An Introduction to Mechanics

For 40 years, Kleppner and Kolenkow's classic text has introduced students to the principles of mechanics. Now brought up-to-date, this revised and improved Second Edition is ideal for classical mechanics courses for first- and second-year undergraduates with foundation skills in mathematics.

The book retains all the features of the first edition, including numerous worked examples, challenging problems, and extensive illustrations, and has been restructured to improve the flow of ideas. It now features

- New examples taken from recent developments, such as laser slowing of atoms, exoplanets, and black holes
- A "Hints, Clues, and Answers" section for the end-of-chapter problems to support student learning
- A solutions manual for instructors at www.cambridge.org/kandk

DANIEL KLEPPNER is Lester Wolfe Professor of Physics, Emeritus, at Massachusetts Institute of Technology. For his contributions to teaching he has been awarded the Oersted Medal by the American Association of Physics Teachers and the Lilienfeld Prize of the American Physical Society. He has also received the Wolf Prize in Physics and the National Medal of Science.

ROBERT KOLENKOW was Associate Professor of Physics at Massachusetts Institute of Technology. Renowned for his skills as a teacher, Kolenkow was awarded the Everett Moore Baker Award for Outstanding Teaching.

Daniel Kleppner

Robert Kolenkow

AN INTRODUCTION TO MECHANICS

SECOND EDITION

CAMBRIDGE
UNIVERSITY PRESS

CAMBRIDGE
UNIVERSITY PRESS

University Printing House, Cambridge CB2 8BS, United Kingdom

One Liberty Plaza, 20th Floor, New York, NY 10006, USA

477 Williamstown Road, Port Melbourne, VIC 3207, Australia

314-321, 3rd Floor, Plot 3, Splendor Forum, Jasola District Centre, New Delhi - 110025, India

79 Anson Road, #06-04/06, Singapore 079906

Cambridge University Press is part of the University of Cambridge.

It furthers the University's mission by disseminating knowledge in the pursuit of education, learning and research at the highest international levels of excellence.

www.cambridge.org
Information on this title: www.cambridge.org/9780521198110

First edition previously published by McGraw-Hill Education 1973

First published by Cambridge University Press 2010
Reprinted 2012

Second edition published by Cambridge University Press 2014

A catalogue record for this publication is available from the British Library

ISBN 978-0-521-19811-0 Hardback

Additional resources for this publication at www.cambridge.org/kandk

CONTENTS

PREFACE

An Introduction to Mechanics grew out of a one-semester course at the Massachusetts Institute of Technology—Physics 8.012—intended for students who seek to understand physics more deeply than the usual freshman level. In the four decades since this text was written physics has moved forward on many fronts but mechanics continues to be a bedrock for concepts such as inertia, momentum, and energy; fluency in the physicist's approach to problem-solving—an underlying theme of this book—remains priceless. The positive comments we have received over the years from students, some of whom are now well advanced in their careers, as well as from faculty at M.I.T. and elsewhere, reassures us that the approach of the text is fundamentally sound. We have received many suggestions from colleagues and we have taken this opportunity to incorporate their ideas and to update some of the discussions.

We assume that our readers know enough elementary calculus to differentiate and integrate simple polynomials and trigonometric functions. We do not assume any familiarity with differential equations. Our experience is that the principal challenge for most students is not with understanding mathematical concepts but in learning how to apply them to physical problems. This comes with practice and there is no substitute for solving challenging problems. Consequently problem-solving takes high priority. We have provided numerous worked examples to help provide guidance. Where possible we try to tie the examples to interesting physical phenomena but we are unapologetic about totally pedagogical problems. A block sliding down a plane is sometimes mocked as the quintessentially dull physics problem but if one allows the plane to accelerate, the system takes on a new complexion.

The problems in the first edition have challenged, instructed, and occasionally frustrated generations of physicists. Some former students have volunteered that working these problems gave them the confidence to pursue careers in science. Consequently, most of the problems in the first edition have been retained and a number of new problems have been added. We continue to respect the wisdom of Piet Hein's aphoristic ditty[1]

> Problems worthy of attack,
>
> Prove their worth by hitting back.

In addition to this inspirational thought, we offer students a few practical suggestions: The problems are meant to be worked with pencil and paper. They generally require symbolic solutions: numerical values, if needed, come last. Only by looking at a symbolic solution can one decide if an answer is reasonable. Diagrams are helpful. Hints and answers are given for some of the problems. We have not included solutions in the book because checking one's approach before making the maximum effort is often irresistible. Working in groups can be instructional for all parties. A separate solutions manual with restricted distribution is however available from Cambridge University Press.

Two revolutionary advances in physics that postdate the first edition deserve mention. The first is the discovery, more accurately the rediscovery, of chaos in the 1970's and the subsequent emergence of chaos theory as a vital branch of dynamics. Because we could not discuss chaos meaningfully within a manageable length, we have not attempted to deal with it. On the other hand, it would have been intellectually dishonest to present evidence for the astounding accuracy of Kepler's laws without mentioning that the solar system is chaotic, though with a time-scale too long to be observable, and so we have duly noted the existence of chaos. The second revolutionary advance is the electronic computer. Computational physics is now a well-established discipline and some level of computational fluency is among the physicist's standard tools. Nevertheless, we have elected not to include computational problems because they are not essential for understanding the concepts of the book, and because they have a seductive way of consuming time.

Here is a summary of the second edition: The first chapter is a mathematical introduction to vectors and kinematics. Vector notation is standard not only in the text but throughout physics and so we take some care to explain it. Translational motion is naturally described using familiar Cartesian coordinates. Rotational motion is equally important but its natural coordinates are not nearly as familiar. Consequently, we put special emphasis on kinematics using polar coordinates. Chapter 2 introduces Newton's laws starting with the decidedly non-trivial concept of inertial systems. This chapter has been converted into two, the first (Chapter 2) discussing principles and the second (Chapter 3) devoted to applying these to various physical systems. Chapter 4 introduces the concepts of momentum, momentum flux, and the conservation of

---

[1] From *Grooks 1* by Piet Hein, copyrighted 1966, The M.I.T. Press.

momentum. Chapter 5 introduces the concepts of kinetic energy, potential energy, and the conservation of energy, including heat and other forms. Chapter 6 applies the preceding ideas to phenomena of general interest in mechanics: small oscillations, stability, coupled oscillators and normal modes, and collisions. In Chapter 7 the ideas are extended to rotational motion. Fixed axis rotation is treated in this chapter, followed by the more general situation of rigid body motion in Chapter 8. Chapter 9 returns to the subject of inertial systems, in particular how to understand observations made in non-inertial systems. Chapters 10 and 11 present two topics that are of general interest in physics: central force motion and the damped and forced harmonic oscillator, respectively. Chapters 12–14 provide an introduction to non-Newtonian physics: the special theory of relativity.

When we created Physics 8.012 the M.I.T. semester was longer than it is today and there is usually not enough class time to cover all the material. Chapters 1–9 constitute the intellectual core of the course. Some combination of Chapters 9–14 is generally presented, depending on the instructor's interest.

We wish to acknowledge contributions to the book made over the years by colleagues at M.I.T. These include R. Aggarwal, G. B. Benedek, A. Burgasser, S. Burles, D. Chakrabarty, L. Dreher, T. J. Greytak, H. T. Imai, H. J. Kendall (deceased), W. Ketterle, S. Mochrie, D. E. Pritchard, P. Rebusco, S. W. Stahler, J. W. Whitaker, F. A. Wilczek, and M. Zwierlein. We particularly thank P. Dourmashkin for his help.

Daniel Kleppner
Robert J. Kolenkow

TO THE TEACHER

This edition of *An Introduction to Mechanics*, like the first edition, is intended for a one-semester course. Like the first edition, there are 14 chapters, though much of the material has been rewritten and two chapters are new. The discussion of Newton's laws, which sets the tone for the course, is now presented in two chapters. Also, the discussion of energy and energy conservation has been augmented and divided into two chapters. Chapter 5 on vector calculus from the first edition has been omitted because the material was not essential and its presence seemed to generate some math anxiety. A portion of the material is in an appendix to Chapter 5.

The discussion of energy has been extended. The idea of heat has been introduced by relating the ideal gas law to the concept of momentum flux. This simultaneously incorporates heat into the principle of energy conservation, and illustrates the fundamental distinction between heat and kinetic energy. At the practical end, some statistics are presented on international energy consumption, a topic that might stimulate thinking about the role of physics in society,

The only other substantive change has been a recasting of the discussion of relativity with more emphasis on the spacetime description. Throughout the book we have attempted to make the math more user friendly by solving problems from a physical point of view before presenting a mathematical solution. In addition, a number of new examples have been provided.

The course is roughly paced to a chapter a week. The first nine chapters are vital for a strong foundation in mechanics: the remainder covers material that can be picked up in the future. The first chapter introduces the language of vectors and provides a background in kinematics that is used throughout the text. Students are likely to return to Chapter 1, using it as a resource for later chapters.

On a few occasions we have been able to illustrate concepts by examples based on relatively recent advances in physics, for instance exoplanets, laser-slowing of atoms, the solar powered space kite, and stars orbiting around the cosmic black hole at the center of our galaxy.

The question of student preparation for Physics 8.012 at M.I.T. comes up regularly. We have found that the most reliable predictor of performance is a quiz on elementary calculus. At the other extreme, occasionally a student takes Physics 8.012 having already completed an AP physics course. Taking a third introductory physics course might be viewed as cruel and unusual, but to our knowledge, these students all felt that the experience was worthwhile.

LIST OF EXAMPLES

Chapter 4   MOMENTUM

Chapter 5   ENERGY

Chapter 6   TOPICS IN DYNAMICS

1 VECTORS AND KINEMATICS

1.1 Introduction

Mechanics is at the heart of physics; its concepts are essential for understanding the world around us and phenomena on scales from atomic to cosmic. Concepts such as momentum, angular momentum, and energy play roles in practically every area of physics. The goal of this book is to help you acquire a deep understanding of the principles of mechanics.

The reason we start by discussing vectors and kinematics rather than plunging into dynamics is that we want to use these tools freely in discussing physical principles. Rather than interrupt the flow of discussion later, we are taking time now to ensure they are on hand when required.

1.2 Vectors

The topic of vectors provides a natural introduction to the role of mathematics in physics. By using vector notation, physical laws can often be written in compact and simple form. Modern vector notation was invented by a physicist, Willard Gibbs of Yale University, primarily to simplify the appearance of equations. For example, here is how Newton's second law appears in nineteenth century notation:

$$F_x = ma_x$$
$$F_y = ma_y$$
$$F_z = ma_z.$$

In vector notation, one simply writes

$$\mathbf{F} = m\mathbf{a},$$

where the bold face symbols $\mathbf{F}$ and $\mathbf{a}$ stand for vectors.

Our principal motivation for introducing vectors is to simplify the form of equations. However, as we shall see in Chapter 14, vectors have a much deeper significance. Vectors are closely related to the fundamental ideas of symmetry and their use can lead to valuable insights into the possible forms of unknown laws.

1.2.1 Definition of a Vector

Mathematicians think of a vector as a set of numbers accompanied by rules for how they change when the coordinate system is changed. For our purposes, a down to earth geometric definition will do: we can think of a vector as a *directed line segment*. We can represent a vector graphically by an arrow, showing both its scale length and its direction. Vectors are sometimes labeled by letters capped by an arrow, for instance $\vec{A}$, but we shall use the convention that a bold face letter, such as $\mathbf{A}$, stands for a vector.

To describe a vector we must specify both its length and its direction. Unless indicated otherwise, we shall assume that parallel translation does not change a vector. Thus the arrows in the sketch all represent the same vector.

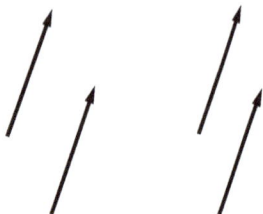

If two vectors have the same length and the same direction they are equal. The vectors **B** and **C** are equal:

$$\mathbf{B} = \mathbf{C}.$$

The magnitude or size of a vector is indicated by vertical bars or, if no confusion will occur, by using italics. For example, the magnitude of **A** is written $|\mathbf{A}|$, or simply A. If the length of **A** is $\sqrt{2}$, then $|\mathbf{A}| = A = \sqrt{2}$. Vectors can have physical dimensions, for example distance, velocity, acceleration, force, and momentum.

If the length of a vector is one unit, we call it a *unit vector*. A unit vector is labeled by a caret; the vector of unit length parallel to **A** is **Â**. It follows that

$$\hat{\mathbf{A}} = \frac{\mathbf{A}}{A}$$

and conversely

$$\mathbf{A} = A\hat{\mathbf{A}}.$$

The physical dimension of a vector is carried by its magnitude. Unit vectors are dimensionless.

1.3 The Algebra of Vectors

We will need to add, subtract, and multiply two vectors, and carry out some related operations. We will not attempt to divide two vectors since the need never arises, but to compensate for this omission, we will define two types of vector multiplication, both of which turn out to be quite useful. Here is a summary of the basic algebra of vectors.

1.3.1 Multiplying a Vector by a Scalar

If we multiply **A** by a simple scalar, that is, by a simple number b, the result is a new vector $\mathbf{C} = b\mathbf{A}$. If $b > 0$ the vector **C** is parallel to **A**, and its magnitude is b times greater. Thus $\hat{\mathbf{C}} = \hat{\mathbf{A}}$, and $C = bA$.

If $b < 0$, then $\mathbf{C} = b\mathbf{A}$ is opposite in direction (antiparallel) to **A**, and its magnitude is $C = |b|\,A$.

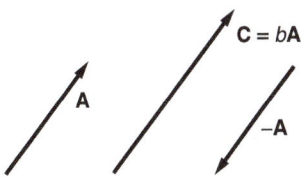

1.3.2 Adding Vectors

Addition of two vectors has the simple geometrical interpretation shown by the drawing. The rule is: to add **B** to **A**, place the tail of **B** at the head of **A** by parallel translation of **B**. The sum is a vector from the tail of **A** to the head of **B**.

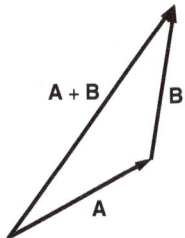

1.3.3 Subtracting Vectors

Because $\mathbf{A} - \mathbf{B} = \mathbf{A} + (-\mathbf{B})$, to subtract **B** from **A** we can simply multiply **B** by -1 and then add. The sketch shows how.

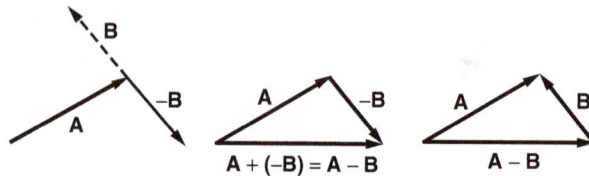

An equivalent way to construct $\mathbf{A} - \mathbf{B}$ is to place the *head* of $\mathbf{B}$ at the *head* of $\mathbf{A}$. Then $\mathbf{A} - \mathbf{B}$ extends from the *tail* of $\mathbf{A}$ to the *tail* of $\mathbf{B}$, as shown in the drawing.

1.3.4 Algebraic Properties of Vectors

It is not difficult to prove the following:

Commutative law

$$\mathbf{A} + \mathbf{B} = \mathbf{B} + \mathbf{A}.$$

Associative law

$$\mathbf{A} + (\mathbf{B} + \mathbf{C}) = (\mathbf{A} + \mathbf{B}) + \mathbf{C}$$
$$c(d\mathbf{A}) = (cd)\mathbf{A}.$$

Distributive law

$$c(\mathbf{A} + \mathbf{B}) = c\mathbf{A} + c\mathbf{B}$$
$$(c + d)\mathbf{A} = c\mathbf{A} + d\mathbf{A}.$$

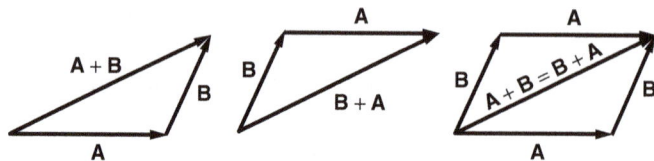

The sketch shows a geometrical proof of the commutative law $\mathbf{A} + \mathbf{B} = \mathbf{B} + \mathbf{A}$; try to cook up your own proofs of the others.

1.4 Multiplying Vectors

Multiplying one vector by another could produce a vector, a scalar, or some other quantity. The choice is up to us. It turns out that two types of vector multiplication are useful in physics.

1.4.1 Scalar Product ("Dot Product")

The first type of multiplication is called the *scalar* product because the result of the multiplication is a scalar. The scalar product is an operation

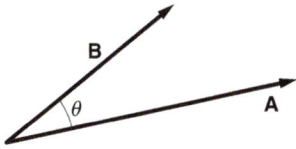

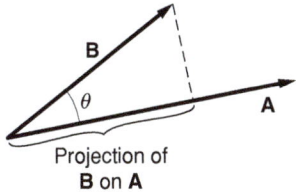

that combines vectors to form a scalar. The scalar product of **A** and **B** is written as **A** · **B**, therefore often called the *dot* product. **A** · **B** (referred to as "A dot B") is defined by

$$\mathbf{A} \cdot \mathbf{B} \equiv AB \cos \theta.$$

Here θ is the angle between **A** and **B** when they are drawn tail to tail. Because $B \cos \theta$ is the projection of **B** along the direction of **A**, it follows that

$$\mathbf{A} \cdot \mathbf{B} = A \text{ times the projection of } \mathbf{B} \text{ on } \mathbf{A}$$
$$= B \text{ times the projection of } \mathbf{A} \text{ on } \mathbf{B}.$$

Note that $\mathbf{A} \cdot \mathbf{A} = |\mathbf{A}|^2 = A^2$. Also, $\mathbf{A} \cdot \mathbf{B} = \mathbf{B} \cdot \mathbf{A}$; the order does not change the value. We say that the dot product is *commutative*.

If either **A** or **B** is zero, their dot product is zero. However, because $\cos \pi/2 = 0$ the dot product of two non-zero vectors is nevertheless zero if the vectors happen to be perpendicular.

A great deal of elementary trigonometry follows from the properties of vectors. Here is an almost trivial proof of the law of cosines using the dot product.

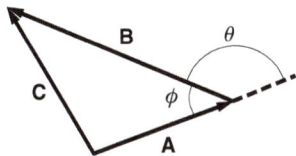

Example 1.1 The Law of Cosines

The law of cosines relates the lengths of three sides of a triangle to the cosine of one of its angles. Following the notation of the drawing, the law of cosines is

$$C^2 = A^2 + B^2 - 2AB \cos \phi.$$

The law can be proved by a variety of trigonometric or geometric constructions, but none is so simple and elegant as the vector proof, which merely involves squaring the sum of two vectors.

$$\mathbf{C} = \mathbf{A} + \mathbf{B}$$
$$\mathbf{C} \cdot \mathbf{C} = (\mathbf{A} + \mathbf{B}) \cdot (\mathbf{A} + \mathbf{B})$$
$$= \mathbf{A} \cdot \mathbf{A} + \mathbf{B} \cdot \mathbf{B} + 2(\mathbf{A} \cdot \mathbf{B})$$
$$C^2 = A^2 + B^2 + 2AB \cos \theta.$$

Recognizing that $\cos \phi = -\cos \theta$ completes the proof.

Example 1.2 Work and the Dot Product

The dot product has an important physical application in describing the work done by a force. As you may already know, the work W done on an object by a force F is defined to be the product of the length of the displacement d and the component of F along the direction of displacement. If the force is applied at an angle θ with respect to the displacement, as shown in the sketch,

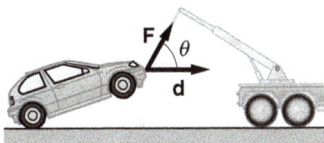

then

$$W = (F \cos \theta) d.$$

Assuming that force and displacement can both be written as vectors, then

$$W = \mathbf{F} \cdot \mathbf{d}.$$

1.4.2 Vector Product ("Cross Product")

The second type of product useful in physics is the *vector* product, in which two vectors **A** and **B** are combined to form a third vector **C**. The symbol for vector product is a cross, so it is often called the *cross product*:

$$\mathbf{C} = \mathbf{A} \times \mathbf{B}.$$

The vector product is more complicated than the scalar product because we have to specify both the magnitude and direction of the vector $\mathbf{A} \times \mathbf{B}$ (called "A cross B"). The magnitude is defined as follows: if

$$\mathbf{C} = \mathbf{A} \times \mathbf{B}$$

then

$$C = AB \sin \theta$$

where θ is the angle between **A** and **B** when they are drawn tail to tail.

To eliminate ambiguity, θ is always taken as the angle smaller than π. Even if neither vector is zero, their vector product is zero if $\theta = 0$ or π, the situation where the vectors are parallel or antiparallel. It follows that

$$\mathbf{A} \times \mathbf{A} = 0$$

for any vector **A**.

Two vectors **A** and **B** drawn tail to tail determine a plane. Any plane can be drawn through **A**. Simply rotate it until it also contains **B**.

We define the direction of **C** to be perpendicular to the plane of **A** and **B**. The three vectors **A**, **B**, and **C** form what is called a *right-hand triple*. Imagine a right-hand coordinate system with **A** and **B** in the x–y plane as shown in the sketch.

A lies on the x axis and **B** lies toward the y axis. When **A**, **B**, and **C** form a right-hand triple, then **C** lies along the positive z axis. We shall always use right-hand coordinate systems such as the one shown.

Here is another way to determine the direction of the cross product. Think of a right-hand screw with the axis perpendicular to **A** and **B**.

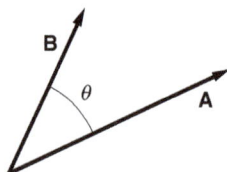

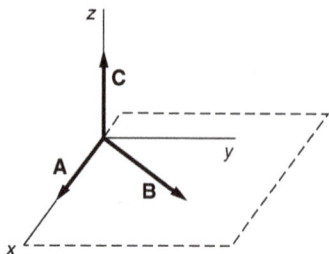

(**A** is into paper)

A

B

C = **A** × **B**

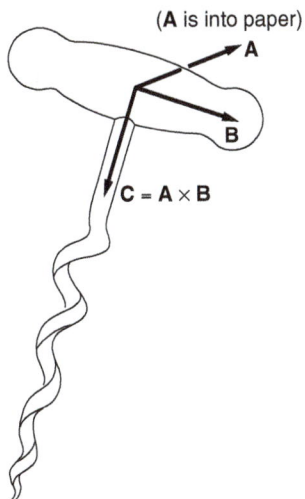

If we rotate it in the direction that swings **A** into **B**, then **C** lies in the direction the screw advances. (Warning: be sure not to use a left-hand screw. Fortunately, they are rare, with hot water faucets among the chief offenders. Your honest everyday wood screw is right-handed.)

A result of our definition of the cross product is that

$$\mathbf{B} \times \mathbf{A} = -\mathbf{A} \times \mathbf{B}.$$

Here we have a case in which the order of multiplication is important. The vector product is *not* commutative. Since reversing the order reverses the sign, it is *anticommutative*.

F

B

q

v

$\tau = \mathbf{r} \times \mathbf{F}$

F

θ

r

Example 1.3 Examples of the Vector Product in Physics

The vector product has a multitude of applications in physics. For instance, if you have learned about the interaction of a charged particle with a magnetic field, you know that the force is proportional to the charge q, the magnetic field B, and the velocity of the particle v. The force varies as the sine of the angle between v and B, and is perpendicular to the plane formed by v and B, in the direction indicated.

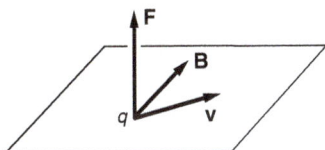

All these rules are combined in the one equation

$$\mathbf{F} = q\mathbf{v} \times \mathbf{B}.$$

Another application is the definition of torque, which we shall develop in Chapter 7. For now we simply mention in passing that the torque vector τ is defined by

$$\tau = \mathbf{r} \times \mathbf{F},$$

$F \sin \theta$

F

θ

r

Top view

where **r** is a vector from the axis about which the torque is evaluated to the point of application of the force **F**. This definition is consistent with the familiar idea that torque is a measure of the ability of an applied force to produce a twist. Note that a large force directed parallel to **r** produces no twist; it merely pulls. Only $F \sin \theta$, the component of force perpendicular to **r**, produces a torque.

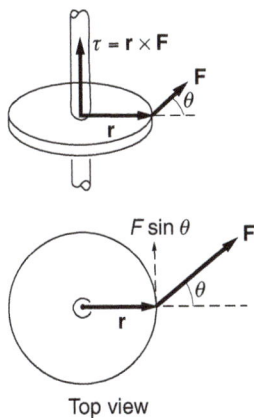

Imagine that we are pushing open a garden gate, where the axis of rotation is a vertical line through the hinges. When we push the gate open, we instinctively apply force in such a way as to make **F** closely perpendicular to **r**, to maximize the torque. Because the torque increases as the lever arm gets larger, we push at the edge of the gate, as far from the hinge line as possible.

As you will see in Chapter 7, the natural direction of τ is along the axis of the rotation that the torque tends to produce. All these ideas are summarized in a nutshell by the simple equation $\tau = \mathbf{r} \times \mathbf{F}$.

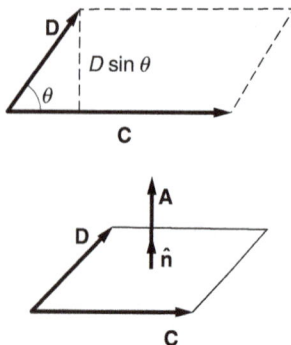

Example 1.4 Area as a Vector

We can use the cross product to describe an area. Usually one thinks of area in terms of magnitude only. However, many applications in physics require that we also specify the orientation of the area. For example, if we wish to calculate the rate at which water in a stream flows through a wire loop of given area, it obviously makes a difference whether the plane of the loop is perpendicular or parallel to the flow. (If parallel, the flow through the loop is zero.) Here is how the vector product accomplishes this:

Consider the area of a quadrilateral formed by two vectors $\mathbf{C}$ and $\mathbf{D}$. The area A of the parallelogram is given by

$$A = \text{base} \times \text{height}$$
$$= CD \sin \theta$$
$$= |\mathbf{C} \times \mathbf{D}| .$$

The magnitude of the cross product gives us the area of the parallelogram, but how can we assign a direction to the area? In the plane of the parallelogram we can draw an infinite number of vectors pointing every which-way, so none of these vectors stands out uniquely. The only unique preferred direction is the *normal* to the plane, specified by a unit vector $\hat{\mathbf{n}}$. We therefore take the vector $\mathbf{A}$ describing the area as parallel to $\hat{\mathbf{n}}$. The magnitude and direction of $\mathbf{A}$ are then given compactly by the cross product

$$\mathbf{A} = \mathbf{C} \times \mathbf{D}.$$

A minor ambiguity remains, because $\hat{\mathbf{n}}$ can point out from either side of the area. We could just as well have defined the area by $\mathbf{A} = \mathbf{D} \times \mathbf{C} = -\mathbf{C} \times \mathbf{D}$, as long as we are consistent once the choice is made.

1.5 Components of a Vector

The fact that we have discussed vectors without introducing a particular coordinate system shows why vectors are so useful; vector operations are defined independently of any particular coordinate system. However, eventually we have to translate our results from the abstract to the concrete, and at this point we have to choose a coordinate system in which to work.

The combination of algebra and geometry, called *analytic geometry*, is a powerful tool that we shall use in many calculations. Analytic geometry has a consistent procedure for describing geometrical objects by a set of numbers, greatly easing the task of performing quantitative calculations. With its aid, students still in school can routinely solve problems that would have taxed the ancient Greek geometer Euclid. Analytic geometry was developed as a complete subject in the first half of the seventeenth

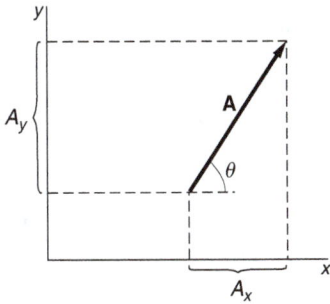

century by the French mathematician René Descartes, and independently by his contemporary Pierre Fermat.

For simplicity, let us first restrict ourselves to a two-dimensional system, the familiar x–y plane. The diagram shows a vector $\mathbf{A}$ in the x–y plane.

The projections of $\mathbf{A}$ along the x and y coordinate axes are called the *components* of $\mathbf{A}$, A_x and A_y, respectively. The magnitude of $\mathbf{A}$ is $A = \sqrt{A_x^2 + A_y^2}$, and the direction of $\mathbf{A}$ makes an angle $\theta = \arctan{(A_y/A_x)}$ with the x axis.

Since its components define a vector, we can specify a vector entirely by its components. Thus

$$\mathbf{A} = (A_x, A_y)$$

or, more generally, in three dimensions,

$$\mathbf{A} = (A_x, A_y, A_z).$$

Prove for yourself that $A = \sqrt{A_x^2 + A_y^2 + A_z^2}$.

If two vectors are equal $\mathbf{A} = \mathbf{B}$, then in the same coordinate system their corresponding components are equal.

$$A_x = B_x \quad A_y = B_y \quad A_z = B_z.$$

The single vector equation $\mathbf{A} = \mathbf{B}$ symbolically represents three scalar equations.

The vector $\mathbf{A}$ has a meaning independent of any coordinate system. However, the components of $\mathbf{A}$ depend on the coordinate system being used. To illustrate this, here is a vector $\mathbf{A}$ drawn in two different coordinate systems.

In the first case,

$$\mathbf{A} = (A, 0) \quad (x, y \text{ system}),$$

while in the second

$$\mathbf{A} = (0, -A) \quad (x', y' \text{ system}).$$

All vector operations can be written as equations for components. For instance, multiplication by a scalar is written

$$c\mathbf{A} = (cA_x, cA_y, cA_z).$$

The law for vector addition is

$$\mathbf{A} + \mathbf{B} = (A_x + B_x, A_y + B_y, A_z + B_z).$$

By writing $\mathbf{A}$ and $\mathbf{B}$ as the sums of vectors along each of the coordinate axes, you can verify that

$$\mathbf{A} \cdot \mathbf{B} = A_x B_x + A_y B_y + A_z B_z.$$

We shall defer evaluating the cross product until the next section.

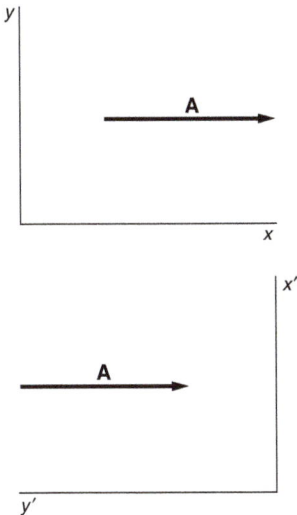

Example 1.5 Vector Algebra

Let

$$\mathbf{A} = (3, \ 5, \ -7)$$
$$\mathbf{B} = (2, 7, \ 1).$$

Find $\mathbf{A} + \mathbf{B}$, $\mathbf{A} - \mathbf{B}$, A, B, $\mathbf{A} \cdot \mathbf{B}$, and the cosine of the angle between $\mathbf{A}$ and $\mathbf{B}$.

$$\mathbf{A} + \mathbf{B} = (3 + 2, 5 + 7, -7 + 1)$$
$$= (5, \ 12, \ -6)$$
$$\mathbf{A} - \mathbf{B} = (3 - 2, 5 - 7, -7 - 1)$$
$$= (1, \ -2, \ -8)$$
$$A = \sqrt{(3^2 + 5^2 + 7^2)}$$
$$= \sqrt{83}$$
$$\approx 9.11$$
$$B = \sqrt{(2^2 + 7^2 + 1^2)}$$
$$= \sqrt{54}$$
$$\approx 7.35$$
$$\mathbf{A} \cdot \mathbf{B} = 3 \times 2 + 5 \times 7 - 7 \times 1$$
$$= 34$$
$$\cos(\mathbf{A}, \mathbf{B}) = \frac{\mathbf{A} \cdot \mathbf{B}}{AB} \approx \frac{34}{(9.11)(7.35)} \approx 0.508.$$

Example 1.6 Constructing a Vector Perpendicular to a Given Vector

The problem is to find a unit vector lying in the x–y plane that is perpendicular to the vector $\mathbf{A} = (3, 5, 1)$.

A vector $\mathbf{B}$ in the x–y plane has components (B_x, B_y). For $\mathbf{B}$ to be perpendicular to $\mathbf{A}$, we must have $\mathbf{A} \cdot \mathbf{B} = 0$:

$$\mathbf{A} \cdot \mathbf{B} = 3B_x + 5B_y$$
$$= 0.$$

Hence $B_y = -\frac{3}{5}B_x$. For $\mathbf{B}$ to be a unit vector, $B_x^2 + B_y^2 = 1$. Combining these gives

$$B_x^2 + \frac{9}{25}B_x^2 = 1,$$

or

$$B_x = \sqrt{\frac{25}{34}}$$
$$\approx \pm 0.858$$
$$B_y = -\frac{3}{5}B_x$$
$$\approx \mp 0.515.$$

There are two solutions, given by the upper and lower signs. Each vector is the negative of the other, so they are equal in magnitude but point in opposite directions.

1.6 Base Vectors

Base vectors are a set of orthogonal (mutually perpendicular) unit vectors, one for each dimension. For example, if we are dealing with the familiar Cartesian coordinate system of three dimensions, the base vectors lie along the $x, y,$ and z axes. We shall designate the x unit vector by $\hat{\mathbf{i}}$, the y unit vector by $\hat{\mathbf{j}}$, and the z unit vector by $\hat{\mathbf{k}}$. (Sometimes the symbols $\hat{\mathbf{x}}, \hat{\mathbf{y}},$ and $\hat{\mathbf{z}}$ are used.)

The base vectors have the following properties, as you can readily verify:

$$\hat{\mathbf{i}} \cdot \hat{\mathbf{i}} = \hat{\mathbf{j}} \cdot \hat{\mathbf{j}} = \hat{\mathbf{k}} \cdot \hat{\mathbf{k}} = 1$$
$$\hat{\mathbf{i}} \cdot \hat{\mathbf{j}} = \hat{\mathbf{j}} \cdot \hat{\mathbf{k}} = \hat{\mathbf{k}} \cdot \hat{\mathbf{i}} = 0$$
$$\hat{\mathbf{i}} \times \hat{\mathbf{j}} = \hat{\mathbf{k}}$$
$$\hat{\mathbf{j}} \times \hat{\mathbf{k}} = \hat{\mathbf{i}}$$
$$\hat{\mathbf{k}} \times \hat{\mathbf{i}} = \hat{\mathbf{j}}.$$

As shown in the drawing, we can write any three-dimensional vector in terms of its components and the base vectors:

$$\mathbf{A} = A_x \hat{\mathbf{i}} + A_y \hat{\mathbf{j}} + A_z \hat{\mathbf{k}}$$

To find the component of a vector in any direction, take the dot product with a unit vector in that direction. For instance, the z component of vector $\mathbf{A}$ is

$$A_z = \mathbf{A} \cdot \hat{\mathbf{k}}.$$

The base vectors are particularly useful in deriving the general rule for evaluating the cross product of two vectors in terms of their components:

$$\mathbf{A} \times \mathbf{B} = (A_x \hat{\mathbf{i}} + A_y \hat{\mathbf{j}} + A_z \hat{\mathbf{k}}) \times (B_x \hat{\mathbf{i}} + B_y \hat{\mathbf{j}} + B_z \hat{\mathbf{k}}).$$

Consider the first term:

$$A_x \hat{\mathbf{i}} \times \mathbf{B} = A_x B_x (\hat{\mathbf{i}} \times \hat{\mathbf{i}}) + A_x B_y (\hat{\mathbf{i}} \times \hat{\mathbf{j}}) + A_x B_z (\hat{\mathbf{i}} \times \hat{\mathbf{k}}).$$

(The associative law holds here.) Because $\hat{\mathbf{i}} \times \hat{\mathbf{i}} = 0, \hat{\mathbf{i}} \times \hat{\mathbf{j}} = \hat{\mathbf{k}},$ and $\hat{\mathbf{i}} \times \hat{\mathbf{k}} = -\hat{\mathbf{j}}$, we find

$$A_x \hat{\mathbf{i}} \times \mathbf{B} = A_x (B_y \hat{\mathbf{k}} - B_z \hat{\mathbf{j}}).$$

The same argument applied to the y and z components gives

$$A_y \hat{\mathbf{j}} \times \mathbf{B} = A_y (B_z \hat{\mathbf{i}} - B_x \hat{\mathbf{k}})$$
$$A_z \hat{\mathbf{k}} \times \mathbf{B} = A_z (B_x \hat{\mathbf{j}} - B_y \hat{\mathbf{i}}).$$

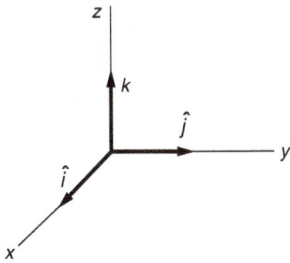

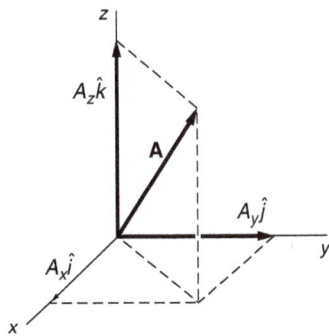

A quick way to derive these relations is to work out the first and then to obtain the others by cyclically permuting x, y, z, and $\hat{\mathbf{i}}, \hat{\mathbf{j}}, \hat{\mathbf{k}}$ (that is, $x \to y, y \to z, z \to x$, and $\hat{\mathbf{i}} \to \hat{\mathbf{j}}, \hat{\mathbf{j}} \to \hat{\mathbf{k}}, \hat{\mathbf{k}} \to \hat{\mathbf{i}}$). A compact mnemonic for expressing this result is to write the base vectors and the components of **A** and **B** as three rows of a determinant, like this:

$$\mathbf{A} \times \mathbf{B} = \begin{vmatrix} \hat{\mathbf{i}} & \hat{\mathbf{j}} & \hat{\mathbf{k}} \\ A_x & A_y & A_z \\ B_x & B_y & B_z \end{vmatrix}$$

$$= (A_y B_z - A_z B_y)\,\hat{\mathbf{i}} - (A_x B_z - A_z B_x)\,\hat{\mathbf{j}} + (A_x B_y - A_y B_x)\,\hat{\mathbf{k}}.$$

For instance, if $\mathbf{A} = \hat{\mathbf{i}} + 3\,\hat{\mathbf{j}} - \hat{\mathbf{k}}$ and $\mathbf{B} = 4\,\hat{\mathbf{i}} + \hat{\mathbf{j}} + 3\,\hat{\mathbf{k}}$, then

$$\mathbf{A} \times \mathbf{B} = \begin{vmatrix} \hat{\mathbf{i}} & \hat{\mathbf{j}} & \hat{\mathbf{k}} \\ 1 & 3 & -1 \\ 4 & 1 & 3 \end{vmatrix}$$

$$= 10\,\hat{\mathbf{i}} - 7\,\hat{\mathbf{j}} - 11\,\hat{\mathbf{k}}.$$

1.7 The Position Vector r and Displacement

So far we have discussed only abstract vectors. However, the reason for introducing vectors is that many physical quantities are conveniently described by vectors, among them velocity, force, momentum, and gravitational and electric fields. In this chapter we shall use vectors to discuss *kinematics*, which is the description of motion without regard for the causes of the motion. *Dynamics*. which we shall take up in Chapter 2, looks at the causes of motion.

Kinematics is largely geometric and perfectly suited to characterization by vectors. Our first application of vectors will be to the description of position and motion in familiar three-dimensional space.

To locate the position of a point in space, we start by setting up a coordinate system. For convenience we choose a three-dimensional Cartesian system with axes x, y, and z, as shown. In order to measure position, the axes must be marked in some convenient unit of length—meters, for instance. The position of the point of interest is given by listing the values of its three coordinates, x_1, y_1, z_1, which we can write compactly as a *position vector* $\mathbf{r}(x_1, y_1, z_1)$ or more generally as $\mathbf{r}(x, y, z)$. This notation can be confusing because we normally label the axes of a Cartesian coordinate system by x, y, z. However, $\mathbf{r}(x, y, z)$ is really shorthand for $\mathbf{r}(x\text{-axis}, y\text{-axis}, z\text{-axis})$. The components of $\mathbf{r}$ are the coordinates of the point referred to the particular coordinate axes.

The three numbers (x, y, z) do not represent the components of a vector according to our previous discussion because they specify only the position of a single point, not a magnitude and direction. Unlike other physical vectors such as force and velocity, $\mathbf{r}$ is tied to a particular coordinate system.

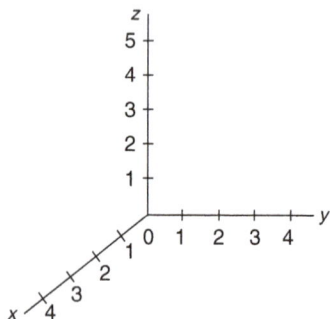

The position of an arbitrary point P at (x, y, z) is written as

$$\mathbf{r} = (x, y, z) = x\hat{\mathbf{i}} + y\hat{\mathbf{j}} + z\hat{\mathbf{k}}.$$

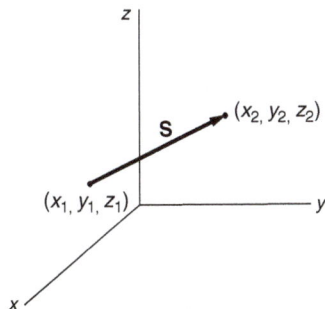

If we move from the point x_1, y_1, z_1 to some new position, x_2, y_2, z_2, then the *displacement* defines a true vector $\mathbf{S}$ with coordinates $S_x = x_2 - x_1$, $S_y = y_2 - y_1$, $S_z = z_2 - z_1$.

$\mathbf{S}$ is a vector from the initial position to the final position—it defines the displacement of a point of interest. Note, however, that $\mathbf{S}$ contains no information about the initial and final positions separately—only about the *relative* position of each. Thus, $S_z = z_2 - z_1$ depends on the *difference* between the final and initial values of the z coordinates; it does not specify z_2 or z_1 separately. Thus $\mathbf{S}$ is a true vector: the values of the co-ordinates of its initial and final points depend on the coordinate system but $\mathbf{S}$ does not, as the sketches indicate.

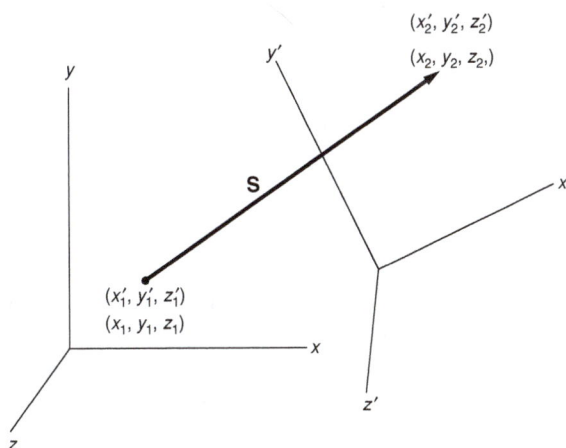

One way in which our displacement vector differs from vectors in pure mathematics is that in mathematics, vectors are usually pure quantities, with components described by simple numbers, whereas the magnitude S has the physical dimension of length associated with it. We will use the convention that the physical dimension of a vector is attached to its magnitude, so that the associated unit vector is dimensionless. Thus, a displacement of 8 m (8 meters) in the x direction is $\mathbf{S} = (8 \text{ m}, 0, 0)$. $S = 8$ m, and $\hat{\mathbf{S}} = \mathbf{S}/S = \hat{\mathbf{i}}$.

The sketch shows position vectors $\mathbf{r}$ and $\mathbf{r}'$ indicating the position of the same point in space but drawn in different coordinate systems. If $\mathbf{R}$ is the vector from the origin of the unprimed coordinate system to the origin of the primed coordinate system, we have $\mathbf{r} = \mathbf{R} + \mathbf{r}'$, or alternatively, $\mathbf{r}' = \mathbf{r} - \mathbf{R}$.

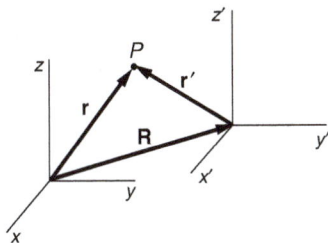

We use these results to show that displacement $\mathbf{S}$, a true vector, is independent of coordinate system. As the sketch indicates,

$$
\begin{aligned}
\mathbf{S} &= \mathbf{r}_2 - \mathbf{r}_1 \\
&= (\mathbf{R} + \mathbf{r}'_2) - (\mathbf{R} + \mathbf{r}'_1) \\
&= \mathbf{r}'_2 - \mathbf{r}'_1.
\end{aligned}
$$

1.8 Velocity and Acceleration

1.8.1 Motion in One Dimension

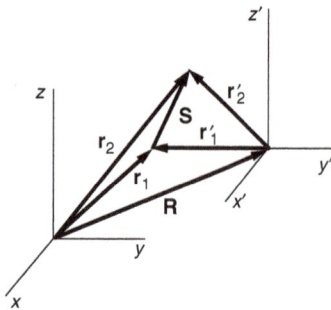

Before employing vectors to describe velocity and acceleration in three dimensions, it may be helpful to review one-dimensional motion: motion along a straight line.

Let x be the value of the coordinate of a particle moving on a line, with x measured in some convenient unit such as meters. We assume that we have a continuous record of position versus time.

The *average velocity* $\bar{v}$ of the point between two times t_1 and t_2 is defined by

$$
\bar{v} = \frac{x(t_2) - x(t_1)}{t_2 - t_1}.
$$

(We shall generally use a bar to indicate the time average of a quantity.)

The *instantaneous velocity* v is the limit of the average velocity as the time interval approaches zero:

$$
v = \lim_{\Delta t \to 0} \frac{x(t + \Delta t) - x(t)}{\Delta t}.
$$

The limit we introduced in defining v is exactly the definition of a derivative in calculus. In the latter half of the seventeenth century Isaac Newton invented calculus to give him the tools he needed to analyze change and motion, particularly planetary motion, one of his greatest achievements in physics. We therefore write

$$
v = \frac{dx}{dt},
$$

using notation due to Gottfried Leibniz, who independently invented calculus. Newton would have written

$$
v = \dot{x}
$$

where the dot stands for d/dt. Following a convention frequently used in physics, we shall use Newton's notation only for derivatives with respect to time. The derivative of a function $f(x)$ can also be written $f'(x) \equiv df(x)/dx$.

In a similar fashion, the *instantaneous acceleration* a is

$$
\begin{aligned}
a &= \lim_{\Delta t \to 0} \frac{v(t + \Delta t) - v(t)}{\Delta t} \\
&= \frac{dv}{dt} = \dot{v}.
\end{aligned}
$$

Using $v = dx/dt$,

$$a = \frac{d^2x}{dt^2} = \ddot{x}.$$

Here d^2x/dt^2 is called the second derivative of x with respect to t.

The concept of speed is sometimes useful. Speed s is simply the magnitude of the velocity: $s = |\mathbf{v}|$. In one dimension, speed and velocity are synonymous.

1.8.2 Motion in Several Dimensions

Our task now is to extend the ideas of velocity and acceleration to several dimensions using vector notation. Consider a particle moving in the x–y plane. As time goes on, the particle traces out a path. We assume that we know the particle's coordinates at every value of time. The instantaneous position of the particle at some time t_1 is

$$\mathbf{r}(t_1) = (x(t_1), y(t_1))$$

or

$$\mathbf{r}(t_1) = (x_1, y_1)$$

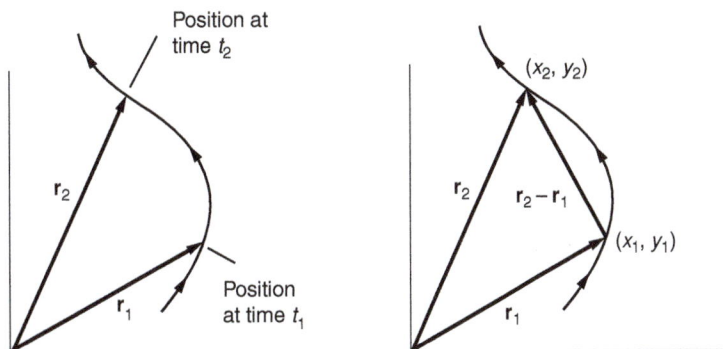

where x_1 is the value of x at $t = t_1$, and so forth. At time t_2 the position is similarly $\mathbf{r}(t_2) = (x_2, y_2)$.

The displacement of the particle between times t_1 and t_2 is

$$\mathbf{r}(t_2) - \mathbf{r}(t_1) = (x_2 - x_1, y_2 - y_1).$$

We can generalize our example by considering the position at some time t and also at some later time $t + \Delta t$. We put no restrictions on the size of Δt—it can be as large or as small as we please.

The displacement of the particle during the interval Δt is

$$\Delta \mathbf{r} = \mathbf{r}(t + \Delta t) - \mathbf{r}(t).$$

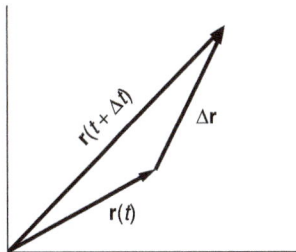

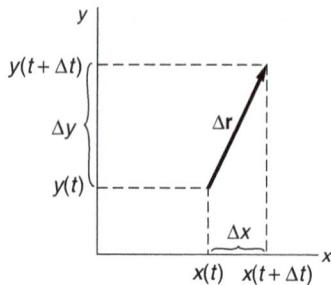

This vector equation is equivalent to the two scalar equations

$$\Delta x = x(t + \Delta t) - x(t)$$
$$\Delta y = y(t + \Delta t) - y(t).$$

The *velocity* **v** of the particle as it moves along the path is

$$\mathbf{v} = \lim_{\Delta t \to 0} \frac{\Delta \mathbf{r}}{\Delta t}$$
$$= \frac{d\mathbf{r}}{dt},$$

which is equivalent to the two scalar equations

$$v_x = \lim_{\Delta t \to 0} \frac{\Delta x}{\Delta t} = \frac{dx}{dt}$$
$$v_y = \lim_{\Delta t \to 0} \frac{\Delta y}{\Delta t} = \frac{dy}{dt}.$$

Extension of the argument to three dimensions is trivial. The third component of velocity is

$$v_z = \lim_{\Delta t \to 0} \frac{z(t + \Delta t) - z(t)}{\Delta t} = \frac{dz}{dt}.$$

Our definition of velocity as a vector is a straightforward generalization of the familiar concept of motion in a straight line. Vector notation allows us to describe motion in three dimensions with a single equation, a great economy compared with the three equations we would need otherwise. The equation $\mathbf{v} = d\mathbf{r}/dt$ expresses concisely the results we have just found.

An alternative approach to calculating the velocity is to start with the definition $\mathbf{r} = x\,\hat{\mathbf{i}} + y\,\hat{\mathbf{j}} + z\,\hat{\mathbf{k}}$, and then differentiate:

$$\frac{d\mathbf{r}}{dt} = \frac{d(x\,\hat{\mathbf{i}} + y\,\hat{\mathbf{j}} + z\,\hat{\mathbf{k}})}{dt}.$$

To evaluate this expression, we use a key property of vectors—they can change with time in magnitude or in direction or in both. But base vectors are unit vectors and therefore have constant magnitude, so they cannot change in magnitude. The Cartesian base vectors also have the special property that they are fixed in direction, and therefore cannot change direction. Hence we can treat the Cartesian base vectors as constants when we differentiate:

$$\frac{d\mathbf{r}}{dt} = \frac{dx}{dt}\,\hat{\mathbf{i}} + \frac{dy}{dt}\,\hat{\mathbf{j}} + \frac{dz}{dt}\,\hat{\mathbf{k}}$$

as before.

Similarly, acceleration **a** is defined by

$$\mathbf{a} = \frac{d\mathbf{v}}{dt} = \frac{dv_x}{dt}\,\hat{\mathbf{i}} + \frac{dv_y}{dt}\,\hat{\mathbf{j}} + \frac{dv_z}{dt}\,\hat{\mathbf{k}}$$
$$= \frac{d^2\mathbf{r}}{dt^2}.$$

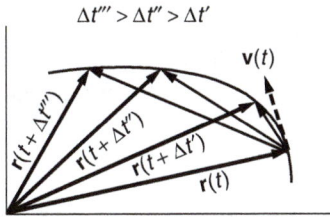

$\Delta t''' > \Delta t'' > \Delta t'$

We could continue to form new vectors by taking higher derivatives of $\mathbf{r}$, but in the study of dynamics it turns out that $\mathbf{r}$, $\mathbf{v}$, and $\mathbf{a}$ are of chief interest.

Let the particle undergo a displacement $\Delta \mathbf{r}$ in time Δt. In the limit $\Delta t \to 0$, $\Delta \mathbf{r}$ becomes tangent to the trajectory, as the sketch indicates. The relation

$$\Delta \mathbf{r} \approx \frac{d\mathbf{r}}{dt} \Delta t$$
$$= \mathbf{v} \Delta t$$

becomes exact in the limit $\Delta t \to 0$, and shows that $\mathbf{v}$ is parallel to $\Delta \mathbf{r}$; the instantaneous velocity $\mathbf{v}$ of a particle is everywhere tangent to the trajectory.

Example 1.7 Finding Velocity from Position

Suppose that the position of a particle is given by

$$\mathbf{r} = A(e^{\alpha t} \, \hat{\mathbf{i}} + e^{-\alpha t} \, \hat{\mathbf{j}}),$$

where A and α are constants. Find the velocity, and sketch the trajectory.

$$\mathbf{v} = \frac{d\mathbf{r}}{dt}$$
$$= A(\alpha e^{\alpha t} \, \hat{\mathbf{i}} - \alpha e^{-\alpha t} \, \hat{\mathbf{j}}) \quad \text{or}$$
$$v_x = A\alpha e^{\alpha t}$$
$$v_y = -A\alpha e^{-\alpha t}.$$

The magnitude of $\mathbf{v}$ is

$$v = \sqrt{v_x{}^2 + v_y{}^2}$$
$$= A\alpha \sqrt{e^{2\alpha t} + e^{-2\alpha t}}.$$

To sketch the trajectory it is often helpful to look at limiting cases. At $t = 0$, we have

$$\mathbf{r}(0) = A \, (\hat{\mathbf{i}} + \hat{\mathbf{j}})$$
$$\mathbf{v}(0) = \alpha A \, (\hat{\mathbf{i}} - \hat{\mathbf{j}}).$$

Note that $\mathbf{v}(0)$ is perpendicular to $\mathbf{r}(0)$.

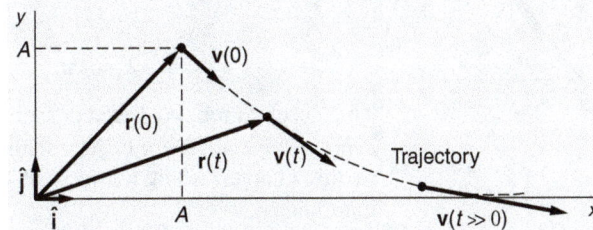

As $t \to \infty$, $e^{\alpha t} \to \infty$ and $e^{-\alpha t} \to 0$. In this limit $\mathbf{r} \to A e^{\alpha t}\,\hat{\mathbf{i}}$, which is a vector along the x axis, and $\mathbf{v} \to \alpha A e^{\alpha t}\,\hat{\mathbf{i}}$; in this unrealistic example, the point rushes along the x axis and the speed increases without limit.

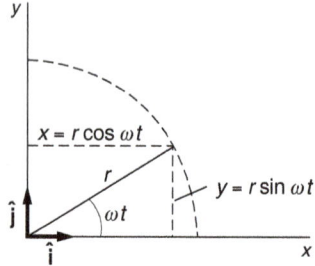

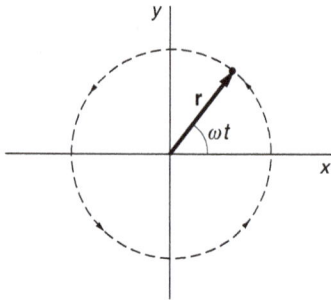

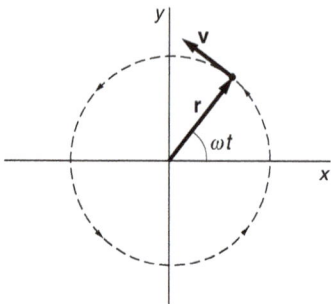

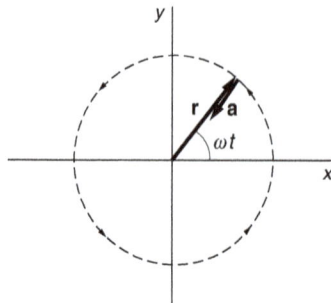

Example 1.8 Uniform Circular Motion

Circular motion plays an important role in physics. Here we look at the simplest and most important case—*uniform* circular motion, which is circular motion at constant speed.

Consider a particle moving in the x–y plane according to $\mathbf{r} = r(\cos \omega t\,\hat{\mathbf{i}} + \sin \omega t\,\hat{\mathbf{j}})$, where r and ω are constants. Find the trajectory, the velocity, and the acceleration.

$$|\mathbf{r}| = \sqrt{r^2 \cos^2 \omega t + r^2 \sin^2 \omega t}.$$

Using the familiar identity $\sin^2 \theta + \cos^2 \theta = 1$,

$$|\mathbf{r}| = r = \text{constant}.$$

The trajectory is a circle.

The particle moves counterclockwise around the circle, starting from $(r, \ 0)$ at $t = 0$. It traverses the circle in a time T such that $\omega T = 2\pi$. ω is called the *angular speed* (or less precisely the *angular velocity*) of the motion and is measured in radians per second. T, the time required to execute one complete cycle, is called the *period*.

$$\mathbf{v} = \frac{d\mathbf{r}}{dt}$$
$$= r\omega(-\sin \omega t\,\hat{\mathbf{i}} + \cos \omega t\,\hat{\mathbf{j}}).$$

We can show that $\mathbf{v}$ is tangent to the trajectory by calculating $\mathbf{v} \cdot \mathbf{r}$:

$$\mathbf{v} \cdot \mathbf{r} = r^2\omega(-\sin \omega t \cos \omega t + \cos \omega t \sin \omega t)$$
$$= 0.$$

Because $\mathbf{v}$ is perpendicular to $\mathbf{r}$, the motion is tangent to the circle, as we expect. It is easy to show that the speed $|\mathbf{v}| = r\omega$ is constant.

$$\mathbf{a} = \frac{d\mathbf{v}}{dt}$$
$$= r\omega^2(-\cos \omega t\,\hat{\mathbf{i}} - \sin \omega t\,\hat{\mathbf{j}})$$
$$= -\omega^2\,\mathbf{r}.$$

The acceleration is directed radially inward and is known as the *centripetal acceleration*. We shall have more to say about it later in this chapter when we look at how motion is described in polar coordinates.

1.9 Formal Solution of Kinematical Equations

Dynamics, which we shall take up in Chapter 2, enables us to find the acceleration of a body if we know the interactions. Once we have the acceleration, finding the velocity and position is a simple matter of integration. Here is the formal integration procedure.

If the acceleration is a known function of time, the velocity can be found from the defining equation

$$\frac{d\mathbf{v}(t)}{dt} = \mathbf{a}(t)$$

by integration with respect to time. Writing this equation in more detail, we have

$$\frac{dv_x}{dt}\hat{\mathbf{i}} + \frac{dv_y}{dt}\hat{\mathbf{j}} + \frac{dv_z}{dt}\hat{\mathbf{k}} = a_x\hat{\mathbf{i}} + a_y\hat{\mathbf{j}} + a_z\hat{\mathbf{k}}.$$

We can separate the corresponding components on each side into separate equations. (To justify this, take the dot product of all the terms with $\hat{\mathbf{i}}$, $\hat{\mathbf{j}}$, or $\hat{\mathbf{k}}$.) For example, the x component is

$$\frac{dv_x}{dt} = a_x.$$

If we know the x velocity at an initial time t_0, then we can integrate this equation with respect to time to find the velocity at a later time t_1:

$$\int_{v_0}^{v_1} dv_x = \int_{t_0}^{t_1} a_x dt, \qquad \text{or}$$

$$v_x(t_1) - v_x(t_0) = \int_{t_0}^{t_1} a_x(t)dt,$$

$$v_x(t_1) = v_x(t_0) + \int_{t_0}^{t_1} a_x(t)dt.$$

Treating the y and z velocity components similarly, we have

$$\mathbf{v}(t_1) = \mathbf{v}(t_0) + \int_{t_0}^{t_1} \mathbf{a}(t)dt.$$

To express the velocity at an arbitrary time t we write

$$\mathbf{v}(t) = \mathbf{v}_0 + \int_{t_0}^{t} \mathbf{a}(t')dt'.$$

The dummy variable of integration has been changed from t to t' to avoid confusion with the upper limit t. The initial velocity $\mathbf{v}(t_0)$ has been written as $\mathbf{v}_0$ to make the notation more compact. When $t = t_0$, $\mathbf{v}(t)$ reduces to $\mathbf{v}_0$, as we expect.

Example 1.9 Finding Velocity from Acceleration

A table-tennis ball is released near the surface of the (airless) moon with velocity $\mathbf{v}_0 = (0, 5, -3)$ m/s. It accelerates (downward) with acceleration $\mathbf{a} = (0, 0, -1.6)$ m/s^2. Find its velocity after 5 s.

The equation

$$\mathbf{v}(t) = \mathbf{v}_0 + \int_{t_0}^{t_1} \mathbf{a}(t')dt'$$

is equivalent to the three component equations

$$v_x(t) = v_{0x} + \int_0^t a_x(t')dt'$$

$$v_y(t) = v_{0y} + \int_0^t a_y(t')dt'$$

$$v_z(t) = v_{0z} + \int_0^t a_z(t')dt'.$$

Taking these equations in turn with the given values of $\mathbf{v}_0$ and $\mathbf{a}$, we obtain at $t = 5$ s:

$$v_x = 0 \, \text{m/s}$$
$$v_y = 5 \, \text{m/s}$$

$$v_z = -3 + \int_0^5 (-1.6)dt' = -11 \, \text{m/s}.$$

Position is found by a second integration. Starting with

$$\frac{d\mathbf{r}(t)}{dt} = \mathbf{v}(t),$$

we find, by an argument identical to the above,

$$\mathbf{r}(t) = \mathbf{r}_0 + \int_0^t \mathbf{v}(t')dt'.$$

A particularly important case is *uniform acceleration*. If we take $\mathbf{a} =$ constant and $t_0 = 0$, we have

$$\mathbf{v}(t) = \mathbf{v}_0 + \mathbf{a}t$$

$$\mathbf{r}(t) = \mathbf{r}_0 + \int_0^t (\mathbf{v}_0 + \mathbf{a}t')dt',$$

or

$$\mathbf{r}(t) = \mathbf{r}_0 + \mathbf{v}_0 t + \frac{1}{2}\mathbf{a}\,t^2.$$

Quite likely you are already familiar with this in its one-dimensional form. For instance, the x component of this equation is

$$x = x_0 + v_{0x}t + \frac{1}{2}a_x t^2,$$

where v_{0x} is the x component of $\mathbf{v}_0$. This expression is so familiar that you may inadvertently apply it to the general case of varying acceleration. Don't! It holds only for *uniform* acceleration. In general, the full procedure described above must be used.

Example 1.10 Motion in a Uniform Gravitational Field

Suppose that an object moves freely under the influence of gravity so that it has a constant downward acceleration g. Choosing the z axis vertically upward, we have

$$\mathbf{a} = -g\,\hat{\mathbf{k}}.$$

If the object is released at $t = 0$ with initial velocity $\mathbf{v}_0$, we have

$$x = x_0 + v_{0x}t$$
$$y = y_0 + v_{0y}t$$
$$z = z_0 + v_{0z}t - \frac{1}{2}g\,t^2.$$

Without loss of generality, we can let $\mathbf{r}_0 = 0$, and assume that $v_{0y} = 0$. (The latter assumption simply means that we choose the coordinate system so that the initial velocity is in the x–z plane.) Then

$$x = v_{0x}t$$
$$z = v_{0z}t - \frac{1}{2}g\,t^2.$$

We can eliminate time from the two equations for x and z to obtain the trajectory, i.e. the path $z(x)$.

$$z = \frac{v_{0z}}{v_{0x}}x - \frac{g}{2v_{0x}^2}x^2.$$

As shown in the sketch, this is the well-known parabola of projectile motion under constant gravity.

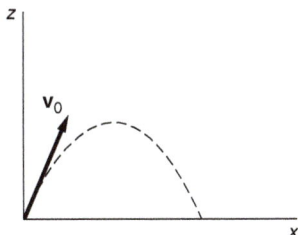

Example 1.11 The Effect of a Radio Wave on an Ionospheric Electron

The ionosphere is a region of electrically neutral gas, composed of positively charged ions and negatively charged electrons, that surrounds the Earth at a height of approximately 200 km (120 mi). If a radio wave passes through the ionosphere, its electric field accelerates the charged particles. Because the electric field oscillates in time, the charged particles tend to jiggle back and forth. The problem is to find the motion of an electron of charge $-e$ and mass m which is initially at rest, and which is suddenly subjected to an electric field $\mathbf{E} = \mathbf{E}_0 \sin \omega t$ (ω is the frequency of oscillation in radians/second).

The force $\mathbf{F}$ on the electron subject to the electric field is $\mathbf{F} = -e\mathbf{E}$, and by Newton's second law we have $\mathbf{a} = \mathbf{F}/m = -e\mathbf{E}/m$. (If the reasoning behind this is not clear, ignore it for now. It will become clear in Chapter 2.) We have

$$\mathbf{a} = \frac{-e\mathbf{E}}{m}$$
$$= \frac{-e\mathbf{E}_0}{m}\sin \omega t.$$

$\mathbf{E}_0$ is a constant vector and we shall choose our coordinate system so that the x axis lies along it. Since there is no acceleration in the y or z directions, we need consider only the x motion. With this understanding, we can drop subscripts and write a for a_x:

$$a(t) = \frac{-eE_0}{m} \sin \omega t = a_0 \sin \omega t$$

where

$$a_0 = \frac{-eE_0}{m}.$$

Then

$$v(t) = v_0 + \int_0^t a(t')dt'$$

$$= v_0 + \int_0^t a_0 \sin \omega t' dt'$$

$$= v_0 - \frac{a_0}{\omega} \cos \omega t' \Big|_0^t = v_0 - \frac{a_0}{\omega}(\cos \omega t - 1)$$

and

$$x(t) = x_0 + \int_0^t v(t')dt'$$

$$= x_0 + \int_0^t \left[v_0 - \frac{a_0}{\omega}(\cos \omega t' - 1)\right] dt'$$

$$= x_0 + \left(v_0 + \frac{a_0}{\omega}\right)t - \frac{a_0}{\omega^2} \sin \omega t.$$

The electron is initially at rest, $x_0 = v_0 = 0$, so we have

$$x(t) = \frac{a_0}{\omega}t - \frac{a_0}{\omega^2} \sin \omega t.$$

The result is interesting: the second term oscillates and corresponds to the jiggling motion of the electron that we predicted. The first term, however, corresponds to motion with uniform velocity, so in addition to the jiggling motion the electron starts to drift away. Can you see why?

1.10 More about the Time Derivative of a Vector

In Section 1.8 we demonstrated how to describe velocity and acceleration by vectors. In particular, we showed how to differentiate the vector $\mathbf{r}$ to obtain a new vector $\mathbf{v} = d\mathbf{r}/dt$. We will want to differentiate other vectors with respect to time on occasion, so it is worthwhile generalizing our discussion.

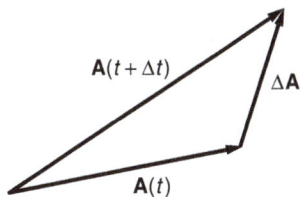

Consider a vector $\mathbf{A}(t)$ that changes with time. The change in $\mathbf{A}(t)$ during the interval from t to $t + \Delta t$ is

$$\Delta \mathbf{A} = \mathbf{A}(t + \Delta t) - \mathbf{A}(t).$$

In complete analogy to the procedure we followed in differentiating $\mathbf{r}$ in Section 1.6, we define the time derivative of $\mathbf{A}$ by

$$\frac{d\mathbf{A}}{dt} = \lim_{\Delta t \to 0} \frac{\mathbf{A}(t + \Delta t) - \mathbf{A}(t)}{\Delta t}.$$

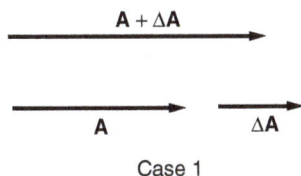

It is important to appreciate that $d\mathbf{A}/dt$ is a new vector that can be large or small, and can point in any direction, depending on the behavior of $\mathbf{A}$.

There is one respect in which $d\mathbf{A}/dt$ differs from the derivative of a simple scalar function. $\mathbf{A}$ can change in both *magnitude* and *direction*—a scalar function can change only in magnitude. This difference is important.

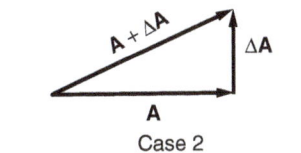

The figures illustrate the addition of an increment $\Delta \mathbf{A}$ to $\mathbf{A}$. In the first case $\Delta \mathbf{A}$ is parallel to $\mathbf{A}$; this leaves the direction unaltered but changes the *magnitude* to $|\mathbf{A}| + |\Delta \mathbf{A}|$.

In the second case, $\Delta \mathbf{A}$ is perpendicular to $\mathbf{A}$. This causes a change of *direction* but leaves the magnitude practically unaltered if $\Delta \mathbf{A}$ is small.

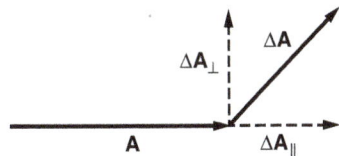

In general, $\mathbf{A}$ will change with time both in magnitude and in direction. It is useful to visualize both types of change taking place simultaneously. In the sketch we show a small increment $\Delta \mathbf{A}$ resolved into a component $\Delta \mathbf{A}_{\parallel}$ parallel to $\mathbf{A}$ and a component $\Delta \mathbf{A}_{\perp}$ perpendicular to $\mathbf{A}$. In the limit $\Delta \mathbf{A} \to 0$, as when we take the derivative, $\Delta \mathbf{A}_{\parallel}$ changes the magnitude of $\mathbf{A}$ but not its direction, while $\Delta \mathbf{A}_{\perp}$ changes the direction of $\mathbf{A}$ but not its magnitude.

Without a clear understanding of the two ways a vector can change it is easy to make an error by neglecting one of them.

1.10.1 Rotating Vectors

If $d\mathbf{A}/dt$ is always perpendicular to $\mathbf{A}$, $\mathbf{A}$ must *rotate*. Because its magnitude cannot change, its time dependence arises solely from change in direction.

The illustrations show how rotation occurs when $\Delta \mathbf{A}$ is always perpendicular to $\mathbf{A}$. The rotational motion is made more apparent by drawing

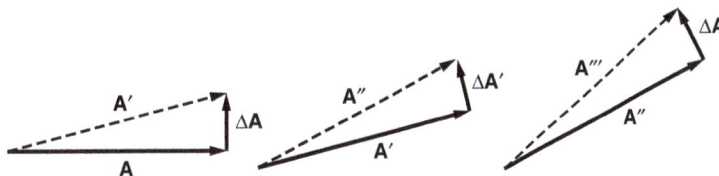

the successive vectors at a common origin.

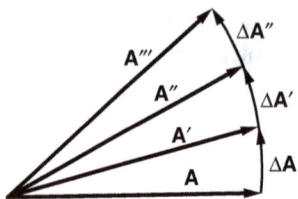

Contrast this with the case where $\Delta\mathbf{A}$ is always parallel to $\mathbf{A}$.

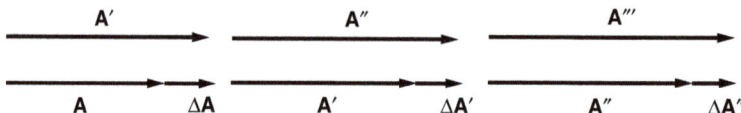

Drawn from a common origin, the vectors look like this:

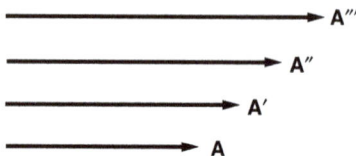

Example 1.12 Circular Motion and Rotating Vectors

This example relates the idea of rotating vectors to circular motion. In Example 1.8 we discussed the motion given by

$$\mathbf{r} = r\,(\cos\omega t\,\hat{\mathbf{i}} + \sin\omega t\,\hat{\mathbf{j}}).$$

The velocity is

$$\mathbf{v} = r\,\omega(-\sin\omega t\,\hat{\mathbf{i}} + \cos\omega t\,\hat{\mathbf{j}}).$$

Because

$$\mathbf{r}\cdot\mathbf{v} = r^2\omega(-\cos\omega t\sin\omega t + \sin\omega t\cos\omega t)$$
$$= 0$$

we see that $d\mathbf{r}/dt$ is perpendicular to $\mathbf{r}$. We conclude that the magnitude of $\mathbf{r}$ is constant. Consequently, the only possible change in $\mathbf{r}$ is a change in its direction, which is to say that $\mathbf{r}$ must rotate and the trajectory is a circle, This is precisely the case: $\mathbf{r}$ rotates about the origin.

We showed earlier that $\mathbf{a} = -\omega^2\mathbf{r}$. Since $\mathbf{r}\cdot\mathbf{v} = 0$, it follows that $\mathbf{a}\cdot\mathbf{v} = -\omega^2\mathbf{r}\cdot\mathbf{v} = 0$ and $\mathbf{a} = d\mathbf{v}/dt$ is perpendicular to $\mathbf{v}$. This means that the velocity vector has constant magnitude, so that it too changes in time only by rotation. That $\mathbf{v}$ indeed rotates is readily seen from the sketch, which shows $\mathbf{v}$ at various positions along the trajectory.

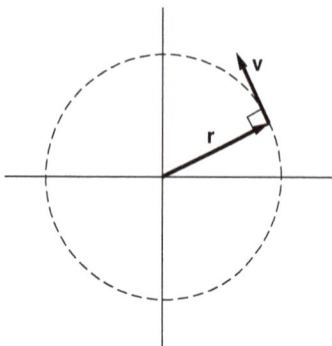

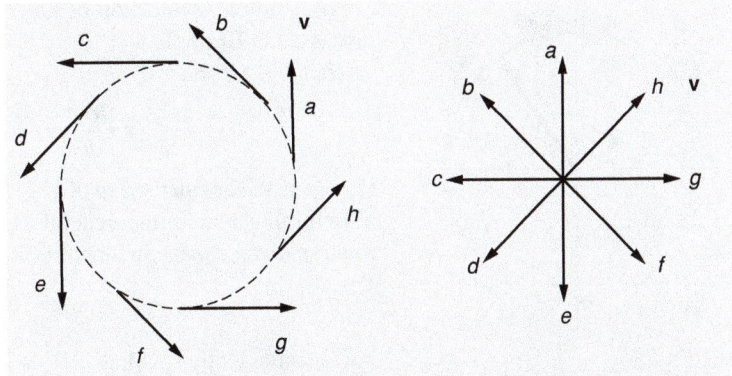

In the sketch the same velocity vectors are drawn from a common origin. It is apparent that each time the particle completes a traversal, the velocity vector has swung around through a full circle.

Perhaps you can show that the acceleration vector also undergoes uniform rotation.

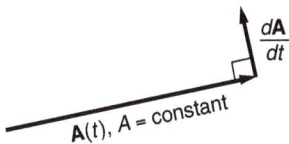

Suppose a vector $\mathbf{A}(t)$ has constant magnitude A. The only way $\mathbf{A}(t)$ can change in time is by rotating, and we shall now develop a useful expression for the time derivative $d\mathbf{A}/dt$ of such a rotating vector. The direction of $d\mathbf{A}/dt$ is always perpendicular to $\mathbf{A}$. The magnitude of $d\mathbf{A}/dt$ can be found by the following geometrical argument.

The change in $\mathbf{A}$ in the time interval t to $t + \Delta t$ is

$$\Delta\mathbf{A} = \mathbf{A}(t + \Delta t) - \mathbf{A}(t).$$

Using the angle $\Delta\theta$ defined in the sketch,

$$|\Delta\mathbf{A}| = 2A \sin\frac{\Delta\theta}{2}.$$

For $\Delta\theta \ll 1$, $\sin \Delta\theta/2 \approx \Delta\theta/2$, as discussed in Note 1.2. We have

$$|\Delta\mathbf{A}| \approx 2A\frac{\Delta\theta}{2}$$
$$= A\Delta\theta$$

and

$$\left|\frac{\Delta\mathbf{A}}{\Delta t}\right| \approx A\frac{\Delta\theta}{\Delta t}.$$

Taking the limit $\Delta t \to 0$

$$\left|\frac{d\mathbf{A}}{dt}\right| = A\frac{d\theta}{dt}.$$

$d\theta/dt$ is called the *angular speed* of $\mathbf{A}$.

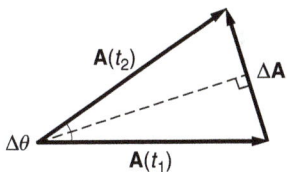

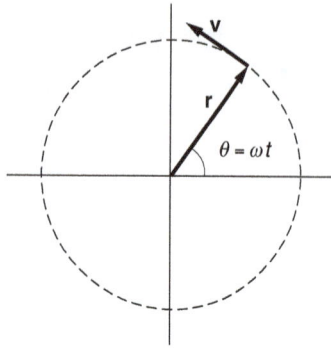

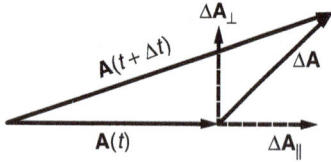

For a simple application of this result, let $\mathbf{A}$ be the rotating vector $\mathbf{r}$ discussed in Example 1.8.

Then $\theta = \omega t$ and

$$\left|\frac{d\mathbf{r}}{dt}\right| = r\frac{d}{dt}(\omega t) = r\omega \quad \text{or} \quad v = r\omega.$$

Here ω is the angular speed of $\mathbf{r}$.

Returning now to the general case, a change in $\mathbf{A}$ is the result of a rotation *and* a change in magnitude:

$$\Delta\mathbf{A} = \Delta\mathbf{A}_\perp + \Delta\mathbf{A}_\parallel.$$

For $\Delta\theta$ sufficiently small,

$$|\Delta\mathbf{A}_\perp| = A\Delta\theta$$
$$|\Delta\mathbf{A}_\parallel| = \Delta A.$$

Dividing by Δt and taking the limit

$$\left|\frac{d\mathbf{A}_\perp}{dt}\right| = A\frac{d\theta}{dt}$$
$$\left|\frac{d\mathbf{A}_\parallel}{dt}\right| = \frac{dA}{dt}.$$

$d\mathbf{A}_\perp/dt$ vanishes if the direction of $\mathbf{A}$ is constant, $(d\theta/dt = 0)$, and $d\mathbf{A}_\parallel/dt$ vanishes if the magnitude of $\mathbf{A}$ is constant.

We conclude this section by stating some formal identities in vector differentiation, with their proofs left as exercises. Let the scalar c and the vectors $\mathbf{A}$ and $\mathbf{B}$ be functions of time. Then

$$\frac{d}{dt}(c\mathbf{A}) = \frac{dc}{dt}\mathbf{A} + c\frac{d\mathbf{A}}{dt}$$
$$\frac{d}{dt}(\mathbf{A} \cdot \mathbf{B}) = \frac{d\mathbf{A}}{dt}\cdot\mathbf{B} + \mathbf{A}\cdot\frac{d\mathbf{B}}{dt}$$
$$\frac{d}{dt}(\mathbf{A} \times \mathbf{B}) = \frac{d\mathbf{A}}{dt}\times\mathbf{B} + \mathbf{A}\times\frac{d\mathbf{B}}{dt}.$$

Consider also $A^2 = \mathbf{A} \cdot \mathbf{A}$. Then

$$\frac{d}{dt}(A^2) = 2\mathbf{A}\cdot\frac{d\mathbf{A}}{dt},$$

and we see again that if $d\mathbf{A}/dt$ is perpendicular to $\mathbf{A}$, the magnitude of $\mathbf{A}$ is constant, $d(A^2)/dt = 0$.

1.11 Motion in Plane Polar Coordinates

The rectangular, or Cartesian, coordinates we have used so far are well suited to describing motion in a straight line. For instance, if we orient the coordinate system so that one axis lies in the direction of motion, then only a single coordinate changes as the point moves. However, rectangular coordinates are cumbersome for describing circular motion. Because circular motion plays a prominent role in physics, it is worth introducing a more convenient coordinate system.

We should emphasize that although in physics one is free to choose any coordinate system one pleases, the proper choice of a coordinate system can *vastly* simplify a problem. The material in this section, which introduces a coordinate system that is beautifully suited for many problems, is very much in the spirit of more advanced physics. Some of this material may be new to you. Be patient if it seems strange at first. Once you have studied the examples and worked a few problems, it will seem natural.

1.11.1 Polar Coordinates

This two-dimensional coordinate system is based on the three-dimensional cylindrical coordinate system, much as the x–y plane of the Cartesian coordinate system is a subset of the three-dimensional x–y–z system. The z axis of the cylindrical system is identical to that of the Cartesian system.

However, position in the x–y plane is described not by x–y coordinates but by r–θ coordinates where r is the distance from the origin and θ is the angle between r and the x axis, as shown in the sketch.

We see that

$$r = \sqrt{x^2 + y^2}$$
$$\theta = \arctan\left(\frac{y}{x}\right).$$

Since we shall be concerned primarily with motion in a plane, we neglect the z axis for now and restrict our discussion to two dimensions. The coordinates r and θ are called *plane polar* coordinates. In the following sections we shall learn how to describe position, velocity, and acceleration in these coordinates.

The contrast between Cartesian and plane polar coordinates is readily seen by comparing drawings of constant coordinate lines for the two systems.

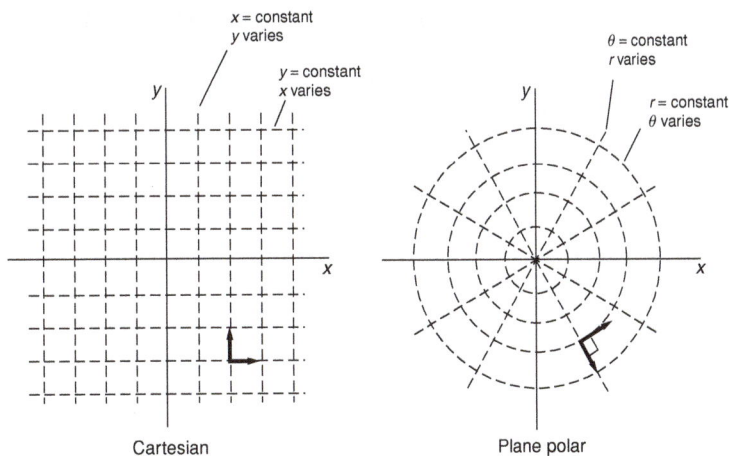

Cartesian

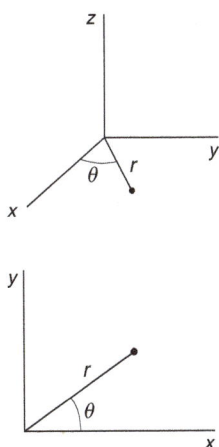

Plane polar

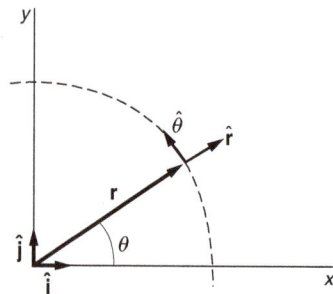

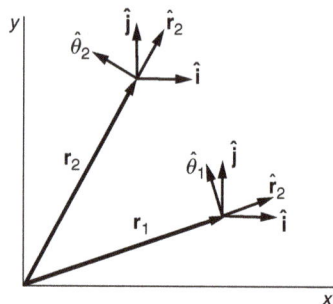

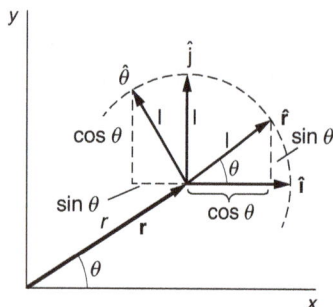

The lines of constant x and of constant y are straight and perpendicular to each other. Lines of constant θ are also straight, directed radially outward from the origin. In contrast, lines of constant r are circles concentric about the origin. Note, however, that the lines of constant θ and constant r are perpendicular wherever they intersect; the Cartesian and plane polar coordinate systems are both *orthogonal* coordinate systems. Orthogonality is important because it makes the base vectors independent of one another, as we have already seen for the $\hat{\mathbf{i}}$, $\hat{\mathbf{j}}$, $\hat{\mathbf{k}}$ triplet in the Cartesian system. In an orthogonal system no base vector has a component along some other base vector.

In Section 1.6 we introduced the base unit vectors $\hat{\mathbf{i}}$ and $\hat{\mathbf{j}}$ which point in the direction of increasing x and increasing y, respectively. In the same spirit we now introduce two new base unit vectors, $\hat{\mathbf{r}}$ and $\hat{\boldsymbol{\theta}}$, that point in the direction of increasing r and increasing θ, respectively.

There is a fundamental difference between polar and Cartesian base vectors: the directions of $\hat{\mathbf{r}}$ and $\hat{\boldsymbol{\theta}}$ vary with position, whereas $\hat{\mathbf{i}}$ and $\hat{\mathbf{j}}$ have fixed directions. The drawing shows this by illustrating both sets of base vectors at two points in space.

Because $\hat{\mathbf{r}}$ and $\hat{\boldsymbol{\theta}}$ vary with position, kinematical formulas can look more complicated in polar coordinates than in the Cartesian system.

Although the directions of $\hat{\mathbf{r}}$ and $\hat{\boldsymbol{\theta}}$ vary with position, the directions depend on θ only, not on r. As a reminder of this θ-dependence, we sometimes show it explicitly by writing $\hat{\mathbf{r}}(\theta)$ and $\hat{\boldsymbol{\theta}}(\theta)$.

The drawing shows the unit vectors $(\hat{\mathbf{i}}, \hat{\mathbf{j}})$ and also $(\hat{\mathbf{r}}(\theta), \hat{\boldsymbol{\theta}}(\theta))$ at a point in the x–y plane. We see that

$$\hat{\mathbf{r}}(\theta) = \cos\theta\,\hat{\mathbf{i}} + \sin\theta\,\hat{\mathbf{j}} \tag{1.1}$$

$$\hat{\boldsymbol{\theta}}(\theta) = -\sin\theta\,\hat{\mathbf{i}} + \cos\theta\,\hat{\mathbf{j}}. \tag{1.2}$$

It is worth convincing yourself that these expressions are reasonable by checking them at a few points, such as $\theta = 0$ and $\pi/2$. Also, you can confirm that $\hat{\mathbf{r}}(\theta)$ and $\hat{\boldsymbol{\theta}}(\theta)$ are indeed orthogonal by showing that $\hat{\mathbf{r}} \cdot \hat{\boldsymbol{\theta}} = 0$.

It is easy to confirm that the vector $\mathbf{r}$ is the same whether we describe it by Cartesian or polar coordinates. In Cartesian coordinates we have

$$\mathbf{r} = x\,\hat{\mathbf{i}} + y\,\hat{\mathbf{j}}$$

while in polar coordinates we have

$$\mathbf{r} = r\,\hat{\mathbf{r}}.$$

If we insert Eq. (1.1) for $\hat{\mathbf{r}}$, we obtain

$$x\,\hat{\mathbf{i}} + y\,\hat{\mathbf{j}} = r\,(\cos\theta\,\hat{\mathbf{i}} + \sin\theta\,\hat{\mathbf{j}}).$$

By orthogonality (or by taking the dot product of this equation with $\hat{\mathbf{i}}$ and $\hat{\mathbf{j}}$, respectively) we have

$$x = r\cos\theta$$
$$y = r\sin\theta,$$

as we expect.

Expressing velocity in polar coordinates requires taking the time derivative of

$$\mathbf{r} = r\,\hat{\mathbf{r}}(\theta),$$

and this requires care. Using the chain rule, we have

$$\frac{d\mathbf{r}}{dt} = \frac{dr}{dt}\hat{\mathbf{r}}(\theta) + r\frac{d\hat{\mathbf{r}}(\theta)}{dt}.$$

The meaning of the first term is transparent: this is the speed in the radial direction. The second term, however, involves a new concept—taking the time derivative of a base vector. So, let us investigate how to do this, both for $\hat{\mathbf{r}}(\theta)$ and for $\hat{\boldsymbol{\theta}}(\theta)$.

1.11.2 $d\hat{\mathbf{r}}/dt$ and $d\hat{\boldsymbol{\theta}}/dt$ in Polar Coordinates

Our goal here is to calculate the time derivatives of $\hat{\mathbf{r}}$ and $\hat{\boldsymbol{\theta}}$. We will need these results to express velocity $\mathbf{v}$ and acceleration $\mathbf{a}$ in polar coordinates.

Using Newton's notation for time derivatives can help make equations easier to read. For example,

$$\frac{d\theta}{dt} = \dot{\theta}$$
$$\frac{d^2\theta}{dt^2} = \ddot{\theta}.$$

Our starting point is Eq. (1.1): $\hat{\mathbf{r}} = \cos\theta\,\hat{\mathbf{i}} + \sin\theta\,\hat{\mathbf{j}}$. Differentiating with respect to time yields

$$\frac{d\hat{\mathbf{r}}}{dt} = \frac{d}{dt}(\cos\theta)\,\hat{\mathbf{i}} + \frac{d}{dt}(\sin\theta)\,\hat{\mathbf{j}}$$
$$= -\sin\theta\,\dot{\theta}\,\hat{\mathbf{i}} + \cos\theta\,\dot{\theta}\,\hat{\mathbf{j}}$$
$$= (-\sin\theta\,\hat{\mathbf{i}} + \cos\theta\,\hat{\mathbf{j}})\,\dot{\theta}.$$

Recalling from Eq. (1.2) that $\hat{\boldsymbol{\theta}} = -\sin\theta\,\hat{\mathbf{i}} + \cos\theta\,\hat{\mathbf{j}}$ we see that

$$\frac{d\hat{\mathbf{r}}}{dt} = \dot{\theta}\,\hat{\boldsymbol{\theta}}.$$

Similarly, by taking the time derivative of Eq. (1.2), we have

$$\frac{d\hat{\boldsymbol{\theta}}}{dt} = (-\cos\theta\,\hat{\mathbf{i}} - \sin\theta\,\hat{\mathbf{j}})\,\dot{\theta}$$
$$= -\dot{\theta}\,\hat{\mathbf{r}}.$$

We will need to call on these results and so we summarize them here:

$$\frac{d\hat{\mathbf{r}}}{dt} = \dot{\theta}\,\hat{\boldsymbol{\theta}} \tag{1.3}$$

$$\frac{d\hat{\boldsymbol{\theta}}}{dt} = -\dot{\theta}\,\hat{\mathbf{r}}. \tag{1.4}$$

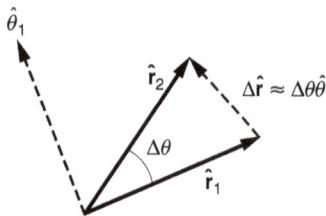

Example 1.13 Geometric Derivation of $d\hat{\mathbf{r}}/dt$ **and** $d\hat{\boldsymbol{\theta}}/dt$
It is always helpful to have alternative ways to understand new concepts. We derived Eqs. (1.3) and (1.4) algebraically, but we can also derive them geometrically by invoking the concept of rotating vectors from Section 1.8.

Because $\hat{\mathbf{r}}(\theta)$ and $\hat{\boldsymbol{\theta}}(\theta)$ are unit vectors, their magnitudes are constant and the only way they can change is by changing their direction, that is, by *rotating*.

The sketch shows $\hat{\mathbf{r}}(\theta)$ at time t where it lies at some angle θ and also $\hat{\mathbf{r}}(\theta + \Delta\theta)$ at time $t + \Delta t$ a little later, when the angle is $\theta + \Delta\theta$. The corresponding change in $\hat{\mathbf{r}}$, denoted by $\Delta\hat{\mathbf{r}}$, is almost perpendicular to $\hat{\mathbf{r}}$:

$$|\Delta\hat{\mathbf{r}}| \approx |\hat{\mathbf{r}}|\,\Delta\theta = \Delta\theta,$$

Dividing by Δt gives

$$\frac{|\Delta\hat{\mathbf{r}}|}{\Delta t} \approx \frac{\Delta\theta}{\Delta t},$$

and taking the limit $\Delta t \to 0$, we have

$$\left|\frac{d\hat{\mathbf{r}}}{dt}\right| = \frac{d\theta}{dt} = \dot{\theta}.$$

Now that we have both the magnitude and direction, we conclude

$$\frac{d\hat{\mathbf{r}}}{dt} = \dot{\theta}\,\hat{\boldsymbol{\theta}},$$

as we expect.

Similarly, the sketch shows that the change in $\hat{\boldsymbol{\theta}}$, denoted by $\Delta\hat{\boldsymbol{\theta}}$, points radially inward along $\hat{\mathbf{r}}$. Thus

$$|\Delta\hat{\boldsymbol{\theta}}| \approx |\hat{\boldsymbol{\theta}}|\,\Delta\theta = \Delta\theta,$$

$$\frac{|\Delta\hat{\boldsymbol{\theta}}|}{\Delta t} \approx \frac{\Delta\theta}{\Delta t}.$$

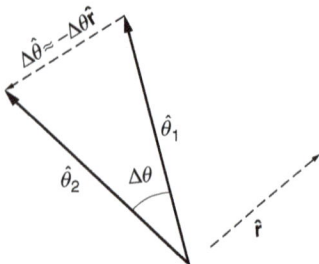

Taking the limit,

$$\left|\frac{d\hat{\boldsymbol{\theta}}}{dt}\right| = \frac{d\theta}{dt} = \dot{\theta},$$

and combining the results gives

$$\frac{d\hat{\boldsymbol{\theta}}}{dt} = -\dot{\theta}\,\hat{\mathbf{r}},$$

as we expect.

1.11.3 Velocity in Polar Coordinates

We now have the means for evaluating velocity using polar coordinates:

$$\mathbf{v} = \frac{d}{dt}(r\,\hat{\mathbf{r}}) = \dot{r}\,\hat{\mathbf{r}} + r\frac{d\hat{\mathbf{r}}}{dt}.$$

Using Eq. (1.3), we obtain

$$\mathbf{v} = \dot{r}\,\hat{\mathbf{r}} + r\dot{\theta}\,\hat{\boldsymbol{\theta}}.$$

We can get insight into the meaning of each term by considering cases where only one component varies at a time.

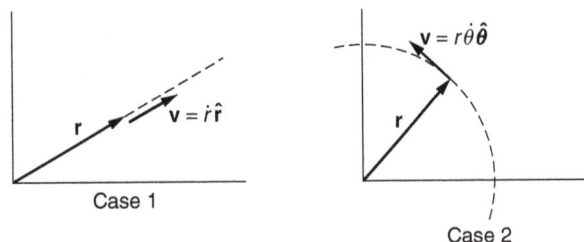

Case 1   Case 2

Case 1: Radial velocity (θ = constant, r varies). If θ is a constant, $\dot{\theta} = 0$, and $\mathbf{v} = \dot{r}\hat{\mathbf{r}}$. We have one-dimensional motion in a fixed radial direction.

Case 2: Tangential velocity (r = constant, θ varies). In this case $\mathbf{v} = r\dot{\theta}\,\hat{\boldsymbol{\theta}}$. Since r is fixed, the motion lies on the arc of a circle, that is, in the *tangential* direction. The speed of the point on the circle is $r\dot{\theta}$, and it follows that $\mathbf{v} = r\dot{\theta}\,\hat{\boldsymbol{\theta}}$.

If r and θ both change, the velocity is a combination of radial and tangential motion.

The next four examples illustrate the use of polar coordinates to describe velocity.

Example 1.14 Circular Motion in Polar Coordinates

A particle moves in a circle of radius b with angular velocity $\dot{\theta} = \alpha t$, where α is a constant. (α has the units rad/s^2.) Describe the particle's velocity in polar coordinates.

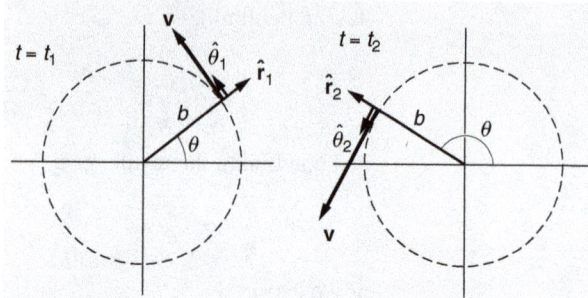

Since $r = b =$ constant, $\mathbf{v}$ is purely tangential and $\mathbf{v} = b\alpha t\hat{\boldsymbol{\theta}}$. The sketches show $\hat{\mathbf{r}}, \hat{\boldsymbol{\theta}}$, and $\mathbf{v}$ at a time t_1 and at a later time t_2.

The particle is located at the position

$$r = b, \quad \theta = \theta_0 + \int_0^t \dot{\theta}\, dt = \theta_0 + \tfrac{1}{2}\alpha t^2.$$

If the particle is on the x axis at $t = 0$, then $\theta_0 = 0$. The particle's position vector is $\mathbf{r} = b\hat{\mathbf{r}}$, but as the sketches indicate, θ must be given to specify the direction of $\hat{\mathbf{r}}$.

Example 1.15 Straight Line Motion in Polar Coordinates

Consider a particle moving with constant velocity $\mathbf{v} = u\hat{\mathbf{i}}$ along the line $y = 2$. Describe $\mathbf{v}$ in polar coordinates:

$$\mathbf{v} = v_r\hat{\mathbf{r}} + v_\theta\hat{\boldsymbol{\theta}}.$$

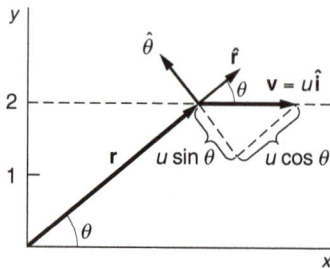

From the sketch,

$$v_r = u\cos\theta$$
$$v_\theta = -u\sin\theta$$
$$\mathbf{v} = u\cos\theta\,\hat{\mathbf{r}} - u\sin\theta\,\hat{\boldsymbol{\theta}}.$$

As the particle moves to the right, θ decreases and $\hat{\mathbf{r}}$ and $\hat{\boldsymbol{\theta}}$ change direction. Ordinarily, of course, we try to use coordinates that make the problem as simple as possible; polar coordinates can be used here, but they are not well suited to this problem.

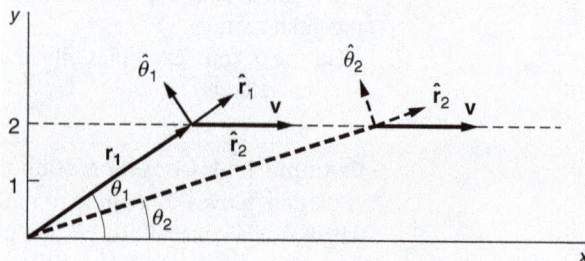

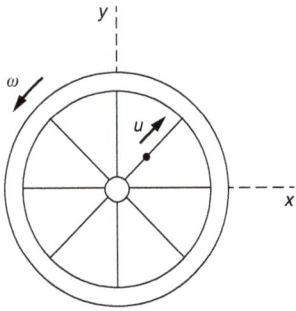

Example 1.16 Velocity of a Bead on a Spoke

A bead moves along the spoke of a wheel at constant speed u meters per second. The wheel rotates with uniform angular velocity $\dot{\theta} = \omega$ radians per second about an axis fixed in space.

At $t = 0$ the spoke is along the x axis, and the bead is at the origin. Find the bead's velocity at time t (a) in polar coordinates; (b) in Cartesian coordinates.

(a) In polar coordinates, $r = ut$, $\dot{r} = u$, $\dot{\theta} = \omega$. Hence

$$\mathbf{v} = \dot{r}\,\hat{\mathbf{r}} + r\dot{\theta}\,\hat{\boldsymbol{\theta}} = u\,\hat{\mathbf{r}} + u\omega t\,\hat{\boldsymbol{\theta}}.$$

At time t, the bead is at radius ut on the spoke, and the spoke makes angle ωt with the x axis.

(b) In Cartesian coordinates, we have

$$v_x = v_r \cos\theta - v_\theta \sin\theta$$
$$v_y = v_r \sin\theta + v_\theta \cos\theta.$$

Since $v_r = u$, $v_\theta = r\omega = u\omega t$, $\theta = \omega t$, we obtain

$$\mathbf{v} = (u\cos\omega t - u\omega t \sin\omega t)\,\hat{\mathbf{i}} + (u\sin\omega t + u\omega t \cos\omega t)\,\hat{\mathbf{j}}.$$

Note how much simpler the result is in plane polar coordinates. (Incidentally, the trajectory of the bead is a figure known as the Archimedean spiral.)

Example 1.17 Motion on an Off-center Circle

A particle moves with constant speed v around a circle of radius b, with the circle offset from the origin of coordinates by distance b so that it is tangential to the y axis. Find the particle's velocity vector in polar coordinates.

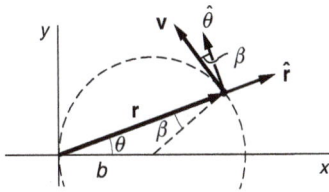

With this origin, $\mathbf{v}$ is no longer purely parallel to $\hat{\boldsymbol{\theta}}$, as the sketch indicates:

$$\mathbf{v} = -v\sin\beta\,\hat{\mathbf{r}} + v\cos\beta\,\hat{\boldsymbol{\theta}}$$
$$= -v\sin\theta\,\hat{\mathbf{r}} + v\cos\theta\,\hat{\boldsymbol{\theta}}.$$

The last step follows because β and θ are the base angles of an isosceles triangle and are therefore equal.

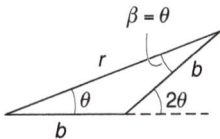

To complete the calculation, we must find θ as a function of time. By geometry, $2\theta = \omega t$ or $\theta = \omega t/2$, where $\omega = v/b$. Hence

$$\mathbf{v} = -v\,\sin(vt/2b)\hat{\mathbf{r}} + v\,\cos(vt/2b)\hat{\boldsymbol{\theta}}.$$

1.11.4 Acceleration in Polar Coordinates

Our remaining task is to express acceleration in polar coordinates. We differentiate **v** to obtain

$$\mathbf{a} = \frac{d\mathbf{v}}{dt}$$

$$= \frac{d}{dt}(\dot{r}\,\hat{\mathbf{r}} + r\dot{\theta}\,\hat{\boldsymbol{\theta}})$$

$$= \ddot{r}\,\hat{\mathbf{r}} + \dot{r}\frac{d}{dt}\hat{\mathbf{r}} + \dot{r}\dot{\theta}\,\hat{\boldsymbol{\theta}} + r\ddot{\theta}\,\hat{\boldsymbol{\theta}} + r\dot{\theta}\frac{d}{dt}\hat{\boldsymbol{\theta}}.$$

Using Eqs. (1.3) and (1.4) for $d\hat{\mathbf{r}}/dt$ and $d\hat{\boldsymbol{\theta}}/dt$, we obtain

$$\mathbf{a} = \ddot{r}\,\hat{\mathbf{r}} + \dot{r}\dot{\theta}\,\hat{\boldsymbol{\theta}} + \dot{r}\dot{\theta}\,\hat{\boldsymbol{\theta}} + r\ddot{\theta}\,\hat{\boldsymbol{\theta}} - r\dot{\theta}^2\,\hat{\mathbf{r}}$$

$$= (\ddot{r} - r\dot{\theta}^2)\,\hat{\mathbf{r}} + (r\ddot{\theta} + 2\dot{r}\dot{\theta})\,\hat{\boldsymbol{\theta}}.$$

This is quite a string of terms, but they will seem more understandable when we give them physical and geometrical interpretations.

Case 1: Radial acceleration. The term $\ddot{r}\,\hat{\mathbf{r}}$ is the acceleration due to a change in radial speed. The second term $-r\dot{\theta}^2\,\hat{\mathbf{r}}$ is the centripetal acceleration that we encountered earlier. The diagram shows how it results from a change in the *direction* of the tangential velocity v_t. From the diagram, $\Delta v_t \approx v_t \Delta\theta$. In the limit, $dv_t/dt = v_t\dot{\theta} = r\dot{\theta}^2$. The direction is radially inward.

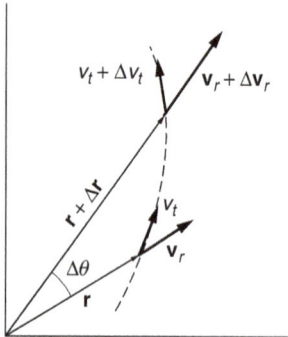

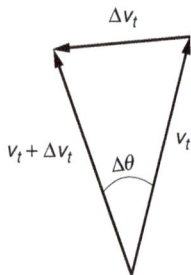

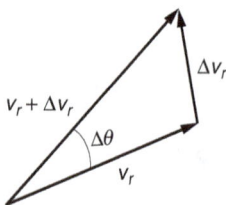

Case 2: Tangential acceleration. The term $r\ddot{\theta}\,\hat{\boldsymbol{\theta}}$ is the acceleration that arises from the changing tangential speed. The next term, $2\dot{r}\dot{\theta}\,\hat{\boldsymbol{\theta}}$, may not be so familiar. This is known as the *Coriolis* acceleration. Perhaps you have heard of the *Coriolis force*. This is a fictitious force that appears to act in a rotating coordinate system, as we shall study in Chapter 9. In contrast, the Coriolis acceleration that we are discussing here is a real acceleration that is present whenever r and θ both change with time. Half of the Coriolis acceleration is due to the change in direction of the radial velocity.

From the diagram, it is evident that the change in time Δt is $\Delta v_r \approx v_r \Delta\theta$, and in the limit, $dv_r/dt = v_r\dot{\theta}$. To see how the other half arises, consider the tangential speed $v_\theta = r\dot{\theta}$. If r changes by Δr, then v_θ changes by $\Delta v_\theta = \Delta r\dot{\theta}$, and the contribution to the tangential acceleration is therefore $\dot{r}\dot{\theta}$, the other half of the Coriolis acceleration.

Example 1.18 Acceleration of a Bead on a Spoke

A bead moves outward with constant speed u along the spoke of a wheel. It starts from the center at $t = 0$. The angular position of the spoke is given by $\theta = \omega t$, where ω is a constant. Find the velocity and acceleration.

$$\mathbf{v} = \dot{r}\,\hat{\mathbf{r}} + r\dot{\theta}\,\hat{\boldsymbol{\theta}}.$$

We are given that $\dot{r} = u$ and $\theta = \omega t$, so that $\dot{\theta} = \omega$. The radial position is given by $r = ut$, and as in Example 1.16 we have

$$\mathbf{v} = u\,\hat{\mathbf{r}} + u\omega t\,\hat{\boldsymbol{\theta}}.$$

The acceleration is

$$\mathbf{a} = (\ddot{r} - r\dot{\theta}^2)\,\hat{\mathbf{r}} + (r\ddot{\theta} + 2\dot{r}\dot{\theta})\,\hat{\boldsymbol{\theta}}$$

$$= -u\omega^2 t\,\hat{\mathbf{r}} + 2u\omega\,\hat{\boldsymbol{\theta}}.$$

The velocity is shown in the sketch for several different positions of the wheel.

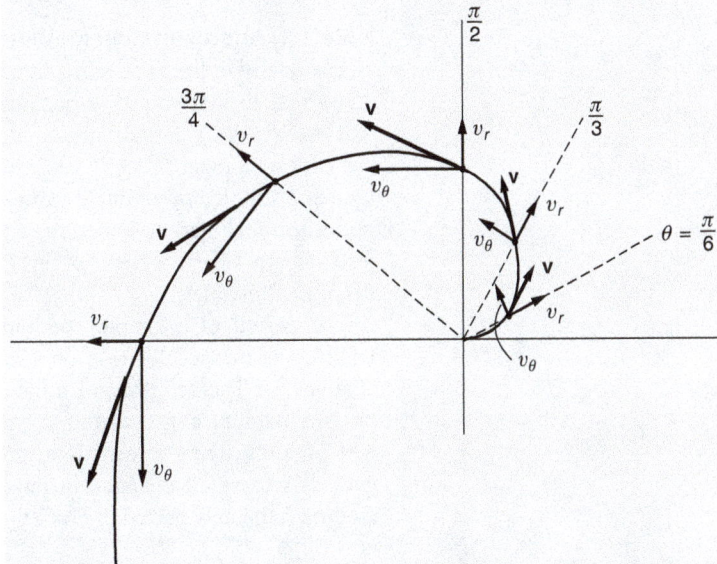

(The trajectory is again an Archimedean spiral.) Note that the radial velocity is constant. The tangential acceleration is also constant—can you visualize this?

Example 1.19 Radial Motion without Acceleration

A particle moves with $\dot{\theta} = \omega = $ constant and $r = r_0 e^{\beta t}$, where r_0 and β are constants. We shall show that for certain values of β, the particle moves with $a_r = 0$.

$$\mathbf{a} = (\ddot{r} - r\dot{\theta}^2)\,\hat{\mathbf{r}} + (r\ddot{\theta} + 2\dot{r}\dot{\theta})\,\hat{\boldsymbol{\theta}}$$

$$= (\beta^2 r_0 e^{\beta t} - r_0 e^{\beta t}\omega^2)\,\hat{\mathbf{r}} + 2\beta r_0 e^{\beta t}\omega\,\hat{\boldsymbol{\theta}}.$$

If $\beta = \pm\omega$, the radial part of $\mathbf{a}$ vanishes.

It is very surprising at first that when $r = r_0 e^{\beta t}$ the particle moves with zero radial acceleration. The error is in thinking that $\ddot{r}$ makes the only contribution to a_r; the term $-r\dot{\theta}^2$ is also part of the radial acceleration, and cannot be neglected.

The paradox is that even though $a_r = 0$, the radial velocity $v_r = \dot{r} = r_0 \omega e^{\beta t}$ is increasing rapidly with time. The answer is that we can be misled by the special case of Cartesian coordinates; in polar coordinates,

$$v_r \neq \int a_r(t)\, dt,$$

because $\int a_r(t)dt$ does not take into account the fact that the unit vectors $\hat{r}$ and $\hat{\theta}$ are functions of time.

Note 1.1 Approximation Methods

Occasionally in the course of solving a physics problem you might suddenly realize that you have become so involved with the mathematics that the physics is largely obscured. In such a case, it is worth stepping back for a moment to see if you can reduce the complexity, for instance by using simple approximate expressions instead of exact formulas. You might feel that there is something essentially wrong with substituting inexact results for exact ones but this is not really the case, as the following example illustrates.

The period of a simple pendulum of length L is $T_0 = 2\pi\sqrt{g/L}$, where g is the acceleration of gravity. (This result will be derived in Chapter 3.) The accuracy of a clock driven by the pendulum depends on L remaining constant, but L can change due to thermal expansion and possibly aging effects. The problem is to find how sensitive the period is to small changes in length. If the length changes by some amount l, the new period is $T = 2\pi\sqrt{g/(L + l)}$. The change in the period is

$$\Delta T = T - T_0 = 2\pi\left(\sqrt{\frac{g}{(L + l)}} - \sqrt{\frac{g}{L}}\right).$$

This equation is exact but not particularly informative. It gives little insight as to how ΔT depends on the change in length l. Also, if $l \ll L$, which is generally the case of interest, ΔT is the small difference of two large numbers, which makes the result very sensitive to numerical errors. However, by recasting the form of ΔT, both of these problems can be solved. The trick is to write ΔT as a power series in the small parameter $x = l/L$. We have

$$\Delta T = T - T_0 = 2\pi\left(\sqrt{\frac{g}{L}\left(\frac{1}{1 + l/L}\right)} - \sqrt{\frac{g}{L}}\right)$$

$$= T_0\left(\sqrt{\frac{1}{1 + x}} - 1\right). \tag{1}$$

Next, we make use of the following identity, which will be derived in the following section:

$$\sqrt{\frac{1}{1+x}} = 1 - \frac{1}{2}x + \frac{3}{8}x^2 - \frac{1}{16}x^3 + \cdots \qquad (2)$$

This expansion is valid provided $x < 1$. Inserting this in Eq. (1) gives

$$\Delta T = T_0 \left(-\frac{1}{2}x + \frac{3}{8}x^2 - \frac{1}{16}x^3 + \cdots \right). \qquad (3)$$

Note that in the limit that $l \to 0$, $\Delta T \to 0$, as we expect. For $x \ll 1$, only the first term on the right may be important, in which case

$$\Delta T \approx -\frac{1}{2}T_0 x = -\frac{1}{2}T_0 \frac{l}{L}.$$

For instance, if $l/L = 0.01$, the period is increased by about $5 \times 10^{-3} T_0$. Although the result is approximate, we can estimate how good it is. The error is less than the first neglected term, which in this case is $(3/8)T_0(l/L)^2 \approx 4 \times 10^{-5} T_0$. If higher accuracy is required, further terms can be included.

Equations in the form of Eqs. (2) and (3) are known as *power series* expansions. Such expansions can be enormously helpful both for finding symbolic solutions to problems, and for calculating numerical results. Here is how they are generated, with some useful examples.

Note 1.2 The Taylor Series
The general form for expressing a function $f(x)$ as a power series in x is

$$f(x) = a_0 + a_1 x + a_2 x^2 + \cdots = \sum_{k=0}^{\infty} a_k x^k, \qquad (1)$$

where $a_0, a_1, a_2, \ldots$ are constants that we find as follows. Evaluating the series at $x = 0$ yields

$$a_0 = f(0).$$

We now differentiate the series, assuming that $f(x)$ is well behaved, in other words, that it is differentiable. This gives

$$\frac{df}{dx} = f'(x) = a_1 + 2a_2 x + 3a_3 x^3 + \cdots$$

We once again evaluate the series at $x = 0$, yielding

$$a_1 = f'(x)\big|_{x=0} \cdots$$

If we differentiate the series k times, we find

$$a_k = \frac{1}{k!} f^{(k)}(x)\big|_{x=0}$$

where $f^{(k)}(x)$ is the kth derivative of $f(x)$. The symbol $k!$, called "k factorial", stands for $k \times (k-1) \times (k-2) \times \cdots \times 1$. To simplify the notation, we often write $f^{(k)}(x)\big|_{x=0}$ as $f^{(k)}(0)$, bearing in mind that $f^{(k)}(0)$ means that first we differentiate $f(x)$ k times and only then set x equal to 0. Combining these results, Eq. (1) becomes

$$f(x) = f(0) + f'(0)x + f''(0)\frac{x^2}{2!} + f'''(0)\frac{x^3}{3!} + \cdots \tag{2}$$

This expansion is known as the *Taylor series*. There is no guarantee that the series converges, but if it does it provides good approximations to $f(x)$ in the vicinity of $x = 0$.

The Taylor series is easily generalized to permit expansion of $f(x)$ as a power series about some other origin, for instance $x = a$:

$$f(a + x) = f(a) + f'(a)x + f''(a)\frac{x^2}{2!} + \cdots$$

This power series expansion is known as the *MacLaurin series*.

For some functions the Taylor series converges for all values of x; for others the range may be limited. We shall simply assume that we are dealing with functions for which the range of convergence is either infinite or is readily apparent.

Note 1.3 Series Expansions of Some Common Functions
A. Trigonometric functions

Using $d\sin x/dx = \cos x$ and $d\cos x/dx = -\sin x$, and the values $\sin(0) = 0$ and $\cos(0) = 1$, Eq. (2) yields

$$\sin x = x - \frac{1}{3!}x^3 + \frac{1}{5!}x^5 - \frac{1}{7!}x^7 + \cdots$$

$$\cos x = 1 - \frac{1}{2!}x^2 + \frac{1}{4!}x^4 - \frac{1}{6!}x^6 + \cdots$$

These series converge for all values of x. For small values of x we have

$$\sin x \approx x$$

$$\cos x \approx 1 - \frac{1}{2}x^2.$$

These expressions, which are sometimes called *the small angle approximation*, are valid up to terms of order x^3, denoted by $O(x^3)$.

B. The exponential function

The exponential function is e^x, where $e = 2.71828\cdots$ is the base of the natural logarithms. The fundamental properties of the exponential are that $de^x/dx = e^x$ and $e^0 = 1$. The Taylor series is

$$e^x = 1 + x + \frac{1}{2}x^2 + \frac{1}{3 \times 2}x^3 + \cdots + \frac{1}{n!}x^n + \cdots$$

This series converges for all values of x.

C. Some algebraic functions

1. $\dfrac{1}{1 \pm x} = 1 \mp x + x^2 \mp x^3 + \cdots \qquad -1 < x < 1$

2. $\dfrac{1}{1 - x} = 1 + x + x^2 + x^3 + \cdots \qquad -1 < x < 1$

3. $\dfrac{1}{\sqrt{1 + x}} = 1 - \dfrac{1}{2}x + \dfrac{1}{8}x^2 + \cdots \qquad -1 < x < 1$

D. The binomial series

The expression $(1 + x)^n$ occurs in many contexts. Its power series expansion, known as the *binomial series*, is easily found as a Taylor series. (If n is an integer, the power series can also be found by direct algebraic expansion.)

$$(1 + x)^n = 1 + nx + \frac{n(n - 1)}{2!}x^2 + \frac{n(n - 1)(n - 2)}{3!}x^3 + \cdots$$
$$+ \frac{n(n - 1)\cdots(n - k + 1)}{k!}x^k + \cdots$$

This result is valid for $-1 < x < 1$, and for both integral and fractional values of n. If n is an integer, the series terminates, the last term being x^n.

In the introduction to this section the expression $f(x) = 1/\sqrt{1 + x} = (1 + x)^{-1/2}$ arose. This is given by the binomial series with $n = 1/2$:

$$(1 + x)^{-\frac{1}{2}} = 1 - \frac{1}{2}x + \frac{3}{8}x^2 - \frac{5}{16}x^3 + \cdots$$

The binomial series can also be employed even if $|x| > 1$ by the following procedure:

$$(1 + x)^n = x^n\left(1 + \frac{1}{x}\right)^n$$
$$= x^n\left[1 + n\frac{1}{x} + \frac{n(n - 1)}{2!}\left(\frac{1}{x}\right)^2 + \cdots\right].$$

Note 1.4 Differentials

Often we need a simple approximation for the change in some function $f(x)$ when x is changed to $x + \Delta x$. We shall denote the change by $\Delta f = f(x + \Delta x) - f(x)$. The Taylor series for $f(x)$ about the point x gives

$$f(x + \Delta x) = f(x) + f'(x)\Delta x + \frac{1}{2!}f''(x)\Delta x^2 + \cdots$$

Omitting terms of order $(\Delta x)^2$ and higher yields the simple linear approximation

$$\Delta f = f(x + \Delta x) - f(x) \approx f'(x)\Delta x.$$

This approximation becomes increasingly accurate the smaller the size of Δx, but for finite values of Δx, the expression

$$\Delta f \approx f'(x)\Delta x$$

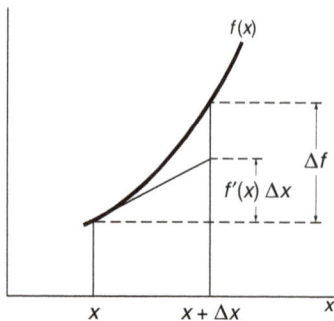

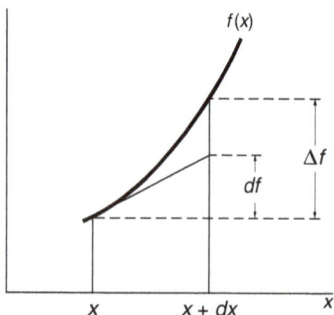

is only an approximation. The sketch shows a comparison of $\Delta f \equiv f(x + \Delta x) - f(x)$ with the linear extrapolation $f'(x)\Delta x$. It is apparent that Δf, the actual change in $f(x)$ as x is changed, is generally not exactly equal to Δf for finite Δx.

We shall use the symbol dx, called the *differential* of x, to stand for Δx. The differential of x can be as large or small as we please. We define df, the *differential of f*, by

$$df \equiv f'(x)dx.$$

This notation is illustrated in the sketches.

The symbols dx and Δx are used interchangeably but df and Δf are different quantities: df is a differential defined by $df = f'(x)dx$, whereas Δf is the actual change $f(x + dx) - f(x)$. Put another way, if we write

$$\frac{df}{dx} \approx \frac{\Delta f}{\Delta x}$$

the term df/dx on the left is the result of taking the limit $\Delta x \to 0$, but the term $\Delta f/\Delta x$ on the right is the ratio of two finite (possibly small) quantities. Nevertheless, when the linear approximation is justified in a calculation, we often use df and dx to represent finite quantities Δf and Δx. We can always do this when a limit will eventually be taken.

Among their uses, differentials provide a shorthand method for changing a variable of integration. Consider the integral

$$\int_a^b xe^{x^2}dx.$$

The exponential is simplified if we introduce the variable $t = x^2$. The procedure is first to solve for x in terms of t,

$$x = \sqrt{t},$$

and then to take differentials:

$$dx = \frac{1}{2}\frac{1}{\sqrt{t}}dt.$$

This result is exact, since we are effectively taking the limit. The original integral can now be written in terms of t:

$$\int_a^b xe^{x^2}dx = \int_{t_1}^{t_2} \sqrt{t}\,e^t\left(\frac{1}{2}\frac{1}{\sqrt{t}}dt\right) = \frac{1}{2}\int_{t_1}^{t_2} e^t dt$$

$$= \frac{1}{2}(e^{t_2} - e^{t_1}),$$

where $t_1 = a^2$ and $t_2 = b^2$.

Note 1.5 Significant Figures and Experimental Uncertainty
When performing a numerical calculation, it is helpful to have a clear guideline as to the accuracy with which the calculation should be carried out. In other words, we should retain only the number of non-trivial digits, or *significant figures*.

Examples of significant figures are: 3.1415 (five significant figures); 9 (one significant figure); 0.00021 (two significant figures); .000210 (three significant figures). Leading zeros don't count, but trailing zeros do.

A useful rule-of-thumb is that the number of significant figures retained in the result of a calculation should equal the smallest number of significant figures of any number in the calculation. For instance, if the acceleration of gravity in a calculation is taken to be 9.8 m/s^2, the result of the calculation should be quoted to no more than two significant figures.

Experimental uncertainty can be expressed in several ways, for example as 72.53±0.20 or concisely as 72.53(20). Another way is the *parts-per notation* based on fractional error. The fractional error in our example is $0.20/72.53 = 2.8\times10^{-3}$, and the error can be stated as 2.8 parts per thousand, alternatively as 2.8 parts in 10^3.

Problems
For problems marked *, refer to page 519 for a hint, clue, or answer.

1.1 *Vector algebra 1**
Given two vectors $\mathbf{A} = (2\hat{\mathbf{i}} - 3\hat{\mathbf{j}} + 7\hat{\mathbf{k}})$ and $\mathbf{B} = (5\hat{\mathbf{i}} + \hat{\mathbf{j}} + 2\hat{\mathbf{k}})$ find:
(a) $\mathbf{A} + \mathbf{B}$;   (b) $\mathbf{A} - \mathbf{B}$;   (c) $\mathbf{A} \cdot \mathbf{B}$;   (d) $\mathbf{A} \times \mathbf{B}$.

1.2 *Vector algebra 2**
Given two vectors $\mathbf{A} = (3\hat{\mathbf{i}} - 2\hat{\mathbf{j}} + 5\hat{\mathbf{k}})$ and $\mathbf{B} = (6\hat{\mathbf{i}} - 7\hat{\mathbf{j}} + 4\hat{\mathbf{k}})$ find:
(a) $\mathbf{A}^2$;   (b) $\mathbf{B}^2$;   (c) $(\mathbf{A} \cdot \mathbf{B})^2$.

1.3 *Cosine and sine by vector algebra**
Find the cosine and the sine of the angle between $\mathbf{A} = (3\hat{\mathbf{i}} + \hat{\mathbf{j}} + \hat{\mathbf{k}})$ and $\mathbf{B} = (-2\hat{\mathbf{i}} + \hat{\mathbf{j}} + \hat{\mathbf{k}})$.

1.4 *Direction cosines*
The direction cosines of a vector are the cosines of the angles it makes with the coordinate axes. The cosines of the angles between the vector and the x, y, and z axes are usually called, in turn, α, β, and γ. Prove that $\alpha^2 + \beta^2 + \gamma^2 = 1$, using either geometry or vector algebra.

1.5 *Perpendicular vectors*
Show that if $|\mathbf{A} - \mathbf{B}| = |\mathbf{A} + \mathbf{B}|$, then $\mathbf{A}$ and $\mathbf{B}$ are perpendicular.

1.6 *Diagonals of a parallelogram*
Show that the diagonals of an equilateral parallelogram are perpendicular.

1.7 *Law of sines**
Prove the law of sines using the cross product. It should only take a couple of lines.

1.8   *Vector proof of a trigonometric identity*
Let $\hat{\mathbf{a}}$ and $\hat{\mathbf{b}}$ be unit vectors in the x–y plane making angles θ and ϕ with the x axis, respectively. Show that $\hat{\mathbf{a}} = \cos\theta\hat{\mathbf{i}} + \sin\theta\hat{\mathbf{j}}, \hat{\mathbf{b}} = \cos\phi\hat{\mathbf{i}} + \sin\phi\hat{\mathbf{j}}$, and using vector algebra prove that

$$\cos(\theta - \phi) = \cos\theta\cos\phi + \sin\theta\sin\phi.$$

1.9   *Perpendicular unit vector**
Find a unit vector perpendicular to $\mathbf{A} = (\hat{\mathbf{i}} + \hat{\mathbf{j}} - \hat{\mathbf{k}})$ and $\mathbf{B} = (2\hat{\mathbf{i}} + \hat{\mathbf{j}} - 3\hat{\mathbf{k}})$.

1.10 *Perpendicular unit vectors**
Given vector $\mathbf{A} = 3\hat{\mathbf{i}} + 4\hat{\mathbf{j}} - 4\hat{\mathbf{k}}$,
(*a*) find a unit vector $\hat{\mathbf{B}}$ that lies in the x–y plane and is perpendicular to $\mathbf{A}$.
(*b*) find a unit vector $\hat{\mathbf{C}}$ that is perpendicular to both $\mathbf{A}$ and $\hat{\mathbf{B}}$.
(*c*) Show that $\mathbf{A}$ is perpendicular to the plane defined by $\hat{\mathbf{B}}$ and $\hat{\mathbf{C}}$.

1.11 *Volume of a parallelepiped*
Show that the volume of a parallelepiped with edges $\mathbf{A}$, $\mathbf{B}$, and $\mathbf{C}$ is given by $\mathbf{A} \cdot (\mathbf{B} \times \mathbf{C})$.

1.12 *Constructing a vector to a point*
Consider two points located at $\mathbf{r}_1$ and $\mathbf{r}_2$, separated by distance $r = |\mathbf{r}_1 - \mathbf{r}_2|$. Find a vector $\mathbf{A}$ from the origin to a point on the line between $\mathbf{r}_1$ and $\mathbf{r}_2$ at distance xr from the point at $\mathbf{r}_1$ where x is some number.

1.13 *Expressing one vector in terms of another*
Let $\mathbf{A}$ be an arbitrary vector and let $\hat{\mathbf{n}}$ be a unit vector in some fixed direction. Show that $\mathbf{A} = (\mathbf{A} \cdot \hat{\mathbf{n}})\hat{\mathbf{n}} + (\hat{\mathbf{n}} \times \mathbf{A}) \times \hat{\mathbf{n}}$.

1.14 *Two points*
Consider two points located at $\mathbf{r}_1$ and $\mathbf{r}_2$, and separated by distance $r = |\mathbf{r}_1 - \mathbf{r}_2|$. Find a time-dependent vector $\mathbf{A}(t)$ from the origin that is at $\mathbf{r}_1$ at time t_1 and at $\mathbf{r}_2$ at time $t_2 = t_1 + T$. Assume that $\mathbf{A}(t)$ moves uniformly along the straight line between the two points.

1.15 *Great circle**
The shortest distance between two points on the Earth (considered to be a perfect sphere of radius R) is the distance along a great circle — the arc of a circle formed where a plane passing through the two points and the center of the Earth intersects the Earth's surface.

The position of a point on the Earth is specified by the point's longitude ϕ and latitude λ. Longitude is the angle between the meridian (a line from pole to pole) passing through the point and the "prime" meridian passing through Greenwich U.K. Longitude is taken to be positive to the east and negative to the west. Latitude

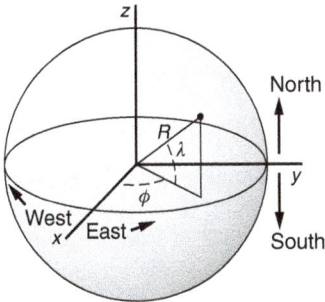

is the angle from the Equator along the point's meridian, taken positive to the north.

Let the vectors from the center of the Earth to the two points be $\mathbf{r}_1$ and $\mathbf{r}_2$. The cosine of the angle θ between them can be found from their dot product, so that the great circle distance between the points is $R\theta$.

Find an expression for θ in terms of the coordinates of the two points. Use a coordinate system with the x axis in the equatorial plane and passing through the prime meridian; let the z axis be on the polar axis, positive toward the north pole, as shown in the sketch.

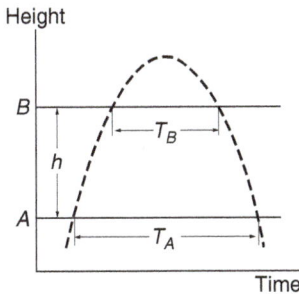

1.16 Measuring g

The acceleration of gravity can be measured by projecting a body upward and measuring the time that it takes to pass two given points in both directions.

Show that if the time the body takes to pass a horizontal line A in both directions is T_A, and the time to go by a second line B in both directions is T_B, then, assuming that the acceleration is constant, its magnitude is

$$g = \frac{8h}{T_A{}^2 - T_B{}^2},$$

where h is the height of line B above line A.

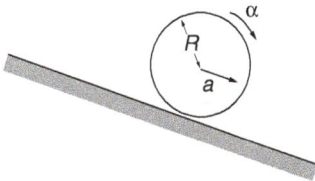

1.17 Rolling drum

A drum of radius R rolls down a slope without slipping. Its axis has acceleration a parallel to the slope. What is the drum's angular acceleration α?

1.18 Elevator and falling marble*

At $t = 0$, an elevator departs from the ground with uniform speed. At time T_1 a child drops a marble through the floor. The marble falls with uniform acceleration $g = 9.8$ m/s^2, and hits the ground at time $t = T_2$. Find the height of the elevator at time T_1.

1.19 Relative velocity*

By relative velocity we mean velocity with respect to a specified coordinate system. (The term velocity, alone, is understood to be relative to the observer's coordinate system.)

(a) A point is observed to have velocity $\mathbf{v}_A$ relative to coordinate system A. What is its velocity relative to coordinate system B, which is displaced from system A by distance $\mathbf{R}$? ($\mathbf{R}$ can change in time.)

(b) Particles a and b move in opposite directions around a circle with angular speed ω, as shown. At $t = 0$ they are both at the point $\mathbf{r} = l\hat{\mathbf{j}}$, where l is the radius of the circle.

Find the velocity of a relative to b.

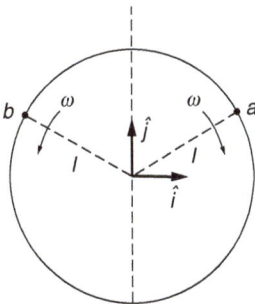

1.20 *Sportscar*

A sportscar, Electro-Fiasco I, can accelerate uniformly to 100 km/h in 3.5 s. Its *maximum* braking rate cannot exceed $0.7g$. What is the minimum time required to go 1.0 km, assuming it begins and ends at rest?

1.21 *Particle with constant radial velocity**

A particle moves in a plane with constant radial velocity $\dot{r} = 4$ m/s, starting from the origin. The angular velocity is constant and has magnitude $\dot{\theta} = 2$ rad/s. When the particle is 3 m from the origin, find the magnitude of (a) the velocity and (b) the acceleration.

1.22 *Jerk*

The rate of change of acceleration is known as "jerk." Find the direction and magnitude of jerk for a particle moving in a circle of radius R at constant angular velocity ω. Draw a vector diagram showing the instantaneous position, velocity, acceleration, and jerk.

1.23 *Smooth elevator ride**

For a smooth ("low jerk") ride, an elevator is programmed to start from rest and accelerate according to

$$a(t) = (a_m/2)[1 - \cos(2\pi t/T)] \qquad 0 \le t \le T$$
$$a(t) = -(a_m/2[(1 - \cos(2\pi t/T)] \qquad T \le t \le 2T$$

where a_m is the maximum acceleration and $2T$ is the total time for the trip.

(a) Draw sketches of $a(t)$ and the jerk as functions of time.

(b) What is the elevator's maximum speed?

(c) Find an approximate expression for the speed at short times near the start of the ride, $t \ll T$.

(d) What is the time required for a trip of distance D?

1.24 *Rolling tire*

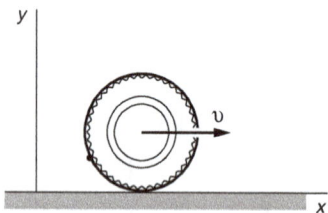

A tire of radius R rolls in a straight line without slipping. Its center moves with constant speed V. A small pebble lodged in the tread of the tire touches the road at $t = 0$. Find the pebble's position, velocity, and acceleration as functions of time.

1.25 *Spiraling particle*

A particle moves outward along a spiral. Its trajectory is given by $r = A\theta$, where A is a constant. $A = (1/\pi)$ m/rad. θ increases in time according to $\theta = \alpha t^2/2$, where α is a constant.

(a) Sketch the motion, and indicate the approximate velocity and acceleration at a few points.

(b) Show that the radial acceleration is zero when $\theta = 1/\sqrt{2}$ rad.

(c) At what angles do the radial and tangential accelerations have equal magnitude?

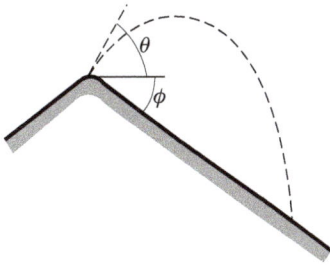

1.26 *Range on a hill**

An athlete stands at the peak of a hill that slopes downward uniformly at angle ϕ. At what angle θ from the horizontal should they throw a rock so that it has the greatest range?

1.27 *Peaked roof**

A peaked roof is symmetrical and subtends a right angle, as shown. Standing at a height of distance h below the peak, with what initial speed must a ball be thrown so that it just clears the peak and hits the other side of the roof at the same height?

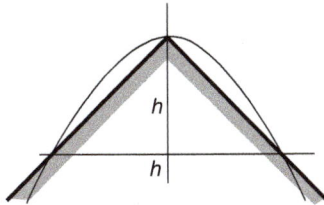

2 NEWTON'S LAWS

2.1 Introduction

Our goal in this chapter is to understand Newton's laws of motion. Newton's laws are simple to state and they are not mathematically complex, so at first glance the task looks modest. As we shall see, Newton's laws combine definitions, observations from nature, partly intuitive concepts, and some unexamined assumptions about space and time. Newton's presentation of his laws of motion in his monumental *Principia* (1687) left some of these points unclear. However, his methods were so successful that it was not until two hundred years later that the foundations of Newtonian mechanics were carefully examined, principally by the Viennese physicist Ernst Mach. Our treatment is very much in the spirit of Mach.

Newton's laws of motion are by no means self-evident. According to Aristotle, the natural state of bodies is rest: bodies move only when a force is applied. Aristotelian mechanics was accepted for two thousand years because it seemed intuitively correct. Careful reasoning from observation and a great leap of imagination were needed to break out of the Aristotelian mold.

Analyzing physical systems from the Newtonian point of view requires effort, but the payoff is handsome. To launch the effort, this chapter is devoted to presenting Newton's laws and showing how to apply them to elementary problems. In addition to deepening our understanding of dynamics, there is an immediate reward for these exercises—the power to analyze physical phenomena that at first sight might seem incomprehensible.

There are alternative approaches to the Newtonian formulation of mechanics. Among these are the formulations of Lagrange and Hamilton, which take energy rather than force as the fundamental concept. However, these formulations are physically equivalent to Newtonian physics. Consequently, a deep understanding of Newton's laws is an invaluable asset to understanding any systematic treatment of mechanics.

2.2 Newtonian Mechanics and Modern Physics

A word about the validity of Newtonian mechanics: Possibly you have had some contact with modern physics—the developments early in the last century of Einstein's theory of relativity and of quantum mechanics. If so, you know that there are important areas of physics where Newtonian mechanics fails while relativity and quantum mechanics succeed. Briefly, Newtonian mechanics breaks down for systems moving with a speed comparable to the speed of light, 3×10^8 m/s, and also for systems of atomic dimensions or smaller where quantum effects are significant. The failure arises because of limitations to the classical concepts of space, time, and the nature of measurement. As you dig deeper into physics, you will learn the boundaries of Newtonian mechanics. Nevertheless, keep in mind that where Newtonian mechanics is valid, it is stunningly successful.

The terms *Newtonian mechanics* and *classical physics* are often used interchangeably, except that classical physics is taken to include Maxwell's theory of electromagnetism. The term *modern physics* typically describes developments in physics after relativity and quantum mechanics appeared on the scene. A natural impulse might be to ignore Newtonian physics and proceed directly to modern theories. This would be a serious error because in major areas of the natural world Newtonian physics works brilliantly while modern theories are of little use. For example, attempting to describe planetary motion using the language of quantum mechanics would lead to a quagmire of impenetrable equations. Analyzing a game of billiards according to the rules of special relativity would quickly lead back to the Newtonian equations as excellent approximations. Rather than make a blanket statement about whether Newtonian physics is right or wrong, we recognize that Newtonian mechanics is exceptionally useful in many areas of physics but is inappropriate in other areas. Newtonian physics enables us to predict eclipses centuries in advance, but it is inadequate for predicting the motions of electrons in atoms. In any case, because classical physics explains so many everyday phenomena, and because many of its fundamental concepts such as momentum, energy, and conservation laws lie at the heart of more advanced formulations, it is an essential tool for all practicing scientists and engineers.

2.3 Newton's Laws

Newton's laws are based on a series of definitions and observations and it is important to understand just which is which. In discussing the laws we must also learn how to apply them, not only because this is the bread and butter of physics but also because this is the only way to obtain a deep understanding of the underlying concepts.

We start by appealing directly to experiment. Unfortunately, simple mechanical experiments can be difficult to carry out because motion in our everyday surroundings is complicated by forces such as gravity and friction. To see the physical essentials, we would like to eliminate all disturbances and examine very simple systems. One way to eliminate the effects of gravity and friction would be to construct a space station laboratory, because in the environment of free fall in space most of the everyday disturbances are negligible. However, lacking the resources to put ourselves in orbit, we settle for second best, a device known as a *linear air track*, which approximates ideal conditions but only in one dimension. (It is not obvious that we can learn about three-dimensional motion from studying motion in one dimension, but happily it turns out that we can.)

The linear air track is a hollow triangular beam, perhaps 2 m long, pierced by many small holes that emit gentle streams of air from a blower. A rider rests on the beam. When the blower is turned on, the rider floats on a thin cushion of air. The friction effects of air are

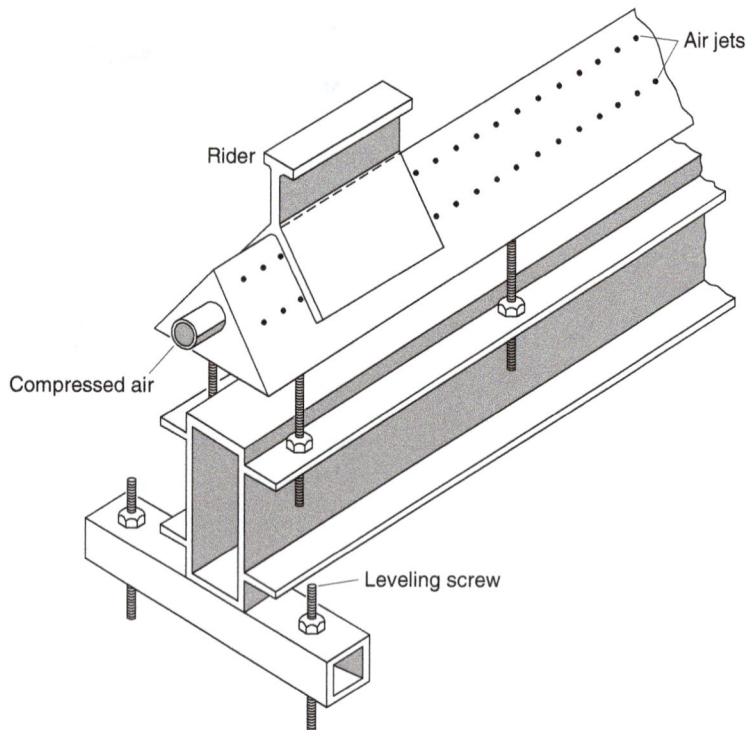

extremely small, typically 5000 times less than for a film of oil, so that friction is negligible compared to the much larger forces we might be applying. If the track is leveled carefully, and if we eliminate stray air currents, the rider moves along the track essentially free of gravity, friction, or any other detectable influences.

Now let's observe how the rider behaves. (It is well worth trying these experiments yourself, if possible.) Suppose that we place the rider on the track and carefully release it from rest. As we might expect, the rider stays at rest, at least until a draft hits it or somebody bumps the apparatus. (This is hardly surprising since we leveled the track until the rider stayed put when left at rest.) Next, we give the rider a slight shove and let it move freely. The motion seems uncanny, for the rider continues to move along slowly and evenly, neither gaining nor losing speed. This is contrary to our everyday experience (not to mention Aristotle's dictum that moving bodies stop moving unless we push them). The rider's behavior seems uncanny because frictionless motion is alien to our normal experience. If the air flow is interrupted, the rider comes to a grinding halt. Evidently friction stops the motion. But we are getting ahead of ourselves; let us return to the properly functioning air track and try to generalize from our experience.

It is possible to make a two-dimensional air table analogous to the one-dimensional air track. (A smooth sheet of glass with a flat piece of dry ice on it does pretty well; the evaporating dry ice provides the gas cushion.) We find again that the undisturbed rider moves with uniform speed. Three-dimensional isolated motion is difficult to observe, short of going into space, but let us for the moment assume that our experience in one and two dimensions also holds in three dimensions. We therefore surmise that an object moves uniformly in space provided external influences are negligible.

2.4 Newton's First Law and Inertial Systems

In describing the air track experiments, we glossed over a fundamental issue. Motion has meaning only when measured with respect to a particular coordinate system. Thus, to describe motion it is essential to specify a coordinate system. For motion along the air track we implicitly used a coordinate system fixed to the track, perhaps with its origin at one end. However, we are free to choose any coordinate systems we please, including systems that are moving with respect to the track. In a coordinate system moving uniformly with respect to the track, the undisturbed rider again moves with constant speed, though a speed different from the one fixed to the track. Such coordinate systems are called *inertial systems*. Not all coordinate systems are inertial. For instance, if viewed from a coordinate system that is accelerating with respect to the track, the speed of the rider would appear to change in time. However, it is always possible to find a coordinate system in which isolated bodies move uniformly. We will return to this subject in Chapter 9.

This is the essence of Newton's first law of motion. The law is often stated in words such as "A uniformly moving body continues to move uniformly unless acted on by a force," but the underlying concept is really the idea of an *isolated* body. Because all interactions decrease with distance, we can take "isolated" to mean a body that has been removed so far from other bodies that interactions are negligible. A coordinate system can always be found in which the body moves uniformly. From this point of view we can state Newton's first law as follows:

Newton's first law of motion is the assertion that inertial systems exist.

Newton's first law is part definition and part experimental fact. Isolated bodies move uniformly in inertial systems by virtue of the definition of an inertial system. In contrast, the assertion that inertial systems exist is a statement about the physical world. Newton's first law raises a number of questions such as what we really mean by an "isolated body," but we defer these for the present.

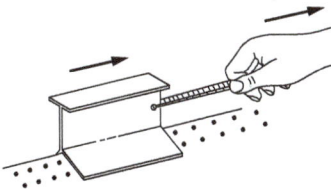

2.5 Newton's Second Law

We now turn to how the rider on the air track behaves when it is no longer isolated. Suppose that we pull the rider with a rubber band. When we start to stretch the rubber band, the rider starts to move. If we move our hand ahead of the rider so that the rubber band's stretch is constant, we find that the rider moves in a wonderfully simple way; its speed increases uniformly with time. The rider moves with *constant acceleration*.

Now suppose that we repeat the experiment but with a different rider, perhaps one a good deal bigger than the first. If the same rubber band is stretched to the same length as before, it causes a constant acceleration, but the acceleration differs from the previous case. Apparently the acceleration depends not only on what we do to the object, since presumably we do the same thing in each case, but also on some property of the object. This property is called *mass*.

2.5.1 Mass

We will use our rubber band experiment to *define* what we mean by mass. We start by arbitrarily asserting that the first body has mass m_1. We can define m_1 to be one unit of mass or x units of mass, where x is any number that is convenient. We then *define* the mass of the second body to be

$$m_2 \equiv m_1 \frac{a_1}{a_2},$$

where a_1 is the acceleration of the first body in our rubber band experiment and a_2 is the acceleration of the second body. In our usage here, the symbol $\equiv$ means "defined to be."

Continuing this procedure, we can assign masses to other objects by measuring their accelerations with the standard stretched rubber band. Thus

$$m_3 \equiv m_1 \frac{a_1}{a_3}$$
$$m_4 \equiv m_1 \frac{a_1}{a_4},$$

etc.

This procedure is straightforward but we have yet to show that it is useful. We could define some other property, call it property Z, such that $Z_2 \equiv Z_1 (a_1/a_2)^2$. However, we shall soon see that mass turns out to be useful but property Z (and most other quantities you might try) does not.

By carrying out further experiments with the air track, for instance causing motion using springs or magnets instead of rubber bands, we find that the ratios of accelerations, hence the mass ratios, are identical no matter how we produce the accelerations, provided that we do the same thing to each body. Thus, mass so defined turns out to be independent of the source of acceleration but appears to be an inherent property of the body. Of course, the particular numerical value of mass that we assign to the body depends on our choice of mass unit. The important thing is that any two bodies have a unique mass ratio.

Our definition of mass is an example of an *operational* definition. By operational we mean that the definition is dominantly in terms of experiments we perform and not in terms of abstract concepts, such as "mass is a measure of the resistance of bodies to a change in motion." Of course, there can be many abstract concepts hidden in apparently simple operations. For instance, when we measure acceleration, we tacitly assume that we have a clear understanding not only of inertial systems but also of space and time. Although our intuitive ideas are adequate for our purposes here, we shall see when we discuss relativity that the behavior of measuring rods and clocks is itself a matter for experiment, and the concept of mass itself needs to be broadened.

Operational definitions may be satisfying theoretically, but they can be useless in practice because in principle they are limited to situations in which the operations can be carried out. However, this is not usually a problem; physics proceeds by constructing a chain of theory and experiment that allows us to employ convenient methods of measurement but that are ultimately based on the operational definitions. For example, the gravitational force on an object turns out to be proportional to its mass. Because weight is proportional to mass, one can compare masses simply by comparing weights. The most practical way to find the mass of a mountain, for instance, is to observe its gravitational pull on a test body, such as a hanging plumb bob, essentially comparing the mass of the mountain to the mass of the Earth. If we had to employ the operational definition of mass, we would need to apply a standard force and measure the mountain's acceleration. This would be impractical, to say the least. Fortunately, the two methods are directly related conceptually.

We defined mass by experiments on laboratory objects; we cannot say *a priori* whether the results are consistent on a much larger or smaller scale. In fact, one of the goals of physics is to find the limitations of such definitions, for the limitations normally reveal new physical laws. Nevertheless, for an operational definition to be useful, it must have a wide field of applications. For instance, our definition of mass holds not only for everyday objects on the Earth but also for planetary, and even galactic, motions on enormously larger scales. It should not surprise us, however, if eventually we find situations in which the operations no longer give the expected results.

Now that we have defined mass, let us turn our attention to *force*.

2.5.2 Force

We describe the operation of acting on the test mass with a stretched rubber band as "applying" a force. Once again we have an operational definition that sidesteps the question of what a force *is*, limiting ourselves to describing how it is produced—namely, by stretching a rubber band a given amount. When we apply the force, the test mass accelerates at some rate a. If we apply two standard stretched rubber bands, side by side, we find that the mass accelerates at the rate $2a$, and if we apply the forces in opposite directions, the acceleration is zero. Thus, the effects of the rubber bands add algebraically, at least for motion in a straight line.

We can establish a force scale by defining the unit force as the force that produces unit acceleration when applied to the unit mass. It follows from our experiments that F units of force accelerate the unit mass by F units of acceleration. From our definition of mass, the force will produce $F \times (1/m)$ units of acceleration when acting on mass m. Hence, the acceleration produced by force F acting on mass m is $a = F/m$. In a more familiar order we write this as $F = ma$.

In the International System of units (SI), the unit of force is the *newton* (N) and the unit of mass is the *kilogram* (kg). There is no special unit for acceleration, which is quoted in units of meters per second per second (m/s^2). Units are discussed in greater detail in Section 2.7.

We have focused our discussion on one-dimensional motion. It is natural to assume that for three-dimensional motion, force, like acceleration, behaves like a vector. Although this turns out to be the case, it is not obviously true. For instance, if mass were different in different directions, acceleration would not be parallel to force and force and acceleration could not be related by a simple vector equation. Although the concept of mass having different values in different directions might sound absurd, it is not impossible. In fact, physicists have carried out very sensitive tests of this hypothesis, without finding any variation. So, we can treat mass as a scalar, i.e. a simple number, and write

$$\mathbf{F} = m\mathbf{a}.$$

This is Newton's second law of motion, which will underlie much of our subsequent discussion.

It is worth emphasizing that force is not merely a matter of definition. For instance, if we observe that an air track rider of mass m starts to accelerate at rate $\mathbf{a}$, it might be tempting to conclude that we have just observed a force $\mathbf{F} = m\mathbf{a}$. Tempting, but wrong. The reason is that forces always arise from real physical *interactions* between systems. Interactions are scientifically significant: accelerations are merely their consequence. Consequently, if we eliminate all interactions by isolating a body sufficiently from its surroundings—an inertial system—we expect it to move uniformly.

You might question whether it is really possible to totally isolate a body from its surroundings. Fortunately, as far as we know, the answer is *yes*. Because interactions decrease with distance, all that is required to make interactions negligible is to move everything else far away. The forces that extend over the greatest distance are the familiar gravitational and Coulomb electric forces. These decrease as $1/r^2$, where r is the distance. Most forces decrease much more rapidly. For example, the force between separated atoms decreases as $1/r^7$. By moving bodies sufficiently far apart, the interactions can be reduced as much as desired.

2.6 Newton's Third Law

That force is necessarily the result of an interaction is made explicit by Newton's third law. The third law states that forces *always* appear in pairs that are equal in magnitude and opposite in direction: if body b exerts force $\mathbf{F}_a$ on body a, then there must be a force $\mathbf{F}_b$ acting on body b, due to body a, such that $\mathbf{F}_b = -\mathbf{F}_a$. There is never a lone force without a partner. As we shall see in Chapter 4, the third law leads directly to a powerful conservation law: the conservation of momentum.

Newton's third law is essential if the second law is to be meaningful: without it, there would be no way to know whether an acceleration results from a real force, or is merely an artifact of being in a non-inertial system. If the acceleration is due to a force, then somewhere in the universe there must be an equal and opposite force acting on some other body.

Example 2.1 Inertial and Non-inertial Systems

Newton's second law $\mathbf{F} = m\mathbf{a}$ holds true only in inertial systems. The concept of inertial systems might seem almost trivial because the Earth provides a reasonably good inertial reference frame, and non-inertial effects are not readily visible on the small scale. However, there is nothing trivial about the concept of an inertial system, as we shall now illustrate.

Aliens infiltrated the intergalactic space police, pilfered a space shuttle, and are making a getaway. Two spaceships set out in pursuit: spaceship A, led by Commander Earhart, and spaceship B, led by Commander Wright. To intercept the shuttle, the commanders must decide whether the shuttle is accelerating or coasting in free flight. For simplicity, we assume that A, B, and the shuttle all move along a straight line.

Commander Earhart measures the distance to the shuttle at a series of times using her super-LIDAR ("Light Detection And Ranging") capabilities. To plot an intercepting course, she sets up a coordinate system along the line of motion with her ship as origin and measures the distance to the space shuttle $x_A(t)$ at a series of times. From $x_A(t)$ she calculates the shuttle's velocity $v_A = \dot{x}_A$ and its acceleration $a_A = \ddot{x}_A$. The results, shown in the sketches, are unambiguous: the distance appears to be varying quadratically in time. She infers that the velocity of the stolen shuttle is varying linearly in time, and that its acceleration is therefore constant.

Earhart calculates from her data that the shuttle is accelerating at the rate of $a_A = 1000$ m/s^2. She concludes that the shuttle's rocket engine must be on and that the force on the shuttle due to the engine is

$$F_A = a_A M_s$$
$$= 1000 \,(\text{m/s}^2) \times M_s(\text{kg}),$$

where M_s is the mass of the shuttle. (Note that the right-hand side of the equation has units kg·m/s^2, exactly as we expect for a force in newtons.)

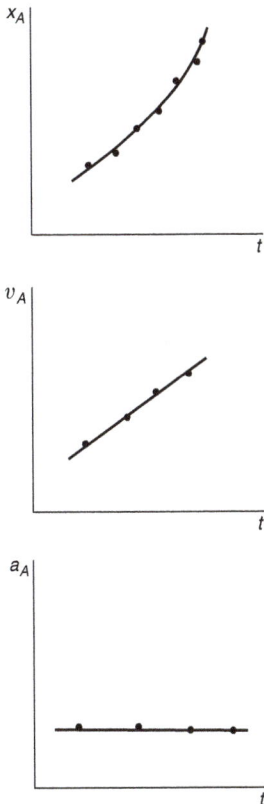

Commander Wright follows the same procedure but finds a different acceleration: $a_B = 950$ m/s^2. He concludes that the force on the shuttle is

$$F_B = a_B M_s$$
$$= 950 \, (\text{m/s}^2) \times M_s(\text{kg}).$$

The disagreement is serious because if different observers obtain different values for the force on a system, at least one of them must be mistaken. Fortunately, both commanders have studied physics, so with confidence in the laws of mechanics they set to work to resolve the discrepancy.

Wright and Earhart recall that Newton's laws hold only in inertial systems. How can they decide whether or not their systems are inertial? Earhart confirms that none of her engines are running and that there are no nearby bodies that could exert a force. She concludes that she is in an isolated system and that it should therefore be inertial. To confirm, she executes a simple but sensitive experiment. She carefully releases her peanut butter sandwich, and observes that it floats in front of her face without motion. Because the sandwich's acceleration is negligible, she concludes that she is indeed in an inertial system. The argument is as follows: as long as Earhart holds the sandwich, it must have the same instantaneous velocity and acceleration as her spaceship. However, once the sandwich is released, no forces act on it, assuming that we can neglect gravitational or electrical interactions with the spaceship, air currents, etc. The sandwich, then, represents an isolated body. If the spaceship were itself accelerating the sandwich would appear to accelerate relative to the cabin. Because the sandwich does not, the spaceship must also define an inertial system.

Earhart's measurement of the force on the shuttle must be correct because she has measured it in an inertial system. But what can we say about measurements made by Commander Wright? To answer this question, we look at the relation between x_A and x_B.

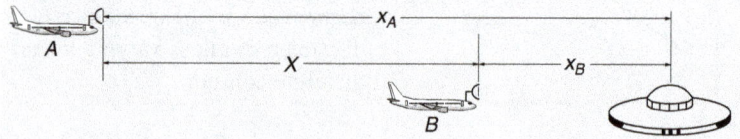

From the sketch,

$$x_A(t) = x_B(t) + X(t),$$

where $X(t)$ is the position of B relative to A. Differentiating twice with respect to time, we have

$$\ddot{x}_A = \ddot{x}_B + \ddot{X}. \qquad (1)$$

Because system A is inertial, Newton's second law applied to the shuttle is

$$F_{\text{true}} = M\ddot{x}_A \qquad (2)$$

where F_{true} is the true force on the shuttle.

The apparent force observed by B is

$$F_{B,\text{apparent}} = M_s\ddot{x}_B.$$

Using the results of Eqs. (1) and (2), we have

$$F_{B,\text{apparent}} = M_s\ddot{x}_A - M_s\ddot{X}$$
$$= F_{\text{true}} - M_s\ddot{X}.$$

Wright can measure the true force only if $\ddot{X} = 0$. However, $\ddot{X} = 0$ only if B moves uniformly with respect to A. Wright suspects that perhaps this is not the case and tries the floating sandwich test. To his embarrassment, he finds that the sandwich will not stay at rest. A check of the ship reveals that an assistant engineer has carelessly left a rocket engine running. Consequently, Wright's system is not inertial but is accelerating with respect to A (and presumably with respect to the rest of the universe) at 50 m/s². When he turns off the engine, Wright finds the same value for the force on the shuttle as Earhart.

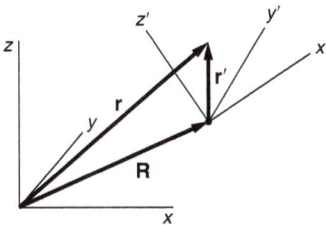

The last example dealt with motion in a straight line, but the results are easily generalized to three dimensions. If **R** is the vector from the origin of an inertial system with coordinates (x, y, z) to the origin of another coordinate system (x', y', z'), then from the sketch, we see that $\mathbf{r}' = \mathbf{r} - \mathbf{R}$. If an acceleration in the (x, y, z) system is $\ddot{\mathbf{r}}$, the force **F** on mass M is

$$\mathbf{F}_{\text{apparent}} = \mathbf{F}_{\text{true}} - M_s\ddot{\mathbf{R}}.$$

If $\ddot{\mathbf{R}} = 0$, then $\mathbf{F}_{\text{apparent}} = \mathbf{F}_{\text{true}}$, which means that the second coordinate system is also inertial. In fact, we have merely proven what we asserted earlier, namely, that any system moving uniformly with respect to an inertial system is also inertial.

2.6.1 Fictitious Forces

Sometimes it is convenient, or possibly essential, to carry out measurements in a non-inertial system. The Earth provides a notable example; the surface of the Earth constitutes a reasonably good inertial system for many purposes, but it is not strictly inertial because of the Earth's rotation. One consequence, to be explained in Chapter 9, is the Coriolis force that causes large weather systems to rotate. Another is the precession of the Foucault pendulum—a circular acceleration with no obvious driving force—on display in many science museums.

What can we do to obtain the correct equations of motion from observations in a non-inertial system? The answer lies in the relation $\mathbf{F}_{apparent} = \mathbf{F}_{true} - M\ddot{\mathbf{R}}$. We can treat the last term like an additional force. Because it is not really a force—no interaction is involved—we shall refer to it as a *fictitious force*. We then have

$$\mathbf{F}_{apparent} = \mathbf{F}_{true} + \mathbf{F}_{fictitious},$$

where $\mathbf{F}_{fictitious} = -M\ddot{\mathbf{R}}$. Here M is the mass of the particle and $\ddot{\mathbf{R}}$ is the acceleration of the non-inertial system with respect to any inertial system.

Fictitious forces are useful in solving certain problems, but they must be treated with care. They generally cause more confusion than they are worth at this stage of study, and for that reason we shall avoid them for the present and agree to use inertial systems only. Later on, in Chapter 9, we shall discuss fictitious forces rigorously and learn how to deal with them.

Some Cautions

Newton's laws can be stated in a clear and consistent fashion but it should be realized that there are fundamental difficulties that cannot be argued away. We shall return to these in later chapters after we have had a chance to become better acquainted with the concepts of Newtonian physics. Some points, however, are well to bear in mind now.

1. You have had to take our word that the experiments we used to define mass and to develop the second law of motion really give the results claimed. It should come as no surprise (although it was a considerable shock when it was first discovered) that this is not always so. For instance, the mass-scale we have set up is no longer consistent when the particles are moving at high speeds where effects predicted by Einstein's special theory of relativity become important. It turns out that instead of the mass we defined, now called the *rest mass* m_0, a more useful quantity is $m = m_0/\sqrt{1 - v^2/c^2}$, where c is the speed of light and v is the speed of the particle. For the case $v \ll c$, m and m_0 differ negligibly. The reason that our table-top experiments did not lead us to the more general expression for mass is that even for the largest everyday velocities, say the velocity of a spacecraft going around the Earth, $v/c \approx 3 \times 10^{-5}$ and m and m_0 differ by only a few parts in 10^{10}.

2. Newton's laws describe the behavior of point masses. If the size of a body is small compared with the interaction distance, this offers no problem. For instance, the Earth and Sun are so small compared with the distance between them that for many purposes their motion can be adequately described by considering the motion of point masses located at the center of each. However, the approximation that we are dealing with ideal point masses is fortunately not essential, and if we wish to describe the motion of large bodies, we can readily generalize Newton's laws, as we shall do in Chapter 4. It turns out to be not much more

difficult to discuss the motion of a rigid body composed of 10^{24} atoms than the motion of a single point mass.

3. Newton's laws deal with particles and are poorly suited for describing a continuous system such as a fluid. We cannot directly apply $\mathbf{F} = m\mathbf{a}$ to a fluid, for both the force and the mass are continuously distributed. However, Newtonian mechanics can be generalized to deal with fluids and provides the underlying principles of fluid mechanics.

One system that is particularly troublesome for our present formulation of Newtonian mechanics is the electromagnetic field. Paradoxes can arise when such a field is present. For instance, two charged bodies which interact electrically actually interact via the electric fields they create. The interaction is not instantaneously transmitted from one particle to the other but propagates at the speed of light. During the propagation time there is an apparent breakdown of Newton's third law; the forces on the particles are not equal and opposite. Similar problems arise in considering gravitational and other interactions. However, the problem lies not so much with Newtonian mechanics as with its misapplication. Simply put, fields possess mechanical properties like momentum and energy that must be included in the analysis. From this point of view there is no such thing as a simple two-particle system. However, for many systems the fields can be taken into account and the paradoxes can be resolved within the Newtonian framework.

2.7 Base Units and Physical Standards

The concepts of length, time, and mass are fundamental to every branch of physics. The units of length, time, and mass are known as the *base units* of physics, and are defined by a set of physical standards that are augmented by descriptions of the procedures for employing them. Base units are not mere matters of practical convenience: because they embody the underlying concepts, they are foundations of physical science. This section presents a brief description of the base units and the systems of units derived from them that are universally used in physics.

The base units play two roles. The precision with which the base units can be defined and reproduced sets a limit to the accuracy of all other metrological standards. In some cases the precision is incredibly high—time, for instance, can be measured to a few parts in 10^{17}. At a deeper level, agreeing to a standard for a physical quantity implies acceptance of an operational definition for that quantity. For example, the modern view of time is that time is what is measured by clocks. Consequently, the properties of time can be understood only by studying the properties of clocks. This is not a trivial point; the rates of all clocks are affected by gravity and by motion (as we shall discuss in Chapters 9 and 12), and unless we are willing to accept the fact that time itself is altered by motion and gravity, we will become enmeshed in conceptual dilemmas and possibly serious practical problems. The global positioning system (GPS), for instance, would not work if relativistic effects were overlooked. Furthermore, nobody knows how clocks work in ultra-intense gravitational fields, for instance in a black hole where the properties of space and time are yet to be understood.

Once a physical quantity has been defined in terms of a measurement procedure, we must appeal to experiment, not to preconceived notions, to understand its properties. To contrast this operational viewpoint with a non-operational approach, consider, for example, Newton's definition of time: "Absolute, true, and mathematical time, of itself, and from its own nature, flows equally without relation to anything external." This may be philosophically and psychologically appealing, but it is difficult to see how to make use of such a definition. Newton's idea of time is metaphysical (beyond physics).

After the operation underlying a physical quantity is agreed upon, the task is to construct the most precise practical standard for defining it. In the past, physical standards were usually particular objects—artifacts—to which all other measurements had to be referred. Thus, the unit of length, the meter, was the distance between two scratches on a platinum-iridium bar, and the unit of mass was a cylinder of platinum-iridium. Such artifacts share some serious disadvantages. Because the standard must be carefully preserved, actual measurements are often done with secondary standards, which causes a loss of accuracy. The precision of an artifact is intrinsically limited. In the case of the standard meter, for example, precision was limited by fuzziness in the scratches that defined the meter interval. Finally, an artifact standard cannot be widely available to the scientific community because it must reside in a single metrological laboratory.

Most physical standards today are not based on artifacts but on atoms and atomic phenomena. These so-called natural or atomic units can be reproduced by anyone with the required apparatus. Such units are often directly connected to fundamental experiments. As the experimental technique improves, the accuracy of the standard increases. Length became the first atomic base unit when it was defined as a specified number of wavelengths of a particular spectral line emitted by a particular atom. The unit of time, the second, became an atomic unit when it was defined as the time for a specified number of oscillations within a particular atom. Mass is the last holdout to natural units. As of 2013, the legal definition of the kilogram is the mass of a platinum-iridium artifact preserved at BIPM (French abbreviation for the International Bureau of Weights and Measures) in Sèvres near Paris. However, this is expected to change in the near future and the kilogram will then be defined in terms of electrical and quantum measurements.

Here is a brief account of the standards of time, length, and mass.

2.7.1 Time

Historically, time was measured in terms of the Earth's rotation. Until 1956 the basic unit, the second, was defined as 1/86 400 of the mean solar day. However, the period of rotation of the Earth turned out to be not as uniform as expected. Variations of up to 1 part in 10^7 per day occur due to atmospheric tides and motions in the Earth's core. Because the movement of the Earth around the Sun is not influenced by these perturbations, in the early 1960's the second was redefined in terms of the mean solar year, which provided accuracy of a few parts in 10^9. However, in the 1950's atomic clocks were developed that made it possible to measure time in terms of a natural microwave frequency in an atom. In 1967 the second was defined to be the time required for $9\,192\,637\,770$ cycles of a hyperfine transition in cesium-133. The initial accuracy was about 1 part in 10^{11} and over the years this has been improved to about 1 part in 10^{15}. A new generation of atomic clocks based on optical transitions promises to provide precision of a few parts in 10^{17} or higher. Time is by far the most accurately determined fundamental quantity.

2.7.2 Length

When the metric system was created in 1795, it was agreed that the meter—the unit of length—should be defined not by an artifact possessed by a single nation but by a standard available to all: the Earth. The meter was defined to be one ten-millionth of the distance from the equator to the pole along the Dunkirk–Barcelona line. Such a distance cannot be measured accurately (in fact it changes due to distortions of the Earth), and in 1889 the meter was redefined to be the separation between two scratches in a platinum-iridium bar that is preserved at the International Bureau of Weights and Measures. However, at about the same time, the physicist Albert A. Michelson devised a method for measuring distance accurately in terms of the wavelength of light, that is, for creating an atomic unit. About seventy years had to pass before the meter was legally adopted as an atomic unit: the distance for $1\,650\,763.73$ wavelengths of the orange-red line of krypton-86. The accuracy of this standard was a few parts in 10^8. The advent of lasers and laser spectroscopy soon rendered this definition obsolete. In measuring the speed of light c, the limiting factor turned out to be the accuracy with which distance can be measured, or, specifically, the accuracy with which wavelengths can be compared. Consequently, the reasoning was turned around, and the speed of light in vacuum was given the assigned value $c = 299\,792\,458$ m/s. The meter was then redefined as the distance traveled by light in $1/c$ seconds. Hence, the meter is now a derived, rather than primary, unit.

2.7.3 Mass

Of the three base units, mass is the only one still defined by an artifact (as of 2013), but it is hoped that this defect will eventually be remedied. The kilogram is the mass of the International Prototype ("*le grand K*"), a platinum-iridium cylinder that is maintained at the International Bureau of Weights and Measures. Secondary standards can be compared with it to an accuracy of about a part in 10^9 but both the prototype and the secondary masses are known to drift in time relative to each other by as much as a part in 10^6. Plans have been advanced to define the kilogram in terms of atomic constants in a system called the *Quantum SI System*. This will not only avoid the ambiguities of a mass prototype, but provide higher accuracy for a number of the fundamental constants of physics.

2.7.4 Systems of Units

Although the standards that define mass, length, and time are accepted by the entire scientific community, a variety of systems of units are in use. The system most widely employed in science, and essentially universally used in physics, is the International System, abbreviated SI (for Système International d'Unités). This is the legal system in most countries with the notable exception of the United States. The SI base units of length, mass, and time are the *meter*, *kilogram*, and *second*. A related system, the CGS system (for centimeter, gram, second), differs from SI only in scaling factors. CGS units appear in older databases and are sometimes used in chemical and biological research. Yet another system of units, the English system, is used for non-scientific measurements in Britain and North America, although Britain also uses the SI system. English units are related to SI units by legally agreed scaling factors; for example, the *inch* is legally defined as 2.54 cm. We shall work chiefly with SI units with occasional lapses into the English system.

The table lists some principal units in the SI, CGS, and English systems.

	SI	CGS	English
Length	1 meter (m)	1 centimeter (cm)	1 inch (in)
Mass	1 kilogram (kg)	1 gram (g)	1 slug
Time	1 second (s)	1 second (s)	1 second (s)
Acceleration	1 m/s^2	1 cm/s^2	1 ft/s^2
Force	1 newton (N) $= 1$ kg·m/s^2	1 dyne $= 1$ g·cm/s^2	1 pound (lb) $= 1$ slug·ft/s^2

Here are some useful relations between these units systems.

1 m = 100 cm	1 m $\approx$ 39.4 in
1 ft = 12 in	1 mile = 5280 ft
1 kg = 1000 g	1 slug $\approx$ 14.6 kg
1 N $= 10^5$ dyne	1 N $\approx$ 0.224 lb

The unit of mass in the English system, the slug, is the mass that a 1-pound force causes to accelerate at a rate of 1 ft/second^2. This unit is archaic. The weight of a slug is about 32 pounds. The word "pound" is sometimes used (incorrectly) as a unit of mass. In this usage it is the mass that experiences a gravitational force of one pound at the surface of the

Earth, a mass of approximately 0.454 kg. We shall avoid this confusing usage.

We often need to deal with quantities that are much larger or much smaller than the base units, so the SI system provides prefixes based on powers of 10 (positive or negative) that can multiply the base units. The most common prefixes are listed in Appendix C.

2.8 The Algebra of Dimensions

Equations in physics are not meaningful unless they are dimensionally consistent. In this context, the term *dimension* refers to the type of physical quantity (ultimately expressed in units of mass, length, and time), in contrast to usage in mathematics, where dimension refers to the number of coordinates needed to specify a point.

A useful check on a calculation is to see whether the units agree on both sides of the final result. If they don't, there is evidently an error somewhere. Sometimes equations may look inconsistent, even though they are physically meaningful, merely because the units need to be reconciled. For instance, the left-hand side of an equation might be in meters and the right-hand side in kilometers.

Fortunately, the units of physics have the pleasing property that they can be treated as algebraic quantities. For example, the relation

$$1 \text{ kilometer} = 1000 \text{ meters}$$

can equally well be written as

$$1 = \frac{1 \text{ kilometer}}{1000 \text{ meters}}.$$

The expression on the right, known as a *conversion factor*, can be manipulated as an ordinary algebraic quantity.

Various units can enter into the solution of a problem and to minimize error it is helpful to have a systematic procedure for reconciling them. Suppose, for instance, that we want to express a length of 3.5 m in cm. We start with the known relation 1 m = 100 cm. The idea is to treat the conversion factor as an algebraic relation in which the numbers and the units both act like algebraic quantities. Writing the conversion factor as a ratio with "unit" value we have

$$(3.5 \text{ m})\left(\frac{100 \text{ cm}}{1 \text{ m}}\right) = 350 \text{ cm}.$$

The units "cancel" symbolically. If we had wanted to change a length in cm to m, we would have written the conversion factor as 1 m/100 cm.

Example 2.2 Converting Units
The mean distance of the Earth from the Sun is 93 million miles. Express this distance in meters, using the conversion factors 1 mile = 5280 feet, 1 foot = 12 inches, 1 inch = 2.54 cm, 100 cm = 1 m.

We can combine the conversion factors to find the answer in one step:

$$(9.3 \times 10^7 \text{ miles}) \left(\frac{5280 \text{ feet}}{1 \text{ mile}} \right) \left(\frac{12 \text{ inches}}{1 \text{ foot}} \right) \left(\frac{2.54 \text{ cm}}{1 \text{ inch}} \right) \left(\frac{1 \text{ m}}{100 \text{ cm}} \right)$$

$$= 1.5 \times 10^{11} \text{ m}.$$

The units "cancel" to give the desired units in the result. Note that the result is expressed to the same number of significant figures as the given data.

We have seen that quantities in mechanics such as velocity and force are measured in units constructed from the base units of mass, length, and time. Regardless of the system of units we use, in Newtonian mechanics every quantity depends on mass, length, and time in a unique way. For example, the units of velocity are m/s in the SI system and cm/s in the CGS system, but both have dimensions length/time.

In analyzing the consistency of units in an equation, the dimension of mass is abbreviated M, the dimension of length is L, and the dimension of time is T. James Clerk Maxwell, who developed the theory of electromagnetism, was the first to use the convenient notation of square brackets to stand for the dimensions of a quantity.

$$[\text{mass}] = M \quad [\text{length}] = L \quad [\text{time}] = T.$$

The dimensions of quantities in mechanics can always be expressed in terms of powers of M, L, and T. For instance,

$$[\text{velocity}] = LT^{-1} \quad [\text{force}] = MLT^{-2}.$$

Units must agree on both sides of an equation and this is only possible if the underlying dimensions also agree. Note that M, L, and T are *independent* quantities; we cannot express mass in terms of time, or length in terms of mass. Consequently, for an equation to be valid, the powers of M, L, and T must separately agree no matter what system of units we choose to employ. To take an example, in Chapter 5 we shall encounter the work–energy theorem, which essentially states that for a body starting from rest,

$$\text{force} \times \text{distance} = \text{kinetic energy} = \tfrac{1}{2}mv^2.$$

We can check this dimensionally: $[\text{force}] = [\text{mass}][\text{acceleration}] = MLT^{-2}$ so that the left-hand side has dimensions ML^2T^{-2}. Kinetic energy has dimensions $M(LT^{-1})^2 = ML^2T^{-2}$. Thus the equation is dimensionally consistent.

2.9 Applying Newton's Laws

Newton's laws are simple to state but become meaningful only when you have learned how to use them. In the remainder of this chapter we shall launch this quest by considering a few problems in which the forces are known and the masses can be treated as particles, rather than as real extended objects. We shall describe a series of steps that, once learned, will seem so natural that the procedure becomes intuitive. A note of reassurance lest you feel that matters are presented too dogmatically: there are different ways to attack most problems, and the procedure described here is certainly not unique. In fact, no cut-and-dried procedure can ever substitute for intelligent analytical thinking. However, the method is worth mastering even if you should later resort to shortcuts or a different approach. Here are the steps for attacking mechanical problems involving systems of a small number of masses acted on by simple forces.

1. Isolate the masses

Mentally divide the system into smaller systems that each contain a single mass. Each mass will be treated as if were a tiny particle. Later, we will generalize the method to real extended bodies.

2. Draw a force diagram for each mass

Force diagrams describe all the important physics in the problem and are the key to understanding. To draw a force diagram:

(a) Represent each body by a point or simple symbol, and label it.

(b) For each mass, draw a force vector starting on the mass, one vector for each force acting *on* it, and label each vector.

This can be tricky. Draw only forces acting *on* the body, not forces exerted *by* the body. The body might be pulled by strings, pushed by other bodies, experiencing the force of gravity or an electric field, etc. In any case, be sure not to omit any. Caution: use symbols on the force diagram, never numerical values. The procedure is to obtain a symbolic solution to the problem before introducing numerical values. Without a symbolic solution, there is no reliable method for determining whether an answer is reasonable.

3. Show a coordinate system on the force diagram

Take the axes along a convenient direction, perhaps along the assumed direction of motion or along an applied force. In any case, the coordinate system must be inertial—that is, it must be fixed to an inertial frame of reference.

4. Write the equations of motion

By equation of motion we mean an equation of the form of Newton's second law, but with the forces and acceleration shown explicitly. Thus, it is of the form $F_{1x} + F_{2x} + \cdots = Ma_x$, where the x component of each force on the body is represented by a term on the left-hand side of the equation. Because force and acceleration are vectors, a separate equation of motion is required for each dimension of interest. The algebraic sign of each force component must be consistent with the force diagram and with the coordinate system. Even if a body is at rest and its acceleration vanishes, it is good practice to write the complete equations of motion before inserting known quantities.

5. Write the constraint equations

In many problems, bodies are constrained to move along certain paths. A pendulum bob, for instance, moves in a circle, and a block sliding on a tabletop is constrained to move in a plane. Each constraint can be described by a geometric equation, known as a *constraint equation*. Furthermore, if two bodies in the same system interact, the forces between them are constrained by Newton's third law to be equal and opposite. Write each constraint equation. Sometimes the constraints are implicit in the statement of the problem. For instance, for a block on a table there is no vertical acceleration a_v, and the constraint equation is simply $a_v = 0$.

6. Solve!

The equations of motion and the constraint equations should provide enough relations to allow every unknown to be found. If an equation is overlooked, however, there will be too few equations for a solution. This is a signal that you need to look further.

Sometimes mechanics is subdivided into statics—the analysis of forces on bodies in equilibrium—and dynamics—the study of motion due to applied forces. However, the distinction is unimportant here: statics can be regarded as a special case of dynamics with vanishing acceleration. This requires that the net force on each body also vanishes.

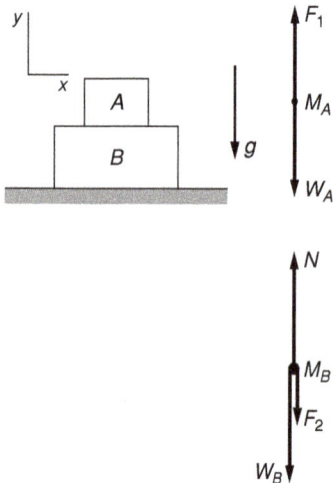

To illustrate our procedure, consider two blocks, A and B, resting on a table, as in the drawing. Force diagrams are shown for each block. The direction of gravity and the coordinate axes are shown by arrows.

The problem is to find all the forces.

Because the forces are vertical, we consider only components along the y axis. The force diagrams show the weights W_A and W_B of the blocks, directed down. F_1 is the force of block B on block A, while F_2 is the force of A on B. We shall refer to such a contact force between two bodies, directed perpendicular to the surface, as a *normal force*. The normal force of the table on B is denoted by N. We have assumed directions for the forces. If we guessed wrong and a force is actually in the opposite direction, it will simply turn out to be negative when we solve.

The equation of motion for block A is

$$F_1 - W_A = m_A a_A$$

and the equation of motion for B is

$$N - F_2 - W_B = m_B a_B.$$

By Newton's third law we have $F_1 = F_2$, according to the directions assumed in the force diagram. Because the system is at rest, $a_A = a_B = 0$. Hence

$$F_1 = W_A,$$

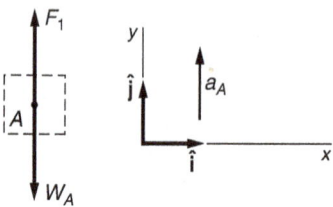

and the equation of motion for B is

$$N = W_B + F_2.$$

The final step in solving a problem is to assure yourself that your answer is reasonable. Here, the normal force N, of the table on block B, is accompanied by an equal and opposite force of the block on the table (not shown). Since $N = W_B + F_2 = W_B + W_A$, the force on the table is the weight of the two blocks, as we expect.

The purpose of this example is to illustrate the difference between the force we apply to an object and the force the object exerts on us. For instance, the gravitational force—the weight $\mathbf{W}_A$—acts on block A only, not on the table or on block B. Of course, the weight of A has an effect on both those objects, but it is through the contact or normal forces that arise to keep the system in equilibrium. Physiologically, forces are easy to confuse. If you push a book across a table, the force you feel is not the force that makes the book move; it is the force the book exerts on you. According to Newton's third law, these two forces are always equal and opposite.

If one of a pair of forces is limited, the other must be limited also, as the following example illustrates.

Example 2.3 Astronauts' Tug-of-War

Two astronauts on a space walk decide to play tug-of-war by pulling on either end of a rope.

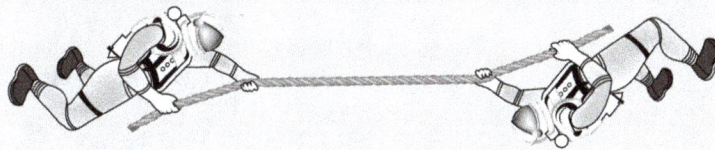

Astronaut Alice, who was a star on her college crew, happens to be much stronger than astronaut Bob, whose passion was video games. As a result, the maximum force F_A with which Alice can pull is much larger than the maximum force F_B with which Bob can pull. The astronauts' masses are M_A and M_B, and the mass of the rope M_r is assumed to be negligible. The problem is to find the motion if each astronaut pulls on the rope as hard as possible.

The figure shows the force diagrams.

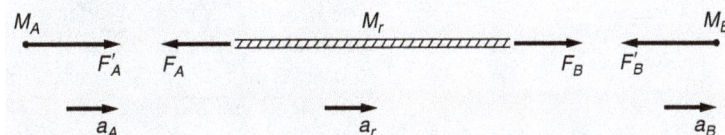

Note that the forces F_A and F_B are exerted *by* the astronauts on the *rope*, not on themselves. The forces exerted by the rope on the astronauts are

F_A' and F_B'. The diagram shows the directions of the forces and the sign convention we have adopted: acceleration to the right is positive.

Only motion along the direction of the rope is of interest. There are no constraints and we proceed to the solution.

With the directions shown in the force diagram, we have from Newton's third law

$$F_A' = F_A$$
$$F_B' = F_B. \tag{1}$$

The equation of motion for the rope is

$$F_B - F_A = M_r a_r. \tag{2}$$

The mass of the rope, M_r, is negligible and so we will take $M_r = 0$ in Eq. (2). This gives $F_B - F_A = 0$ or

$$F_B = F_A.$$

This illustrates a general principle: because a finite force acting on zero mass would produce an infinite acceleration, the total force on any body of negligible mass must be vanishingly small.

Since $F_B = F_A$, Eq. (1) gives $F_A' = F_A = F_B = F_B'$. Hence

$$F_A' = F_B'.$$

Consequently, the astronauts must pull with the same force. No matter how much stronger Alice is than Bob, she is unable to pull harder than he can pull. Physically, if Alice pulls too hard, Bob cannot hold on and the rope slips. Thus the force Alice can exert is limited by the strength of Bob's grip.

The accelerations of the astronauts are

$$a_A = \frac{F_A'}{M_A}$$

$$a_B = \frac{-F_B'}{M_B}$$

$$= \frac{-F_A'}{M_B}.$$

The negative sign means that a_B is to the left. Often the direction of an acceleration or force component is initially unknown. In writing the equations of motion, any choice of direction is valid, as long as it is consistent with the directions shown in the force diagram. If the solution yields a negative sign, the acceleration or force is opposite to the direction assumed.

The next example shows that for a system of several masses (a compound system) to accelerate, there must be a net force on each mass in the system.

Example 2.4 Multiple Masses: a Freight Train

Three freight cars each of mass M are pulled with force F by a locomotive. Friction is negligible. Find the forces on each car.

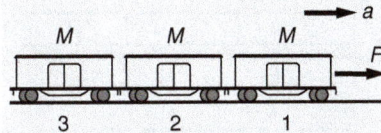

Before drawing the force diagram, it is worth thinking about the system as a whole. The cars are joined and are thus constrained to have the same acceleration. Because the total mass is $3M$, their acceleration is

$$a = \frac{F}{3M}.$$

The force diagram for the end car, #3, is shown, where W is the weight, N is the upward force exerted by the track, and F_3 is the force exerted on car #3 by car #2.

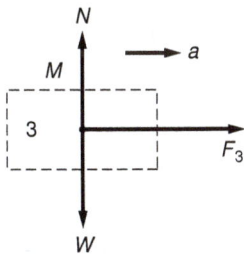

The vertical acceleration is zero, so that $N = W$. The horizontal equation of motion is

$$F_3 = Ma$$
$$= M\left(\frac{F}{3M}\right)$$
$$= \frac{F}{3}.$$

Now let us consider the middle car, #2. The vertical forces are as before, and we omit them. F_3' is the force exerted by the last car, and F_2 is the force exerted by car #1.

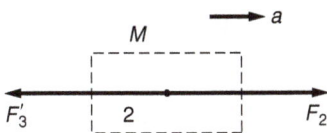

The equation of motion is

$$F_2 - F_3' = Ma.$$

By Newton's third law, $F_3' = F_3$, so that $F_3' = F/3$. Since $a = F/3M$, we have

$$F_2 = M\left(\frac{F}{3M}\right) + \frac{F}{3}$$
$$= \frac{2F}{3}.$$

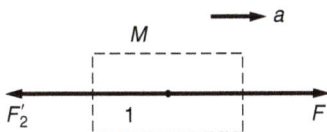

The horizontal forces on the first car, #1, are F, to the right, and

$$F_2' = F_2 = \frac{2F}{3},$$

to the left. Each car experiences a net force $F/3$ to the right.

Here is a more general way to look at the problem. Consider a string of N cars, each of mass M, pulled by a force F. The acceleration is $a = F/(NM)$.

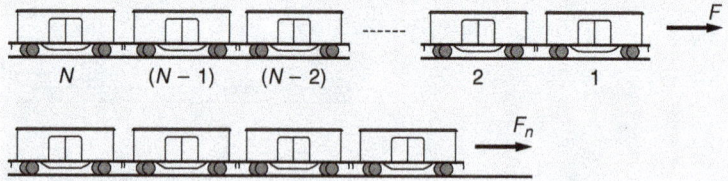

To find the force F_n pulling the last n cars, note that F_n must give the mass nM an acceleration $F/(NM)$. Hence

$$F_n = nM\left(\frac{F}{NM}\right)$$

$$= \frac{n}{N}F.$$

The force by which one car pulls another, in other words, the force between the cars, is proportional to the number of cars pulled.

Let us turn now to systems more complicated than single bodies at rest. An apple falling under the force of gravity is the archetype for Newtonian dynamics, and whether or not the legend is true, the system encompasses the law of universal gravitation and must be counted among the greatest intellectual syntheses in the history of science. We shall get to that later but start here with a system that some cynics view as the dullest problem in all of physics: a block sliding on a plane.

However, we shall permit the plane to slide, which makes the problem more interesting. As a first look at that system, let us investigate how the acceleration of the block and the plane are related by the constraint that the block must stay on the plane. Their accelerations are related by a geometrical constraint equation, and we shall look at two examples of these as a prelude to attacking some real dynamical problems.

Example 2.5 Examples of Constrained Motion
1. wedge and block
A block slides on a wedge (a planar surface) which in turns slides on a horizontal table, as shown in the sketch. The angle of the wedge is θ and its height is h. How are the accelerations of the block and the wedge related? Neglect friction.

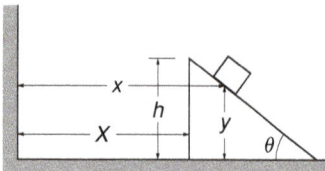

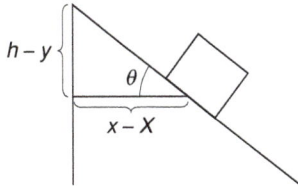

Because the wedge is in contact with the table, we have the trivial constraint that the vertical acceleration of the wedge is zero. To find the less obvious constraint that the block slides on the wedge, let X be the horizontal coordinate of the end of the wedge and let x and y be the horizontal and vertical coordinates of the block, as shown.

From the geometry, we see that

$$\frac{h - y}{x - X} = \tan\theta$$

$$h - y = (x - X)\tan\theta.$$

Differentiating twice with respect to time and rearranging, we obtain the constraint equation for the accelerations:

$$\ddot{y} = (\ddot{X} - \ddot{x})\tan\theta. \tag{1}$$

A few comments: Note that the coordinates are inertial. We would have trouble using Newton's second law if we measured the position of the block with respect to the wedge, because the wedge is accelerating and cannot specify an inertial system. Second, unimportant geometric parameters, like the height of the wedge, disappear when we take time derivatives, but they can be useful in setting up the geometry. Finally, constraint equations are independent of applied forces. For example, even if friction between the block and wedge affects their accelerations, the constraint equation (1) is still valid (but only as long as the bodies remain in contact).

2. pulley system
The pulley system shown is used to hoist a block. How does the acceleration of the end of the rope relate to the acceleration of the block?

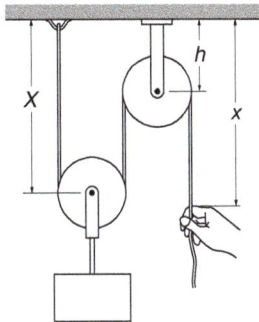

Using the coordinates indicated, the length of the rope is given by

$$l = X + \pi R + (X - h) + \pi R + (x - h),$$

where R is the radius of each pulley. Hence

$$\ddot{X} = -\frac{1}{2}\ddot{x}.$$

The block accelerates half as fast as the hand, and in the opposite direction.

Example 2.6 Masses and Pulley
Two masses, M_1 and M_2, are connected by a string that passes over a pulley. The pulley is accelerating upward at rate A, as shown, and the gravitational force on each mass is $W_i = M_i\, g$. The problem is to find the rate at which the masses accelerate and the tension T in the string.

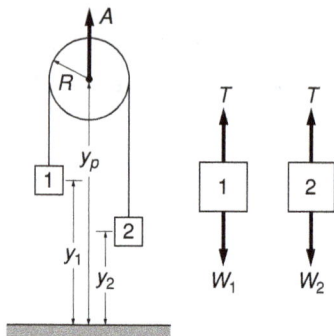

From the force diagrams, we have

$$T - W_1 = M_1\ddot{y}_1 \tag{1}$$

$$T - W_2 = M_2\ddot{y}_2. \tag{2}$$

We have two equations but three unknowns: y_1, y_2, and T. We need a third equation, which is the constraint equation that relates the accelerations. The coordinates are shown in the drawing. If y_p is measured to the center of the pulley of radius R, then if l is the length of the string, we have

$$l = (y_p - y_1) + \pi R + (y_p - y_2).$$

Differentiating twice with respect to time, we find

$$0 = 2\ddot{y}_p - \ddot{y}_1 - \ddot{y}_2.$$

Using $A = \ddot{y}_p$, we have the constraint condition

$$\ddot{y}_1 + \ddot{y}_2 = 2A. \tag{3}$$

Equations (1)–(3) are easily solved:

$$T = 2(A + g)\frac{M_1 M_2}{M_1 + M_2}$$

$$\ddot{y}_1 = \frac{(2A + g)M_2 - M_1 g}{M_1 + M_2}$$

$$\ddot{y}_2 = \frac{(2A + g)M_1 - M_2 g}{M_1 + M_2}.$$

Note that the result is reasonable. If the masses are identical, they accelerate equally and the tension is $M(A + g)$. If either mass vanishes, the other mass falls freely under gravity. If $A = 0$, the apparatus is known as "Atwood's machine," a standby in lecture demonstrations and elementary labs. The purpose of Atwood's machine is to "decrease gravity"—the larger mass descends more slowly than if it were in free fall.

Our examples so far have involved only linear motion. Let us look at the dynamics of rotational motion.

2.10 Dynamics Using Polar Coordinates

To set the stage for understanding dynamics using polar coordinates, we start with the simple example of circular motion. The basic feature of circular motion is that it undergoes radial acceleration. This elementary property is often a source of confusion because our intuitive idea of acceleration usually relates to a change in speed whereas in circular motion radial acceleration arises from a change in the direction of motion.

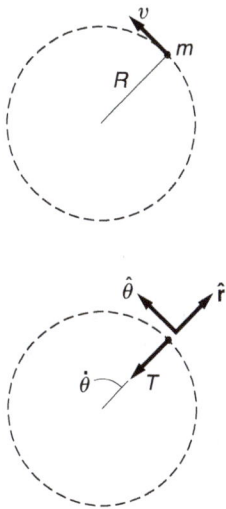

Example 2.7 Block and String 1

A mass m on the end of a string of length R whirls in free space in a horizontal plane, with constant speed v. Find the force on m.

The only force on m is the string force T, which acts toward the center, as shown in the diagram. It is natural to use polar coordinates. According to the derivation in Section 1.11, radial acceleration is $a_r = \ddot{r} - r\dot{\theta}^2$ where $\dot{\theta}$ is the angular velocity. (a_r is positive outward.)

Since $\mathbf{T}$ is directed toward the origin, $\mathbf{T} = -T\hat{\mathbf{r}}$, and the radial equation of motion is

$$-T = m\,a_r$$
$$= m(\ddot{r} - r\dot{\theta}^2).$$

The constraint equation is that $r = R$, so that $\ddot{r} = \ddot{R} = 0$. Because $\dot{\theta} = v/R$, we obtain

$$a_r = -R\,(v/R)^2 = -v^2/R$$
$$T = \frac{mv^2}{R}.$$

Note that T is directed toward the origin; there is no outward force on m. If you whirl a pebble at the end of a string, *you* feel an outward force. However, the force you feel does not act on the pebble, it acts on you. This force is equal in magnitude and opposite in direction to the force with which you pull the pebble, assuming the string's mass to be negligible.

In the next example both radial and tangential motion play a role in circular motion.

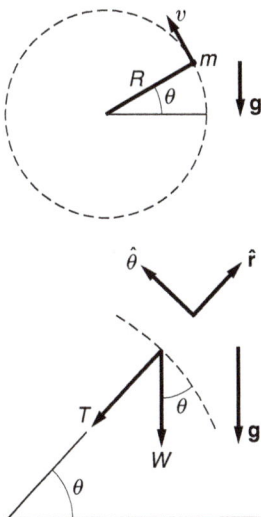

Example 2.8 Block and String 2

Mass m is whirled at instantaneous speed v on the end of a string of length R. The motion is in a vertical plane in the gravitational field of the Earth. The forces on m are the weight $W = mg$ down and the string force T toward the center. The string makes instantaneous angle θ with the horizontal. Find T and the tangential acceleration at any instant.

The diagram shows the forces and unit vectors $\hat{\mathbf{r}}$ and $\hat{\boldsymbol{\theta}}$.

The radial force is $-T - W\sin\theta$, so the radial equation of motion is

$$-(T + W\sin\theta) = m\,a_r$$
$$= m(\ddot{r} - r\dot{\theta}^2). \tag{1}$$

The tangential force is $-W \cos \theta$. Hence

$$-W \cos \theta = m \, a_\theta$$
$$= m(r\ddot{\theta} + 2\dot{r}\dot{\theta}). \tag{2}$$

Since $r = R = $ constant, $a_r = -R(\dot{\theta}^2) = -v^2/R$, and Eq. (1) gives

$$T = \frac{mv^2}{R} - W \sin \theta = m\frac{v^2}{R}\left(1 - \frac{gR}{v^2}\sin\theta\right).$$

A string can pull but not push, so that T cannot be negative. This requires that $mv^2/R \geq W \sin\theta$. The maximum value of $W \sin\theta$ occurs when the mass is vertically up; in this case $mv^2/R > W$. If this condition is not satisfied, the mass does not follow a circular path but starts to fall; $\ddot{r}$ is no longer zero.

The tangential acceleration is given by Eq. (2). Since $\dot{r} = 0$ we have

$$a_\theta = R\ddot{\theta}$$
$$= -\frac{W \cos\theta}{m}$$
$$= -g \cos\theta.$$

The whirling block has tangential acceleration that varies between $+g$ and $-g$, no matter what the value of v. On the downswing the tangential speed increases, on the upswing it decreases. If we wanted to swing the mass with constant velocity, we would need to make $a_\theta = 0$ by giving T a tangential component $W \cos\theta$.

The next example involves rotational motion, translational motion, and constraints.

Example 2.9 The Whirling Block

A horizontal frictionless table has a small hole in its center. Block A on the table is connected to block B hanging beneath by a string of negligible mass which passes through the hole.

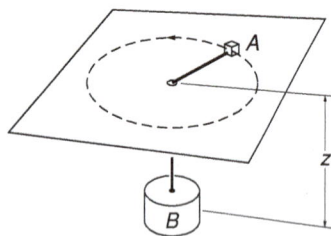

Initially, B is held stationary and A rotates at constant radius r_0 with steady angular velocity ω_0. If B is released at $t = 0$, what is its acceleration immediately afterward?

The force diagrams for A and B after the moment of release are shown in the sketches.

For the block on the table we need consider only horizontal forces. The only such force acting on A is the string force T. The forces on B are the string force T and its weight W_B.

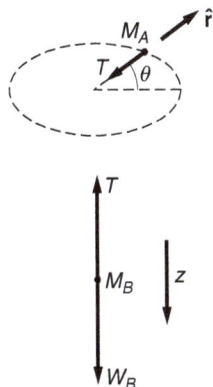

For A it is natural to use polar coordinates r, θ, while for B the linear coordinate z is sufficient, as shown in the force diagrams.

As usual, the unit vector $\hat{r}$ is radially outward. For convenience, we have taken z to be positive in the downward direction. The equations of motion are

$$-T = M_A(\ddot{r} - r\dot{\theta}^2) \qquad \text{radial,} \, A \qquad (1)$$
$$0 = M_A(r\ddot{\theta} + 2\dot{r}\dot{\theta}) \qquad \text{tangential,} \, A \qquad (2)$$
$$W_B - T = M_B\ddot{z} \qquad \text{vertical,} \, B. \qquad (3)$$

Because the length of the string, l, is constant, we have

$$r + z = l. \qquad (4)$$

Differentiating Eq. (4) twice with respect to time gives the constraint equation

$$\ddot{r} = -\ddot{z}. \qquad (5)$$

The negative sign means that if mass A moves outward, mass B would rise. Combining Eqs. (1), (3), and (5), we find

$$\ddot{z} = \frac{W_B - M_A r\dot{\theta}^2}{M_A + M_B}.$$

Immediately after B is released, $r = r_0$ and $\dot{\theta} = \omega_0$. Hence

$$\ddot{z}(0) = \frac{W_B - M_A r_0\omega_0^2}{M_A + M_B}. \qquad (6)$$

$\ddot{z}(0)$ can be positive, negative, or zero depending on the value of the numerator in Eq. (6); if ω_0 is large enough, block B will begin to rise after release. Before release, $\ddot{r} = 0$, but immediately after, the acceleration has a finite value. It is evident that because forces can be applied suddenly, acceleration can change abruptly—acceleration can be discontinuous in time. In contrast, position and velocity are time integrals of acceleration and are therefore continuous in time.

The apparently simple problem in the next example has some unexpected subtleties.

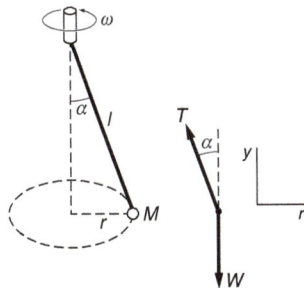

Example 2.10 The Conical Pendulum
Mass M is fixed to the end of a rod of length l and negligible mass that is pivoted to swing from the end of a hub that rotates at constant angular frequency ω, as shown in the drawing. The mass moves with steady speed in a circular path of constant radius. The problem is to find α, the angle the rod makes with the vertical.

We start with the force diagram. T is the tension with which the rod pulls the mass and W is the weight of the mass. Note that there are no other forces on M. If this is not clear, you are most likely confusing an acceleration with a force—a serious error. Because y is constant and $\ddot{y}$ is zero, the vertical equation of motion is

$$T \cos \alpha - W = 0. \tag{1}$$

To find the horizontal equation of motion note that the bob is accelerating in the $\hat{\mathbf{r}}$ direction at rate $a_r = -\omega^2 r$. Then

$$-T \sin \alpha = -M r \omega^2. \tag{2}$$

Because $r = l \sin \alpha$ we have

$$T \sin \alpha = M l \omega^2 \sin \alpha \tag{3}$$

which gives

$$T = M l \omega^2. \tag{4}$$

Combining Eqs. (1) and (3) gives

$$M l \omega^2 \cos \alpha = W. \tag{5}$$

The weight is $W = Mg$. Consequently, Eq. (5) gives

$$\cos \alpha = \frac{g}{l \omega^2}.$$

As $\omega \to \infty$, $\cos \alpha \to 0$ and $\alpha \to \pi/2$. This is reasonable because at high speeds we expect the bob to fly outward, which is equivalent to expecting that $\alpha \to \pi/2$. However, at low speeds the solution becomes unreasonable. As $\omega \to 0$, our solution predicts $\cos \alpha \to \infty$, which is nonsense because $\cos \alpha \leq 1$. Something has gone seriously wrong.

Here is the trouble. Our solution predicts $\cos \alpha > 1$ for $\omega < \sqrt{g/l}$. When $\omega = \sqrt{g/l}$, $\cos \alpha = 1$ and $\sin \alpha = 0$; the bob simply hangs vertically. In going from Eq. (2) to Eq. (3) we divided both sides of Eq. (2) by $\sin \alpha$ and, in this case, we divided by 0, which is not permissible. However, we see that we have overlooked a second possible solution: $\sin \alpha = 0$, $T = W$. This solution is true for all values of ω. In this solution, the bob hangs straight down. The drawing shows a plot of the complete solution.

Physically, for $\omega \leq \sqrt{g/l}$ the only acceptable solution is $\alpha = 0$, $\cos \alpha = 1$. For $\omega > \sqrt{g/l}$ there are two possible solutions:

$$\cos \alpha = 1 \tag{A}$$

$$\cos \alpha = \frac{g}{l \omega^2}. \tag{B}$$

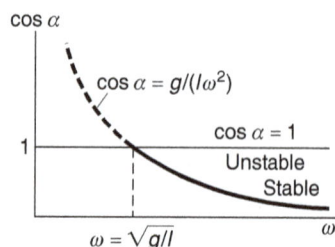

Solution (A) corresponds to the mass rotating rapidly but hanging vertically. Solution (B) corresponds to the mass flying in a circular path with the rod at an angle with the vertical. For $\omega > \sqrt{g/l}$, solution (A) is unstable—if the system is in that state and is slightly perturbed, it will jump outward. We expect the bifurcation point, the point where the equation of motion suddenly has two solutions, has some special significance. As we shall see in Example 3.10, $\omega = \sqrt{g/l}$ is the oscillation frequency, in radians per second, of a simple pendulum of length l. If the rotation frequency is less than this value, the pendulum can hang vertically. (It could also swing back and forth, executing pendulum motion.) However, for a rotation frequency higher than the pendulum frequency, the pendulum will fly outward, unless it is precisely vertical. Can you see why this is so?

The moral of this example is that it is important to check that a mathematical solution makes good physical sense.

Problems

*For problems marked *, refer to page 520 for a hint, clue, or answer.*

2.1 *Time-dependent force**
A 5-kg mass moves under the influence of a force $\mathbf{F} = (4t^2\hat{\mathbf{i}} - 3t\hat{\mathbf{j}})$ N, where t is the time in seconds (1 N = 1 newton). It starts at rest from the origin at $t = 0$. Find: (*a*) its velocity; (*b*) its position; and (*c*) $\mathbf{r} \times \mathbf{v}$, for any later time.

2.2 *Two blocks and string**
The two blocks M_1 and M_2 shown in the sketch are connected by a string of negligible mass. If the system is released from rest, find how far block M_1 slides in time t. Neglect friction.

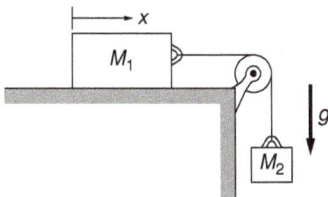

2.3 *Two blocks on table*
Two blocks m_1 and m_2 are in contact on a horizontal table. A horizontal force is applied to one of the blocks, as shown in the drawing. If $m_1 = 2$ kg, $m_2 = 1$ kg, and $F = 3$ N, find the force of contact between the two blocks.

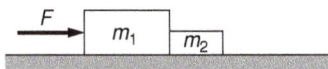

2.4 *Circling particle and force*
Two particles of mass m and M undergo uniform circular motion about each other at a separation R under the influence of an attractive constant force F. The angular velocity is ω radians per second. Show that $R = (F/\omega^2)(1/m + 1/M)$.

2.5 *Concrete mixer**
In a concrete mixer, cement, gravel, and water are mixed by tumbling action in a slowly rotating drum. If the drum spins too fast the ingredients stick to the drum wall instead of mixing.
    Assume that the drum of a mixer has radius $R = 0.5$ m and that it is mounted with its axle horizontal. What is the fastest the drum can rotate without the ingredients sticking to the wall all the time? Assume $g = 9.8$ m/s^2.

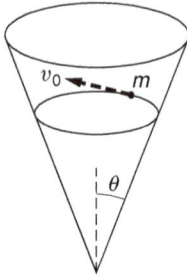

2.6 *Mass in cone*

A particle of mass m slides without friction on the inside of a cone. The axis of the cone is vertical, and gravity is directed downward. The apex half-angle of the cone is θ, as shown.

The path of the particle happens to be a circle in a horizontal plane. The speed of the particle is v_0. Draw a force diagram and find the radius of the circular path in terms of v_0, g, and θ.

2.7 *Leaning pole*

A pole of negligible mass leans against a wall, at angle θ with the horizontal. Gravity is directed down.

(a) Find the constraint relating the vertical acceleration of one end to the horizontal acceleration of the other.

(b) Now suppose that each end carries a pivoted mass M. Find the initial vertical and horizontal components of acceleration as the pole just begins to slide on the frictionless wall and floor. Assume that at the beginning of the motion the forces exerted by the rod are along the line of the rod. (As the motion progresses, the system rotates and the rod exerts sidewise forces.)

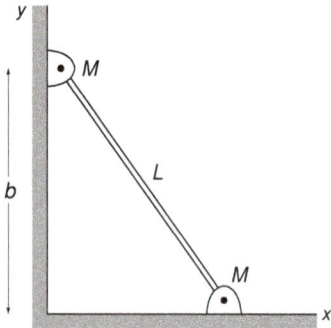

2.8 *Two masses and two pulleys**

Masses M_1 and M_2 are connected to a system of strings and pulleys as shown. The strings are massless and inextensible, and the pulleys are massless and frictionless. Find the acceleration of M_1.

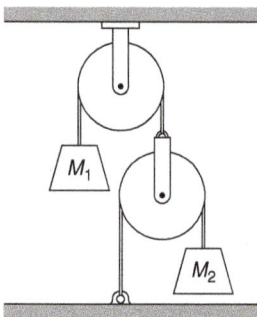

2.9 *Masses on table*

Two masses, A and B, lie on a frictionless table, as shown.

They are attached to either end of a light rope of length l which passes around a pulley of negligible mass. The pulley is attached to a rope connected to a hanging mass, C. Find the acceleration of each mass. (You can check whether or not your answer is reasonable by considering special cases—for instance, the cases $M_A = 0$, or $M_A = M_B = M_c$.)

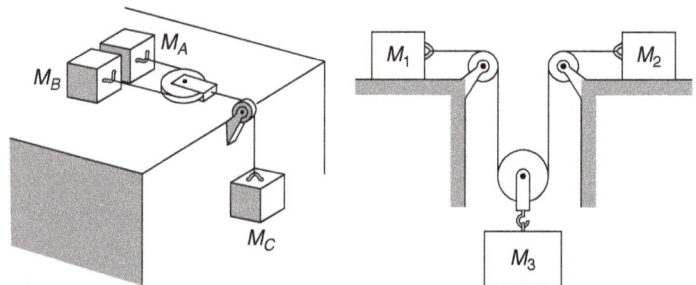

2.10 *Three masses*

The system of masses M_1, M_2, and M_3 in the sketch uses massless pulleys and ropes. The horizontal table is frictionless. Gravity is directed downward.

(a) Draw force diagrams, and show all relevant coordinates.

(b) How are the accelerations related?

2.11 *Mass on wedge**

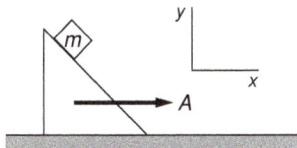

A 45° wedge is pushed along a table with constant acceleration A. A block of mass m slides without friction on the wedge. Find the block's acceleration. Gravity is directed down.

2.12 *Painter on scaffold**

A painter of mass M stands on a scaffold of mass m and pulls himself up by two ropes which hang over pulleys, as shown.

He pulls each rope with force F and accelerates upward with a uniform acceleration a. Find a—neglecting the fact that no one could do this for long.

2.13 *Pedagogical machine**

A "pedagogical machine" is illustrated in the sketch. All surfaces are frictionless. What force F must be applied to M_1 to keep M_3 from rising or falling?

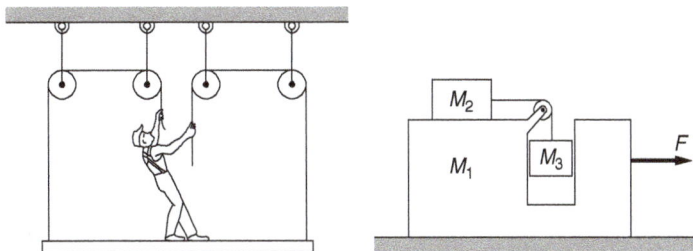

2.14 *Pedagogical machine 2**

Consider the "pedagogical machine" of the previous problem in the case where F is zero. Find the acceleration of M_1.

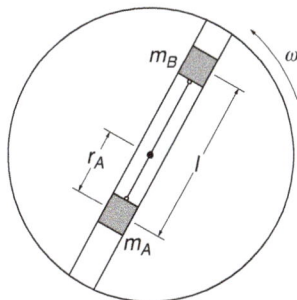

2.15 *Disk with catch*

A disk rotates with constant angular velocity ω, as shown. Two masses, m_A and m_B, slide without friction in a groove passing through the center of the disk. They are connected by a light string of length l, and are initially held in position by a catch, with mass m_A at distance r_A from the center. Neglect gravity. At $t = 0$ the catch is removed and the masses are free to slide.

Find $\ddot{r}_A$ immediately after the catch is removed, in terms of m_A, m_B, l, r_A, and ω.

2.16 *Planck units**

Max Planck introduced a constant h, now called Planck's constant, to relate the energy of an oscillator to its frequency. $h = 6.6 \times 10^{-34}$ J $\cdot$ s, where 1 joule (J) = 1 newton-meter. (h is engraved on Planck's tombstone in Göttingen, Germany.)

Planck pointed out that if one takes h and Newton's gravitational constant $G = 6.7 \times 10^{-11}$ m^{3}kg^{-1}s^{-2} and the speed of light $c = 3.0 \times 10^8$ m/s as fundamental quantities, it is possible to combine them to form three new independent quantities to replace the customary units of mass, length, and time. The three new quantities are called the Planck units.

(*a*) Planck length L_p

(*b*) Planck mass M_p

(*c*) Planck time T_p.

The Planck units have a natural role in modern cosmology, particularly the cosmology of the early universe.

Find the SI values of the Planck units, as for example 1 $L_p = $ (?) m. (Note: published results may differ from yours because they are often evaluated using $\hbar = h/2\pi$.)

2.17 *Block on accelerating wedge*

A block rests on a wedge on a horizontal surface. The coefficient of friction of the block on the wedge is μ. Gravity is directed down. The wedge angle θ obeys $\tan \theta > \mu$. The wedge is accelerated horizontally at rate a. Find the maximum and minimum values of a for the block to stay fixed on the plane.

3 FORCES AND EQUATIONS OF MOTION

3.1 Introduction

The concept of force is central in Newtonian physics. This chapter describes the gravitational force and the electrostatic force, two of the fundamental forces of nature. We also discuss several phenomenological forces, for example friction. Such forces are commonly encountered in "everyday" physics and are approximately described by empirical equations. Because the concept of force is meaningful only if one knows how to solve problems involving forces, this chapter includes many examples in which Newton's laws are put into practice.

The problem of calculating motion from known forces frequently occurs in physics. For instance, a physicist who sets out to design a particle accelerator employs the laws of mechanics and knowledge of electric and magnetic forces to calculate how the particles will move in the accelerator. Equally important, however, is the converse process of deducing the physical interaction from observations of the motion, which is how new laws are discovered. The classic example is Newton's deduction of the inverse-square law of gravitation from Kepler's laws of planetary motion. A contemporary example is the effort to elucidate the interactions between elementary particles from high energy scattering experiments at the Large Hadron Collider at CERN in Geneva and at other high energy laboratories.

Unscrambling experimental observations to find the underlying forces can be complicated. In a facetious mood, the British cosmologist Arthur Eddington once said that force is the mathematical expression we put into the left-hand side of Newton's second law to obtain results that agree with observed motions. Fortunately, force has a more concrete physical reality.

Much of our effort in the following chapters will be to understand how systems behave under applied forces. The task would be hopeless if every pair of particles in the universe had their own unique interaction. Fortunately, nature is kinder than this. As far as we know, there are only four fundamentally different types of interactions in the universe: gravity, electromagnetic interactions, the so-called weak interaction, and the strong interaction. At one time electricity and magnetism were regarded as different forces but the work of James Clerk Maxwell in the 1870's unified them, revealing them to be different aspects of a single force field, called the electromagnetic field. In another great synthesis, the weak interaction and the electromagnetic interaction were unified in the 1970's by Steven Weinberg, Sheldon Glashow, and Abdus Salaam, for which they received the Nobel Prize in 1979.

3.2 The Fundamental Forces of Physics

The most familiar fundamental forces are gravity and electromagnetic forces, both of which act over a long range. Their strengths decrease only as the inverse square of the distance between particles. In spite of this similarity, they play totally different roles in nature. The gravitational force always attracts, whereas electrical forces can either attract or repel. However, the major difference is that gravity is incredibly weak compared to electromagnetic interactions. For instance, the gravitational force between an electron and a proton in a hydrogen atom is smaller than the electric force by a factor of about 10^{-30}. In large systems, however, electrical attraction and repulsion cancel to a high degree, and gravity alone is left. Gravitational forces therefore dominate the cosmic scale of our universe. In contrast, the world immediately around us is dominated by electrical forces, which are far stronger than gravity on the atomic scale. Electrical forces are responsible for the structure of atoms, molecules, and more complex forms of matter, as well as the existence of light.

There are two other fundamental forces: the weak and the strong interactions. They have such a short range that they are important only at nuclear distances, typically 10^{-15} m. These interactions are negligible even at atomic distances, 10^{-10} m. As its name implies, the strong interaction is very strong, much stronger than the electromagnetic force at nuclear distances. It is the "glue" that binds protons and neutrons together in the atomic nucleus, but aside from this it has little effect in the everyday world. The weak interaction plays a less dramatic role; it mediates in the creation and destruction of neutrinos—particles of no charge and almost no mass that are essential to our understanding of matter but that can be detected only by the most arduous experiments.

The forces of gravity and electromagnetism are regarded as fundamental because they cannot easily be explained in simpler terms. The phenomenological forces that we shall discuss, such as friction, the contact force, and the viscous force, can be described by relatively simple empirical mathematical expressions, but when examined in detail they can be explained as the macroscopic manifestation of complicated interatomic forces.

3.3 Gravity

Gravity, the most familiar of the fundamental forces, played an honored role in the development of mechanics; Newton discovered the law of universal gravitation in 1666, the same year that he formulated his laws of motion. By calculating the motion of two gravitating particles, Newton was able to derive Kepler's empirical laws of planetary motion. (And by accomplishing all this by age 26, Newton established a tradition that still maintains—that great advances are often made by young physicists.)

According to Newton's law of gravitation, two particles attract each other with a force that is proportional to the product of their masses and inversely proportional to the square of the distance between them. Gravity is always attractive.

This verbal description of the gravitational force is essentially correct but not useful for solving problems, for which we need a mathematical expression. Consider two particles, a and b, with masses M_a and M_b,

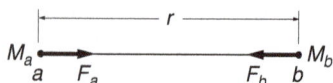

respectively, separated by distance r. Let $\mathbf{F}_{b,a}$ be the force exerted on particle b by particle a. Our verbal description of the magnitude of the gravitational force can be expressed mathematically as

$$|\mathbf{F}_{b,a}| = \frac{GM_aM_b}{r^2}.$$

G is known as the *gravitational constant*. The value of G can be found by measuring the force between masses in a known geometry, a procedure first carried out by Henry Cavendish in 1771 using a torsion balance. The mass of the Earth can be found from G, the acceleration due to gravity g, and the radius of the Earth, as we shall see in a later section. Cavendish became famous as the scientist who "weighed the Earth."

The value of G is $6.673(10) \times 10^{-11} \mathrm{m}^3\mathrm{kg}^{-1}\mathrm{s}^{-2}$. G is difficult to measure because of the weakness of gravity, and at a relative uncertainty of 10^{-4}, it is the least accurately known of the "fundamental" constants in physics. In contrast, other constants are typically known with a relative uncertainty of 10^{-8} or better.

The gravitational force between two particles is a *central force* because it is directed along the line joining them. Vector notation is ideally suited for describing these properties mathematically. By convention, we introduce the vector $\mathbf{r}_{b,a}$ that extends *from* the particle exerting the force, particle a in this case, *to* the particle experiencing the force, particle b. It is evident that $\mathbf{r}_{b,a} = -\mathbf{r}_{a,b}$. Note that $|\mathbf{r}_{b,a}| = r$. Introducing the unit vector $\hat{\mathbf{r}}_{b,a} = \mathbf{r}_{b,a}/r$, we have

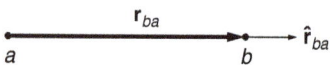

$$\mathbf{F}_{b,a} = -\frac{GM_aM_b}{r^2}\hat{\mathbf{r}}_{b,a}.$$

The negative sign indicates that the force on particle b is directed toward particle a, that is, the force is attractive. The force on a due to b is

$$\mathbf{F}_{a,b} = -\frac{GM_aM_b}{r^2}\hat{\mathbf{r}}_{b,a} = +\frac{GM_aM_b}{r^2}\hat{\mathbf{r}}_{a,b} = -\mathbf{F}_{b,a}$$

where we have used $\hat{\mathbf{r}}_{b,a} = -\hat{\mathbf{r}}_{a,b}$. Thus the forces on the two particles are equal and opposite, as Newton's third law requires.

3.3.1 The Gravitational Force of a Sphere

Newton's law of gravitation describes the interaction between point particles. How can we find the gravitational force on a particle due to a real extended body like the Earth? Because force obeys the law of superposition, the force due to a collection of particles is the vector sum of the forces exerted by the particles individually. This allows us to mentally divide the body into a collection of small elements that can be treated as particles. We can then sum the forces from all the particles using standard methods from integral calculus. This approach is applied in Note 3.1 to calculate the force between a particle of mass m and a uniform thin spherical shell of mass M and radius R. The result is

$$\mathbf{F} = -G\frac{Mm}{r^2}\hat{\mathbf{r}} \qquad r > R$$

$$\mathbf{F} = 0 \qquad r < R,$$

where r is the distance from the center of the shell to the particle. If the particle lies outside the shell $r > R$, the force is the same as if all the mass of the shell were concentrated at its center. If the particle lies inside, the force vanishes.

The reason why gravitational force vanishes inside a spherical shell can be seen by a simple argument due to Newton. Consider the two small mass elements marked out by a conical surface with its apex at m.

The amount of mass in each element is proportional to its surface area. The area increases as (distance)^2. However, the strength of the force varies as $1/$(distance)^2, where the distance is measured from the apex to the shell. Thus the forces of the two mass elements are equal and opposite, and cancel. We can pair up all the elements of the shell this way and so the total force on m is zero.

A uniform solid sphere can be regarded as a succession of thin spherical shells, so for particles outside the sphere, the sphere behaves gravitationally as if its mass were concentrated at its center. This result also holds if the density of the sphere varies with radius, provided the mass distribution is spherically symmetric.

For example, although the Earth has a dense core, the mass distribution is nearly spherically symmetric, so to good approximation the gravitational force of the Earth on a mass m at distance r is

$$\mathbf{F} = -\frac{GM_e m}{r^2}\hat{\mathbf{r}} \qquad r \geq R_e$$

where M_e is the mass of the Earth and R_e its radius.

3.3.2 The Acceleration Due to Gravity

At the surface of the Earth, the gravitational force on mass m is

$$\mathbf{F} = -\frac{GM_e m}{R_e^2}\hat{\mathbf{r}},$$

and the acceleration due to gravity is

$$\mathbf{a} = \frac{\mathbf{F}}{m}$$

$$= -\frac{GM_e}{R_e^2}\hat{\mathbf{r}}.$$

As we expect, the acceleration is independent of m. The acceleration due to the Earth's gravity, GM_e/R_e^2, is universally designated by g. When g is written as a vector, the vector is directed "down" toward the center of the Earth:

$$\mathbf{g} = -\frac{GM_e}{R_e^2}\hat{\mathbf{r}}.$$

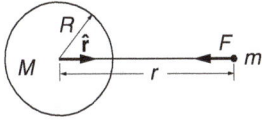

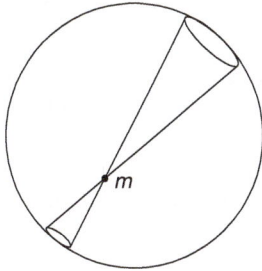

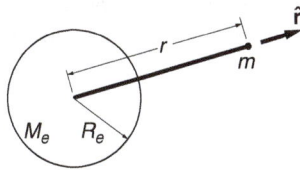

This result justifies our earlier claim that M_e can be found from G, g, and R_e.

The value of g varies slightly over the surface of the Earth, but if high accuracy is not required it can be taken to have a nominal value of $9.80 \text{ m/s}^2 = 980 \text{ cm/s}^2 \approx 32 \text{ ft/s}^2$.

By convention, g stands for the acceleration of an object measured with respect to the Earth's surface. This differs slightly from the true gravitational acceleration—the acceleration measured in an inertial system—due to the rotation of the Earth, a point to which we shall return in Chapter 9. Furthermore, g increases by about 5 parts per thousand from the Equator to the poles. About half this variation is due to the slight flattening of the Earth, and the remainder arises from the Earth's rotation. Local mass concentrations, for instance ocean and atmospheric tides, affect g; a variation in g of 10 parts per million is typical.

The acceleration of gravity also varies with altitude h. This effect is easily calculated for values of h that are small compared to the Earth's radius. We can write

$$\Delta g = g(R_e + h) - g(R_e) \approx h\frac{dg}{dr}$$

where the derivative is evaluated at R_e. The result is

$$\Delta g \approx -2GM_e\frac{h}{R_e{}^3} = -2g\frac{h}{R_e}.$$

It is good practice in physics to express the variation of some variable as a fractional change, since this immediately sets the scale for the size of effects. The fractional change in g with altitude is

$$\frac{\Delta g}{g} = -\frac{2h}{R_e}.$$

The Earth's radius is approximately 6×10^6 m and so g decreases by roughly 1 part per million for each increase in altitude of 3 m.

3.3.3 Weight

The weight of a body is the gravitational force exerted on it by the Earth. At the surface of the Earth the weight of a mass m is

$$\mathbf{W} = -G\frac{M_e m}{R_e{}^2}\hat{\mathbf{r}}$$

$$= m\mathbf{g}.$$

The unit of weight is the newton (SI), dyne (CGS), or, in the United States and a few other countries, the pound (English). In everyday affairs, "weight" and "mass" are used interchangeably, for the most part without ambiguity. Thus one hears such statements as "1 kg equals 2.2 lbs," which strictly speaking means that the weight of a 1 kg mass is 2.2 lbs. Similarly, reference to a "10-lb mass" means a mass that weighs 10 lbs.

Our definition of weight is unambiguous. According to this definition, the weight of a body is not affected by its motion. However, weight is often used in another sense in which it is taken to be the magnitude of the force on a body that must be exerted by its surroundings to support the body under the influence of gravity. The following example illustrates the difference between these two definitions.

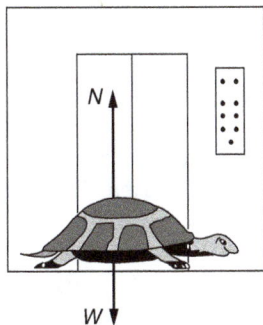

Example 3.1 Turtle in an Elevator

An amiable turtle of mass M stands in an elevator accelerating up at rate a. Find N, the force exerted on the turtle by the floor of the elevator.

The forces acting on the turtle are the normal force N and the weight, the true gravitational force $W = Mg$. Taking the positive direction to be up, we have

$$N - W = Ma$$
$$N = Mg + Ma$$
$$= M(g + a).$$

This result illustrates the two senses in which weight is used. In the sense that weight is the gravitational force, the weight of the turtle, Mg, is independent of the motion of the elevator. However, if the weight is taken to be the magnitude of the force exerted by the elevator on the turtle, for example the reading on a scale on which the turtle is standing, then the scale would indicate weight $N = M(g+a)$. With this definition, the turtle's weight increases when the elevator accelerates up and decreases if the elevator accelerates down. If the downward acceleration equals g, N becomes zero, and the turtle "floats" in the elevator. The turtle is then said to be in a state of weightlessness.

Although the two definitions of weight in the previous example are both commonly used and are both acceptable, we shall generally consider weight to mean the true gravitational force on mass m : $W = mg$, as measured in an inertial system. This is consistent with our resolve to refer all motion to inertial systems and helps us to keep the real forces on a body distinct.

Our definition of weight has one minor drawback. As we saw in the last example, a scale does not read mg in an accelerating system. As we have pointed out, systems at rest on the Earth's surface have a small acceleration due to the Earth's rotation, so that the reading of a scale fixed to the Earth's surface is not the true gravitational force on a mass. However, the effect is small and we shall treat the surface of the Earth as if it were an inertial system for the present.

3.3.4 The Principle of Equivalence

The gravitational force displays a profoundly mysterious behavior. Consider the equation of motion of particle b under the gravitational attraction of particle a. Let $\mathbf{a}_b$ denote the acceleration of particle b. Then

$$\mathbf{F}_b = -\frac{GM_b M_a}{r^2}\hat{\mathbf{r}}_{b,a} \tag{3.1}$$

$$= M_b\,\mathbf{a}_b \tag{3.2}$$

or

$$\mathbf{a}_b = -\frac{GM_a}{r^2}\hat{\mathbf{r}}_{a,b}.$$

The acceleration of a particle under gravity is independent of its mass! This is why all bodies fall with the same acceleration, in the absence of friction. We have, however, glossed over a subtle point in canceling M_b on both sides when combining Eqs. (3.1) and (3.2). The "mass" in the law of gravitation (*gravitational* mass) measures the strength of gravitational interaction and is operationally distinct from the "mass" (*inertial* mass) that characterizes inertia in Newton's second law. Why gravitational mass is proportional to inertial mass is a deep mystery. Newton recognized the mystery and confirmed the fact experimentally to an accuracy of about 1% by observing that the period of a pendulum does not depend on the material of the pendulum.

3.3.5 The Electrostatic Force

We discuss the electrostatic force only briefly because its full description is better left to a systematic study of electricity and magnetism. The salient feature of the electrostatic force between two particles is that the force, like gravity, is an inverse-square central force. The force depends upon a fundamental property of the particle called its *electric charge q*. There are two types of electric charge: experimentally, like charges repel, unlike charges attract.

The electrostatic force $F_{b,a}$ on charge q_b due to charge q_a is given by Coulomb's law:

$$\mathbf{F}_{b,a} = k\frac{q_a q_b}{r^2}\hat{\mathbf{r}}_{b,a}.$$

k is a constant of proportionality and $\hat{\mathbf{r}}_{b,a}$ is a unit vector that points from a to b. Following an inspired idea of Benjamin Franklin, we distinguish the two types of charge by assigning an algebraic sign to q, either positive or negative. If q_a and q_b are both negative or both positive, the force is repulsive, but if the charges have different signs, $\mathbf{F}_{b,a}$ is attractive.

In the SI system the unit of charge is the *coulomb* (C), which is defined in terms of electric currents and magnetic forces. In this system, k is now a defined quantity

$$k = 10^{-7}c^2 \approx 8.99 \times 10^9 \text{ N·m}^2/\text{C}^2,$$

where c is the defined speed of light.

The concept of the *electric field* is fundamental in electromagnetic theory. For our present purposes, we shall simply define the electric field

E to be the electric force on a body divided by its charge. The electric field at **r** due to a charge q at the origin is

$$\mathbf{E} = k\frac{q}{r^2}\hat{\mathbf{r}}.$$

3.4 Some Phenomenological Forces

3.4.1 Contact Forces

Contact forces are forces transmitted between bodies by short-range atomic or molecular interactions. Examples include the pull of a string, friction, and the force of viscosity between a moving body and a fluid. When examined on the atomic or molecular scale, such forces are found to originate primarily in electrostatic interactions between particles. However, our interest here is in the dynamics of particles subject to applied forces, so we will treat these forces phenomenologically by describing them with approximate empirical formulas, generally ignoring their microscopic origins.

3.4.2 Tension—The Force of a String

We usually take the "string" force for granted, having some primitive idea of how it behaves. The following example is intended to bring this force into sharper focus.

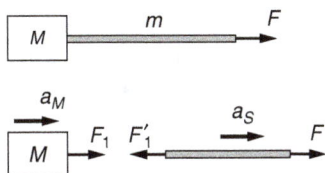

Example 3.2 Block and String

A string of mass m attached to a block of mass M is pulled with force F. Neglect gravity. What is the force F_1 on the block due to the string?

The sketch shows the force diagrams. F_1 is the force of the string on the block and F_1' is the force of the block on the string, The acceleration of the block is a_M and the acceleration of the string is a_s. The equations of motion are

$$F_1 = Ma_M$$
$$F - F_1' = ma_s.$$

Assuming that the string doesn't stretch, the string must accelerate at the same rate as the block so the constraint equation is $a_s = a_M$. Furthermore, $F_1 = F_1'$ by Newton's third law. Solving for the acceleration $a = a_M = a_s$, we find that

$$a = \frac{F}{M+m},$$

as we expect, and

$$F_1 = F_1'$$
$$= \frac{M}{M+m}F.$$

The force on the block is less than F; the string does not transmit the full applied force. The solution is reasonable because we expect that if $m \ll M$, then $F_1 \approx F$. In the opposite extreme $M \ll m$, we expect $F_1 \approx 0$, since there is practically no load for the string to pull.

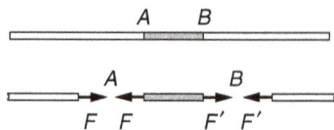

We can think of a string as composed of short sections interacting by contact forces. Each section pulls the sections to either side of it, and by Newton's third law, it is pulled by the adjacent sections. The magnitude of the force acting between adjacent sections is called *tension*. There is no direction associated with tension. In the sketch, the tension at A is F and the tension at B is F'.

Although a string may be under considerable tension, for example a string on a guitar, the net string force on each small section is zero if the tension is uniform, and the section remains at rest unless external forces act on it. If there are external forces on the section, or if the string is accelerating, the tension varies along the string, as Example 3.3 shows.

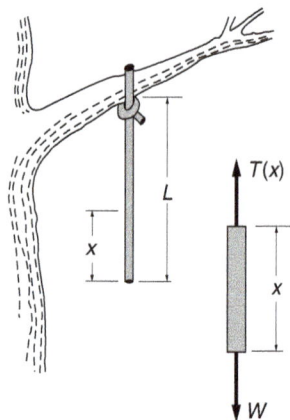

Example 3.3 Dangling Rope

A uniform rope of mass M and length L hangs from the limb of a tree. Find the tension in the rope at distance x from the bottom.

(The force diagram for the section of length x of the rope is shown in the sketch.) The section is pulled up by a force of magnitude $T(x)$, where $T(x)$ is the tension at x. The downward force on the section is its weight $W = Mg(x/L)$. The total force on the section is zero since it is at rest. Hence

$$T(x) = \frac{Mg}{L}x.$$

At the bottom of the rope the tension is zero, while at the top where $x = L$ the tension equals the total weight of the rope Mg.

Tension and Atomic Forces

The total force on each segment of a string in equilibrium must be zero. Nevertheless, the string will break if the tension is too large. We can understand this qualitatively by looking at a string from the atomic viewpoint. An idealized model of a string is a single long chain of molecules bound together by intermolecular forces. Suppose that force F is applied to molecule 1 at the end of the string. The force diagrams for molecules 1, 2, and 3 are shown in the sketch. In equilibrium, $F = F'$, $F' = F''$, and $F'' = F'''$ so that $F''' = F$. We see that the string "transmits" the force F.

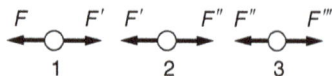

To understand how this comes about, we need to look at the nature of intermolecular forces.

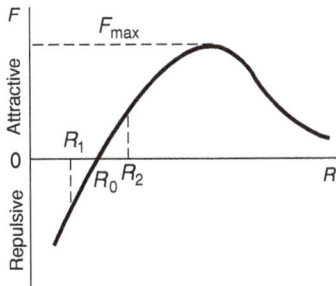

Qualitatively, the force between two molecules—which in this model is the tension in the string—depends on the distance R between them as shown in the drawing. At large distances, all molecules attract each other with a force known as the van der Waals force. The intermolecular force is repulsive at small distances, vanishes at some separation R_0, and is attractive for $R > R_0$. For large values of R the force must fall to zero because molecules do not interact over long distances. There are no scales on the sketch, but R_0 is typically a few angstroms (1 Å = 0.1 nm = 10^{-10} m).

In the absence of an applied force, the separation between adjacent molecules must be R_0; otherwise the intermolecular forces would cause the string to contract or expand. As we pull on the string, the molecules move apart slightly, say to $R = R_2$, where the intermolecular attractive force just balances the applied force so that the total force on each molecule is zero. If the string were stiff like a metal rod, we could push as well as pull. A push makes the molecules move slightly together, say to $R = R_1$, where the intermolecular repulsive force balances the applied force. The change in the length depends on the slope of the interatomic force curve at r_0. The steeper the curve, the less the stretch for a given pull.

The attractive intermolecular force has a maximum value F_{max}, as shown in the sketch. If the applied pull is greater than F_{max}, the intermolecular force is too weak to restore balance—the molecules continue to separate and the string breaks.

For a real string or rod, the intermolecular forces act on a three-dimensional lattice of atoms. The breaking strength of most materials is considerably less than the limit set by F_{max}. Breaks occur at points of weakness, or "defects," in the lattice, where the molecular arrangement departs from regularity. Microscopically thin metal "whiskers" and carbon "nanotubes" seem to be nearly free from defects, and they exhibit breaking strengths close to the theoretical maximum.

3.4.3 The Normal Force

If a body is in contact with a surface, the force on the body due to the surface can be resolved into two components, one perpendicular and the other tangential to the surface. The perpendicular component is called the *normal* force; the tangential component is called *friction*.

The origin of the normal force is similar to the origin of tension in a string. When we put a book on a table, the molecules of the book exert downward forces on the molecules of the table. The tabletop moves downward until the repulsion of the molecules in the table balances the force applied by the book. The more rigid the surface, the smaller the deflection. Because no surface is perfectly rigid, compression always occurs. However, the compression is usually too slight to notice and for most purposes we can assume ideal surfaces that are perfectly rigid.

In equilibrium, the normal force N exerted on a body by a surface is equal and opposite to the resultant of all other forces that act on the body in the perpendicular direction. When you stand, the normal force exerted by the ground is equal to your weight. When you walk, the normal force fluctuates as the surface accelerates up and down.

3.4.4 Friction

Friction is a force that opposes the relative motion of bodies in contact. In mechanical design friction is usually regarded as a problem but in fact life without friction would be hopeless: we could not walk and cars could not drive. Astronauts on a space walk, for instance, move in a frictionless environment. To return to the space ship they must rely on contact forces by pulling a tether line or perhaps by employing the rocket force produced by a jet pack.

Friction arises when the surface of one body moves, or tries to move, along the surface of a second body. Friction depends on detailed structure at the molecular level and is generally too complicated to be analyzed from basic principles. Consequently, friction must be treated phenomenologically, described by empirical rules. The magnitude of the force of friction varies in a complicated way with the nature of the surfaces and their relative velocity. In fact, the only thing we can always say about friction is that it opposes the motion that would occur in its absence. For instance, suppose that we try to push a book across a table. If we push gently, the book does not move and the force of friction assumes whatever value is needed to keep the book at rest. However, this force cannot increase indefinitely. If we push hard enough, the book starts to slide. Once the book is sliding, the friction force is approximately constant at low speeds.

In many cases the maximum value of the friction between two surfaces is found to be essentially independent of the area of contact, which may seem strange. The reason is that the actual area of contact on an atomic scale is a minute fraction of the total surface area. This fraction is proportional to the pressure, that is, the force per unit area. If the area is doubled while the normal force is held constant, then the pressure is halved. Thus there is twice as much contact area, but only half as much microscopic force on each. The resultant total friction force, the product of contact area and microscopic force, is unchanged. This is an oversimplified explanation and non-rigid bodies like automobile tires are more complicated. A wide tire is generally better than a narrow one for good acceleration and braking.

Empirically, the friction force f is proportional to the normal force N, which we express as $f = \mu N$, where μ is an empirical dimensionless constant called the *coefficient of friction*. Typical values of μ are in the range of 0.3 to 0.6. For Teflon®, which is exceptionally slippery, $\mu = 0.04$, and for some surfaces μ can exceed 1.

When a body slides on a surface, the friction force has its maximum value μN and is directed so as to oppose the motion. Experimentally, the force of friction decreases slightly when the body begins to slide. The terms "static friction" and "kinetic friction" (or "sliding friction") are sometimes employed to distinguish the cases of rest and motion, but we shall generally neglect any difference between the coefficients of static and kinetic friction.

In summary, we take f to behave as follows:

1. For bodies not in relative motion (static friction),
   $0 \leq f \leq \mu N$.
   f opposes the motion that would occur in its absence.
2. For bodies in relative motion (kinetic friction),
   $f = \mu N$.
   f is directed opposite to the relative velocity.

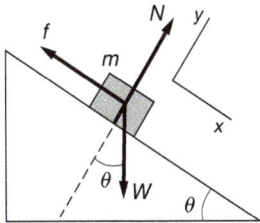

Example 3.4 Block and Wedge with Friction

A block of mass m rests on a fixed wedge of angle θ. The coefficient of friction is μ. (For wooden blocks, μ is typically in the range of 0.2 to 0.5.) Find the value of θ at which the block starts to slide, and the acceleration $\ddot{x}$ when it slides.

In the absence of friction, the block would slide down the plane; hence the friction force f points up the plane. With the coordinates shown, we have

$$m\ddot{x} = W \sin\theta - f$$

and

$$m\ddot{y} = N - W \cos\theta$$
$$= 0.$$

When sliding is just about to start, f has its maximum value μN and $\ddot{x} = 0$. The equations then give

$$W \sin\theta_{\text{max}} = \mu N$$
$$W \cos\theta_{\text{max}} = N.$$

Hence

$$\tan\theta_{\text{max}} = \mu.$$

As the wedge angle is gradually increased from zero, the friction force grows in magnitude from zero to its maximum value μN, since before the block begins to slide we have

$$f = W \sin\theta \qquad \theta \leq \theta_{\text{max}}.$$

Once the block is sliding, its acceleration is given by

$$m\ddot{x} = W \sin\theta - \mu W \cos\theta$$
$$\ddot{x} = g(\sin\theta - \mu\cos\theta).$$

The physical dimension of the right-hand side is that of g, acceleration, while on the left it is $\ddot{x}$, which is also acceleration, so the equation is dimensionally correct. However, it is good practice to look at the behavior for various values of the parameters, in this case, for various values of μ. When μ is small, the acceleration is somewhat less than in the absence of friction, which is to be expected. However, if μ is large enough for the expression $\sin\theta - \mu\cos\theta$ to become negative, then $\ddot{x}$ is negative, indicating that the block is sliding uphill, contrary to the assumption that it starts from rest. The contradiction between the solution and reality arises because if the block were to slide uphill, the friction force would actually point downhill, contrary to the force diagram. The dilemma arises because of the assumption in the analysis that friction has its maximum value. This, however, is true only if the block is sliding. In the case that $\mu > \tan\theta$, the block never starts to slide.

Example 3.5 The Spinning Terror

The Spinning Terror is an amusement park ride—a large vertical drum that spins so fast that everyone inside stays pinned against the wall when the floor drops away. What is the minimum steady angular velocity ω that allows the floor to be dropped safely?

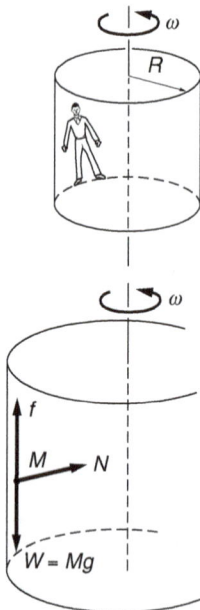

Suppose that the radius of the drum is R and the mass of the body is M. Let μ be the coefficient of friction between the drum and the body. The forces on M are the weight W, the friction force f, and the normal force exerted by the wall, N, as shown.

The radial acceleration is $R\omega^2$ toward the axis, and the radial equation of motion is

$$N = MR\omega^2.$$

For M to be in vertical equilibrium

$$f = Mg.$$

By the law of static friction,

$$f \le \mu N = \mu MR\omega^2,$$

and we have

$$Mg \le \mu MR\omega^2$$

or

$$\omega^2 \geq \frac{g}{\mu R}.$$

Consequently, the smallest value of ω for safety is

$$\omega_{\min} = \sqrt{\frac{g}{\mu R}}.$$

As installed in amusement parks under names like "Gravitron" and "Starship 2000", the drum's radius is $R \approx 7$ m, large enough to hold more than 40 people at a time. The coefficient of friction between clothing and the padded cloth backing in the drum is at least 0.5. Then $\omega_{\min} = \sqrt{9.8 \text{ m/s}^2/(0.5 \times 7 \text{ m})} = 1.7$ rad/s. The drum must rotate $\geq \omega/2\pi = 0.27$ turns per second = 16 rpm (revolutions per minute). In practice, the drum usually spins at 24 rpm.

3.5 A Digression on Differential Equations

Differential equations frequently arise in the course of solving physical problems. Often we can analyze how a system behaves in a short time interval or are able to write all the forces acting on a small element of a system. Finding the behavior at all times or the behavior of the whole system then comes down to an integration, or, essentially, solving a differential equation.

The term *differential equation* can create anxiety because differential equations can be mathematically challenging. The study of differential equations constitutes a lively branch of applied mathematics, but the systems of interest in this book are usually described by simple equations that lead to physically intuitive solutions. These same equations arise in many different contexts and are an essential part of the "working vocabulary" of physics, so it is worth making their acquaintance.

There are two aspects to dealing with differential equations in mechanics. First, framing the differential equation, which requires application of physical laws. Second, solving the differential equation, which requires mathematics.

If different parts of a system accelerate at different rates, Newton's laws still hold true but they must be applied with care. The next example deals with such a system. By considering each small section of a rope separately, we can obtain a simple differential equation from which the tension can be found at any point along the rope.

Example 3.6 Whirling Rope

A uniform rope of mass M and length L is pivoted at one end and whirls in a horizontal plane with constant angular speed ω. What is the tension in the rope at distance r from the pivot? Neglect gravity.

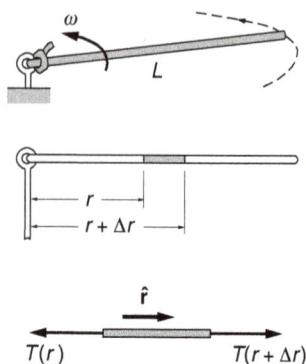

Let's find the equation of motion for a small section of rope between r and $r + \Delta r$. The length of the section is Δr and its mass is $\Delta m = M\Delta r / L$. Because the motion is circular, the section is undergoing radial acceleration. This requires a net radial force, which is possible only if the forces pulling the ends of the section are not equal. Consequently, the tension must vary with r, so we write the tension as $T(r)$.

The inward force on the section at distance r from the center is $T(r)$ while the outward force is $T(r + \Delta r)$. If we treat the small section as a particle of mass Δm, its inward radial acceleration is $r\omega^2$. (This point could be confusing; it would be just as reasonable to take the acceleration to be $(r + \Delta r)\omega^2$. However, when we take the limit $\Delta r \to 0$ the two expressions give the same result.)

The equation of motion for the section is

$$T(r + \Delta r) - T(r) = -(\Delta m)\, r\, \omega^2$$

$$\Delta T = -\frac{M\, r\, \omega^2 \Delta r}{L}.$$

Dividing the last equation by Δr gives $\Delta T / \Delta r$, which is the rate of change of T with r. Strictly speaking, this result is only approximate, but it becomes exact when we take the limit $\Delta r \to 0$ to give dT/dr:

$$\frac{dT}{dr} = \lim_{\Delta r \to 0} \frac{T(r + \Delta r) - T(r)}{\Delta r}$$

$$\frac{dT}{dr} = -\frac{Mr\omega^2}{L}.$$

This is a differential equation, an equation that gives us not the tension but only how tension varies with position. Because the right-hand side is < 0, the tension decreases as r increases, a reasonable behavior.

To find the tension as a function of r, we integrate:

$$dT = -\frac{M\omega^2}{L} r\, dr$$

$$\int_{T_0}^{T(r)} dT = -\int_0^r \frac{M\omega^2}{L} r\, dr,$$

where T_0, the tension at $r = 0$, is a constant that remains to be determined.

$$T(r) - T_0 = -\frac{M\omega^2}{L}\frac{r^2}{2}$$

or

$$T(r) = T_0 - \frac{M\omega^2}{2L} r^2.$$

To evaluate T_0 we need an additional piece of information called a *boundary condition*. Because the end of the rope at $r = L$ is free, the tension there must be zero. Consequently

$$T(L) = 0 = T_0 - \tfrac{1}{2}M\omega^2 L.$$

Hence $T_0 = \tfrac{1}{2}M\omega^2 L$, and the final result can be written

$$T(r) = \frac{M\omega^2}{2L}(L^2 - r^2).$$

The result is reasonable: the tension $M\omega^2 L/2$ at the origin is the same as if the mass of the rope were concentrated at a point whirling in a circle of radius half the length of the rope.

Next we consider a rope in equilibrium, but in a situation where the forces on the rope lie in different directions. The problem is to find the force when a pulley is used to change the direction of a rope. As every sailor knows, this force depends on the tension T and also the angle $2\theta_0$ through which the rope is deflected. It may be obvious from the diagram that the total force on the pulley is $2T \sin\theta_0$. This is the correct solution, but it is instructive to derive the result from first principles by finding the forces due to each element of the rope and then adding them vectorially.

Example 3.7 Pulleys

A string with constant tension T is deflected through angle $2\theta_0$ by a smooth fixed pulley. What is the force on the pulley?

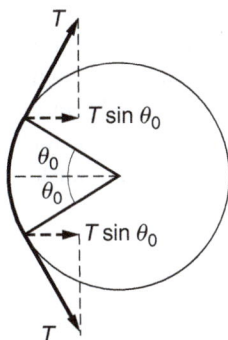

Consider a short segment of string between angles θ and $\theta + \Delta\theta$. Let ΔF be the outward force on the segment due to the pulley.

We shall shortly take the limit $\Delta\theta \to 0$, so we can treat the segment like a particle. The total force on the segment is zero at equilibrium. We have

$$\Delta F - 2T \sin\frac{\Delta\theta}{2} = 0.$$

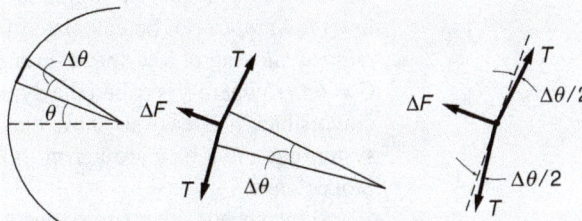

For small $\Delta\theta$, $\sin(\Delta\theta/2) \approx \Delta\theta/2$ and therefore

$$\Delta F = 2T\frac{\Delta\theta}{2} = T\Delta\theta.$$

Thus the segment exerts an inward radial force $T\Delta\theta$ on the pulley.

Finding the total force on the pulley requires adding (integrating) the contribution from each segment. It might be tempting to simply integrate each side of the equation by writing

$$\int dF = T \int d\theta$$

but blindly plugging in symbols would give the absurd answer $F = T\theta$. Although the force would vanish as the deflection angle θ is decreased, it would increase without limit as the angle is made larger.

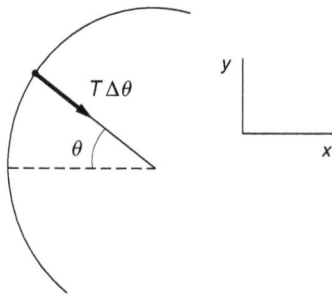

The essential point is that the forces at either end of the segment are not parallel and the force on the pulley due to each segment must be added vectorially. By symmetry, the net force $\mathbf{F}$ lies in the x direction, so we only need to calculate its x component. Using $\Delta F_x = T\Delta\theta \cos\theta$, the sum of contributions from each segment becomes an integral in the limit $\Delta\theta \to 0$. We have

$$F_x = \int dF_x = T \int_{-\theta_0}^{\theta_0} \cos\theta \, d\theta = 2T \sin\theta_0,$$

as we expect.

3.6 Viscosity

A body moving through a liquid or gas is retarded by *viscosity*—a force due to the fluid. Viscosity arises because a body moving through a medium exerts forces that set the nearby fluid into motion. By Newton's third law the fluid exerts a reaction force on the body. The viscous force $\mathbf{F}_v$ is along the line of motion, opposes the motion, and is proportional to the velocity, so we can write it in the vector form

$$\mathbf{F}_v = -C\mathbf{v},$$

where C is a constant that depends on the fluid and the geometry of the body. For objects of simple shape moving slowly through a gas at low pressure, C can be calculated from first principles. For a sphere of radius r moving at low speed through a common fluid like water or air, $C = 6\pi\eta r$, where η is called the dynamic viscosity of the fluid. Checking dimensions, it is easy to show that $[\eta] = \mathrm{M}\,\mathrm{L}^{-1}\,\mathrm{T}^{-1}$, so that in the SI system the units of η are kg/(m · s). The relation $F = 6\pi\eta r v$ is called Stokes' law.

At high speeds other forces due to turbulence occur and the total friction force, often called the drag force, can have a more complicated velocity dependence. (Sports car designers assume a force proportional to the square of the speed to account for the drag forces.) However, in many practical cases at low speeds viscosity is the only important drag force.

Example 3.8 Terminal Velocity

The Earth's atmosphere normally contains a large number of very small particles, for instance water droplets and carbon soot, generally called aerosols. Consider a spherical water droplet in still air falling under gravity. If the radius of the droplet is 5 microns, what is its maximum speed of fall (the terminal velocity)? The density of water is $\rho_w = 1000\,\text{kg/m}^3$ and the dynamic viscosity of air at $20\,°\text{C}$ is $1.8 \times 10^{-5}\,\text{kg/m} \cdot \text{s}$.

If m is the mass of the droplet and v its instantaneous speed, the equation for vertical motion is

$$m\frac{dv}{dt} = -6\pi\eta\, r\, v + mg$$

or

$$\frac{dv}{dt} = -\frac{6\pi\eta\, r\, v}{m} + g$$

where positive is downward.

Using $m = (4/3)\pi r^3 \rho_w$ gives a differential equation for the speed:

$$\frac{dv}{dt} = -\frac{9}{2}\left(\frac{\eta\, v}{\rho_w r^2}\right) + g.$$

If the droplet starts from rest with $v = 0$ the droplet initially begins to fall downward with acceleration g. But as the droplet accelerates, the viscous retarding force increases until finally the acceleration becomes 0. From then on the droplet falls at the constant terminal velocity v_t,

$$0 = -\frac{9}{2}\left(\frac{\eta\, v_t}{\rho_w r^2}\right) + g$$

which gives

$$
\begin{aligned}
v_t &= \frac{2}{9}\left(\frac{g\rho_w r^2}{\eta}\right) \\
&= \frac{2}{9}\left(\frac{(9.8\,\text{m/s}^2)(1000\,\text{kg/m}^3)(5 \times 10^{-6}\text{m})^2}{1.8 \times 10^{-5}\,\text{kg/m} \cdot \text{s}}\right) \\
&= 3 \times 10^{-3}\,\text{m/s}.
\end{aligned}
$$

The very small droplet practically floats in the air. For a larger droplet, such as a raindrop with $r = 1\,\text{mm} = 10^{-3}\,\text{m}$, the terminal velocity is much higher, 120 m/s.

If the only force on a body is the viscous retarding force, the equation of motion is

$$m\frac{d\mathbf{v}}{dt} = -C\mathbf{v}.$$

Once again we have a differential equation, in this case an equation for the variable **v**. Because the force is along the line of motion, only the magnitude of **v** changes and the vector equation reduces to the scalar equation

$$m\frac{dv}{dt} = -Cv.$$

We can rewrite this in an even simpler form:

$$\frac{dv}{dt} = -\frac{1}{\tau}v$$

where we have introduced the parameter $\tau = m/C$.

A few differential equations, such as this, are so simple and occur so frequently that it is worth becoming thoroughly familiar with their solutions whatever the context or the symbols. For instance, the same form of equation governs the loss of a nuclear species due to radioactive decay, the temperature of a body coming into thermal equilibrium, and the decrease of charge of an electrical capacitor attached to a resistor.

The equation $dv/dt = -v/\tau$ tells us that the velocity decreases at a rate proportional to its value. If you are familiar with the exponential function e^x, then you know that this is actually its fundamental property: $(d/dx)e^x = e^x$. This suggests that we try a solution of the form $v = e^{bt}$, where b is a constant that is required to make the argument bt dimensionless. However, there is an obvious defect in this trial solution: it is dimensionally impossible. The left-hand side has the dimension of velocity, LT^{-1}, but the right-hand side is dimensionless. To repair this defect we can try a solution of the form $v = Be^{bt}$ where B is a constant that has the dimensions of velocity. This trial solution yields

$$\frac{dv}{dt} = bBe^{bt},$$

or

$$\frac{dv}{dt} = bv.$$

This agrees with $dv/dt = -v/\tau$ provided that $b = -1/\tau$, so that our solution takes the form

$$v = Be^{-t/\tau}.$$

B could be any value, but this presents a dilemma because we would like a solution that has no arbitrary constant. To evaluate B we make use of what is called an *initial condition*, that is, a specific piece of information known about the motion at some particular time. The problem stated that at $t = 0$ the body had speed $v = v_0$. Hence

$$v(t = 0) = v_0 = Be^0.$$

Because $e^0 = 1$, it follows that $B = v_0$, and the full solution is

$$v = v_0e^{-t/\tau}.$$

We solved our equation by simply guessing the form of the solution. This common sense approach can be a good way to solve such a problem, but the equation can also be solved formally:

$$\frac{dv}{dt} = -\frac{v}{\tau}$$

$$\frac{dv}{v} = -\frac{dt}{\tau}$$

$$\int_{v_0}^{v} \frac{dv}{v} = -\int_{0}^{t} \frac{1}{\tau} dt.$$

Note the correspondence between the limits: v is the velocity at time t and v_0 is the velocity at time 0. The initial condition is built into the limits.

$$\ln\left(\frac{v}{v_0}\right) = -\frac{1}{\tau} t$$

$$\frac{v}{v_0} = e^{-t/\tau}$$

$$v = v_0 e^{-t/\tau}.$$

Before leaving this problem, let us look at the solution in a little more detail. We introduced the parameter $\tau = m/C$, which has the physical dimension of time. The velocity decreases exponentially in time and τ is a *characteristic time* for the system; it is the time for the velocity to drop by a factor of $e^{-1} \approx 0.37$ of its original velocity. Mathematical parameters that arise in the solution to a physical problem generally have a physical meaning. It is not difficult to show that although the body theoretically never comes to rest, it travels only a finite distance $v_o \tau$.

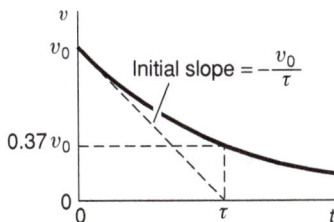

Example 3.9 Falling Raindrop

In Example 3.8 we found the terminal velocity of a droplet falling under gravity with a viscous retarding force. Here we shall solve the equation of motion to find the speed at any time after the droplet is released from rest.

Taking the coordinate system as positive downwards, the equation of motion can be written

$$\frac{dv}{dt} = -\frac{1}{\tau} v + g. \tag{1}$$

In this notation, the terminal velocity v_t is

$$v_t = \tau g. \tag{2}$$

To solve Eq. (1), we convert it to the form we have studied by changing the dependent variable from v to $u = v + \alpha$ where α is a constant we shall determine. Because the derivative of a constant is zero, the left-hand side of Eq. (1) is then $dv/dt = du/dt$. In terms of u, Eq. (1) becomes

$$\frac{du}{dt} = -\frac{1}{\tau}u + \frac{\alpha}{\tau} + g. \qquad (3)$$

Taking $\alpha = -\tau g$ puts Eq. (3) in the recognizable form

$$\frac{du}{dt} = -\frac{1}{\tau}u$$

which has the general solution

$$u = B\,e^{-t/\tau}.$$

Thus

$$v = u + \tau g = \tau g + B\,e^{-t/\tau}.$$

Using the initial condition $v = 0$ at $t = 0$, we find $B = -\tau g$ and the solution becomes

$$v = \tau g\left(1 - e^{-t/\tau}\right)$$
$$= v_t\left(1 - e^{-t/\tau}\right)$$

from Eq. (2). If $t \gg \tau$, the droplet falls at essentially its terminal velocity. For a 2 mm diameter raindrop in air, $\tau \approx 12\,\text{s}$.

3.7 Hooke's Law and Simple Harmonic Motion

In the mid-seventeenth century Robert Hooke (a contemporary of Newton) discovered that the extension of a spring is proportional to the applied force, for both positive and negative displacements. The force F_S exerted by a spring is given by Hooke's law

$$F_S = -k(x - x_0),$$

where k is a constant called the *spring constant* and $x - x_0$ is the displacement of the end of the spring from its equilibrium position x_0.

The important features of Hooke's law are that the force varies linearly with the displacement and that it is always directed toward the equilibrium position. When the spring is stretched, $x > x_0$ and F_S is negative, directed toward x_0. When the spring is compressed, $x < x_0$ and F_S is positive, once again directed toward x_0. For this reason, a Hooke's law force is sometimes called a *linear restoring force*.

Hooke's law is essentially empirical and breaks down for large displacements. Taking a frivolous view of affairs, we could rephrase Hooke's law as "extension is proportional to force, as long as it is." However, for sufficiently small displacements Hooke's law is remarkably

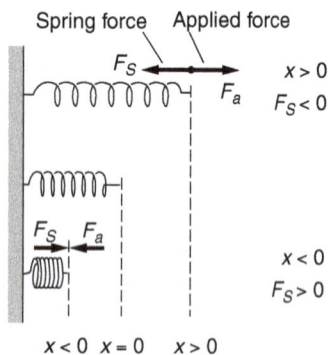

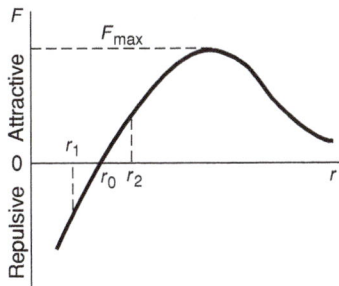

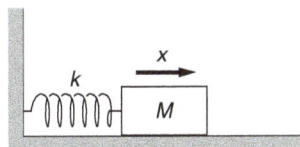

accurate, not only for springs but also for practically every system near equilibrium. A look at the intermolecular force curve again tells why the spring force is so common in nature.

If the force curve is linear in the neighborhood of the equilibrium point, then the force is proportional to the displacement from equilibrium. This is almost always the case in physics: a sufficiently short segment of force vs. displacement is generally linear to a good approximation. Only in pathological cases does a force near equilibrium lack a linear component. It is to be expected, however, that the linear approximation will inevitably break down for a large displacement.

The motion of a mass on a spring that is displaced from equilibrium and released is called *simple harmonic motion (SHM)*. This behavior is ubiquitous in physics, arising in all sorts of mechanical and electromagnetic contexts, in all wave phenomena, and in phenomena ranging down to the atomic scale, such as the vibration of nuclei in a molecule.

To derive the equation for simple harmonic motion, consider the motion of a block of mass M attached to one end of a spring, the other end of which is fixed.

The block rests on a horizontal frictionless surface. We will take the zero of the coordinate system to lie at the equilibrium position. The equation of motion is

$$M \frac{d^2x}{dt^2} = -kx$$

or

$$\frac{d^2x}{dt^2} + \frac{k}{M}x = 0.$$

Introducing the variable $\omega = \sqrt{k/M}$ puts the equation for SHM in the standard form

$$\frac{d^2x}{dt^2} + \omega^2 x = 0.$$

Any system that obeys an equation of this form is called a *harmonic oscillator*. Before looking at the solution of the harmonic oscillator equation, it is useful to see what the equation tells us physically. The displacement and the acceleration always have opposite signs. As the mass heads outward $x > 0$, the negative acceleration eventually brings the mass to rest and accelerates it back toward the equilibrium position. After the mass speeds through equilibrium, the acceleration changes sign and the mass is pulled back. We therefore expect the mass to oscillate about the equilibrium position.

Motion that repeats regularly is called *periodic motion*, and we might guess that simple harmonic motion is periodic. Sine and cosine functions are periodic, repeating themselves whenever their arguments increase by

2π radians (360°). We shall derive the solution to the SHM equation formally in Example 5.2, but the solution is so straightforward that we can guess its form:

$$x = A \sin \omega t + B \cos \omega t,$$

where A and B are constants, or equivalently

$$x = C \sin(\omega t + \phi)$$
$$= C(\cos \phi) \sin \omega t + C(\sin \phi) \cos \omega t.$$

Our two ways of writing the solution hold at any time t only if $A = C \cos \phi$ and $B = C \sin \phi$. Conversely, $C = \sqrt{A^2 + B^2}$ and $\tan \phi = B/A$. It is easy to show that both these solutions satisfy the equation of motion, where A and B, alternatively C and ϕ, are constants to be determined by initial conditions, for instance the position and the velocity at $t = 0$.

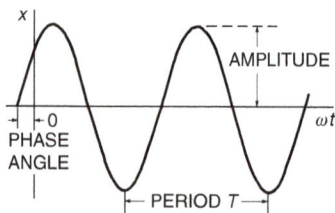

The motion is periodic in time, going through one cycle in time T given by $\omega T = 2\pi$. T (not to be confused with tension) is known as the *period* of the motion, and ω is known as the *angular frequency* of the motion. The maximum excursion C is called the *amplitude* of the motion and the angle ϕ is called the *phase angle*.

A fundamental property of simple harmonic motion is that the frequency of the motion, and hence the period, does not depend on the amplitude. If the spring is stretched farther before releasing the mass, the amplitude of the motion increases, which means that the mass travels farther during each cycle. However, because the force also increases with amplitude, the acceleration increases, and the mass moves faster as it passes through the origin. The effect of larger distance is compensated by the effect of higher acceleration, and the time for a complete cycle remains constant.

A note of caution: The unit of frequency in cycles per second (sometimes called the circular frequency) is the hertz: one cycle per second is 1 hertz (Hz). Circular frequency is often denoted by the symbol f or ν. The natural unit of angle in physics (and mathematics) is the radian, and the natural unit of frequency in physics, sometimes called the angular frequency, is *radians per second*. Angular frequency is often denoted by ω, but the unit has no special name. Circular and angular frequency both have the same physical dimension $[f] = T^{-1}$ and $[\omega] = T^{-1}$, but the two quantities differ by a factor of 2π: $\omega = 2\pi f$.

Example 3.10 Pendulum Motion

We show here that a simple pendulum—a point mass hanging from a massless string—is among the many physical systems that execute simple harmonic motion. Later, in Chapter 7, we shall drop the assumptions that the mass is a particle and the string massless, and

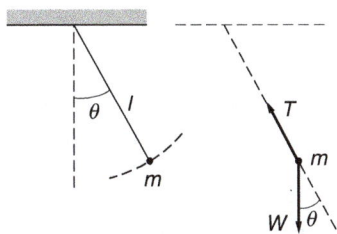

analyze what is called a physical pendulum. It, too, displays simple harmonic motion to good approximation if the amplitude of swing is a small angle.

The sketch shows a simple pendulum of length l, with mass M, and corresponding weight $W = Mg$.

The mass moves in a circular arc in a vertical plane. Denoting the angle from the vertical by θ, we see that the velocity is $l\,d\theta/dt$ and the acceleration is $l\,d^2\theta/dt^2$. The tangential force is $-W \sin \theta$. Thus the equation of motion is

$$Ml\frac{d^2\theta}{dt^2} = -Mg\sin\theta$$

or

$$\frac{d^2\theta}{dt^2} + \frac{g}{l}\sin\theta = 0.$$

This is not the equation for SHM because of the sine function, and it cannot be solved in terms of familiar functions. However, if the pendulum never swings far from the vertical so that $\theta \ll 1$, we can make the approximation $\sin\theta \approx \theta$, giving

$$\frac{d^2\theta}{dt^2} + \frac{g}{l}\theta = 0.$$

This is the equation for simple harmonic motion. To put it in standard form, take $\omega = \sqrt{g/l}$:

$$\frac{d^2\theta}{dt^2} + \omega^2\theta = 0.$$

It is important not to confuse ω, the angular frequency at which the pendulum oscillates, with $d\theta/dt$, which is the angular speed of the pendulum as it moves.

The solution is $\theta = A\sin\omega t + B\cos\omega t$, where $\omega = \sqrt{g/l}$ and A and B are constants. If at time $t = 0$ the pendulum is released from rest at angle θ_0, the solution is

$$\theta = \theta_0\cos\omega t.$$

The motion is *periodic*, which means it occurs identically over and over again. The period T, the time between successive repetitions of the motion, is given by $\omega T = 2\pi$, or

$$T = \frac{2\pi}{\sqrt{g/l}}$$

$$= 2\pi\sqrt{\frac{l}{g}}.$$

The maximum angle θ_0 is the amplitude of this motion. For small angles, the period is nearly independent of the amplitude, which is why the pendulum is so well suited to regulating the rate of a clock. However, this feature of the motion is a consequence of the approximation $\sin\theta \approx \theta$. A more accurate solution shows that the period lengthens slightly with increasing amplitude. Nevertheless, a pendulum clock can be an adequate timekeeper for household purposes if its mechanism can hold the amplitude nearly constant. A typical pendulum clock will not gain or lose more than 5 s/day if the amplitude (assumed to be 5 degrees) can be kept constant to within about 0.3 degree.

The general solution for the motion of a simple pendulum, without making the small angle approximation, is calculated in Note 5.1 using energy methods.

Example 3.11 Spring Gun and Initial Conditions

A spring gun fires a marble of mass M by means of a spring and piston in a barrel, as shown.

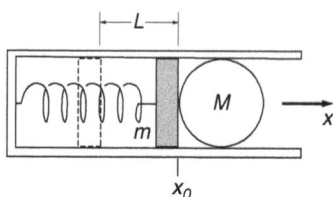

The piston has mass m and is attached to the end of a spring having spring constant k. The piston and marble are pulled back a distance L from equilibrium and released. The problem is to find the speed of the marble just as it loses contact with the piston. Gravity and friction are neglected.

We take the x axis to be along the direction of motion and the origin to be at the spring's unstretched position. The position of the piston obeys the equation for SHM

$$(M + m)\frac{d^2x}{dt^2} + kx = 0,$$

or

$$\frac{d^2x}{dt^2} + \frac{k}{M+m}x = 0,$$

which has the general solution

$$x(t) = A\sin\omega t + B\cos\omega t \tag{1}$$

where $\omega = \sqrt{k/(M+m)}$. The velocity is

$$v(t) = \dot{x}(t)$$
$$= \omega A\cos\omega t - \omega B\sin\omega t. \tag{2}$$

To evaluate the constants A and B we use the initial conditions $x(t = 0) = -L$ and $v(t = 0) = 0$. Substituting in Eqs. (1) and (2), we have

$$x(0) = -L = A\sin(\omega \cdot 0) + B\cos(\omega \cdot 0) = B$$
$$v(0) = 0 = \omega A\cos(\omega \cdot 0) - \omega B\sin(\omega \cdot 0) = A$$

so the solution can be written

$$x(t) = -L \cos \omega t \qquad (3)$$
$$v(t) = \omega L \sin \omega t. \qquad (4)$$

Our solution holds until the marble and piston lose contact. The piston can only push on the marble, not pull, and when the piston begins to slow down, contact is lost and the marble moves on at a constant velocity. From Eq. (4), we see that the time t_m at which the velocity reaches a maximum is given by

$$\omega t_m = \frac{\pi}{2}.$$

Substituting this in Eq. (3), we find

$$x(t_m) = -L \cos \frac{\pi}{2} = 0.$$

The marble loses contact as the spring passes its equilibrium point, as we expect, since the spring force retards the piston for $x > 0$.

From Eq. (4), the final speed of the marble is

$$v_{\max} = v(t_m)$$
$$= \omega L \sin \frac{\pi}{2}$$
$$= \sqrt{\frac{k}{M + m}} L.$$

For high speed, k and L should be large and $M + m$ should be small. In other words, to get high speed use a small projectile and pull on the spring gun as hard as you can. The answer is reasonable.

Note 3.1 The Gravitational Force of a Spherical Shell

In this note we calculate the gravitational force between a uniform thin spherical shell of mass M and a particle of mass m located a distance r from its center. We shall show that the magnitude of the force is GMm/r^2 if the particle is outside the shell and zero if the particle is inside.

To attack the problem, we divide the shell into narrow rings and add their forces using integral calculus. Let R be the radius of the shell and t its thickness, $t \ll R$. The ring at angle θ, which subtends angle $d\theta$, has circumference $2\pi R \sin \theta$, width $R \, d\theta$, and thickness t.

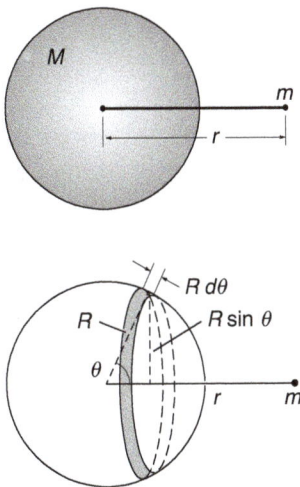

Its volume is

$$dV = 2\pi R^2 t \, \sin \theta \, d\theta$$

and its mass is

$$\rho \, dV = 2\pi R^2 t \rho \, \sin \theta \, d\theta$$
$$= \frac{M}{2} \sin \theta \, d\theta.$$

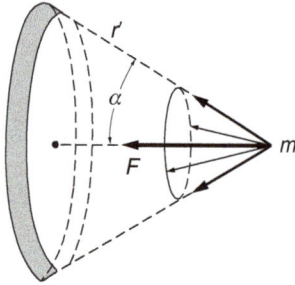

The total volume of the shell is $4\pi R^2 t$, so its mass density is $\rho = M/(4\pi R^2 t)$.

Each part of the ring is the same distance r' from m. The force on m due to a small section of the ring points toward that section. By symmetry, the transverse force components for the whole ring add vectorially to zero.

Since the angle α between the force vector and the line of centers is the same for all sections of the ring, the force components along the line of centers add to give the total force of the ring on the mass:

$$dF = \frac{Gm\rho \, dV}{r'^2} \cos \alpha$$

for the whole ring.

The force due to the entire shell is

$$F = \int dF$$
$$= \int \frac{Gm\rho \, dV}{r'^2} \cos \alpha.$$

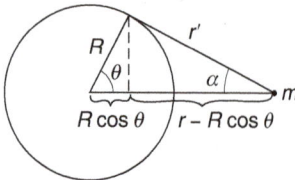

The problem now is to express all the quantities in the integrand in terms of one variable, say the polar angle θ. From the sketch, $\cos \alpha = (r - R \cos \theta)/r'$, and $r' = \sqrt{r^2 + R^2 - 2rR \cos \theta}$.

Since

$$\rho \, dV = M \sin \theta \, d\theta / 2,$$

we have

$$F = \left(\frac{GMm}{2}\right) \int_0^\pi \frac{(r - R \cos \theta) \sin \theta \, d\theta}{(r^2 + R^2 - 2rR \cos \theta)^{3/2}}.$$

A convenient substitution for evaluating this integral is $u = r - R \cos \theta$, $du = R \sin \theta \, d\theta$. Hence

$$F = \left(\frac{GMm}{2R}\right) \int_{r-R}^{r+R} \frac{u \, du}{(R^2 - r^2 + 2ru)^{3/2}}. \tag{1}$$

This integral is listed in tables, but we can evaluate it directly using integration by parts. For arbitrary functions f and g, the rule is

$$d(fg) = f \, dg + g \, df.$$

Integrating from a to b, we have

$$fg\big|_a^b = \int_a^b f \, dg + \int_a^b g \, df.$$

To apply, we choose f and g so that $\int f \, dg$ is the integral we want to evaluate, and also so that $\int g \, df$ is simpler to do. A good choice in this problem is

$$f = u$$

$$g = \frac{(-1/r)}{\sqrt{R^2 - r^2 + 2ru}}$$

so that

$$\int f\,dg = \int \frac{u\,du}{(R^2 - r^2 + 2ru)^{3/2}}$$

and

$$\int g\,df = -\int \frac{du/r}{\sqrt{R^2 - r^2 + 2ru}}$$

$$= \frac{-1}{r^2}\sqrt{R^2 - r^2 + 2ru}.$$

Combining, the result is

$$F = \frac{GMm}{2R}\frac{1}{r^2}\left(\frac{-ur}{\sqrt{R^2 - r^2 + 2ru}} + \sqrt{R^2 - r^2 + 2ru}\right)\Bigg|_{r-R}^{r+R}$$

$$= \frac{GMm}{2R}\frac{1}{r^2}\left[\left(\frac{-r(r+R)}{r+R} + r + R\right) - \left(\frac{-r(r-R)}{r-R} + r - R\right)\right]$$

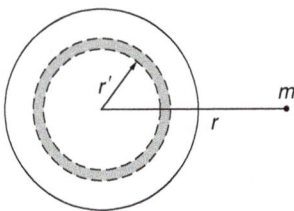

$$= \frac{GMm}{r^2}\quad r > R.$$

For $r > R$, the shell acts gravitationally as though all its mass were concentrated at its center.

There is one subtlety in our evaluation of the integral. The term $\sqrt{r^2 + R^2 - 2rR}$ is inherently positive, and we must take $\sqrt{r^2 + R^2 - 2rR} = r - R$, since $r > R$. If the particle is inside the shell, the magnitude of the force is still given by Eq. (1). However, in this case $r < R$, so we take $\sqrt{r^2 + R^2 - 2rR} = R - r$ in the evaluation. We find

$$F = \frac{GMm}{2R}\frac{1}{r^2}\left[\left(\frac{-r(r+R)}{r+R} + r + R\right) - \left(\frac{-r(r-R)}{R-r} + R - r\right)\right]$$

$$= 0\quad r < R.$$

A solid sphere can be thought of as a succession of spherical shells. It is not hard to extend our results to the case when the density of the sphere $\rho(r')$ is a function only of radial distance r' from the center of the sphere. The mass of a spherical shell of radius r' and thickness dr' is $\rho(r')4\pi r'^2 dr'$. The force it exerts on m is

$$dF = \frac{Gm}{r^2}\rho(r')4\pi r'^2 dr'.$$

Since the force exerted by every shell is directed toward the center of the sphere, the total force is

$$F = \frac{Gm}{r^2}\int_0^R \rho(r')4\pi r'^2 dr'.$$

However, the integral is simply the total mass of the sphere, and we find that for $r > R$, the force between m and the sphere is identical to the force between two particles m and M separated a distance r.

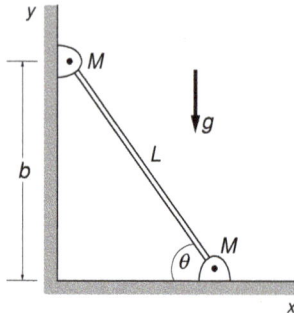

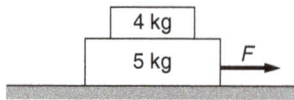

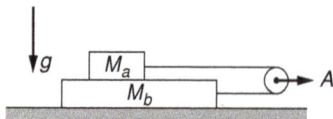

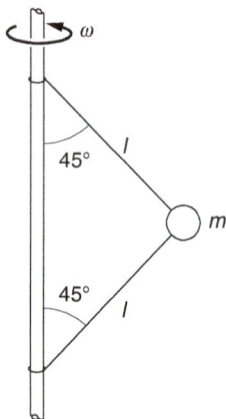

Problems

For problems marked *, refer to page 520 for a hint, clue, or answer.

3.1 *Leaning pole with friction*
Two identical masses M are pivoted at each end of a massless pole of length L. The pole is held leaning against a frictionless wall at angle θ, as shown. The coefficient of friction with the floor is μ. The pole is released and starts to slide. Find the initial acceleration of each mass.

3.2 *Sliding blocks with friction**
Mass $M_A = 4$ kg rests on top of mass $M_B = 5$ kg that rests on a frictionless table. The coefficient of friction between the two blocks is such that the blocks just start to slip when the horizontal force F applied to the lower block is 27 N. Suppose that now a horizontal force is applied to the upper block. What is its maximum value for the blocks to slide without slipping relative to each other?

3.3 *Stacked blocks and pulley*
Mass M_a lies on top of mass M_b, as shown. Assume $M_b > M_a$. The two blocks are pulled from rest by a massless rope passing over a pulley. The pulley is accelerated at rate A. Block M_b slides on the table without friction, but there is a constant friction force f between M_a and M_b due to their relative motion. Find the tension in the rope.

3.4 *Synchronous orbit**
Find the radius of the orbit of a synchronous satellite that circles the Earth. (A synchronous satellite goes around the Earth once every 24 h, so that its position appears stationary with respect to a ground station.) The simplest way to find the answer and give your results is by expressing all distances in terms of the Earth's radius R_e.

3.5 *Mass and axle**
A mass m is connected to a vertical revolving axle by two strings of length l, each making an angle of 45° with the axle, as shown. Both the axle and mass are revolving with angular velocity ω. Gravity is directed downward.

(*a*) Draw a clear force diagram for m.

(*b*) Find the tension in the upper string, T_{up}, and lower string, T_{low}.

3.6 *Tablecloth trick*
If you have courage and a tight grip, you can yank a tablecloth out from under the dishes on a table. What is the longest time in which the cloth can be pulled out so that a glass 6 in from the edge comes to rest before falling off the table? Assume that the coefficient of friction of the glass sliding on the tablecloth or sliding on the table-top is 0.5. (For the trick to be effective the cloth should be pulled out so rapidly that the glass does not move appreciably.)

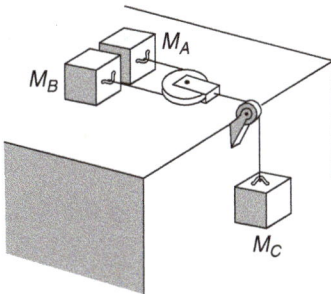

3.7 *Pulleys and rope with friction*
The system shown uses massless pulleys and rope. The coefficient of friction between the masses and horizontal surfaces is μ. Assume that M_A and M_B are sliding. Gravity is directed downward.
(*a*) Draw a force diagram for each mass, showing all relevant coordinates.
(*b*) How are the accelerations related?
(*c*) Find the tension in the rope.

3.8 *Block and wedge**
A block rests on a wedge inclined at angle θ. The coefficient of friction between the block and plane is μ.
(*a*) Find the maximum value of θ for the block to remain motionless on the wedge when the wedge is fixed in position.
(*b*) The wedge is given horizontal acceleration a, as shown. Assuming that $\tan\theta > \mu$, find the minimum acceleration for the block to remain on the wedge without sliding.
(*c*) Repeat part (*b*), but find the maximum value of the acceleration.

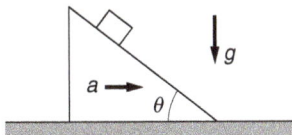

3.9 *Tension in a rope*
A uniform rope of mass m and length l is attached to a block of mass M. The rope is pulled with force F. Find the tension at distance x from the end of the rope. Neglect gravity.

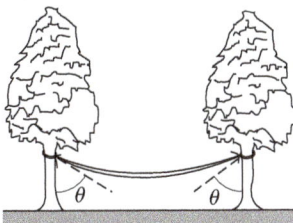

3.10 *Rope and trees**
A uniform rope of weight W hangs between two trees. The ends of the rope are the same height, and they each make angle θ with the trees. Find
(*a*) The tension at either end of the rope.
(*b*) The tension in the middle of the rope.

3.11 *Spinning loop**
A piece of string of length l and mass M is fastened into a circular loop and set spinning about the center of a circle with uniform angular velocity ω. Find the tension in the string. Suggestion: Draw a force diagram for a small piece of the loop subtending a small angle, $\Delta\theta$.

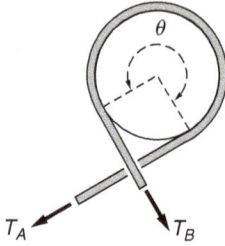

3.12 Capstan

A device called a capstan is used aboard ships in order to control a rope which is under great tension. The rope is wrapped around a fixed drum, usually for several turns (the drawing shows about a three-quarter turn). The load on the rope pulls it with a force T_A, and the sailor holds it with a much smaller force T_B. Show that $T_B = T_A e^{-\mu\theta}$, where μ is the coefficient of friction and θ is the total angle subtended by the rope on the drum.

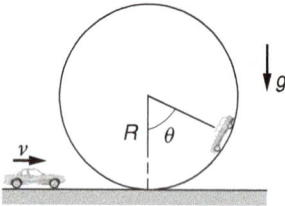

3.13 Incomplete loop-the-loop

An automobile of mass M drives onto a loop-the-loop, as shown. The minimum speed for going completely around the loop without falling off is v_0. However, the automobile drives at constant speed v, where $v < v_0$. The coefficient of friction between the auto and the track is μ.

Find an equation for the angle θ where the auto starts to slip. There is no need to solve the equation.

3.14 Orbiting spheres

Find the shortest possible period of revolution of two identical gravitating solid spheres that are in circular orbit in free space about a point midway between them. The spheres are made of platinum, density 21.5 g/cm^3.

3.15 Tunnel through the Earth

The gravitational force on a body located at distance R from the center of a uniform spherical mass is due solely to the mass lying at distance $r \le R$, measured from the center of the sphere. This mass exerts a force as if it were a point mass at the origin.

Use the above result to show that if you drill a hole through the Earth and then fall in, you will execute simple harmonic motion about the Earth's center. Find the time it takes you to return to your point of departure and show that this is the time needed for a satellite to circle the Earth in a low orbit with $r \approx R_e$. In deriving this result, treat the Earth as a uniformly dense sphere, neglect friction, and neglect any effects due to the Earth's rotation.

3.16 Off-center tunnel

As a variation of the previous problem, show that you will also execute simple harmonic motion with the same period even if the straight hole passes far from the Earth's center.

3.17 Turning car*

A car enters a turn whose radius is R. The road is banked at angle θ, and the coefficient of friction between the wheels and the road is μ. Find the maximum and minimum speeds for the car to stay on the road without skidding sideways.

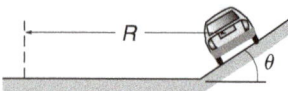

3.18 *Car on rotating platform*

A car is driven on a large revolving platform which rotates with constant angular speed ω. At $t = 0$ a driver leaves the origin and follows a line painted radially outward on the platform with constant speed v_0. The total weight of the car is W, and the coefficient of friction between the car and stage is μ.

(*a*) Find the acceleration of the car as a function of time using polar coordinates. Draw a clear vector diagram showing the components of *acceleration* at some time $t > 0$.

(*b*) Find the time at which the car just starts to skid.

(*c*) Find the direction of the friction force with respect to the instantaneous position vector **r** just before the car starts to skid. Show your result on a clear diagram.

3.19 *Mass and springs**

Find the frequency of oscillation of mass m suspended by two springs having constants k_1 and k_2, in each of the configurations shown.

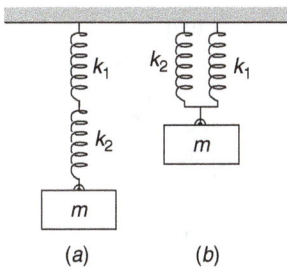

(a)   (b)

3.20 *Wheel and pebble*

A wheel of radius R rolls along the ground with velocity V. A pebble is carefully released on top of the wheel so that it is instantaneously at rest on the moving wheel.

(*a*) Show that the pebble will immediately fly off the wheel if $V > \sqrt{Rg}$.

(*b*) Show that in the case where $V < \sqrt{Rg}$, and the coefficient of friction is $\mu = 1$, the pebble starts to slide when it has rotated through an angle given by $\theta = \arccos [(1/\sqrt{2})(V^2/Rg)] - \pi/4$.

3.21 *Bead on rod*

A small bead of mass m is free to slide on a thin rod.

The rod rotates in a plane about one end at constant angular velocity ω. Show that the motion is given by $r = Ae^{-\gamma t} + Be^{+\gamma t}$, where γ is a constant which you must find and A and B are arbitrary constants. Neglect gravity.

Show that for a particular choice of initial conditions [that is, $r(t = 0)$ and $v(t = 0)$] it is possible to obtain a solution such that r decreases continually in time, but that for any other choice r will eventually increase. (Exclude cases where the bead hits the origin.)

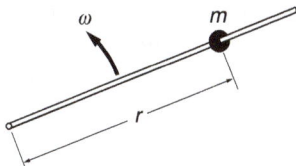

3.22 *Mass, string, and ring**

A mass m whirls around on a string which passes through a ring, as shown. Neglect gravity. Initially the mass is distance r_0 from the center and is revolving at angular velocity ω_0. The string is pulled with constant velocity V starting at $t = 0$ so that the radial distance to the mass decreases. Draw a force diagram and obtain a differential equation for ω. This equation is quite simple and can be solved either by inspection or by formal integration. Find

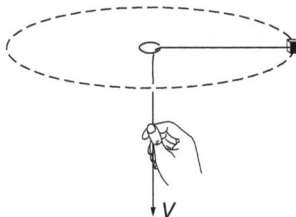

(a) $\omega(t)$.

(b) The force needed to pull the string.

3.23 *Mass and ring*

This problem involves solving a simple differential equation.

A block of mass m slides on a frictionless table. It is constrained to move inside a ring of radius l that is fixed to the table. At $t = 0$, the block is moving along the inside of the ring (in the tangential direction) with velocity v_0. The coefficient of friction between the block and the ring is μ.

(a) Find the velocity of the block at later times.

(b) Find the position of the block at later times.

3.24 *Retarding force*

This problem involves a simple differential equation. You should be able to integrate it after a little "playing around."

A particle of mass m moving along a straight line is acted on by a retarding force (one always directed against the motion) $F = be^{av}$, where b and a are constants and v is the velocity. At $t = 0$ it is moving with velocity v_0. Find the velocity at later times.

3.25 *Hovercraft*

The Eureka Hovercraft Corporation wanted to hold hovercraft races as an advertising stunt. The hovercraft supports itself by blowing air downward, and has a big fixed propeller on the top deck for forward propulsion. Unfortunately, it has no steering equipment, so that the pilots found that making high speed turns was very difficult. The company decided to overcome this problem by designing a bowl-shaped track in which the hovercraft, once up to speed, would coast along in a circular path with no need to steer.

When the company held their first race, they found to their dismay that the craft took exactly the same time T to circle the track, no matter what its speed. Find the equation for the cross-section of the bowl in terms of T.

3.26 *v^2 retarding force*

A body of mass m is retarded by a force $-Cv^2$, where C is a constant and v is its speed. Find the time required for it to travel distance s if it is initially moving with speed v_0. Show that the result is reasonable for short times.

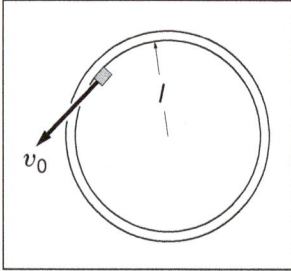

4 MOMENTUM

4.1 Introduction

So far we have viewed nature as if it were composed of ideal particles rather than real bodies. Sometimes such a simplification is justified—for instance in the study of planetary motion, where the size of the planets is of little consequence compared with the vast distances of our solar system, or in the case of elementary particles moving through an accelerator, where the size of the particles, about 10^{-15} m, is minute compared with the size of the machine. However, most of the time we deal with large bodies that may have elaborate structure. For example, consider the landing of an explorer vehicle on Mars. Even if we could calculate the gravitational field of such an irregular and inhomogeneous body as Mars, the explorer itself hardly resembles a particle—it has wheels, gawky antennas, extended solar panels, and a lumpy body.

Furthermore, the methods of the last chapter fail when we try to analyze systems such as rockets in which there is a flow of mass. Rockets accelerate forward by ejecting mass backward; it is not obvious how we can apply $\mathbf{F} = \mathbf{Ma}$ to such a system.

In this chapter we shall generalize the laws of motion to overcome these difficulties. We begin by restating Newton's second law in a slightly modified form. In Chapter 2 we wrote the law in the familiar form

$$\mathbf{F} = \mathbf{Ma}.$$

Newton, however, wrote it in the form

$$\mathbf{F} = \frac{d}{dt}(\mathbf{Mv}).$$

In Newtonian mechanics, the mass M of a particle is a constant and $(d/dt)(\mathbf{Mv}) = M(d\mathbf{v}/dt) = \mathbf{Ma}$, as before. The quantity $\mathbf{Mv}$ plays a prominent role in mechanics and is called *momentum*, or sometimes *linear momentum*, to distinguish it from angular momentum. Momentum is a vector because it is the product of a vector $\mathbf{v}$ and a scalar M. Denoting momentum by $\mathbf{P}$, Newton's second law becomes

$$\mathbf{F} = \frac{d\mathbf{P}}{dt}.$$

This form is preferable to $\mathbf{F} = \mathbf{Ma}$ because it is readily generalized to complex systems, as we shall soon see, and because momentum turns out to be more fundamental than mass or velocity separately. The units of momentum are kg · m/s in the SI system, and g · cm/s in CGS. There are no special names for these units.

4.2 Dynamics of a System of Particles

To generalize the laws of motion to extended bodies consider a system of interacting particles, for instance the Sun and the planets. The bodies of our solar system are so far apart compared with their diameters that they can be treated as particles to an excellent approximation. All particles in the solar system interact gravitationally. Planets interact primarily with the Sun, although their interactions with one another also influence their motion. In addition, the entire solar system is attracted by distant matter.

On a much smaller scale, the system could be a billiard ball resting on a table. Here the particles are atoms (disregarding for now that atoms are not particles but are themselves composed of smaller particles) and the interactions are primarily interatomic electric forces. The external forces on the billiard ball include the gravitational force of the Earth and the contact force of the tabletop.

We shall now prove some simple properties of physical systems. It is important to be clear about what we mean by "system." We are free to choose the boundaries of a system as we please, but once the choice is made, we must be consistent about which particles are included in the system and which are not.

We suppose that the particles in the system interact with particles outside the system as well as with each other. To make the argument general, consider a system of N interacting particles with masses $m_1, m_2, m_3, \cdots, m_N$. The position of the jth particle is $\mathbf{r}_j$, the force on it is $\mathbf{f}_j$, and its momentum is $\mathbf{p}_j = m_j \dot{\mathbf{r}}_j$. The equation of motion for the jth particle is

$$\mathbf{f}_j = \frac{d\mathbf{p}_j}{dt}.$$

The force on particle j can be split into two terms:

$$\mathbf{f}_j = \mathbf{f}_j^{\text{int}} + \mathbf{f}_j^{\text{ext}}.$$

Here $\mathbf{f}_j^{\text{int}}$, the *internal* force on particle j, is the force due to all other particles in the system, and $\mathbf{f}_j^{\text{ext}}$, the *external* force on particle j, is the force due to sources outside the system. The equation of motion of particle j can therefore be written

$$\mathbf{f}_j^{\text{int}} + \mathbf{f}_j^{\text{ext}} = \frac{d\mathbf{p}_j}{dt}.$$

Now let us focus on the system as a whole by the following stratagem: add the equations of motion of all the particles in the system.

$$\mathbf{f}_1^{\text{int}} + \mathbf{f}_1^{\text{ext}} = \frac{d\mathbf{p}_1}{dt}$$
$$\mathbf{f}_2^{\text{int}} + \mathbf{f}_2^{\text{ext}} = \frac{d\mathbf{p}_2}{dt}$$
$$\cdots\cdots\cdots\cdots\cdots$$
$$\mathbf{f}_N^{\text{int}} + \mathbf{f}_N^{\text{ext}} = \frac{d\mathbf{p}_N}{dt}.$$

The result of adding these equations can be written

$$\sum_{j=1}^{N} \mathbf{f}_j^{\,\text{int}} + \sum_{j=1}^{N} \mathbf{f}_j^{\,\text{ext}} = \sum_{j=1}^{N} \frac{d\mathbf{p}_j}{dt}. \tag{4.1}$$

The summations extend over all particles, $j = 1, \ldots, N$.

The first term in Eq. (4.1), $\Sigma \mathbf{f}_j^{\,\text{int}}$, is the sum of all internal forces acting on all the particles. According to Newton's third law, the forces between any two particles are equal and opposite so that their sum is zero. It follows that the internal forces all cancel in pairs so that the sum of all the forces between all the particles is also zero. Hence

$$\sum_{j=1}^{N} \mathbf{f}_j^{\,\text{int}} = 0.$$

The second term, $\sum_{j=1}^{N} \mathbf{f}_j^{\,\text{ext}}$, is the sum of all external forces acting on all the particles. It is the *total external force* $\mathbf{F}_{\text{ext}}$ acting on the system:

$$\sum_{j=1}^{N} \mathbf{f}_j^{\,\text{ext}} \equiv \mathbf{F}_{\text{ext}}.$$

Equation (4.1) then simplifies to

$$\mathbf{F}_{\text{ext}} = \sum_{j=1}^{N} \frac{d\mathbf{p}_j}{dt}.$$

The right-hand side, $\Sigma(d\mathbf{p}_j/dt)$, can be written $(d/dt)\Sigma \mathbf{p}_j$, because the derivative of a sum is the sum of the derivatives. $\Sigma \mathbf{p}_j$ is the *total momentum* of the system, which we designate by $\mathbf{P}$:

$$\mathbf{P} \equiv \sum_{j=1}^{N} \mathbf{p}_j.$$

With this substitution, Eq. (4.1) becomes

$$\mathbf{F}_{\text{ext}} = \frac{d\mathbf{P}}{dt}.$$

In words, the total force applied to a system equals the rate of change of the system's momentum. This is true regardless of the details of the interaction; $\mathbf{F}_{\text{ext}}$ could be a single force acting on a single particle, or it could be the resultant of many tiny interactions involving each particle of the system.

Example 4.1 The Bola

The bola is used by gauchos for entangling their cattle. It consists of three balls of stone or iron connected by thongs. The gaucho whirls the bola in the air and hurls it at the animal. What can we say about its motion?

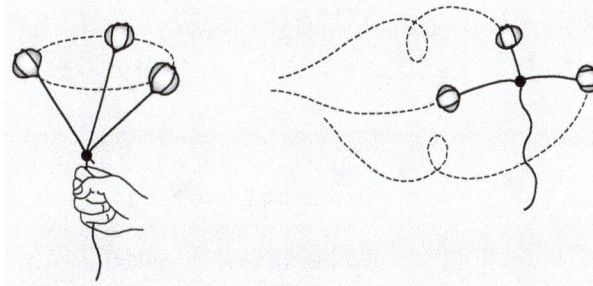

Consider a bola with masses m_1, m_2, and m_3. Each ball is pulled by its binding thong and by gravity. (We neglect air resistance.) Since the constraining forces depend on the instantaneous positions of all three balls, it is a real problem even to write the equation of motion of one ball. However, the total momentum obeys the simple equation

$$\frac{d\mathbf{P}}{dt} = \mathbf{F}_{\text{ext}} = \mathbf{f}_1{}^{\text{ext}} + \mathbf{f}_2{}^{\text{ext}} + \mathbf{f}_3{}^{\text{ext}}$$

$$= m_1\mathbf{g} + m_2\mathbf{g} + m_3\mathbf{g}$$

or

$$\frac{d\mathbf{P}}{dt} = M\mathbf{g},$$

where M is the total mass. This equation represents an important first step in finding the detailed motion. The equation is identical to that of a single particle of mass M with momentum $\mathbf{P}$. This is instinctively clear to the gaucho when he hurls the bola; although it is a complicated system, he need only aim it like a single mass.

4.3 Center of Mass

According to Eq. (4.1),

$$\mathbf{F} = \frac{d\mathbf{P}}{dt}, \tag{4.2}$$

where we have dropped the subscript "ext" with the understanding that $\mathbf{F}$ stands for the external force. This result is identical to the equation of motion of a single particle, although it may in fact refer to a system of several particles. It is tempting to push the analogy between Eq. (4.2) and single-particle motion even further by writing

$$\mathbf{F} = M\ddot{\mathbf{R}}, \tag{4.3}$$

where M is the total mass of the system and $\mathbf{R}$ is a vector yet to be defined. Because $\mathbf{P} = \sum_{j=1}^{N} m_j \dot{\mathbf{r}}_{\mathbf{j}}$, Eqs. (4.2) and (4.3) give

$$M\ddot{\mathbf{R}} = \frac{d\mathbf{P}}{dt} = \sum_{j=1}^{N} m_j \ddot{\mathbf{r}}_j,$$

which is true if

$$\mathbf{R} = \frac{1}{M} \sum_{j=1}^{N} m_j \mathbf{r}_j. \tag{4.4}$$

$\mathbf{R}$ is a vector from the origin to a point called *the center of mass*. The motion of a system's center of mass behaves as if all the mass were concentrated there and all the external forces act at that point.

We are often interested in the motion of comparatively rigid bodies like baseballs or automobiles. Such a body is merely a system of particles that are fixed relative to each other by strong internal forces. Equation (4.4) shows that with respect to external forces, the body behaves as if it were a single particle. In Chapters 2 and 3, we casually treated every body as if it were a particle; we see now that this is justified provided that we focus attention on the center of mass.

You may wonder whether this description of center of mass motion isn't also an oversimplification—experience tells us that an extended body like a plank behaves differently from a compact body like a rock, even if the masses are the same and we apply the same force. Indeed, center of mass motion is only part of the story. The relation $\mathbf{F} = M\ddot{\mathbf{R}}$ describes only the translation of the body (the motion of its center of mass); it does not describe the body's orientation in space. In Chapters 7 and 8 we shall investigate the rotation of extended bodies. It turns out that, as we expect, the rotational motion of a body depends on its shape and where the forces are applied. Nevertheless, as far as translation of the center of mass is concerned, $\mathbf{F} = M\ddot{\mathbf{R}}$ is true for any system of particles, not just for those fixed in rigid objects, as long as the forces between the particles obey Newton's third law. It is immaterial whether or not the particles move relative to each other and whether or not there happens to be any matter at the center of mass.

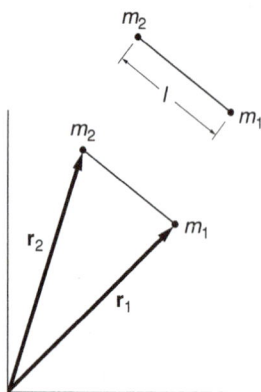

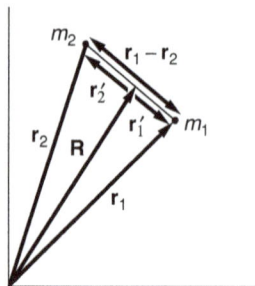

Example 4.2 Drum Major's Baton

A drum major's baton consists of two masses m_1 and m_2 separated by a thin rod of length l. The baton is thrown into the air. Find the baton's center of mass and the equation of motion for the center of mass.

Let the position vectors of m_1 and m_2 be $\mathbf{r}_1$ and $\mathbf{r}_2$, respectively. The position vector of the center of mass, measured from the same origin, is

$$\mathbf{R} = \frac{m_1 \mathbf{r}_1 + m_2 \mathbf{r}_2}{m_1 + m_2} \tag{1}$$

where we have neglected the mass of the thin rod. The center of mass lies on the line joining m_1 and m_2; the proof is left as a problem.

Assuming that air resistance is negligible, the external force on the baton is

$$\mathbf{F} = m_1 \mathbf{g} + m_2 \mathbf{g}.$$

The equation of motion of the center of mass is

$$(m_1 + m_2)\ddot{\mathbf{R}} = (m_1 + m_2)\mathbf{g}$$

or

$$\ddot{\mathbf{R}} = \mathbf{g}.$$

The center of mass follows the parabolic trajectory of a single mass in a uniform gravitational field. With the methods to be developed in Chapter 8, we shall be able to find the motion of m_1 and m_2 about the center of mass, completing the solution.

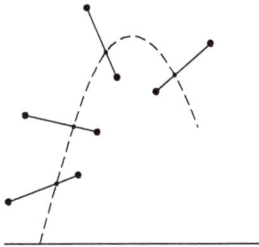

Although it is a simple matter of algebra to find the center of mass of a system of particles, finding the center of mass of an extended body normally requires integration. We proceed by dividing the body of mass M into N mass elements. If $\mathbf{r}_j$ is the position of the jth element, and m_j is the element's mass, then

$$\mathbf{R} = \frac{1}{M} \sum_{j=1}^{N} m_j \mathbf{r}_j. \qquad (4.5)$$

In the limit where N approaches infinity, the size of each element approaches zero and the approximation becomes exact:

$$\mathbf{R} = \lim_{N \to \infty} \frac{1}{M} \sum_{j=1}^{N} m_j \mathbf{r}_j.$$

This limiting process defines an integral. Formally

$$\lim_{N \to \infty} \sum_{j=1}^{\infty} m_j \mathbf{r}_j = \int \mathbf{r} \, dm,$$

where dm is a differential mass element at position $\mathbf{r}$. Then

$$\mathbf{R} = \frac{1}{M} \int_V \mathbf{r} \, dm.$$

To visualize this integral, think of dm as the mass in an element of volume dV located at position $\mathbf{r}$. If the mass density at the element is ρ, then $dm = \rho \, dV$ and

$$\mathbf{R} = \frac{1}{M} \int_V \mathbf{r}\rho \, dV.$$

This integral is called a volume integral. It is sometimes written with three integral signs (a triple integral) to emphasize that the integration proceeds over all three space coordinates:

$$\mathbf{R} = \frac{1}{M} \iiint_V \mathbf{r}\rho \, dV.$$

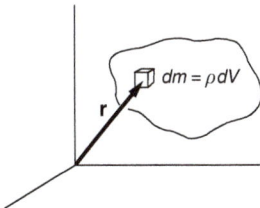

The integral of a sum of terms equals the sum of the integrals over each term. We can use this fundamental property of integrals to express the center of mass of several extended bodies in terms of the centers of mass of the individual bodies. Consider extended body 1 with mass M_1 and body 2 with mass M_2. Let $\mathbf{R}_1$ and $\mathbf{R}_2$ be the position vector of each center of mass. From Eq. (4.5)

$$\mathbf{R}_1 = \frac{1}{M_1} \int_{V_1} \mathbf{r}_1 \, dm$$

$$\mathbf{R}_2 = \frac{1}{M_2} \int_{V_2} \mathbf{r}_2 \, dm.$$

The position vector $\mathbf{R}$ of the system's center of mass can then be written

$$(M_1 + M_2)\mathbf{R} = \int_{V_1} \mathbf{r}_1 \, dm + \int_{V_2} \mathbf{r}_2 \, dm$$

$$= M_1 \, \mathbf{R}_1 + M_2 \, \mathbf{R}_2.$$

In other words, to find the center of mass of a system of several extended bodies, treat each body as if its mass were concentrated at its center of mass.

We shall only be concerned with a few simple cases of calculating the center of mass of extended bodies, as illustrated by the following examples. Further examples are given in Note 4.1 at the end of the chapter.

Example 4.3 Center of Mass of a Non-uniform Rod

A rod of length L has a non-uniform density. The mass per unit length of the rod, λ, varies as $\lambda = \lambda_0(x/L)$, where λ_0 is a constant and x is the distance from the end marked 0. Find the center of mass.

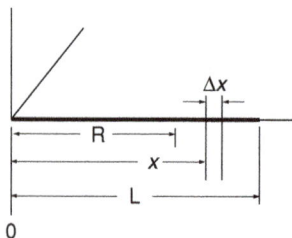

It is apparent that $\mathbf{R}$ lies on the rod. Let the x axis lie along the rod, with the origin $x = 0$ at the end of the rod. The mass in an element of length dx is $dm = \lambda dx = \lambda_0 x \, dx/L$. The rod extends from $x = 0$ to $x = L$ and the total mass is

$$M = \int dm$$

$$= \int_0^L \lambda \, dx$$

$$= \int_0^L \frac{\lambda_0 x \, dx}{L}$$

$$= \tfrac{1}{2}\lambda_0 L.$$

The center of mass is at

$$\mathbf{R} = \frac{1}{M} \int \lambda \, \mathbf{r} \, dx$$

$$= \frac{2}{\lambda_0 L} \int_0^L (x\hat{\mathbf{i}} + 0\hat{\mathbf{j}} + 0\hat{\mathbf{k}}) \frac{\lambda_0 x \, dx}{L}$$

$$= \frac{2}{L^2} \left. \frac{x^3}{3} \right|_0^L \hat{\mathbf{i}}$$

$$= \frac{2}{3} L \hat{\mathbf{i}}.$$

Example 4.4 Center of Mass of a Triangular Plate

Calculating the center of mass is straightforward if the object can be subdivided into parts with known centers of mass. Consider the two-dimensional case of a uniform right triangular plate of mass M, base b, height h, and small thickness t.

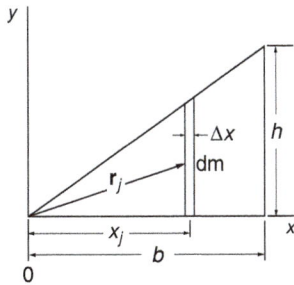

Divide the triangle into strips of width Δx parallel to the y axis, as shown.

The jth strip at x_j has its center of mass halfway up, because the plate is uniform, and the total height of the jth strip is $x_j h / b$ by similar triangles. The position vector to the strip's center of mass is therefore

$$\mathbf{r}_j = x_j \hat{\mathbf{i}} + \frac{x_j h}{2b} \hat{\mathbf{j}}.$$

The center of mass of the plate is located at $\mathbf{R}$, and in the limit of very narrow strips,

$$\mathbf{R} = \frac{1}{M} \int \mathbf{r} \, dm \qquad (1)$$

where

$$M = \rho A t = \rho t b h / 2$$

$$dm = \rho t y dx = \rho t \frac{xh}{b} \, dx.$$

Then Eq. (1) can be written

$$\mathbf{R} = \left(\frac{2}{\rho t b h} \right) \int \mathbf{r} \rho t \frac{xh}{b} \, dx$$

$$= \frac{2}{b^2} \int_0^b x \mathbf{r} \, dx$$

$$= \frac{2}{b^2} \int_0^b \left(x^2 \hat{\mathbf{i}} + \frac{x^2 h}{2b} \hat{\mathbf{j}} \right) dx$$

$$= \tfrac{2}{3} b \hat{\mathbf{i}} + \tfrac{1}{3} h \hat{\mathbf{j}}.$$

To find the center of mass if the plate is not uniform, we would need to use multiple integrals, as discussed in Note 4.1.

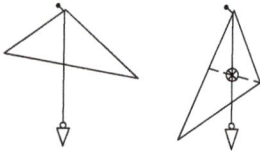

Physical arguments are sometimes able to take the place of complicated calculations. Suppose we want to find the center of mass of a thin irregular non-uniform plate. Let it hang from a pivot and draw a plumb line from the pivot. The center of mass will hang directly below the pivot (this may be intuitively obvious, and can easily be proved with the methods of Chapter 7), so the center of mass is somewhere on the plumb line. Repeat the procedure with a different pivot point. The two lines intersect at the center of mass.

Example 4.5 Center of Mass Motion

A rectangular crate is held with one corner resting on a frictionless table. The crate is gently released and falls in a complex tumbling motion. We are not yet prepared to predict the full motion because it involves rotation, but there is no difficulty in finding the trajectory of the center of mass.

The external forces acting on the box are gravity and the normal force of the table. Both of these are vertical, so the center of mass must accelerate vertically. If the box is released from rest its center falls straight down.

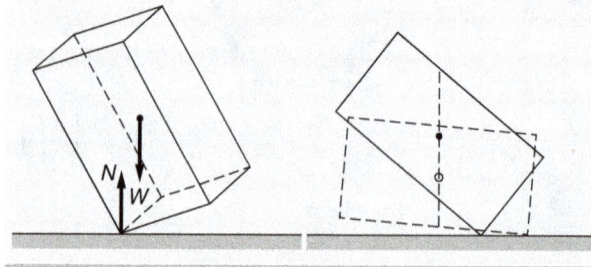

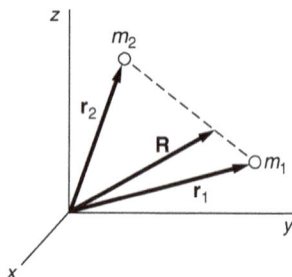

4.4 Center of Mass Coordinates

Often a problem can be simplified by a clever choice of coordinates. The center of mass coordinate system, in which the origin lies at the center of mass, is particularly useful. Consider the case of a two-particle system with masses m_1 and m_2.

In the initial coordinate system x, y, z, the particles are located at $\mathbf{r}_1$ and $\mathbf{r}_2$ and their center of mass is at

$$\mathbf{R} = \frac{m_1\mathbf{r}_1 + m_2\mathbf{r}_2}{m_1 + m_2}.$$

We now set up the center of mass coordinate system, x', y', z', with its origin at the center of mass. The origins of the old and new system are

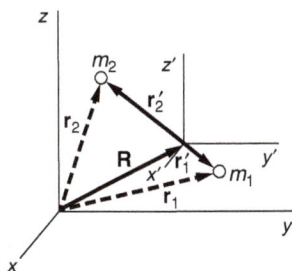

displaced by $\mathbf{R}$. The center of mass coordinates of the two particles are

$$\mathbf{r}_1' = \mathbf{r}_1 - \mathbf{R}$$
$$\mathbf{r}_2' = \mathbf{r}_2 - \mathbf{R}.$$

Center of mass coordinates are the natural coordinates for an isolated two-body system. Such a system has no external forces, so the motion of the center of mass is trivial—it moves uniformly. Furthermore, $m_1\mathbf{r}_1' + m_2\mathbf{r}_2' = 0$ by the definition of center of mass, so that if the motion of one particle is known, the motion of the other particle follows directly. Here are two examples.

Example 4.6 Exoplanets

For many centuries people have wondered if there might be life on other planets. Searching for life on the other planets and moons of our solar system is an active field of inquiry, and has been extended to other stars to discover orbiting planets that might be able to sustain life. Planets not members of our own solar system are called exoplanets (Greek $\epsilon\chi o$, "outside of"). In a few favorable cases, telescopes have seen planets orbiting a nearby star, but for distant stars, a small dark planet is undetectable in the star's bright glare. This example shows how exoplanets can be detected using the concept of center of mass.

Newton was the first to calculate the motion of two gravitating bodies. As we shall show in Chapter 10, two bodies bound by gravity move so that the vector joining them traces out an ellipse with its focus at the center of mass. Consider a single planet of mass m orbiting a star of mass M. Let $\mathbf{r}_p$ and $\mathbf{r}_s$ be the position vectors of the planet and star, respectively. Taking the origin at the center of mass,

$$m\mathbf{r}_p + M\mathbf{r}_s = 0$$

which gives

$$\mathbf{r}_s = -\left(\frac{m}{M}\right)\mathbf{r}_p. \tag{1}$$

As the planet swings around the star, Eq. (1) shows that the center of the star also moves in an orbit, but a much smaller one, because $m \ll M$, as shown schematically in the sketch. The line joining the star and the planet always passes through the center of mass as they orbit. As seen edge on to the orbits, the star is advancing toward the observer as the planet completes half its revolution, and the star is receding the other half.

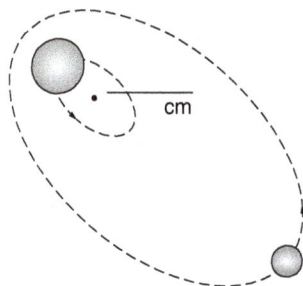

Another relation comes from considering the dynamics. The Earth's orbit is very nearly circular, only modestly elliptical, probably a good situation for life because the temperature on such a planet would not vary greatly. A highly elliptical elongated orbit could lead to large temperature swings, with water going between freezing and boiling.

Assuming a circular orbit, the angular velocity $\dot{\theta}$ is constant, and

$$mr_p\dot{\theta}^2 = \frac{GmM}{(r_p + r_s)^2}$$

$$r_p\dot{\theta}^2 \approx \frac{GM}{r_p^2}$$

$$\dot{\theta} = \sqrt{\frac{GM}{r_p^3}}. \tag{2}$$

Integrating,

$$\theta = \sqrt{\frac{GM}{r_p^3}}t.$$

The planet makes a complete orbit as θ goes from 0 to 2π, so the time T for a complete orbit (the planet's "year") is

$$T = 2\pi\sqrt{\frac{r_p^3}{GM}}.$$

Incidentally, this result is a special case of one of the seventeenth-century astronomer Johannes Kepler's laws of planetary motion, that the square of the "year" is proportional to the cube of the orbital radius. We shall derive the general case in Chapter 10. One consequence of this law is that a planet orbiting close to a star has a much shorter period than a planet farther away.

Ever since large telescopes became available in the late eighteenth century, astronomers have attempted to observe the "wobble" of stars as a means of detecting exoplanets. Although the wobble is typically too small to detect directly, many exoplanets have been discovered from the effect of the wobble on a star's spectrum.

The Kepler space satellite telescope has detected hundreds of exoplanets using a different, but related, technique. If the planet's orbit is nearly edge-on as seen from the Earth, the wobbling star will move periodically toward and away from the Earth. The velocity of this motion can be detected by the *Doppler shift* of the star's light, using the same principle underlying police or baseball radar guns. (We shall discuss the Doppler shift in Chapter 12.) For the edge-on case, Eq. (2) gives the variation in velocity of the star's center as

$$r_s \dot{\theta} = \pm \sqrt{\frac{GM r_s^2}{r_p^3}}$$

$$= \pm \sqrt{\frac{G m^2}{M r_p}}$$

where the last step follows from Eq. (1). For the Earth–Sun system, this is ± 0.09 m/s. With the method's sensitivity of a few m/s, our Earth would not be readily detectable from a distant solar system, but a planet as massive as Jupiter or Saturn could be. A planet of moderate mass could be detected if it orbits close to the star, but then it might be too hot to support life as we know it.

The exoplanet Gliese 876d (upper figure) has a period of 1.9 Earth days, and Gliese 876b (lower figure) has a period of 61 days. The figures are from NSF press release 05-097 (2005), based on data published by Eugenio Rivera *et al.*, *The Astrophysical J.* **634**(1):625–640 (2005).

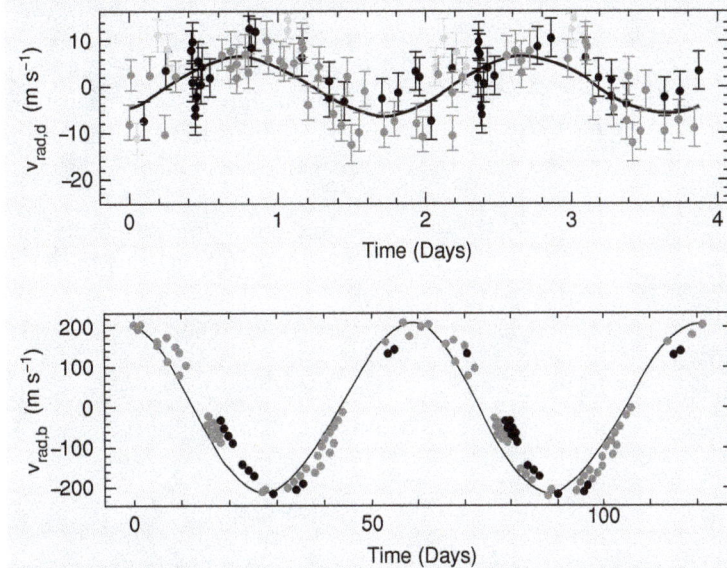

The figures show measured radial speeds inferred from the Doppler shift for two different exoplanets orbiting the same star "Gliese 876" located 15.3 light years $= 1.4 \times 10^{17}$ m from Earth. The upper figure is for exoplanet "d" close to the star; we see that the "year" for this planet is only a few days, and the "wobble" it induces in the star gives a radial speed variation of only a few m/s. The lower figure is for exoplanet "b" farther from the star. It has a much longer "year" than exoplanet "d" but it induces a much larger "wobble" because of its much greater mass.

We would like to know the exoplanet's mass and orbital radius to see if it might be suitable for sustaining life. We have three unknowns, m, M, and r_p (or r_s) but only two measured values, the Doppler shift velocity, and the period T from the time variation of the Doppler shift.

The third quantity we need comes from estimating the star's mass M using stellar models based on color and brightness.

One weakness of this approach is its lack of sensitivity to planets of smaller mass, such as the Earth. Another weakness is our incomplete understanding of the different forms life might take. Even on the Earth life takes on unexpected forms. For instance, tube worms live without light near thermal vents in the deep ocean, sustained by chemicals from the vent, with the help of bacteria. The field of astrobiology is concerned with expanding our conception of life that might exist on exoplanets.

Example 4.7 The Push Me–Pull You

Two identical blocks a and b each of mass m slide without friction on a straight track. They are attached by a spring with unstretched length l and spring constant k; the mass of the spring is negligible compared to the mass of the blocks. Initially the system is at rest. At $t = 0$, block a is hit sharply, giving it an instantaneous velocity v_0 to the right. Find the velocity of each block at later times. (Try this yourself if there is a linear air track available—the motion is unexpected.)

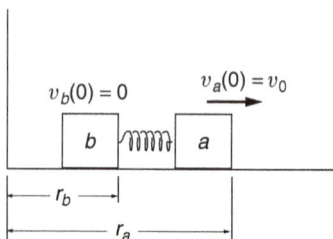

Since the system slides freely after the collision, the center of mass moves uniformly and therefore defines an inertial frame.

Let us transform to center of mass coordinates. The center of mass lies at

$$R = \frac{mr_a + mr_b}{m + m}$$

$$= \frac{1}{2}(r_a + r_b).$$

R always lies halfway between a and b, which is hardly surprising. The center of mass coordinates of a and b are

$$r'_a = r_a - R$$

$$= \tfrac{1}{2}(r_a - r_b)$$

$$r'_b = r_b - R$$

$$= -\tfrac{1}{2}(r_a - r_b)$$

$$= -r'_a.$$

The sketch shows these coordinates.

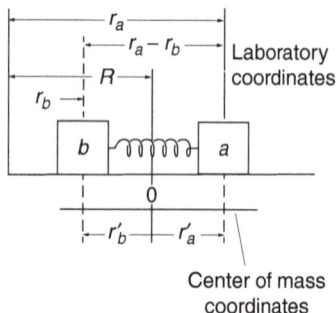

The instantaneous length of the spring is $r_a - r_b = r'_a - r'_b$. The instantaneous departure of the spring from its equilibrium length l is $r_a - r_b - l = r'_a - r'_b - l$. The equations of motion in the center of mass system are

$$m\ddot{r}'_a = -k(r'_a - r'_b - l)$$
$$m\ddot{r}'_b = +k(r'_a - r'_b - l).$$

The form of these equations suggests that we subtract them, obtaining

$$m(\ddot{r}'_a - \ddot{r}'_b) = -2k(r'_a - r'_b - l).$$

It is natural to introduce the departure of the spring from its equilibrium length as a variable. Letting $u = r'_a - r'_b - l$, we have

$$m\ddot{u} + 2ku = 0.$$

This is the equation for simple harmonic motion that we discussed in Chapter 3. The general solution is

$$u = A \sin \omega t + B \cos \omega t,$$

where $\omega = \sqrt{2k/m}$. Since the spring is unstretched at $t = 0$, $u(0) = 0$, which requires $B = 0$. Then $u = A \sin \omega t$ so that $\dot{u} = A\omega \cos \omega t$. At $t = 0$

$$\dot{u}(0) = A\omega \cos(0)$$

and since $u = r'_a - r'_b - l = r_a - r_b - l$

$$\dot{u}(0) = v_a(0) - v_b(0)$$
$$= v_0,$$

so that

$$A = v_0/\omega.$$

Therefore

$$u = (v_0/\omega) \sin \omega t$$

and

$$\dot{u} = v_0 \cos \omega t.$$

Since $v'_a - v'_b = \dot{u}$, and $v'_a = -v'_b$, we have

$$v'_a = -v'_b = \tfrac{1}{2} v_0 \cos \omega t.$$

The laboratory velocities are

$$v_a = \dot{R} + v'_a$$
$$v_b = \dot{R} + v'_b.$$

Since $\dot{R}$ is constant, it is always equal to its initial value

$$\dot{R} = \tfrac{1}{2}(v_a(0) + v_b(0))$$
$$= \tfrac{1}{2} v_0.$$

Putting these results together gives

$$v_a = \frac{v_0}{2}(1 + \cos \omega t)$$

$$v_b = \frac{v_0}{2}(1 - \cos \omega t).$$

The masses move to the right on the average, but they alternately come to rest in a push me–pull you fashion.

4.5 Conservation of Momentum

In Section 4.2, we found that the total external force $\mathbf{F}$ acting on a system is related to the total momentum $\mathbf{P}$ of the system by

$$\mathbf{F} = \frac{d\mathbf{P}}{dt}.$$

Consider the implications of this for an isolated system. In this case $\mathbf{F} = 0$, and $d\mathbf{P}/dt = 0$. The total momentum of an isolated system is constant, no matter how strong the interactions among its constituents, and no matter how complicated the motions. This is the *law of conservation of momentum*. As we shall show, this apparently simple law can provide powerful insights into complex systems.

Example 4.8 Spring Gun Recoil

A loaded spring gun, initially at rest on a horizontal frictionless surface, fires a marble at angle of elevation θ. The mass of the gun is M, the mass of the marble is m, and the muzzle velocity of the marble (the speed with which the marble is ejected, relative to the muzzle) is v_0. What is the final motion of the gun?

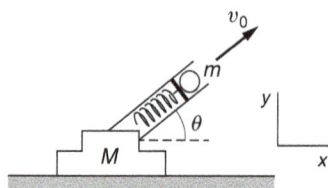

Take the physical system to be the gun and marble. Gravity and the normal force of the table act on the system. These external forces are both vertical. Because there are no horizontal external forces, the x component of the vector equation $\mathbf{F} = d\mathbf{P}/dt$ is

$$0 = \frac{dP_x}{dt}. \tag{1}$$

According to Eq. (1), P_x is conserved:

$$P_{x,\text{initial}} = P_{x,\text{final}}. \tag{2}$$

Let the initial time be prior to firing the gun. Because the system is initially at rest, $P_{x,\text{initial}} = 0$, After the marble has left the muzzle, the gun recoils to the left with some speed V_f, and its final horizontal momentum is $-MV_f$.

Finding the final velocity of the marble involves a subtle point, however. Physically, the marble's acceleration is due to the force of the gun,

$v_0 \sin \theta$

$v_0 \cos \theta - V_f$

V_f

θ

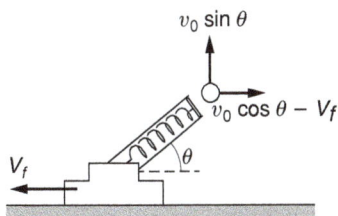

and the gun's recoil is due to the reaction force of the marble. The gun stops accelerating once the marble leaves the barrel, so at the instant the marble and the gun part company, the gun has its final speed $-V_f$. At that same instant the speed of the marble *relative to the gun* is v_0. Hence, the final horizontal speed of the marble relative to the table is $v_0 \cos \theta - V_f$. By conservation of horizontal momentum, we therefore have

$$0 = m(v_0 \cos \theta - V_f) - MV_f$$

or

$$V_f = \frac{mv_0 \cos \theta}{M + m}.$$

The law of conservation of momentum follows directly from Newton's third law, so that the conservation of momentum appears to be a natural consequence of Newtonian mechanics. However, conservation of momentum turns out to hold true even in the realms of quantum mechanics and relativity where Newtonian mechanics proves inadequate. Conservation of momentum can also be generalized to apply to light, because light can be thought of as a stream of particles called *photons* that are massless but nevertheless possess momentum. For these reasons, the law of conservation of momentum is generally regarded as being more fundamental than the laws of Newtonian mechanics. In this view, Newton's third law is a simple consequence of the conservation of momentum for interacting particles. For our present purposes it is purely a matter of taste whether we wish to regard Newton's third law or conservation of momentum as more fundamental.

4.6 Impulse and a Restatement of the Momentum Relation

The relation between force and momentum is

$$\mathbf{F} = \frac{d\mathbf{P}}{dt}. \tag{4.6}$$

As a general rule, any law of physics that can be expressed in terms of derivatives can also be written in an integral form. The integral form of the force–momentum relationship is

$$\int_0^t \mathbf{F}\, dt = \mathbf{P}(t) - \mathbf{P}(0). \tag{4.7}$$

The change in momentum of a system is given by the integral of force with respect to time. Equation (4.7) contains essentially the same physical information as Eq. (4.6), but it gives a new way of looking at the effect of a force: the change in momentum is the time integral of the force. To produce a given change in the momentum in time interval t

requires only that $\int_0^t \mathbf{F}\,dt$ have the appropriate value; we can use a small force acting for much of the time or a large force acting for only part of the interval.

The integral $\int_0^t \mathbf{F}\,dt$ is called the *impulse*. The word impulse calls to mind a short, sharp shock, as in Example 4.7, where a blow to a mass at rest gave it a velocity v_0. However, the physical definition of impulse can just as well apply to a weak force acting for a long time. Change of momentum depends only on $\int \mathbf{F}dt$, independent of the detailed time dependence of the force.

Here are three examples involving impulse and momentum.

Example 4.9 Measuring the Speed of a Bullet

Faced with the problem of measuring the speed of a bullet, our first thought might be to turn to a raft of high-tech equipment—fast photodetectors, fancy electronics, whatever. In this example we show that a simple mechanical system can make the measurement, with the aid of conservation of momentum.

We take a simplified model to emphasize the fundamental principles. Consider a block of soft wood on a horizontal frictionless surface. A compression spring with spring constant k and uncompressed length l connects the block to a wall. The block has mass M, and the spring has negligible mass.

At $t = 0$ a gun fires a bullet of mass m and speed v_0 into the block, which moves back at initial speed V_i due to the impulse. Our system is the bullet and the block, and by conservation of momentum applied at very short times after the collision

$$mv_0 = (M + m)V_i. \tag{1}$$

Conservation of momentum is accurate here, because during the very short time of the collision the horizontal force of the spring has very little time to act. During the ensuing time, however, the spring force has plenty of time to act—it brings the system momentarily to rest. Measuring how far the system moves can tell us the speed of the bullet.

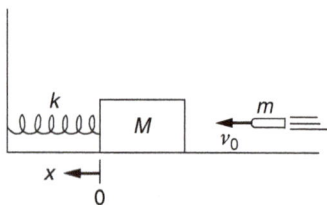

After the initial impulse, the equation of motion of the system is

$$(M + m)\ddot{x} = -kx.$$

The spring length does not appear in the equation of motion because we have taken a coordinate system with $x = 0$ when the spring is uncompressed. We recognize the equation for simple harmonic motion, which has the general solution

$$x = A \sin \omega t + B \cos \omega t \tag{2}$$

where

$$\omega = \sqrt{\frac{k}{M+m}}.$$

Using the initial conditions $x(0) = 0$, $\dot{x}(0) = V_i$ in Eq. (2) we find $A = V_i/\omega$ and $B = 0$. The position and speed of the system are then

$$x = \frac{V_i}{\omega} \sin \omega t \qquad (3)$$

$$\dot{x} = V_i \cos \omega t.$$

The system first comes to rest at $t = t_f$ when $\omega t_f = \pi/2$. Using Eqs. (1), (2), and (3),

$$x(t_f) = \frac{V_i}{\omega}$$

$$= \frac{mv_0}{\sqrt{k(M+m)}}$$

so that

$$v_0 = \frac{\sqrt{k(M+m)}}{m} x(t_f).$$

Example 4.10 Rubber Ball Rebound

A rubber ball of mass 0.2 kg falls to the floor. The ball hits with a speed of 8 m/s and rebounds with approximately the same speed. High speed photographs show that the ball is in contact with the floor for $\Delta t = 10^{-3}$ s. What can we say about the force exerted on the ball by the floor?

The momentum of the ball just before it hits the floor is $\mathbf{P}_a = -1.6\,\hat{\mathbf{k}}$ kg $\cdot$ m/s and its momentum 10^{-3} s later is $\mathbf{P}_b = +1.6\,\hat{\mathbf{k}}$ kg $\cdot$ m/s. Using $\int_{t_a}^{t_b} \mathbf{F}\, dt = \mathbf{P}_b - \mathbf{P}_a$ gives $\int_{t_a}^{t_b} \mathbf{F}\, dt = 1.6\,\hat{\mathbf{k}} - (-1.6\,\hat{\mathbf{k}}) = 3.2\,\hat{\mathbf{k}}$ kg $\cdot$ m/s.

Although the exact variation of $\mathbf{F}$ with time is not known, it is easy to find the average force. If the collision time is $\Delta t = t_b - t_a$, the average force $\mathbf{F}_{av}$ acting during the collision is

$$\mathbf{F}_{av}\Delta t = \int_{t_a}^{t_a + \Delta t} \mathbf{F}\, dt.$$

Since $\Delta t = 10^{-3}$ s,

$$\mathbf{F}_{av} = \frac{3.2\,\hat{\mathbf{k}}\ \text{kg} \cdot \text{m/s}}{10^{-3}\ \text{s}} = 3200\,\hat{\mathbf{k}}\ \text{N}.$$

The average force is directed upward, as we expect. In English units, 3200 N $\approx$ 720 lb—a sizable force. The instantaneous force on the ball is even larger at the peak, as the sketch implies.

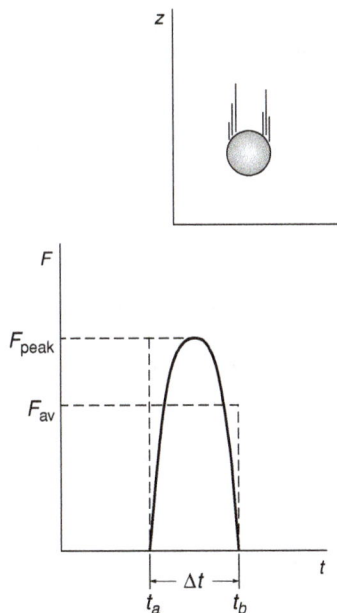

If the ball hits a softer surface the collision time is longer, and the peak and average forces are less.

Actually, there is a weakness in our treatment of the rubber ball rebound. In calculating the impulse $\int \mathbf{F}\, dt$, $\mathbf{F}$ is the total force. This includes the gravitational force, which we have neglected. Proceeding more carefully, we write

$$\mathbf{F} = \mathbf{F}_{\text{floor}} + \mathbf{F}_{\text{grav}}$$
$$= \mathbf{F}_{\text{floor}} - Mg\hat{\mathbf{k}}.$$

The impulse equation then becomes

$$\int_0^{10^{-3}} \mathbf{F}_{\text{floor}}\, dt - \int_0^{10^{-3}} Mg\,\hat{\mathbf{k}}\, dt = 3.2\,\hat{\mathbf{k}}\ \text{kg} \cdot \text{m/s}.$$

The impulse due to the gravitational force is

$$-\int_0^{10^{-3}} Mg\,\hat{\mathbf{k}}\, dt = -Mg\,\hat{\mathbf{k}} \int_0^{10^{-3}} dt = -(0.2)(9.8)(10^{-3})\,\hat{\mathbf{k}}$$
$$= -1.96 \times 10^{-3}\,\hat{\mathbf{k}}\ \text{kg} \cdot \text{m/s}.$$

This is less than one-thousandth of the total impulse, and we can neglect it with little error. Over a long period of time, gravity can produce a large change in the ball's momentum (the ball gains speed as it falls, for example). In the short time of contact, however, gravity contributes little momentum change compared with the tremendous force exerted by the floor. Contact forces during a short collision are generally so huge that we can neglect the impulse due to other forces of moderate strength, such as gravity or friction.

The rubber ball rebound example shows why a quick collision is more violent than a slow collision, even when the initial and final velocities are identical. This is the reason that a hammer can produce a force far greater than the carpenter could produce on his own; the hard steel hammerhead rebounds in a time short compared to the time of the hammer swing, and the force driving the hammer is correspondingly amplified. Contrariwise, pounding a nail into a tall fence picket can be difficult, because the thin picket can spring back under the blow, increasing the collision time and therefore decreasing the force of the hammer.

Many devices to prevent bodily injury in accidents are based on prolonging the time of the collision, which is the design basis for bicycle helmets and automobile airbags. The following example shows what can happen in even a relatively mild collision, as when you jump to the ground.

Example 4.11 How to Avoid Broken Ankles

Animals, including humans, instinctively reduce the force of impact with the ground by flexing while running or jumping. Consider what happens to someone who hits the ground with legs rigid.

Suppose a person of mass M jumps to the ground from height h, and that their center of mass moves downward a distance s during the time of collision with the ground. The average force during the collision is

$$F = \frac{Mv_0}{\Delta t}, \tag{1}$$

where Δt is the collision time and v_0 is the velocity with which they hit the ground. As a reasonable approximation, we can take the acceleration due to the force of impact to be constant, so that the person comes uniformly to rest. In this case the collision time is given by $v_0 = 2s/\Delta t$, or

$$\Delta t = \frac{2s}{v_0}.$$

Inserting this in Eq. (1) gives

$$F = \frac{Mv_0^2}{2s}. \tag{2}$$

For a body in free fall under gravity through height h, $v_0^2 = 2gh$. Inserting this in Eq. (2) gives

$$F = Mg\frac{h}{s}.$$

If the person hits the ground rigidly in a vertical position, their center of mass will not move far during the collision. Suppose that the center of mass moves downward by only 1 cm. If they jump from a height of 2 m, the force is 200 times their weight! If this person has mass 90 kg (≈ 200 lb), the force on them is

$$F = 90 \, \text{kg} \times 9.8 \, \text{m/s}^2 \times 200$$

$$= 1.8 \times 10^5 \, \text{N}.$$

Where is a bone fracture most likely to occur? Because the mass above a horizontal plane through the body decreases with height, the force is maximum at the feet. Thus the ankles will break, not the neck. If the area of contact of bone at each ankle is 5 cm^2, then the force per unit area is

$$\frac{F}{A} = \frac{1.8 \times 10^5 \, \text{N}}{10 \, \text{cm}^2}$$

$$= 1.8 \times 10^4 \, \text{N/cm}^2.$$

This is approximately the compressive strength of human bone, and so there is a good probability that the ankles will snap.

Of course, no one would be so rash as to jump rigidly. We instinctively cushion the impact when jumping by flexing as we hit the ground, in the extreme case collapsing to the ground. If the center of mass drops 50 cm instead of 1 cm during the collision, the force is only four times their weight, and there is no danger of compressive fracture.

4.7 Momentum and the Flow of Mass

Analyzing the forces on a system in which there is a flow of mass can be totally confusing if you try to apply Newton's laws blindly. A rocket provides the most dramatic example of such a system, although there are many other everyday problems where the same considerations apply— for instance, the problem of calculating the reaction force on a fire hose.

There is no fundamental difficulty in handling any of these problems provided that we keep clearly in mind exactly what is included in the system. Recall that $\mathbf{F} = d\mathbf{P}/dt$ was established for a system composed of a certain set of particles. When we apply this equation in integral form,

$$\int_{t_a}^{t_b} \mathbf{F}\, dt = \mathbf{P}(t_b) - \mathbf{P}(t_a),$$

it is essential to deal with the same set of particles throughout the time interval t_a to t_b; we must keep track of all the particles that were originally in the system. Consequently, the integral form applies correctly only to systems defined so that the system's mass does not change during the time of interest.

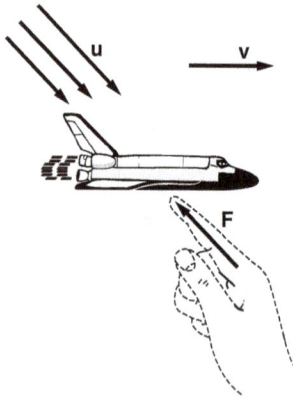

Example 4.12 Mass Flow and Momentum

A spacecraft moves through space with constant velocity $\mathbf{v}$. The space-craft encounters a stream of dust particles that embed themselves in the hull at rate dm/dt. The dust has velocity $\mathbf{u}$ just before it hits.

At time t the total mass of the spacecraft is $M(t)$. The problem is to find the external force $\mathbf{F}$ necessary to keep the spacecraft moving uniformly. (In practice, $\mathbf{F}$ would most likely come from the spacecraft's own rocket engines. Until we discuss rocket motion in Section 4.8, we can for simplicity visualize the source $\mathbf{F}$ to be completely external—an "invisible hand".)

Let us focus on the short time interval between t and $t + \Delta t$. The draw-ings show the system at the beginning and end of the interval. The system consists of $M(t)$ and the mass increment Δm added to the craft during Δt. The initial momentum is

$$\mathbf{P}(t) = M(t)\mathbf{v} + (\Delta m)\mathbf{u}.$$

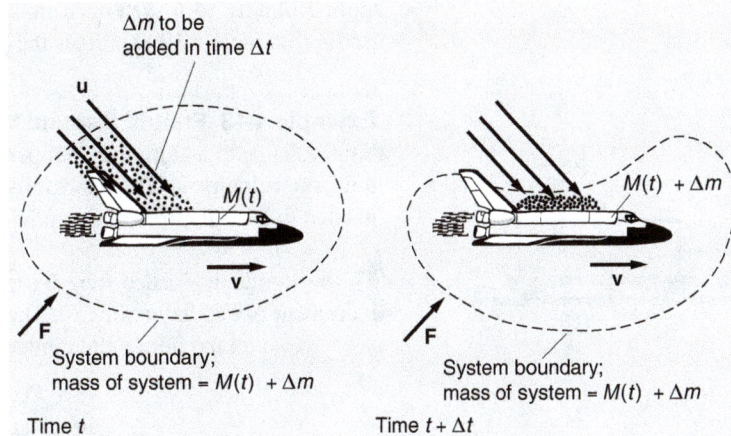

The final momentum is

$$\mathbf{P}(t + \Delta t) = M(t)\mathbf{v} + (\Delta m)\mathbf{v}.$$

The change in momentum is

$$\Delta \mathbf{P} = \mathbf{P}(t + \Delta t) - \mathbf{P}(t)$$
$$= (\mathbf{v} - \mathbf{u})\Delta m.$$

The rate of change of momentum is approximately

$$\frac{\Delta \mathbf{P}}{\Delta t} = (\mathbf{v} - \mathbf{u})\frac{\Delta m}{\Delta t}.$$

In the limit $\Delta t \rightarrow 0$, the result is exact:

$$\frac{d\mathbf{P}}{dt} = (\mathbf{v} - \mathbf{u})\frac{dm}{dt}.$$

Since $\mathbf{F} = d\mathbf{P}/dt$, the required external force is

$$\mathbf{F} = (\mathbf{v} - \mathbf{u})\frac{dm}{dt}.$$

Note that $\mathbf{F}$ can be either positive or negative, depending on the direction of the stream of mass. If $\mathbf{u} = \mathbf{v}$, the momentum of the system is constant, and $\mathbf{F} = 0$.

The procedure of isolating the system, focusing on differentials, and taking the limit may appear a trifle formal but it helps to avoid errors in a subject where it is easy to become confused. For instance, a common mistake is to argue that $\mathbf{F} = (d/dt)(m\mathbf{v}) = m(d\mathbf{v}/dt) + \mathbf{v}(dm/dt)$. In the last example this would have led to the incorrect result that $\mathbf{F} = \mathbf{v}(dm/dt)$ rather than $(\mathbf{v} - \mathbf{u})(dm/dt)$. The origin of the error is that the expression for the momentum of a single particle $\mathbf{p} = m\mathbf{v}$ cannot be applied blindly to a system of many particles. The limiting procedure used in Example 4.12 expresses the physical situation correctly.

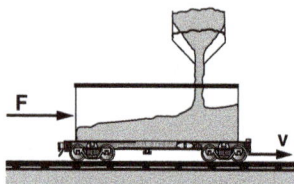

Example 4.13 Freight Car and Hopper
Sand falls from a stationary hopper onto a freight car that moves with uniform velocity v. The sand falls at the rate dm/dt. What force is needed to keep the freight car moving at the speed v?

The system is the loaded freight car of mass M and the incoming mass increment Δm added in time Δt. The initial horizontal speed of the sand is $v = 0$, so taking horizontal components of momentum we have

$$P(t) = Mv + 0$$
$$P(t + \Delta t) = (M + \Delta m)v$$
$$P(t + \Delta t) - P(t) = v\Delta m.$$

Dividing by Δt and taking the limit $\Delta t \to 0$, the required force is $F = dP/dt = v\,dm/dt$.

Example 4.14 Leaky Freight Car
Now consider a case related to Example 4.13. A freight car leaks sand at the rate dm/dt. What force is needed to keep the freight car moving uniformly with speed $\mathbf{v}$?

Here the mass is decreasing. However, the velocity of the sand just after leaving the freight car is identical to its initial velocity, and its momentum does not change

$$P(t) = (M + \Delta m)v$$
$$P(t + \Delta t) = Mv + v\Delta m.$$

Since $dP/dt = 0$, no force is required.

4.8 Rocket Motion
We can readily explain the principle of rocket motion by focusing on momentum. During a time interval Δt the engine exerts a force that accelerates some of the fuel Δm, expelling it from the rocket with exhaust velocity $\mathbf{u}$. By Newton's third law, there is an equal and opposite force on the rocket, propelling the rocket in the opposite direction. Another way to look at this is that the center of mass of the expelled mass and the rocket moves at constant velocity. Hence, if Δm is accelerated backward, the rocket must be accelerated forward.

Suppose that a rocket coasts in deep space with its engines turned off and that external forces are negligible. Let the mass of the rocket be $M + \Delta m$, and let the rocket coast at velocity $\mathbf{v}$ with respect to our coordinate system. At time t a thruster engine fires and expels mass Δm

in the time interval Δt. What is the velocity $\mathbf{v} + \Delta\mathbf{v}$ of the rocket body at time $t + \Delta t$?

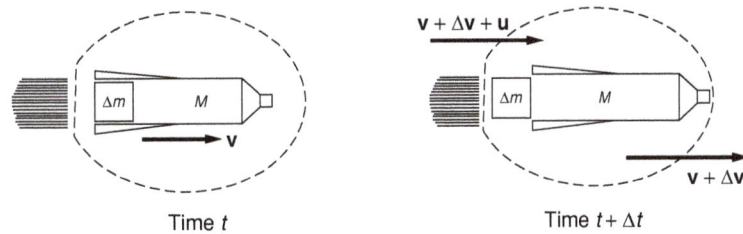

Time t Time $t + \Delta t$

Comparing the momentum at the initial time t and at a slightly later time $t + \Delta t$, when mass Δm has been expelled with velocity $\mathbf{u}$ with respect to the rocket, we have

$$\mathbf{P}(t) = (M + \Delta m)\mathbf{v}$$

$$\mathbf{P}(t + \Delta t) = M(\mathbf{v} + \Delta\mathbf{v}) + \Delta m(\mathbf{v} + \Delta\mathbf{v} + \mathbf{u})$$

$$\Delta\mathbf{P} = M\Delta\mathbf{v} + \Delta m\,(\Delta\mathbf{v} + \mathbf{u}).$$

Hence the rate of change of the system's momentum is

$$\frac{d\mathbf{P}}{dt} = \lim_{\Delta t \to 0}\left(M\frac{\Delta\mathbf{v}}{\Delta t} + \frac{\Delta m}{\Delta t}(\Delta\mathbf{v} + \mathbf{u})\right)$$

$$= M\frac{d\mathbf{v}}{dt} + \mathbf{u}\frac{dm}{dt}$$

$$= M\frac{d\mathbf{v}}{dt} - \mathbf{u}\frac{dM}{dt}.$$

In this last equation we used the identity $dm/dt = -dM/dt$, because the expelled mass decreases the total mass of the rocket. The equation of rocket motion in free space is therefore

$$M\frac{d\mathbf{v}}{dt} - \mathbf{u}\frac{dM}{dt} = 0.$$

If an external force $\mathbf{F}$ such as gravity acts on the system, the general equation for rocket motion becomes

$$\mathbf{F} = M\frac{d\mathbf{v}}{dt} - \mathbf{u}\frac{dM}{dt}.$$

Example 4.15 Center of Mass and the Rocket Equation
In this example we shall derive the rocket equation using center of mass considerations. The general expression for the velocity $\dot{\mathbf{R}}$ of the center of mass of a system of two masses m_1 and m_2 is

$$\dot{\mathbf{R}} = \frac{m_1\dot{\mathbf{r}}_1 + m_2\dot{\mathbf{r}}_2}{(m_1 + m_2)}.$$

If no external forces act on the system, $\ddot{\mathbf{R}} = 0$, so the center of mass moves with constant velocity. In the inertial frame moving along with the rocket, $\dot{\mathbf{R}} = 0$.

Using the same notation as in our previous derivation (Example 4.14),

$$\dot{\mathbf{R}} = \frac{M\Delta v + \Delta m(\mathbf{u} + \Delta v)}{M + \Delta m} = 0.$$

Hence

$$M\Delta v + \Delta m(\mathbf{u} + \Delta v) = 0.$$

Dividing by Δt and taking the limit $\Delta t \to 0$ gives

$$M\frac{d\mathbf{v}}{dt} + \mathbf{u}\frac{dm}{dt} = 0.$$

The second-order term $\Delta m\Delta\mathbf{v}$ does not contribute in the limit. Then, with the identity $dm/dt = -dM/dt$,

$$M\frac{d\mathbf{v}}{dt} - \mathbf{u}\frac{dM}{dt} = 0$$

as before.

Our approach to rocket motion illustrates a powerful method for analyzing physical problems. It is easy to become confused trying to take into account the detailed acceleration of Δm and the rocket body while they separate. But these details vanish when taking the limit; their effect is actually included in the final equation of motion. The correct equation of motion results from taking the limit and including only the non-vanishing "first-order" terms. Terms beyond first order, such as $\Delta m\Delta\mathbf{v}$, vanish in the limit $\Delta t \to 0$.

Here are three examples on rocket motion.

Example 4.16 Rocket in Free Space

If there is no external force on a rocket, $\mathbf{F} = 0$, and the rocket's motion is given by

$$M\frac{d\mathbf{v}}{dt} = \mathbf{u}\frac{dM}{dt}$$

or

$$\frac{d\mathbf{v}}{dt} = \frac{\mathbf{u}}{M}\frac{dM}{dt}.$$

Checking signs—always useful—we expect the rocket to accelerate ($d\mathbf{v}/dt > 0$) while its mass decreases ($dM/dt < 0$). To make both sides of the last equation positive, $\mathbf{u} < 0$, which means that the mass is expelled in the backward direction, as expected.

The exhaust velocity **u** is usually constant, in which case it is easy to integrate the equation of motion:

$$\int_{t_0}^{t_f} \frac{d\mathbf{v}}{dt}\, dt = \mathbf{u} \int_{t_0}^{t_f} \frac{1}{M} \frac{dM}{dt}\, dt$$

$$\int_{\mathbf{v}_0}^{\mathbf{v}_f} d\mathbf{v} = \mathbf{u} \int_{M_0}^{M_f} \frac{dM}{M}$$

or

$$\mathbf{v}_f - \mathbf{v}_0 = \mathbf{u} \ln \frac{M_f}{M_0}$$

$$= -\mathbf{u} \ln \frac{M_0}{M_f}.$$

If $\mathbf{v}_0 = 0$, then

$$\mathbf{v}_f = -\mathbf{u} \ln \frac{M_0}{M_f}.$$

The final velocity is independent of how the mass is released—the fuel can be expended rapidly or slowly without affecting $\mathbf{v}_f$. The only important quantities are the exhaust velocity and the ratio of initial to final mass.

The situation is quite different if a gravitational field is present, as shown by the next example.

Example 4.17 Rocket in a Constant Gravitational Field

If a rocket takes off in a constant gravitational field $\mathbf{F} = M\mathbf{g}$, the equation for rocket motion becomes

$$M\mathbf{g} = M\frac{d\mathbf{v}}{dt} - \mathbf{u}\frac{dM}{dt},$$

where **u** and **g** are directed down and are assumed to be constant

$$\frac{d\mathbf{v}}{dt} = \frac{\mathbf{u}}{M}\frac{dM}{dt} + \mathbf{g}.$$

Integrating with respect to time we obtain

$$\mathbf{v}_f - \mathbf{v}_0 = \mathbf{u} \ln\left(\frac{M_f}{M_0}\right) + \mathbf{g}(t_f - t_0).$$

Let $\mathbf{v}_0 = 0$ and $t_0 = 0$, with velocity **v** positive upward:

$$v_f = -u \ln\left(\frac{M_0}{M_f}\right) - gt_f.$$

Now there is a premium attached to burning the fuel rapidly. The shorter the burn time, the greater the final velocity. This is why the takeoff of a large rocket is so spectacular—it is essential to burn the fuel as quickly as possible.

Example 4.18 Saturn V

The Saturn V ("Five") three-stage rocket, one of the most powerful expendable launch vehicles ever constructed, fulfilled its purpose by sending Apollo astronauts to land on the Moon on six different missions. The first stage was powered by five enormous F-1 rocket engines (each nearly 6 m tall and 4 m diameter at the outlet). The F-1 engines burned a hydrocarbon similar to kerosene, with liquid oxygen as the oxidizer. All of these materials had to be carried by the rocket; a fully fueled Saturn V had a total mass of 3.0×10^6 kg, of which 2.1×10^6 kg was the fuel for the first stage. All the first-stage fuel was expended in 168 seconds.

The rocket equation with constant gravity is

$$M\frac{d\mathbf{v}}{dt} = \mathbf{u}\frac{dM}{dt} + M\mathbf{g}. \tag{1}$$

The first term on the right-hand side of Eq. (1) is called the "thrust." The second term is the weight of the rocket, directed vertically downward. At launch, the weight was 2.9×10^7 N, but this decreased rapidly as the first stage burned its fuel. The five F-1 engines in the first stage produced a total thrust of 3.4×10^7 N, somewhat greater than the initial weight. The initial upward acceleration, as you can easily verify, was only about 0.17 g.

Where does the thrust come from? Because $\mathbf{u}\Delta M$ is the momentum carried off by the expelled gases in time Δt, the thrust is the rate at which momentum is carried off by the burning fuel. Because both $\mathbf{u}$ and dM/dt are negative, the thrust is positive, opposite to $\mathbf{g}$.

Fuel is a precious commodity on rockets. To minimize the fuel mass required for a given thrust, the exhaust velocity must be as large as possible. The exhaust velocity for the first stage F-1 engines was 2600 m/s, but the second and third stages used liquid hydrogen and liquid oxygen, giving an exhaust velocity of 4100 m/s.

Evaluating the right-hand side of Eq. (1) for the first stage gives $(2600 \, \text{m/s})(2.1 \times 10^6 \, \text{kg})/168 \, \text{s} = 3.4 \times 10^7$ N, in good agreement with the thrust.

Rocket data tables often do not list the exhaust velocity but instead a quantity called the "specific impulse," which is the exhaust velocity divided by g. Specific impulse has units of seconds, and is therefore independent of whether we use SI, CGS, or English units.

4.9 Momentum Flow and Force

When catching a ball one expects to feel a recoil force or, more precisely, to experience an impulse. The concepts of momentum and impulse are reasonably intuitive: the recoil that we experience is merely the reaction to the impulse we must deliver to the ball to bring it to rest. Closely related to these concepts, though perhaps less intuitive, is the concept of momentum flow. Anyone who has been on the receiving end of a stream of water from a hose knows that a stream can exert a force. If the stream is intense, as in the case of a fire hose, the push can be dramatic—a jet of high pressure water can break through the wall of a burning building.

How can a column of water flying through the air exert a force that is every bit as real as a force transmitted by a rigid steel rod? The origin of the force can be visualized by picturing the stream as a series of small uniform droplets each of mass m traveling with velocity v. Let the droplets be distance l apart. Assume that the drops collide with your hand without rebounding, with final velocity $v_f = 0$, and then simply fall to the ground. Consider the force exerted by your hand on the stream. As each drop hits your hand there is a large force for a short time. Although we do not know the instantaneous force, we can find the impulse $I_{droplet}$ given *to* each drop *by* your hand:

$$I_{droplet} = \int_{1\ collision} F\ dt$$
$$= \Delta p$$
$$= m(v_f - v)$$
$$= -mv.$$

By Newton's third law, the impulse delivered to your hand by the droplet is equal and opposite to the impulse delivered to the droplet by your hand:

$$I_{hand} = mv.$$

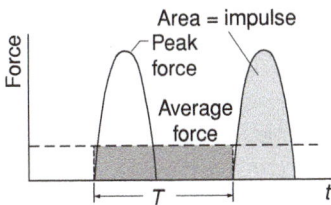

The positive sign means that the impulse to your hand is in the same direction as the velocity of the droplet. The impulse equals the area under one of the peaks of the instantaneous force shown in the drawing.

If there are many collisions per second, you feel the average force F_{av} (indicated by the dashed line in the drawing) rather than the shock of individual drops. If the average time between collisions is T, then the area under F_{av} during the time T is identical to the impulse due to one droplet

$$F_{av}T = \int_{1\ collision} F\ dt = mv.$$

The average distance between droplets is $l = vT$ and so the average force exerted by the stream can be written

$$F_{av} = \frac{mv}{T} = \frac{mv^2}{l}. \tag{4.8}$$

Momentum transfer by a stream is the physics underlying the force on a wind turbine blade and the lift on an airplane wing.

This description of successive collisions generating an average force employed an idealized model of a stream of water, but the model is pretty accurate for a related scenario: laser slowing of atoms. Just as a stream of water can exert a force on a hand, a stream of light can exert a force on an atom. The force can be so large that atoms are brought to near rest almost instantaneously. This process is the first step in the creation of ultra-cold atomic gases, in which atoms are cooled into the sub-microkelvin temperature regime using laser light.

Example 4.19 Slowing Atoms with Laser Light

Understanding how laser light can slow atoms requires a few facts from quantum physics that we will simply state without stopping to explain them.

The starting point is the classical description of light. According to the electromagnetic theory of James Clerk Maxwell, light is a wave of oscillating electric and magnetic fields that carries energy. The speed of a light wave, c, its wavelength λ, and its frequency ν are related by the familiar wave condition $c = \lambda\nu$.

Einstein put forth an alternate picture of light that seems at first sight to be totally incompatible with Maxwell's: light energy is received in discrete bundles or *quanta*, now called *photons*, with particle-like properties. Einstein argued that the energy of a photon associated with light of frequency ν is $h\nu$, where the constant h, known as Planck's constant, has the numerical value of 6.63×10^{-34} kg $\cdot$ m^2/s. Einstein also argued that each photon carries momentum $h\nu/c$ or, equivalently, h/λ.

If a gas of atoms is heated or excited by an electrical discharge, the atoms radiate light at characteristic wavelengths. They can also absorb light at those wavelengths. Niels Bohr proposed that the energy of atoms cannot vary arbitrarily, as we expect in classical physics, but that atoms exist only in certain states that he called *stationary states*. If the lowest-lying state, called the *ground state*, has energy E_a and an excited state has energy E_b, then an excited atom can get rid of its energy by creating a photon. Conservation of energy requires that $h\nu = E_b - E_a$. Thus, the different colors radiated by atoms reflect their particular stationary states.

An excited atom rapidly jumps back to its ground state by emitting a photon, a process called *spontaneous emission*. The process is similar to radioactive decay with a characteristic decay time τ that is typically tens of nanoseconds. To complete the description, we need one further concept, also proposed by Einstein, *stimulated emission*. An atom in an excited state will in time spontaneously emit a photon, but if it is illuminated by that frequency, the emission will occur sooner. Stimulated emission is the fundamental process in the generation of laser light. (The term "LASER" is an acronym for Light Amplification by Stimulated Emission of Radiation.)

In a laser-cooling apparatus, a stream of atoms initially in their ground state flows into a high vacuum through a small aperture. Laser light tuned to one of the atom's spectral lines is directed toward the aperture. Lasers are so intense that their light causes absorption and stimulated emission in times much shorter than the decay time τ. In such a situation, the atom can be viewed as being in the ground state half the time, and in its excited state the other half.

Because the laser light is directed against the motion of the atoms, every time an atom absorbs a photon it recoils with a momentum kick, or impulse, $\Delta p = h/\lambda$. The atom also experiences a momentum kick when it emits a photon, causing it to recoil in the opposite direction from the emission. However, spontaneous emission occurs in random directions and its momentum kicks average out. Consequently, the atom experiences a series of kicks that retard its motion. The time-average force is given by

$$F_{av} = \frac{1}{2}\frac{\Delta p}{\tau} = \frac{1}{2}\frac{h}{\lambda\tau},$$

where the factor of 1/2 takes into account that the atom is in the excited state only half the time. If the mass of the atom is M, then the average acceleration is

$$a_{av} = \frac{F_{av}}{M} = \frac{1}{2M}\frac{h}{\lambda\tau}.$$

The first experiments in laser slowing were to a stream of sodium atoms. For sodium, the wavelength of the excited state transition is $\lambda = 589 \times 10^{-9}$ m, $M = 3.85 \times 10^{-26}$ kg, and $\tau = 15 \times 10^{-9}$ s. Using $h = 6.6 \times 10^{-34}$ kg $\cdot$ m^2/s,

$$a_{av} = 9.7 \times 10^5 \text{ m/s}^2.$$

This acceleration is about 10^5 times larger than the acceleration of gravity! The average speed of sodium atoms at room temperature is about 560 m/s, and laser slowing can bring them essentially to rest in less than a meter.

4.10 Momentum Flux

In Section 4.9 we found that the average force on a surface due to a perpendicular stream of droplets of mass m moving with velocity v and separated by distance l is given by Eq. (4.8):

$$F_{av} = \frac{m}{l}v^2.$$

This expression has a natural interpretation. The quantity mv/l is the average momentum per unit length in the stream of particles. If we multiply this by v, the distance per second that the droplets travel, we obtain the momentum per second carried by the stream past any point, that is, the *rate* of momentum transport. Thus, the average force exerted on a surface is the rate at which the stream transports momentum to the surface.

More realistic than a hypothetical stream of particles is a real stream of matter, for instance a stream of water in a hose with cross-section A, flowing with speed v in the direction given by the unit vector $\hat{\mathbf{v}}$. If the water has mass density $\rho_m(\mathrm{kg/m^3})$, then the mass per unit length in the stream is $\rho_m A$ and the momentum per unit length is $\rho_m vA$. The rate at which momentum flows through a hypothetical surface across the stream is

$$\dot{\mathbf{P}} = \rho_m v^2 A\, \hat{\mathbf{v}}.$$

If the stream is brought to a halt by striking a solid surface, the force exerted by the surface must cause the stream to lose momentum at the same rate the stream transports momentum to the surface. The force of the surface on the stream is therefore $-\dot{\mathbf{P}}$. The reaction force of the stream on the surface is

$$\mathbf{F}_{\text{on surface}} = +\dot{\mathbf{P}} = \rho_m v^2 A\, \hat{\mathbf{v}}.$$

As expected, the direction of the force on the surface is in the direction of flow $\hat{\mathbf{v}}$. If the stream does not come to rest at the surface but is reflected straight back, then the surface must exert the force needed not only to cancel the incoming momentum but also to generate the outgoing momentum, doubling the force on the surface. On the other hand, if the surface is transparent so that the matter simply passes through, then momentum is carried to and away from the surface at the same rate; the net rate of momentum transfer to the surface is zero and hence there is no force.

If the surface is not perpendicular to the flow, but tilted at angle θ as shown in the drawing, then the momentum flow to the surface is

$$\dot{P} = \rho_m v^2 A \cos\theta$$

if the momentum is cancelled at the surface.

It is useful to introduce the vector $\mathbf{J}$ with magnitude $\rho_m v^2$, and directed along the flow $\hat{\mathbf{v}}$:

$$\mathbf{J} = \rho_m v^2 \hat{\mathbf{v}}.$$

The vector $\mathbf{J}$ is called the *flux density* of the stream.

It is also useful to describe the area by a vector $\mathbf{A}$. The magnitude of $\mathbf{A}$ is numerically equal to the area and its direction is perpendicular to the surface, as described by a unit vector $\hat{\mathbf{n}}$ normal to the surface. The normal vector can lie in either of two directions. We choose the following convention: in evaluating the momentum transfer through a surface into a system, $\hat{\mathbf{n}}$ is positive if it points inward.

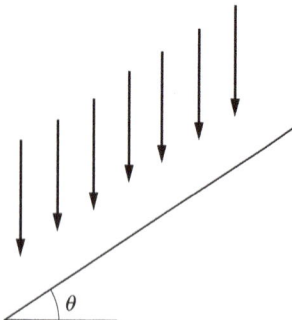

Because the momentum transfer by a stream is in the direction of flow $\hat{\mathbf{v}}$, the rate of momentum transfer to a surface—consequently, the force on the system—is

$$\dot{\mathbf{P}} = (\mathbf{J} \cdot \mathbf{A})\,\hat{\mathbf{v}}.$$

The quantity $\dot{\mathbf{P}} = (\mathbf{J} \cdot \mathbf{A})\,\hat{\mathbf{v}}$ is called the *flux* (or *flow*) of momentum to the surface.

Momentum is a vector, so momentum flux is a vector. Flux as a vector occurs often in physics, particularly in fluid dynamics and electromagnetic theory.

In situations such as a stream of water rebounding from a surface, momentum can be transported both to the surface and away from it. According to our convention with $\hat{\mathbf{n}}$ pointing inward through a surface surrounding a system, flux is positive $\mathbf{J} \cdot \mathbf{A} > 0$ if the momentum flows in and negative $\mathbf{J} \cdot \mathbf{A} < 0$ if it flows out. The total force on a system due to a number of sources of momentum flow can be written

$$\mathbf{F}_{\text{tot}} = \sum_k \dot{\mathbf{P}}_k = \sum (\mathbf{J}_k \cdot \mathbf{A}_k)\hat{\mathbf{v}}_k$$

where the sum is over all the surface elements through which momentum flows. As the number of elements is increased, the sum becomes what is called a *surface integral*. However, there is no need for us to take that limit now because we shall be concerned here only with surfaces of simple geometry.

A helpful way to calculate the force on a system in which there are several sources of momentum flow is to sum the inward flow terms into a total inward flow $\dot{\mathbf{P}}_{\text{in}}$, and all the outgoing flow terms into a total outward flow $\dot{\mathbf{P}}_{\text{out}}$. The total force on the system can then be written

$$\mathbf{F}_{\text{tot}} = \dot{\mathbf{P}}_{\text{in}} - \dot{\mathbf{P}}_{\text{out}}.$$

Example 4.20 Reflection from an Irregular Object

A stream is reflected from an object, as shown in the drawing.

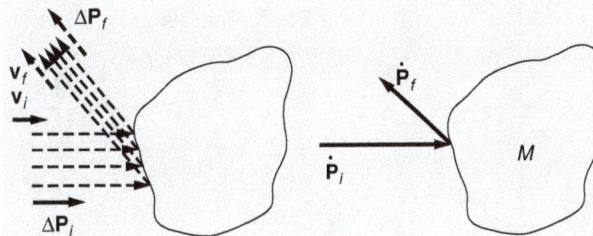

The incident momentum flux is $\dot{\mathbf{P}}_i$ and the reflected (outgoing) flux is $\dot{\mathbf{P}}_o$. Hence, the total force on the system is

$$\mathbf{F} = \dot{\mathbf{P}}_i - \dot{\mathbf{P}}_o.$$

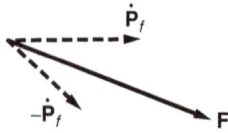

Example 4.21 Solar Sail Spacecraft

Interplanetary exploration has many applications for a lightweight fuel-saving spacecraft capable of carrying a small instrument package across the solar system. One attractive design is the solar sail craft, which sports a large sail made of thin plastic sheet; propulsion is the force exerted by sunlight—the momentum carried by photons.

Light from the Sun arrives at the Earth with an energy flux density called the *solar constant*, which has the value $S_{sun} = 1370\,\text{watts/m}^2$. The solar constant can be viewed as a flux of photons with energies $h\nu$ that vary over a wide frequency range. Details of the Sun's spectrum are unimportant because every photon carries momentum $h\nu/c$. Consequently, the momentum flux density of sunlight at the Earth is simply $S_{sun}/c = 1370/(3 \times 10^8) = 4.6 \times 10^{-6}\,\text{kg/(m s}^2)$, where we have used 1 watt = 1 joule/s = 1 kg m^2/s^3. This can also be written $S_{sun}/c = 4.6 \times 10^{-6}\,\text{(kg m/s)/(m}^2\text{s)}$, showing that it has the units of momentum per unit area per unit time.

In 2010, Japan launched a solar sail craft called IKAROS ("Interplanetary Kite-craft Accelerated by Radiation Of the Sun"). The sail was very thin polyimide (Kapton®), only 7.5×10^{-6} m thick. The sail's area was $A = 150\,\text{m}^2$, and the craft's total mass was $M = 1.6\,\text{kg}$. The craft was lifted by a chemical rocket into space, where the sail unfurled. Rotation at 25 revolutions/minute about the axis kept the sail flat, eliminating the need for struts that would have added to the mass. IKAROS traveled to the orbit of Venus, the first craft to demonstrate solar sail technology in deep space.

We shall calculate the initial acceleration of IKAROS when it started out near the Earth. Suppose that the solar sail is a perfect reflector, and that all the sunlight is reflected back. The total force on the sail is

$$\mathbf{F} = \dot{\mathbf{P}}_{in} - \dot{\mathbf{P}}_{out} = \frac{2S_{sun}\mathbf{A}}{c}.$$

Near Earth orbit, the magnitude of the acceleration a_{photon} due to photons is therefore

$$a_{photon} = \frac{F}{M} = \left(\frac{2S_{sun}}{c}\right)\frac{A}{M}$$

$$= 2\left(4.6 \times 10^{-6}\,\text{kg/(m s}^2)\right)\left(\frac{150\,\text{m}^2}{1.6\,\text{kg}}\right)$$

$$= 8.6 \times 10^{-4}\,\text{m/s}^2.$$

The Sun exerts an inward gravitational acceleration g_{sun}. Near Earth orbit,

$$g_{sun} = \frac{GM_{sun}}{R_E^2}$$

where $R_E = 1.50 \times 10^{11}$ m is the mean distance of the Earth from the Sun, known as the *astronomical unit (AU)*

$$g_{sun} = \frac{(6.7 \times 10^{-11} \text{m}^3/(\text{kg s}^2))(2.0 \times 10^{30} \text{ kg})}{(1.5 \times 10^{11} \text{ m})^2}$$
$$= 5.9 \times 10^{-3} \text{ m/s}^2.$$

The net acceleration is

$$a_{net} = a_{photon} - g_{sun}$$
$$= (0.86 - 5.9) \times 10^{-3} \text{ m/s}^2$$
$$= -5.0 \times 10^{-3} \text{ m/s}^2.$$

The craft falls inward toward the Sun. Because the solar intensity and the solar gravity both vary as the inverse square, the acceleration increases as the craft moves toward the Sun, but is always directed inward. However, during IKAROS' flight the radiation force slowed the craft enough to allow it to come near Venus. A craft with a much larger sail would be needed to travel outward toward Jupiter.

Example 4.22 Pressure of a Gas

The pressure of a gas arises from momentum flow to and from the enclosing surfaces due to the random motion of the particles in the gas. The tiny gas molecules exert a real force, because of their momentum. Consider a thin aluminum can containing a carbonated beverage. Before opening, the can feels strong and rigid because of the outward pressure of the gas on the walls. When the tab is popped and the excess pressure is released, the can is weak and easily crushed.

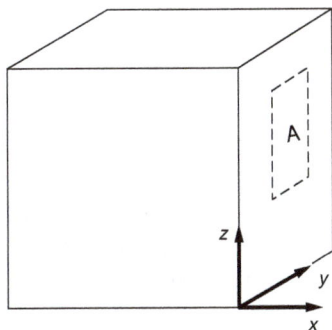

Consider a gas of n particles per unit volume, each having mass m. The mass density is $\rho = nm$ kg/m^3. Let us find the momentum transport to a surface of the container having area A, oriented in the y–z plane, as shown.

Although particles move in all directions, we shall be concerned only with motion in the x direction. We suppose for the moment that the particles have only a single x velocity, v_x, but they are just as likely to move in the $-x$ direction as the $+x$ direction. Hence, at any instant the density of particles moving toward the wall is $\rho/2$. They transport momentum in the $+x$ direction at the rate

$$\dot{P}_x = \frac{\rho}{2}v_x^2 A.$$

Particles must leave the wall at the same rate as they approach; otherwise, they would accumulate at the surface. Hence, the momentum

flow away from the wall is

$$\dot{P}_{-x} = \frac{\rho}{2}v_x^2 A$$

and the total force on the wall is

$$F_x = \dot{P}_x - \dot{P}_{-x} = \rho v_x^2 A.$$

Let us now drop the simplistic assumption that atoms move only in the positive or negative x directions, with a single speed v_x. We expect atoms to move in random directions with speeds that change as they collide. If we calculate the contribution to the pressure for particles moving in a small range of velocities, we would obtain the above result, with the understanding that it is for the *average* force. Hence

$$\overline{F_x} = \overline{\dot{P}_x} - \overline{\dot{P}_{-x}} = \rho \overline{v_x^2} A$$

where the bar indicates an average over all the particles.

The pressure of a gas is the force per unit area on the surface. Consequently, the pressure on the area normal to the $x-y$ plane is

$$\mathcal{P}_x = F_x/A = \rho \overline{v_x^2}.$$

(Here we use the symbol $\mathcal{P}$ for pressure, to distinguish it from P, the symbol for total momentum.) Because the pressure of a gas is the same in all directions, we expect a similar result for pressure on surfaces that are normal to the y and z axes. This would be the case if

$$\overline{v_x^2} = \overline{v_y^2} = \overline{v_z^2} = \frac{1}{3}\overline{v^2},$$

where $\overline{v^2} = \overline{v_x^2} + \overline{v_y^2} + \overline{v_z^2}$. Consequently,

$$\mathcal{P} = (1/3)\rho\overline{v^2}.$$

This result provides a crucial link connecting the concepts of heat, energy, and microscopic motion, topics we shall pursue in Chapter 5.

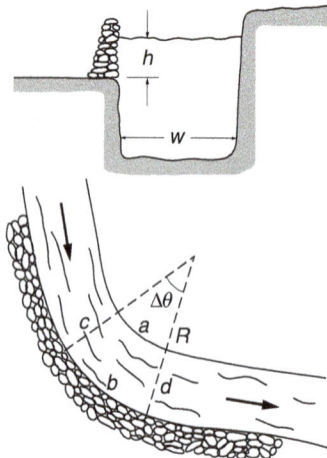

Example 4.23 Dike at the Bend of a River

The problem is to build a dike at the bend of a river to prevent flooding when the river rises. The dike must be strong enough to withstand the static pressure of the river $\rho g h$, where ρ is the density of the water and h is the height from the base of the dike to the surface of the water. We can understand from Newton's laws that the outer bank and the dike must exert a sideways force on the stream to deflect it from straight-line flow. The dike must therefore withstand the dynamic pressure due to the deflection of the flow in addition to the static pressure. How do the dynamic and static pressures compare?

We approximate the bend by a circular curve with radius R, and focus our attention on a short length of the curve subtending angle $\Delta\theta$. We need concern ourselves only with the height h of the river above the base of the dike. Let us calculate the momentum flux to the volume bounded by the river bank a, by the dike b, and by the fictitious surfaces across the river c and d. The river flows with velocity $\mathbf{v}$; because the cross-sectional area is constant, the magnitude of $\mathbf{v}$ is constant.

Momentum flows through surfaces c and d at rates $\dot{\mathbf{P}}_{\text{in,c}} = \rho v^2 A \hat{\mathbf{v}}_{\text{c}}$ and $\dot{\mathbf{P}}_{\text{out,d}} = \rho v^2 A \hat{\mathbf{v}}_{\text{d}}$, respectively. Here $A = hw$ is the cross-sectional area of the river lying above the base of the dike. The total rate of momentum transfer to the bounded volume is

$$\dot{\mathbf{P}} = \dot{\mathbf{P}}_{\text{in,c}} - \dot{\mathbf{P}}_{\text{out,d}} = \rho v^2 A (\hat{\mathbf{v}}_{\text{c}} - \hat{\mathbf{v}}_{\text{d}}).$$

From the drawing, the magnitude of the momentum transfer is

$$\dot{P} = \rho v^2 A (2 \sin \Delta\theta/2).$$

The momentum transfer points to the center of the bending circle. The dike must provide a force to account for this momentum transfer, and the reaction to that force gives rise to the dynamic pressure on the dike.

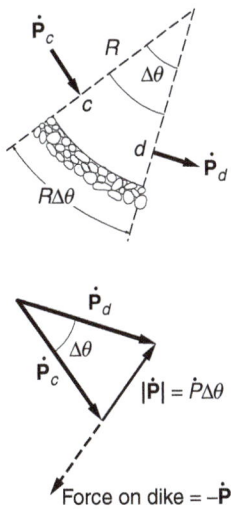

To calculate the pressure we can take the small angle limit $\sin \Delta\theta/2 \approx \Delta\theta/2$, and consider the force arising from a section of the dike subtending angle $\Delta\theta$, with area $(R\Delta\theta)h$. The dynamic force on the dike is radially outward, and has magnitude $\dot{P} \approx \rho v^2 A \Delta\theta$. The force is exerted over the area $(R\,\Delta\theta)h$, and the dynamic pressure is therefore

$$
\begin{aligned}
\text{dynamic pressure} &= \frac{\dot{P}}{R\,\Delta\theta\,h} \\
&= \frac{\rho v^2 A \Delta\theta}{R\,\Delta\theta h} \\
&= \frac{\rho v^2 A}{Rh} \\
&= \frac{\rho v^2 w}{R}.
\end{aligned}
$$

The ratio of dynamic to static pressure is

$$
\begin{aligned}
\frac{\text{dynamic pressure}}{\text{static pressure}} &= \frac{\rho v^2 w}{R}\frac{1}{\rho g h} = \frac{w}{h}\frac{v^2}{Rg} \\
&= \frac{\text{width}}{\text{height}} \times \frac{\text{centripetal acceleration}}{g}.
\end{aligned}
$$

For a river in flood with a speed of 10 mph (approximately 15 ft/s), a radius of 2000 ft, a flood height of 3 ft, and a width of 200 ft, the ratio is 0.22, so that the dynamic pressure is by no means negligible.

Note 4.1 Center of Mass of Two- and Three-dimensional Objects

In this note we shall find the center of mass of some multidimensional objects. These examples are straightforward if you have had experience evaluating two- or three-dimensional integrals. Otherwise, read on.

1. Uniform Triangular Plate

Consider the two-dimensional case of a uniform right triangular plate of mass M, base b, height h, and small thickness t. We treated this problem in Example 4.4, reducing it to a single integral. Here we shall treat it in a more general way using a double integral, an approach that would also apply to cases where the density is not uniform.

Divide the plate into small rectangular areas of sides Δx and Δy, as shown. The volume of each element is $\Delta V = t\,\Delta x\,\Delta y$, and

$$\begin{aligned}\mathbf{R} &\approx \frac{\sum m_j \mathbf{r}_j}{M} \\ &= \frac{\sum \rho_j\, t\, \Delta x\, \Delta y\, \mathbf{r}_j}{M},\end{aligned} \tag{1}$$

where j is the label of a volume element and ρ_j is its density. Because the plate is uniform,

$$\rho_j = \text{constant} = \frac{M}{V} = \frac{M}{At},$$

where $A = bh/2$ is the area of the plate.

We can evaluate the sum in Eq. (1) by summing first over the Δx's and then over the Δy's, instead of over the single index j. This gives a double sum that can be converted to a double integral by taking the limit, as follows:

$$\begin{aligned}\mathbf{R} &= \lim_{\substack{\Delta x \to 0 \\ \Delta y \to 0}} \left(\frac{M}{At}\right)\left(\frac{t}{M}\right) \sum \sum \mathbf{r}_j\, \Delta x\, \Delta y \\ &= \frac{1}{A} \iint \mathbf{r}\, dx\, dy.\end{aligned}$$

Let $\mathbf{r} = x\hat{\mathbf{i}} + y\hat{\mathbf{j}}$ be the position vector of the element $dx\,dy$. Then, writing $\mathbf{R} = X\hat{\mathbf{i}} + Y\hat{\mathbf{j}}$, we have

$$\begin{aligned}\mathbf{R} &= X\hat{\mathbf{i}} + Y\hat{\mathbf{j}} \\ &= \frac{1}{A}\iint (x\hat{\mathbf{i}} + y\hat{\mathbf{j}})\, dx\, dy \\ &= \frac{1}{A}\left(\iint x\, dx\, dy\right)\hat{\mathbf{i}} + \frac{1}{A}\left(\iint y\, dx\, dy\right)\hat{\mathbf{j}}.\end{aligned}$$

Hence the coordinates of the center of mass are

$$X = \frac{1}{A}\iint x\, dx\, dy$$

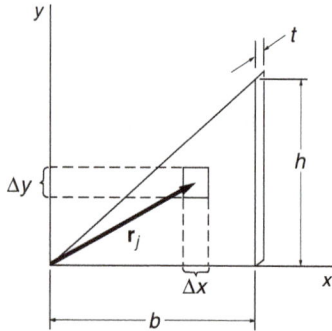

$$Y = \frac{1}{A}\iint y\, dx\, dy.$$

The double integrals may look strange, but they are easily evaluated. Consider first the double integral for X:

$$X = \frac{1}{A}\iint x\, dx\, dy.$$

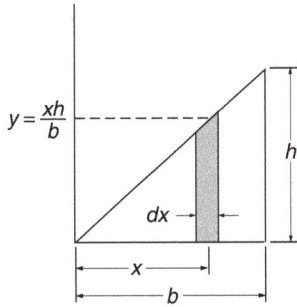

This integral instructs us to take each element, multiply its area by its x coordinate, and sum the results. We can do this in stages by first considering the elements in a strip parallel to the y axis. The strip runs from $y = 0$ to $y = xh/b$. (By similar triangles, any point on the slanted boundary obeys the relation $y/x = h/b$.)

Each element in the strip has the same x coordinate, and the contribution of the strip to the double integral is

$$\frac{1}{A} x\,dx \int_0^{xh/b} dy = \frac{h}{bA} x^2\,dx.$$

Finally, we sum the contributions of all such strips $x = 0$ to $x = b$ to find

$$X = \frac{h}{bA} \int_0^b x^2\,dx = \frac{h}{bA}\frac{b^3}{3}$$

$$= \frac{hb^2}{3A}.$$

Since $A = \frac{1}{2}bh$,

$$X = \tfrac{2}{3}b.$$

Similarly for Y,

$$Y = \frac{1}{A} \int_0^b \left(\int_0^{xh/b} y\,dy \right) dx$$

$$= \frac{h^2}{2Ab^2} \int_0^b x^2\,dx = \frac{h^2 b}{6A}$$

$$= \tfrac{1}{3}h.$$

Hence

$$\mathbf{R} = \tfrac{2}{3}b\hat{\mathbf{i}} + \tfrac{1}{3}h\hat{\mathbf{j}}.$$

Although the coordinates of $\mathbf{R}$ depend on the particular coordinate system we choose, the position of the center of mass with respect to the triangular plate is, of course, independent of the coordinate system.

2. Non-uniform Rectangular Plate

Find the center of mass of a thin non-uniform rectangular plate with sides of length a and b, whose mass per unit area σ varies as $\sigma = \sigma_0(xy/ab)$, where σ_0 is a constant.

$$\mathbf{R} = \frac{1}{M} \iint (x\hat{\mathbf{i}} + y\hat{\mathbf{j}})\sigma\,dx\,dy.$$

We first find M, the mass of the plate:

$$M = \int_0^b \int_0^a \sigma\,dx\,dy$$

$$= \int_0^b \int_0^a \sigma_0 \frac{x}{a}\frac{y}{b}\,dx\,dy.$$

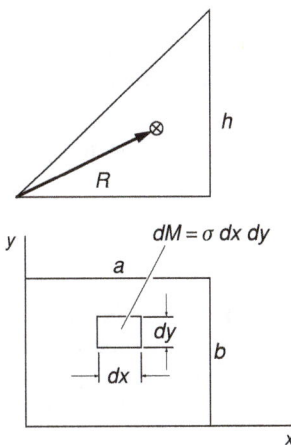

Integrate over x, treating y as a constant

$$M = \int_0^b \sigma_0 \frac{y}{b} \left(\int_0^a \frac{x}{a} \, dx \right) dy$$

$$= \int_0^b \sigma_0 \frac{y}{b} \left(\frac{x^2}{2a} \Big|_{x=0}^{x=a} \right) dy$$

$$= \int_0^b \sigma_0 \frac{y}{b} \frac{a}{2} \, dy.$$

Then integrate over y

$$= \frac{\sigma_0 a}{2} \frac{y^2}{2b} \Big|_{y=0}^{y=b} = \frac{1}{4} \sigma_0 ab.$$

Using the same approach, the x component of **R** is

$$X = \frac{1}{M} \iint x\sigma \, dx \, dy$$

$$= \frac{1}{M} \int_0^b \frac{\sigma_0}{ab} y \left(\int_0^a x^2 \, dx \right) dy$$

$$= \frac{1}{M} \int_0^b \frac{\sigma_0}{ab} y \left(\frac{x^3}{3} \Big|_0^a \right) dy$$

$$= \frac{1}{M} \frac{\sigma_0}{ab} \int_0^b \frac{ya^3}{3} \, dy$$

$$= \frac{1}{M} \frac{\sigma_0}{ab} \frac{a^3}{3} \frac{b^2}{2}$$

$$= \frac{4}{\sigma_0 ab} \frac{\sigma_0 a^2 b}{6}$$

$$= \frac{2}{3} a.$$

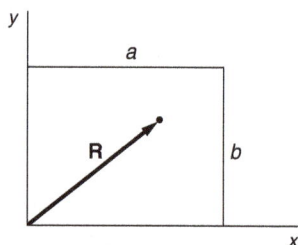

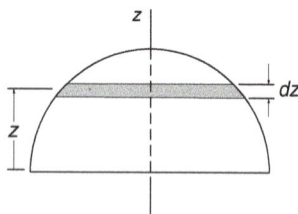

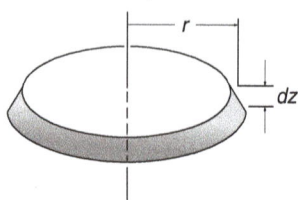

Similarly, $Y = \frac{2}{3}b$.

3. Uniform Solid Hemisphere

Find the center of mass of a uniform solid hemisphere of radius R and mass M.

From symmetry it is apparent that the center of mass lies on the z axis. Its height above the equatorial plane is

$$Z = \frac{1}{M} \int z \, dM.$$

The integral is over three dimensions, but the symmetry of the situation lets us treat it as a one-dimensional integral. We mentally subdivide the hemisphere into a pile of thin disks. Consider a circular disk of radius r and thickness dz. Its volume is $dV = \pi r^2 dz$, and its mass is $dM = \rho \, dV = (M/V)(dV)$, where $V = \frac{2}{3}\pi R^3$.

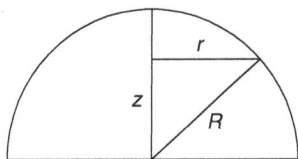

Hence

$$Z = \frac{1}{M} \int \frac{M}{V} z \, dV$$

$$= \frac{1}{V} \int_{z=0}^{R} \pi r^2 z \, dz.$$

To evaluate the integral we need to find r in terms of z.
Since $r^2 = R^2 - z^2$, we have

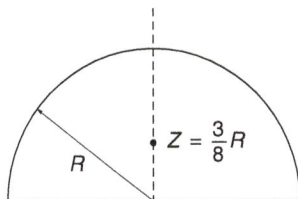

$$Z = \frac{\pi}{V} \int_0^R z(R^2 - z^2) \, dz$$

$$= \frac{\pi}{V} \left(\frac{1}{2} z^2 R^2 - \frac{1}{4} z^4 \right) \Big|_0^R$$

$$= \frac{\pi}{V} \left(\frac{1}{2} R^4 - \frac{1}{4} R^4 \right)$$

$$= \frac{\frac{1}{4} \pi R^4}{\frac{2}{3} \pi R^3}$$

$$= \frac{3}{8} R.$$

Problems

For problems marked *, refer to page 521 for a hint, clue, or answer.

4.1 *Center of mass of a non-uniform rod**
    The mass per unit length of a non-uniform rod of length l is given
    by $\lambda = A \cos(\pi x / 2l)$, where x is position along the rod, $0 \le x \le l$.
    (a) What is the mass M of the rod?
    (b) What is the coordinate X of the center of mass?

4.2 *Center of mass of an equilateral triangle*
    Find the center of mass of a thin uniform plate in the shape of an
    equilateral triangle with sides a.

4.3 *Center of mass of a water molecule*
    A water molecule H_2O consists of a central oxygen atom bound to
    two hydrogen atoms. The two hydrogen–oxygen bonds subtend an
    angle of $104.5°$, and each bond has a length of 0.097 nm.
    Find the center of mass of the water molecule.

4.4 *Failed rocket*
    An instrument-carrying rocket accidentally explodes at the top of
    its trajectory. The horizontal distance between the launch point and
    the point of explosion is L. The rocket breaks into two pieces that
    fly apart horizontally. The larger piece has three times the mass
    of the smaller piece. To the surprise of the scientist in charge, the
    smaller piece returns to Earth at the launching station. How far

away does the larger piece land? Neglect air resistance and effects due to the Earth's curvature.

4.5 *Acrobat and monkey*

A circus acrobat of mass M leaps straight up with initial velocity v_0 from a trampoline. As he rises up, he takes a trained monkey of mass m off a perch at a height h above the trampoline.

What is the maximum height attained by the pair?

4.6 *Emergency landing*

A light plane weighing 2500 lb makes an emergency landing on a short runway. With its engine off, it lands on the runway at 120 ft/s. A hook on the plane snags a cable attached to a 250-lb sandbag and drags the sandbag along. If the coefficient of friction between the sandbag and the runway is 0.4, and if the plane's brakes give an additional retarding force of 300 lb, how far does the plane go before it comes to a stop?

4.7 *Blocks and compressed spring*

A system is composed of two blocks of mass m_1 and m_2 connected by a massless spring with spring constant k. The blocks slide on a frictionless plane. The unstretched length of the spring is l. Initially m_2 is held so that the spring is compressed to $l/2$ and m_1 is forced against a stop, as shown. m_2 is released at $t = 0$.

Find the motion of the center of mass of the system as a function of time.

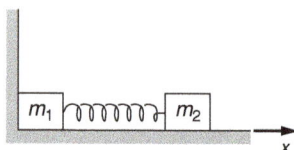

4.8 *Jumper*

A 50 kg woman jumps straight into the air, rising 0.8 m from the ground. What impulse does she receive from the ground to attain this height?

4.9 *Rocket sled*

A rocket sled moves along a horizontal plane, and is retarded by a friction force $f_{\text{friction}} = \mu W$, where μ is constant and W is the weight of the sled.

The sled's initial mass is M, and its rocket engine expels mass at constant rate $dM/dt \equiv \gamma$; the expelled mass has constant speed v_0 relative to the rocket.

The rocket sled starts from rest and the engine stops when half the sled's total mass is gone. Find an expression for the maximum speed.

4.10 *Rolling freight car with sand*

A freight car of mass M contains a mass of sand m. At $t = 0$ a constant horizontal force F is applied in the direction of rolling and at the same time a port in the bottom is opened to let the sand flow out at constant rate dm/dt. Find the speed of the freight car when all the sand is gone. Assume the freight car is at rest at $t = 0$.

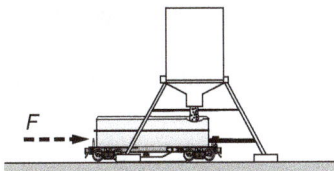

4.11 *Freight car and hopper**

An empty freight car of mass M starts from rest under an applied force F. At the same time, sand begins to run into the car at steady rate b from a hopper at rest along the track.

Find the speed when a mass of sand m has been transferred.

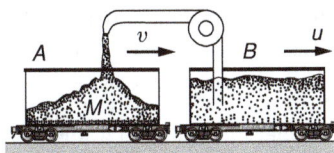

4.12 *Two carts and sand*

Material is blown into cart A from cart B at a rate b kilograms per second, as shown. The material leaves the chute vertically downward, so that it has the same horizontal velocity u as cart B. At the moment of interest, cart A has mass M and velocity v. Find dv/dt, the instantaneous acceleration of A.

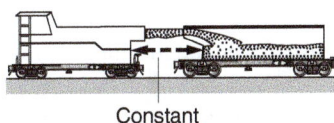

4.13 *Sand sprayer*

A sand-spraying locomotive sprays sand horizontally into a freight car as shown in the sketch. The locomotive and freight car are not attached. The engineer in the locomotive maintains his speed so that the distance to the freight car is constant. The sand is transferred at a rate $dm/dt = 10$ kg/s with a velocity 5 m/s relative to the locomotive. The freight car starts from rest with an initial mass of 2000 kg. Find its speed after 100 s.

4.14 *Ski tow*

A ski tow consists of a long belt of rope around two pulleys, one at the bottom of a slope and the other at the top. The pulleys are driven by a husky electric motor so that the rope moves at a steady speed of 1.5 m/s. The pulleys are separated by a distance of 100 m, and the angle of the slope is 20°.

Skiers take hold of the rope and are pulled up to the top, where they release the rope and glide off. If a skier of mass 70 kg takes the tow every 5 s on the average, what is the average force required to pull the rope? Neglect friction between the skis and the snow.

4.15 *Men and flatcar*

N men, each with mass m, stand on a railway flatcar of mass M. They jump off one end of the flatcar with velocity u relative to the car. The car rolls in the opposite direction without friction.

(*a*) What is the final velocity of the flatcar if all the men jump off at the same time?

(*b*) What is the final velocity of the flatcar if they jump off one at a time? (The answer can be left in the form of a sum of terms.)

(*c*) Does case (*a*) or case (*b*) yield the larger final velocity of the flatcar? Can you give a simple physical explanation for your answer?

4.16 *Rope on table**

A rope of mass M and length l lies on a frictionless table, with a short portion, l_0, hanging through a hole. Initially the rope is at rest.

(a) Find a general equation for $x(t)$, the length of rope through the hole.

(b) Find the particular solution so that the initial conditions are satisfied.

4.17 *Solar sail 1*

With reference to Example 4.21, what is the maximum film thickness for a space sail like IKAROS to be accelerated outward away from the Sun? Take the density of Kapton® to be 1.4 g/cm³.

4.18 *Solar sail 2*

With reference to Example 4.21, consider the design of a solar sail intended to reach escape velocity from the Earth $\sqrt{2gR_e} = 11.2$ km/s using only the pressure due to sunlight. The sail is made of a Kapton® film 0.0025 cm thick with a density 1.4 g/cm³. Take the solar constant to be 1370 watts/m², assumed to be constant during the acceleration.

(a) What is the acceleration near the Earth due to sunlight pressure alone?

(b) How far from the Earth, as measured in units of the Earth's radius, R_e, would the sail have to be launched so that it could escape from the Earth?

(c) What area of sail would be needed to accelerate a 1 kg payload at half the rate of the sail alone?

4.19 *Tilted mirror*

On the Earth, a mirror of area 1 m² is held perpendicular to the Sun's rays.

(a) What is the force on the mirror due to photons from the Sun, assuming that the mirror is a perfect reflector? The momentum flux density from the Sun's photons is $J_{sun} = 4.6 \times 10^{-6}$ kg/(m · s²).

(b) Find how the force varies with angle if the mirror is tilted at angle α from the perpendicular.

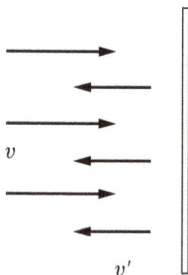

4.20 *Reflected particle stream**

A one-dimensional stream of particles of mass m with density λ particles per unit length, moving with speed v, reflects back from a surface, leaving with a different speed v', as shown. Find the force on the surface.

4.21 *Force on a fire truck*

A fire truck pumps a stream of water on a burning building at a rate K kg/s. The stream leaves the truck at angle θ with respect to the horizontal and strikes the building horizontally at height h above the nozzle, as shown. What is the magnitude and direction of the force on the truck due to the ejection of the water stream?

4.22 *Fire hydrant*

Water shoots out of a fire hydrant having nozzle diameter D, with nozzle speed V_0. What is the reaction force on the hydrant?

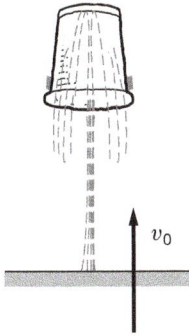

4.23 Suspended garbage can*

An inverted garbage can of weight W is suspended in air by water from a geyser. The water shoots up from the ground with a speed v_0, at a constant rate K kg/s. The problem is to find the maximum height at which the garbage can rides. Neglect any effect of the falling water after it rebounds elastically from the garbage can.

4.24 Growing raindrop

A raindrop of initial mass M_0 starts falling from rest under the influence of gravity. Assume that the drop gains mass from the cloud at a rate proportional to the product of its instantaneous mass and its instantaneous velocity:

$$\frac{dM}{dt} = kMV,$$

where k is a constant.

Show that the speed of the drop eventually becomes effectively constant, and give an expression for the terminal speed. Neglect air resistance.

4.25 Bowl of water

A bowl full of water is sitting out in a pouring rainstorm. Its surface area is 500 cm^2. The rain is coming straight down at 5 m/s at a rate of 10^{-3}g/cm^{2}s. If the excess water drips out of the bowl with negligible velocity, find the force on the bowl due to the falling rain.

What is the force if the bowl is moving uniformly upward at 2 m/s?

4.26 Rocket in interstellar cloud

A cylindrical rocket of diameter $2R$ and mass M is coasting through empty space with speed v_0 when it encounters an interstellar cloud. The number density of particles in the cloud is N particles/m^3. Each particle has mass $m \ll M$, and they are initially at rest.

(*a*) Assume that each cloud particle bounces off the rocket elastically, and that the collisions are so frequent they can be treated as continuous. Prove that the retarding force has the form bv^2, and determine b. Assume that the front cone of the rocket subtends angle $\alpha = \pi/2$, as shown.

(*b*) Find the speed of the rocket in the cloud.

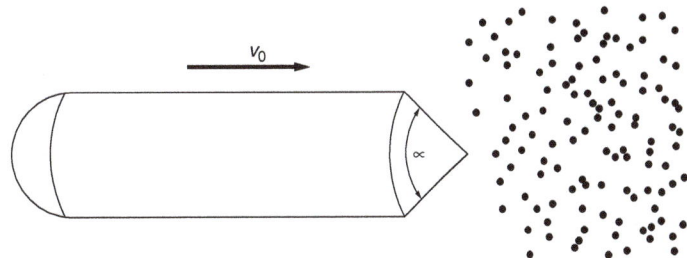

4.27　*Exoplanet detection*

The data plots in Example 4.6 show that with the methods then in use, a shift of 1 m/s in radial velocity of a star is just barely detectable. Could an astronomer on a far-off planet using these same methods detect that our Sun has a planet? The biggest effect would be due to Jupiter.

　　Use only the following data:

mass of the Sun = 1.99×10^{30} kg

mass of Jupiter = 1.90×10^{27} kg

mean radius of Jupiter's orbit = 7.8×10^8 km

period of Jupiter's orbit = 4330 days.

5 ENERGY

5.1 Introduction

In this chapter we make another attack on the fundamental problem of classical mechanics—predicting the motion of a system under known interactions. We shall encounter two important new concepts, energy and work, which first appear to be mere computational aids, mathematical crutches so to speak, but which turn out to possess deep physical significance.

At first glance there seems to be no problem in finding the motion of a particle if we know the force. Newton's second law tells us the acceleration, which we can integrate to find the velocity, and we can then integrate the velocity to find the position. This sounds simple but there is a problem: to carry out these calculations we need to know the force as a function of time, but force is usually known as a function of position as, for example, the spring force or the gravitational force. The problem is serious because physicists are generally interested in interactions between systems, which means knowing how the force varies with position, not how it varies with time.

The task, then, is to find $\mathbf{v}(t)$ from the equation

$$\frac{d\mathbf{v}(t)}{dt} = \mathbf{F}(\mathbf{r}),\qquad(5.1)$$

where the notation emphasizes that $\mathbf{F}$ is a known function of position. A physicist with a penchant for mathematical formalism might stop here and point out that what we are dealing with is a problem in differential equations and that what we ought to do now is study the methods available, including numerical methods, for solving such equations. This is perfectly reasonable from a calculational point of view but such an approach is too narrow to give us much physical understanding.

Fortunately, the solution to Eq. (5.1) is simple for the important case of one-dimensional motion in a single variable. The general case is more complex, but we shall see that it is nevertheless possible to integrate Eq. (5.1) for three-dimensional motion provided we are content with less than a complete solution. This will lead us to a very helpful physical relation, the work–energy theorem; its generalization, the law of conservation of energy, is one of the most useful conservation laws in physics.

5.2 Integrating Equations of Motion in One Dimension

A large class of important problems involves only a single variable to describe the motion, for example the one-dimensional harmonic oscillator. For such problems the equation of motion reduces to

$$m\frac{d^2x}{dt^2} = F(x)$$

or

$$m\frac{dv}{dt} = F(x).\qquad(5.2)$$

We can solve this equation for v by a mathematical trick. First, formally integrate $m\, dv/dt = F(x)$ with respect to x:

$$m \int_{x_a}^{x_b} \frac{dv}{dt} dx = \int_{x_a}^{x_b} F(x)\, dx. \tag{5.3}$$

The integral on the right can be evaluated by standard methods since $F(x)$ is known. The integral on the left is intractable as it stands, but it can be integrated by changing the variable from x to t, using differentials as discussed in Note 1.4:

$$dx = \left(\frac{dx}{dt}\right) dt$$
$$= v\, dt.$$

Then

$$m \int_{x_a}^{x_b} \frac{dv}{dt} dx = m \int_{t_a}^{t_b} \frac{dv}{dt} v\, dt$$
$$= m \int_{t_a}^{t_b} \frac{d}{dt}\left(\frac{1}{2}v^2\right) dt$$
$$= m \int_{t_a}^{t_b} d\left(\frac{1}{2}v^2\right)$$
$$= \frac{1}{2}mv^2 \Big|_{t_a}^{t_b}$$
$$= \tfrac{1}{2}mv_b^2 - \tfrac{1}{2}mv_a^2,$$

where $x_a \equiv x(t_a)$, $v_a \equiv v(t_a)$, etc.

Putting these results in Eq. (5.3) yields

$$\tfrac{1}{2}mv_b^2 - \tfrac{1}{2}mv_a^2 = \int_{x_a}^{x_b} F(x)\, dx. \tag{5.4}$$

Alternatively, we can use indefinite upper limits in Eq. (5.4):

$$\tfrac{1}{2}mv^2 - \tfrac{1}{2}mv_a^2 = \int_{x_a}^{x} F(x)\, dx, \tag{5.5}$$

where v is the speed of the particle when it is at position x. Equation (5.5) gives us v as a function of x and, as we shall see, this is enough for finding x as a function of t. Before carrying this through, let us look at how to solve a familiar problem using Eq. (5.4).

Example 5.1 Mass Thrown Upward Under Constant Gravity

A mass m is thrown vertically upward with initial speed v_0. How high does it rise, assuming the gravitational force to be constant, and neglecting air friction?

Taking the z axis to be directed vertically upward, $F = -mg$, Eq. (5.4) gives

$$\tfrac{1}{2}mv_1{}^2 - \tfrac{1}{2}mv_0{}^2 = \int_{z_0}^{z_1} F \, dz$$

$$= -mg \int_{z_0}^{z_1} dz$$

$$= -mg(z_1 - z_0).$$

At the peak, $v_1 = 0$, so

$$mg(z_1 - z_0) = \tfrac{1}{2}mv_0^2,$$

which gives

$$z_1 = z_0 + \frac{v_0{}^2}{2g}.$$

Our solution makes no explicit reference to time. Of course, we could have easily solved the problem by applying Newton's second law, but the solution would have been less elegant because we would have had to eliminate t to obtain the final answer.

Note that the result depends on two *initial conditions*, v_0 at z_0 and v_1 at z_1. This is a general property of Newton's second law calculations. In the language of differential equations, Newton's second law is a "second order" equation in the position; the highest order derivative it involves is the acceleration d^2x/dt^2. The theory of differential equations shows that the complete solution of a differential equation of nth order must involve n initial conditions.

Here is an example that is much simpler to solve by the energy method than by direct application of Newton's second law.

Example 5.2 Solving the Equation for Simple Harmonic Motion

In Section 3.7 we discussed the equation of simple harmonic motion and more or less pulled the solution out of a hat without proof. Now we shall derive the solution using Eq. (5.5).

Consider a mass M attached to a spring. Using the coordinate x measured from the equilibrium position, the spring force is $F = -kx$. Then Eq. (5.5) becomes

$$\tfrac{1}{2}Mv^2 - \tfrac{1}{2}Mv_0{}^2 = -k \int_{x_0}^{x} x \, dx$$

$$= -\tfrac{1}{2}kx^2 + \tfrac{1}{2}kx_0{}^2.$$

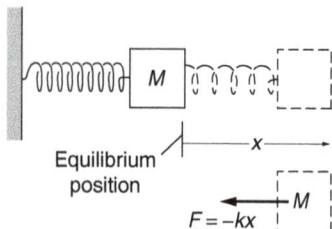

Equilibrium position

x

$F = -kx$

M

To complete the solution for x and v, we must specify initial conditions, because physically the equation of motion by itself cannot completely specify the motion for any given situation. We are free to choose any initial conditions we wish (as long as they are independent); a useful choice here is the position x_0 and the velocity v_0 at some time t_0. Let us consider the case where at $t = 0$ the mass is released from rest, so that $v_0 = 0$. If the mass is released at a distance x_0 from the origin,

$$v^2 = -\frac{k}{M}x^2 + \frac{k}{M}x_0{}^2.$$

Because $v = dx/dt$, we have

$$\frac{dx}{dt} = \sqrt{\frac{k}{M}}\sqrt{x_0{}^2 - x^2}.$$

This equation gives the velocity as a function of position, but what we really want in this problem is the position as a function of time. To accomplish this, we rearrange the equation and integrate again

$$\int_{x_0}^{x} \frac{dx}{\sqrt{x_0{}^2 - x^2}} = \sqrt{\frac{k}{M}} \int_{0}^{t} dt$$

$$= \sqrt{\frac{k}{M}}\, t.$$

The integral on the left-hand side is $\arcsin(x/x_0)$. The integral is listed in standard tables. It can also be generated by symbolic mathematical routines, or by the time-honored methods of "guesswork" and "playing around." Using these methods is as respectable for a physicist as consulting a dictionary is for a writer. Of course, in both cases one hopes that experience gradually reduces dependence.

Denoting $\sqrt{k/M}$ by ω, we obtain

$$\arcsin\left(\frac{x}{x_0}\right)\Big|_{x_0}^{x} = \omega t$$

or

$$\arcsin\left(\frac{x}{x_0}\right) - \arcsin(1) = \omega t.$$

Because $\arcsin(1) = \pi/2$, we obtain

$$x = x_0 \sin\left(\omega t + \frac{\pi}{2}\right)$$

$$= x_0 \cos \omega t.$$

Note that the solution indeed satisfies the given initial conditions: at $t = 0$, $x = x_0 \cos(0) = x_0$, and $v_0 = \dot{x} = x_0\,\omega\,\sin(0) = 0$. For these particular initial conditions our result agrees with the general solution $A \sin \omega t + B \cos \omega t$ given in Section 3.7.

5.3 Work and Energy

5.3.1 The Work–Energy Theorem in One Dimension

In Section 5.2 we demonstrated the formal procedure for integrating Newton's second law with respect to position. We shall now interpret the result

$$\tfrac{1}{2}mv_b{}^2 - \tfrac{1}{2}mv_a{}^2 = \int_{x_a}^{x_b} F(x)\,dx$$

in physical terms.

The quantity $\tfrac{1}{2}mv^2$ is called the *kinetic energy K*, and the left-hand side can be written $K_b - K_a$. The integral $\int_{x_a}^{x_b} F(x)\,dx$ is called the *work* W_{ba} by the force F on the particle as the particle moves from a to b. Our relation now takes the form

$$W_{ba} = K_b - K_a. \tag{5.6}$$

The result in Eq. (5.6) is known as the *work–energy theorem* or, more precisely, the work–energy theorem in one dimension. (We shall generalize to three dimensions shortly.) The unit of work and energy in the SI system is the *joule* (J):

$$1\,\text{N} \cdot \text{m} = 1\,\text{J} = 1\,\text{kg} \cdot \text{m}^2/\text{s}^2.$$

The unit of work and energy in the CGS system is the *erg*:

$$1\,\text{dyne} \cdot \text{cm} = 1\,\text{erg} = 1\,\text{gm} \cdot \text{cm}^2/\text{s}^2$$
$$= 10^{-7}\,\text{J}.$$

The unit of work in the English system is the *foot-pound:*

$$1\,\text{ft} \cdot \text{lb} \approx 1.356\,\text{J}.$$

A table of various other units employed to measure energy is given in Section 5.11.

Example 5.3 Vertical Motion in an Inverse Square Field

A mass m is shot vertically upward from the surface of the Earth with initial speed v_0. Assuming that the only force is gravity, find its maximum altitude and the minimum value of v_0 for the mass to escape the Earth completely.

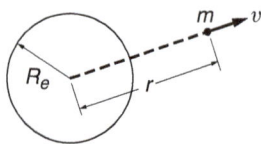

The force on m is

$$F = -\frac{GM_e m}{r^2}.$$

The problem is one dimensional in the variable r, and it is simple to find the kinetic energy at distance r by the work–energy theorem.

Let the particle start at $r = R_e$ with initial velocity v_0:

$$K(r) - K(R_e) = \int_{R_e}^{r} F(r)\,dr$$

$$= -GM_e m \int_{R_e}^{r} \frac{dr}{r^2}$$

or

$$\tfrac{1}{2}mv(r)^2 - \tfrac{1}{2}mv_0{}^2 = GM_e m\left(\frac{1}{r} - \frac{1}{R_e}\right).$$

We can immediately find the maximum height of m. At the highest point, $v(r) = 0$ and we have

$$v_0{}^2 = 2GM_e\left(\frac{1}{R_e} - \frac{1}{r_{\max}}\right).$$

It is good practice to simplify expressions by writing them in terms of familiar constants whenever possible. Because $g = GM_e/R_e{}^2$, we can write

$$v_0{}^2 = 2gR_e{}^2\left(\frac{1}{R_e} - \frac{1}{r_{\max}}\right)$$

$$= 2gR_e\left(1 - \frac{R_e}{r_{\max}}\right)$$

or

$$r_{\max} = \frac{R_e}{1 - v_0{}^2/2gR_e}. \tag{1}$$

The escape velocity from the Earth is the minimum initial velocity needed to move $r_{\max}$ to infinity. The escape velocity is therefore

$$v_{\text{escape}} = \sqrt{2gR_e}$$

$$= \sqrt{(2)(9.8 \text{ m/s}^2)(6.4 \times 10^6 \text{ m})}$$

$$= 1.1 \times 10^4 \text{ m/s}.$$

With this expression for v_{escape}, Eq. (1) can be written

$$r_{\max} = \frac{R_e}{1 - v_0{}^2/v_{\text{escape}}^2}. \tag{2}$$

For $v_0 = v_{\text{escape}}$, $r_{\max} \to \infty$, as we expect. But if $v_0 > v_{\text{escape}}$, Eq. (2) gives $r_{max} < 0$. The reason for this absurd result is that we assumed that the final speed of m is zero. If $v_0 > v_{\text{escape}}$, the mass never comes to rest.

From the work–energy theorem, Eq. (5.6), the minimum energy needed to send a 50-kg spacecraft from the surface of the Earth to infinity is

$$W = \tfrac{1}{2}Mv_{\text{escape}}^2$$

$$= \tfrac{1}{2}(50)(1.1 \times 10^4)^2 = 3.0 \times 10^9 \text{ J.}$$

5.3.2 Integrating Equations of Motion in Several Dimensions

Returning to the central problem of this chapter, let us try to integrate the equation of motion of a particle acted on by a force that depends on

position

$$\mathbf{F}(\mathbf{r}) = m\frac{d\mathbf{v}}{dt}. \qquad (5.7)$$

In the case of one-dimensional motion we integrated with respect to position. To generalize this, consider what happens when the particle moves a short distance $\Delta\mathbf{r}$.

We assume that $\Delta\mathbf{r}$ is so small that $\mathbf{F}$ is effectively constant over this displacement. If we take the scalar product of Eq. (5.7) with $\Delta\mathbf{r}$, we obtain

$$\mathbf{F} \cdot \Delta r = m\frac{d\mathbf{v}}{dt} \cdot \Delta\mathbf{r}. \qquad (5.8)$$

The sketch shows the trajectory and the force at some point along the trajectory. At this point, $\mathbf{F} \cdot \Delta\mathbf{r} = F\,\Delta r\,\cos\theta$.

Perhaps you are wondering how we know $\Delta\mathbf{r}$, since this requires knowing the trajectory, which is what we are trying to find. Let us overlook this problem for a few moments and pretend we know the trajectory.

In Eq. (5.8) the right-hand side is $m(d\mathbf{v}/dt) \cdot \Delta\mathbf{r}$. We can transform this by noting that $\mathbf{v}$ and $\Delta\mathbf{r}$ are not independent; for a sufficiently short length of path, $\mathbf{v}$ is approximately constant. Hence $\Delta\mathbf{r} = \mathbf{v}\,\Delta t$, where Δt is the time the particle requires to travel $\Delta\mathbf{r}$. Consequently,

$$m\frac{d\mathbf{v}}{dt} \cdot \Delta\mathbf{r} = m\frac{d\mathbf{v}}{dt} \cdot \mathbf{v}\Delta t. \qquad (5.9)$$

We can transform Eq. (5.9) with the vector identity $2\mathbf{A} \cdot d\mathbf{A}/dt = dA^2/dt$ that we proved in Section 1.10:

$$\mathbf{v} \cdot \frac{d\mathbf{v}}{dt} = \frac{1}{2}\frac{d}{dt}(v^2).$$

Equation (5.9) becomes

$$\mathbf{F} \cdot \Delta\mathbf{r} = \frac{m}{2}\frac{d}{dt}(v^2)\Delta t. \qquad (5.10)$$

The next step is to divide the entire trajectory from the initial position $\mathbf{r}_a$ to the final position $\mathbf{r}_b$ into N short segments of length $\Delta\mathbf{r}_j$, where j is an index numbering the segments. (When we take the limit $\Delta\mathbf{r} \to 0$ it will make no difference whether all the pieces have the same length.) For each segment we can write a relation similar to Eq. (5.10):

$$\mathbf{F}(\mathbf{r}_j) \cdot \Delta\mathbf{r}_j = \frac{m}{2}\frac{d}{dt}(v_j{}^2)\Delta t_j,$$

where $\mathbf{r}_j$ is the location of segment j, $\mathbf{v}_j$ is the velocity the particle has there, and Δt_j is the time it spends in traversing it. If we add together the equations of all the segments, we have

$$\sum_{j=1}^{N} \mathbf{F}(\mathbf{r}_j) \cdot \Delta\mathbf{r}_j = \sum_{j=1}^{N} \frac{m}{2}\frac{d}{dt}(v_j{}^2)\Delta t_j.$$

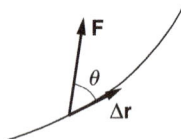

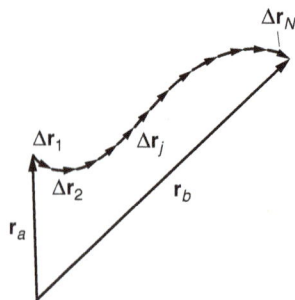

Next we take the limit where the length of each segment approaches zero, and the number of segments approaches infinity. In the limit the sums become integrals, and we have

$$\int_{\mathbf{r}_a}^{\mathbf{r}_b} \mathbf{F}(\mathbf{r}) \cdot d\mathbf{r} = \int_{t_a}^{t_b} \frac{m}{2} \frac{d}{dt}(v^2) \, dt, \qquad (5.11)$$

where t_a and t_b are the times corresponding to $\mathbf{r}_a$ and $\mathbf{r}_b$. In converting the sum to an integral, we dropped the numerical index j. The location of the first segment $\Delta \mathbf{r}_1$ is indicated by $\mathbf{r}_a$, and the location of the last section $\Delta \mathbf{r}_N$ by $\mathbf{r}_b$.

In Eq. (5.11) we can evaluate the integral on the right by a now-familiar procedure.

$$\begin{aligned}
\int_{\mathbf{r}_a}^{\mathbf{r}_b} \mathbf{F}(\mathbf{r}) \cdot d\mathbf{r} &= \int_{t_a}^{t_b} \frac{m}{2} \frac{d}{dt}(v^2) \, dt, \\
&= \frac{m}{2} \int_{t_a}^{t_b} \frac{d}{dt}(v^2) \, dt \\
&= \tfrac{1}{2} m v^2 \Big|_{t_a}^{t_b} \\
&= \tfrac{1}{2} m v_b^2 - \tfrac{1}{2} m v_a^2.
\end{aligned}$$

This represents a simple generalization of the result we found for one dimension, Eq. (5.4). Here, however, $v^2 = v_x^2 + v_y^2 + v_z^2$, while for the one-dimensional case we had $v^2 = v_x^2$.

Equation (5.11) becomes

$$\int_{\mathbf{r}_a}^{\mathbf{r}_b} \mathbf{F} \cdot d\mathbf{r} = \tfrac{1}{2} m v_b^2 - \tfrac{1}{2} m v_a^2. \qquad (5.12)$$

The integral on the left is called a *line integral*, because the integration is carried out along a path. We shall see how to evaluate line integrals in the next two sections, and we shall also see how to interpret Eq. (5.12) physically. However, before proceeding, let's pause for a moment to summarize.

Our starting point was $\mathbf{F}(\mathbf{r}) = m \, d\mathbf{v}/dt$. All we have done is to integrate this equation with respect to distance, but because we described each step carefully, it looks like many operations are involved. This is not really the case; the whole argument can be stated in a few lines as follows:

$$\mathbf{F} = m \frac{d\mathbf{v}}{dt}$$

$$\begin{aligned}
\int_a^b \mathbf{F} \cdot d\mathbf{r} &= \int_a^b m \frac{d\mathbf{v}}{dt} \cdot d\mathbf{r} \\
&= \int_a^b m \frac{d\mathbf{v}}{dt} \cdot \mathbf{v} \, dt \\
&= \int_a^b \frac{m}{2} \frac{d}{dt}(v^2) \, dt \\
&= \tfrac{1}{2} m v_b^2 - \tfrac{1}{2} m v_a^2.
\end{aligned}$$

5.3.3 The Work–Energy Theorem

We now state Eq. (5.12) in the language of physics. The quantity $\frac{1}{2}mv^2$ is called the *kinetic energy K*, so the right-hand side of Eq. (5.12) can be written as $K_b - K_a$. The integral $\int_{\mathbf{r}_a}^{\mathbf{r}_b} \mathbf{F} \cdot d\mathbf{r}$ is called the *work W_{ba}* by the force $\mathbf{F}$ on the particle as the particle moves from a to b. Equation (5.12) now takes the form

$$W_{ba} = K_b - K_a. \tag{5.13}$$

This result is the general statement of the *work–energy theorem* which we met in restricted form, Eq. (5.6), in our discussion of one dimensional motion.

Recall that the work ΔW by a force $\mathbf{F}$ in a small displacement $\Delta \mathbf{r}$ is

$$\Delta W = \mathbf{F} \cdot \Delta \mathbf{r} = F \cos \theta \Delta r = F_\parallel \Delta r,$$

where $F_\parallel = F \cos \theta$ is the component of $\mathbf{F}$ along the direction of $\Delta \mathbf{r}$. The component of $\mathbf{F}$ perpendicular to $\Delta \mathbf{r}$ does no work. For a finite displacement from $\mathbf{r}_a$ to $\mathbf{r}_b$, the work on the particle, $\int_a^b \mathbf{F} \cdot d\mathbf{r}$, is the sum of the contributions $\Delta W = F_\parallel \Delta r$ from each segment of the path, in the limit where the size of each segment approaches zero.

In the work–energy theorem Eq. (5.13), W_{ba} is the work on the particle by the total force $\mathbf{F}$. If $\mathbf{F}$ is the sum of several forces $\mathbf{F} = \Sigma \mathbf{F}_i$, we can write

$$W_{ba} = \sum_i (W_i)_{ba}$$
$$= K_b - K_a,$$

where

$$(W_i)_{ba} = \int_{\mathbf{r}_a}^{\mathbf{r}_b} \mathbf{F}_i \cdot d\mathbf{r}$$

is the work by the ith force $\mathbf{F}_i$.

Our discussion so far has been restricted to the case of a single particle. We showed in Section 4.3 that the center of mass of an extended system moves according to the equation of motion

$$\mathbf{F} = M\ddot{\mathbf{R}}$$
$$= M\frac{d\mathbf{V}}{dt}, \tag{5.14}$$

where $\mathbf{V} = \dot{\mathbf{R}}$ is the velocity of the center of mass. Integrating Eq. (5.14) with respect to position gives

$$\int_{\mathbf{R}_a}^{\mathbf{R}_b} \mathbf{F} \cdot d\mathbf{R} = \frac{1}{2}MV_b^2 - \frac{1}{2}MV_a^2, \tag{5.15}$$

where $d\mathbf{R} = \mathbf{V}\,dt$ is the displacement of the center of mass in time dt.

Equation (5.15) is the work–energy theorem for the translational motion of an extended system. Later, we shall see several ways of doing work on a system, for instance work that causes the system to rotate or

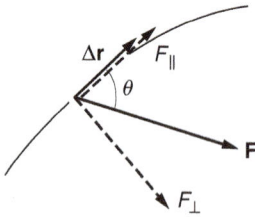

to get warmer. Nevertheless, Eq. (5.15) always holds for the center of mass motion.

Example 5.4 The Conical Pendulum

We discussed the motion of the conical pendulum in Example 2.10. Since the mass moves with constant angular velocity ω in a circle of constant radius R, the kinetic energy of the mass, $\frac{1}{2}m(R\omega)^2$, is constant. The work–energy theorem then tells us that no net work is being done on the mass.

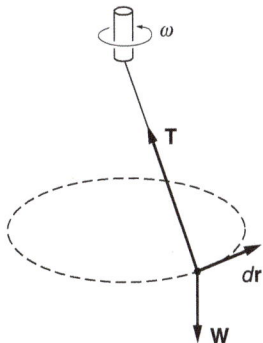

Analyzing the situation in more detail, the string force and the weight force both act on m. However, each of these forces is perpendicular to the circular trajectory, making the integrand of the work integral zero. Consequently, the total work on m is zero, and the kinetic energy is constant.

It is important to realize that in the work integral $\int \mathbf{F} \cdot d\mathbf{r}$, the vector $d\mathbf{r}$ is along the path of the particle. Since $\mathbf{v} = d\mathbf{r}/dt$, $d\mathbf{r} = \mathbf{v}\,dt$ and $d\mathbf{r}$ is always parallel to $\mathbf{v}$.

Example 5.5 Escape Velocity—the General Case

In Example 5.3 we discussed the one-dimensional motion of a mass m projected vertically upward from the Earth. We found that if the initial speed is greater than $v_0 = \sqrt{2gR_e}$, the mass will escape from the Earth. Now we look at the problem once again, but allow the mass to be projected at angle α from the vertical.

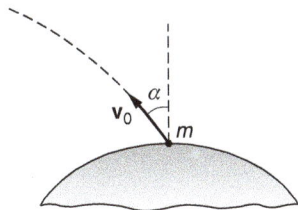

The force on m, neglecting air resistance, is

$$\mathbf{F} = -\frac{GM_e m}{r^2}\hat{\mathbf{r}}$$
$$= -mg\frac{R_e^2}{r^2}\hat{\mathbf{r}},$$

where $g = GM_e/R_e^2$ is the acceleration due to gravity at the Earth's surface. We do not know the trajectory of the particle without solving the problem in detail, but for any element of the path the displacement $d\mathbf{r}$ can be written

$$d\mathbf{r} = dr\,\hat{\mathbf{r}} + r\,d\theta\,\hat{\boldsymbol{\theta}}.$$

Because $\hat{\mathbf{r}} \cdot \hat{\boldsymbol{\theta}} = 0$, we have

$$\mathbf{F} \cdot d\mathbf{r} = -mg\frac{R_e^2}{r^2}\hat{\mathbf{r}} \cdot (dr\,\hat{\mathbf{r}} + r\,d\theta\,\hat{\boldsymbol{\theta}})$$
$$= -mg\frac{R_e^2}{r^2}dr.$$

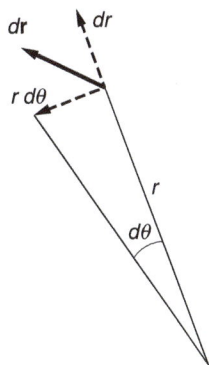

The work–energy theorem becomes

$$\tfrac{1}{2}mv^2 - \tfrac{1}{2}mv_0^2 = -mgR_e^2 \int_{R_e}^{r} \frac{dr}{r^2}$$

$$= -mgR_e^2 \left(\frac{1}{r} - \frac{1}{R_e} \right).$$

The escape velocity is the minimum value of v_0 for which $v = 0$ when $r \to \infty$. We find

$$v_{escape} = \sqrt{2gR_e}$$

$$= 1.1 \times 10^4 \text{ m/s},$$

the same result as in Example 5.3. In the absence of air friction, the escape velocity is independent of the launch direction, a result that may not be intuitively obvious.

We have neglected the Earth's rotation in our analysis. In the absence of air resistance the projectile should be fired horizontally to the east, since the rotational speed of the Earth's surface is then added to the launch velocity. This is the reason satellites in the U.S. are usually launched in a trajectory toward the east from Florida, the part of the U.S. nearest the Equator, where the tangential speed is greatest. Similarly, European satellites are often launched from French Guiana in South America, a location only a few degrees north of the Equator.

5.3.4 Power

Power is the rate at which work is done. The work ΔW by force $\mathbf{F}$ on a system as it moves through a short distance $\Delta \mathbf{r}$ is $\mathbf{F} \cdot \Delta \mathbf{r}$. If the displacement takes place during time Δt, then the rate of work is

$$\frac{\Delta W}{\Delta t} \approx \mathbf{F} \cdot \frac{\Delta \mathbf{r}}{\Delta t}.$$

In the limit $\Delta t \to 0$, $\Delta \mathbf{r}/\Delta t \to \mathbf{v}$, so we have

$$\frac{dW}{dt} = \mathbf{F} \cdot \mathbf{v}. \tag{5.16}$$

Power can be either positive or negative depending on whether the work is *on* or *by* the system. The SI unit of power is the *watt* (W); 1 W = 1 joule/s = 1 kg $\cdot$ m^2/s^3. Many other units are used for non-scientific purposes, for example the *horsepower* to describe the power of machinery and automobiles. There are several slightly different definitions of the horsepower, depending on the application, but it is generally taken to be 746 W. Some other units of power are summarized in Section 5.11.

Example 5.6 Empire State Building Run-Up

Hundreds of athletes compete in early February to be the fastest climber up 1576 stairs to the 86th floor deck of the Empire State Building in New York City, a vertical distance of $h = 320$ m. In this event, first held in 1978, the winning time is typically about 10 minutes. Assuming a body mass $m = 75$ kg, what is the average power the winner exerts?

The work W lifting 75 kg through a height of 320 m is

$$W = mgh = 75 \text{ kg} \times 9.80 \text{ m/s}^2 \times 320 \text{ m} = 2.35 \times 10^5 \text{ J}$$

If this work is done in 10 minutes = 600 s, the average power is

$$\overline{\text{power}} = (2.35 \times 10^5 \text{ J})/600 \text{ s} = 392 \text{ J/s} = 392 \text{ W}$$

The stair climber exerted an average power of 392 W/746 W/hp = 0.53 hp. A human being in good condition can exert close to 1 hp for a short time, for example racing up a few flights of stairs.

5.3.5 Applying the Work–Energy Theorem

In the last section we derived the work–energy theorem

$$W_{ba} = K_b - K_a$$

and applied it to a few simple cases. In this section we shall use it to tackle more complicated problems. But before we start, a few comments about the theorem may be helpful.

To begin, we should emphasize that the work–energy theorem is a mathematical consequence of Newton's second law; we have introduced no new physical ideas. The work–energy theorem is merely the statement that the change in kinetic energy is equal to the net work. This should not be confused with the general law of conservation of energy, an independent physical law that we shall discuss in Sections 5.9 and 5.10.

Possibly you are troubled by the following problem: to apply the work–energy theorem, we have to evaluate the work along some possibly curved path:

$$W_{ba} = \oint_a^b \mathbf{F} \cdot d\mathbf{r}$$

Such an integral is known as a *line integral*, because the integral is to be evaluated along a specific curve, or path, from a to b. The C on the integral sign is the symbol for a line integral. But evaluating this integral requires knowing the path the particle actually follows. We would seem to need to know the solution in order to apply the theorem, making it difficult to see how the work–energy theorem could be useful.

The work–energy theorem is indeed not particularly useful if the work actually depends on the path. Fortunately, the theorem is extremely useful in two cases that happen to be of considerable importance. For many forces of interest, the work integral does not depend on the particular path but only on the end points. Such forces are called *conservative* forces; they include many of the important forces in physics. As we shall see, the work–energy theorem assumes a marvelously simple form when the forces are conservative.

The work–energy theorem is also useful in cases where the path is known because the motion is *constrained*. By constrained motion, we mean motion in which external constraints act to keep the particle on a predetermined trajectory. The roller coaster is a perfect example. A roller coaster follows the track because it is held on by wheels both below and above the track. There are many other examples of constrained motion—the conical pendulum, for example, is constrained by the fixed length of the pendulum—but all have one feature in common—the constraining force does no work. This is because the effect of the constraint force is to assure that the direction of the velocity is always tangential to the predetermined path. Hence the constraint forces change only the direction of **v**. Thus the constraint force $\mathbf{F_c}$ is normal to the velocity **v**. On the other hand, the displacement $\Delta\mathbf{r}$ is parallel to **v**. Consequently, $\mathbf{F} \cdot \Delta\mathbf{r} = \mathbf{0}$ and the constraint force does no work.

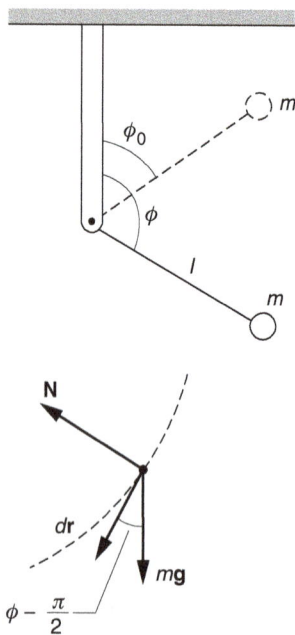

Example 5.7 The Inverted Pendulum

A pendulum consists of a light rigid rod of length l, pivoted at one end, with mass m attached at the other end. The pendulum is released from rest at angle ϕ_0, as shown. What is the velocity of m when the rod is at angle ϕ?

The work–energy theorem gives

$$\tfrac{1}{2}mv(\phi)^2 - \tfrac{1}{2}mv_0^2 = W_{\phi,\phi_0}.$$

Because $v_0 = 0$, we have

$$v(\phi) = \sqrt{\frac{2W_{\phi,\phi_0}}{m}}. \tag{1}$$

To evaluate W_{ϕ,ϕ_0}, the work by gravity as the bob swings from ϕ_0 to ϕ, we see that $d\mathbf{r}$ lies along the circle of radius l.

The forces acting are gravity, directed down, and the force of the rod, **N**. Since **N** lies along the radius, $\mathbf{N} \cdot d\mathbf{r} = 0$, and **N** does no work. The work by gravity is

$$m\mathbf{g} \cdot d\mathbf{r} = mgl \, \cos\left(\phi - \frac{\pi}{2}\right)d\phi$$

$$= mgl \sin\phi \, d\phi$$

where we have used $|d\mathbf{r}| = l\,d\phi$:

$$W_{\phi,\phi_0} = \int_{\phi_0}^{\phi} mgl \, \sin \, \phi \, d\phi$$

$$= -mgl \, \cos \phi \Big|_{\phi_0}^{\phi}$$

$$= mgl \, (\cos \phi_0 - \cos \phi).$$

From Eq. (1), the speed at ϕ is therefore

$$v(\phi) = \sqrt{2gl \, (\cos \, \phi_0 - \cos \, \phi)}.$$

The maximum velocity is obtained by letting the pendulum fall from the top $\phi_0 = 0$ to the bottom $\phi = \pi$:

$$v_{\max} = 2\sqrt{gl}.$$

This is the same speed attained by a mass falling through the same vertical distance $2l$. However, the mass on the pendulum is traveling horizontally at the bottom of its path, not vertically.

To convince yourself of the utility of the work–energy theorem, you might try solving this example by integrating the equation of motion. You will find that using the work–energy theorem is much easier.

Example 5.7 illustrates not only the utility but also one of the short-comings of the method: although we found a simple solution for the speed of the mass at any point on the circle, we have no information on *when* the mass gets there. For instance, if the pendulum is released at $\phi_0 = 0$, in principle the mass balances there forever, never reaching the bottom. Fortunately, in many problems we are not interested in time. When time is important, the work–energy theorem can provide a valuable first step toward a complete solution, as we shall see in the next section.

Next we turn to the general problem of evaluating work by a known force over a given path, which involves evaluating line integrals. We start by looking at the case of a constant force.

Example 5.8 Work by a Uniform Force

The case of a constant force is particularly simple. Here is how to find the work by a force $\mathbf{F} = F_0\hat{\mathbf{n}}$, where F_0 is a constant and $\hat{\mathbf{n}}$ is a unit vector in some given direction, as the particle moves from $\mathbf{r}_a$ to $\mathbf{r}_b$ along an arbitrary path. All the steps are put in to make the procedure clear, but with any practice this problem can be solved by

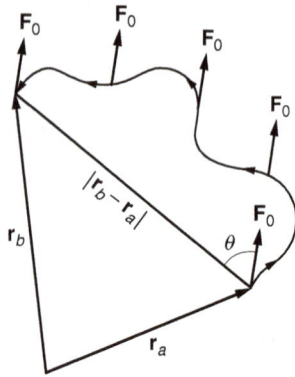

inspection.

$$W_{ba} = \oint_{\mathbf{r}_a}^{\mathbf{r}_b} \mathbf{F} \cdot d\mathbf{r}$$

$$= \oint_{\mathbf{r}_a}^{\mathbf{r}_b} F_0 \hat{\mathbf{n}} \cdot d\mathbf{r}$$

$$= F_0 \hat{\mathbf{n}} \cdot \oint_{\mathbf{r}_a}^{\mathbf{r}_b} d\mathbf{r}$$

$$= F_0 \hat{\mathbf{n}} \cdot \left(\hat{\mathbf{i}} \int_{x_a,y_a,z_a}^{x_b,y_b,z_b} dx + \hat{\mathbf{j}} \int_{x_a,y_a,z_a}^{x_b,y_b,z_b} dy + \hat{\mathbf{k}} \int_{x_a,y_a,z_a}^{x_b,y_b,z_b} dz \right)$$

$$= F_0 \hat{\mathbf{n}} \cdot [\hat{\mathbf{i}}(x_b - x_a) + \hat{\mathbf{j}}(y_b - y_a) + \hat{\mathbf{k}}(z_b - z_a)]$$

$$= F_0 \hat{\mathbf{n}} \cdot (\mathbf{r}_b - \mathbf{r}_a)$$

$$= F_0 \cos\theta \, |\mathbf{r}_b - \mathbf{r}_a| \, .$$

This result shows that for a constant force the work depends only on the net displacement, $\mathbf{r}_b - \mathbf{r}_a$, not on the particular path followed. Such a simple result is not always the case, but it holds true for an important class of forces, conservative forces.

We can use the results of this example to illustrate a characteristic feature of conservative forces. Suppose we return from b to a, but along a different path. The work is W_{ab}, and proceeding as above we find

$$W_{ab} = F_0 \hat{\mathbf{n}} \cdot (\mathbf{r}_a - \mathbf{r}_b)$$

$$= -W_{ba}.$$

It follows that $W_{ba} + W_{ab} = 0$; the work by F_0 around a closed path is zero. We shall have more to say about this property later.

As the next example shows, the work by a central force also depends only on the end points, and not on the particular path followed.

Example 5.9 Work by a Central Force

A *central force* is a radial force that depends only on the distance from the origin. Let us find the work by the central force $\mathbf{F} = f(r)\hat{\mathbf{r}}$ on a particle that moves from $\mathbf{r}_a$ to $\mathbf{r}_b$. For simplicity we shall consider motion in a plane, for which $d\mathbf{r} = dr\,\hat{\mathbf{r}} + r\,d\theta\,\hat{\boldsymbol{\theta}}$. Then

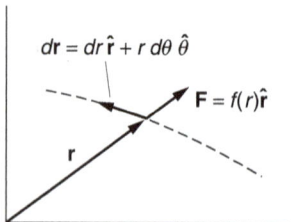

$$W_{ba} = \oint_a^b \mathbf{F} \cdot d\mathbf{r}$$

$$= \oint_a^b f(r)\hat{\mathbf{r}} \cdot (dr\,\hat{\mathbf{r}} + r\,d\theta\,\hat{\boldsymbol{\theta}})$$

$$= \int_a^b f(r)\,dr.$$

The work is given by a simple one-dimensional integral over the variable r. Because θ has disappeared from the problem, it should be obvious that the work for a given $f(r)$ depends only on the initial and final radial distances and not on the particular path. It follows that $W_{ba} + W_{ab} = 0$; the work by a central force around a closed path is zero.

For some forces, the work depends on the particle's path between the initial and final points. A familiar example is work by the force of sliding friction. Here the force always opposes the motion, so that the work by friction in moving through distance dS is $dW = -f\,dS$, where f is the magnitude of the friction force. If we assume that f is constant, then the work by friction in going from $\mathbf{r}_a$ to $\mathbf{r}_b$ along some path is

$$W_{ba} = -\oint_{\mathbf{r}_a}^{\mathbf{r}_b} f\,dS$$
$$= -fS,$$

where S is the total length of the path. The work is negative because the force always retards the particle. W_{ba} is never smaller in magnitude than fS_0, where S_0 is the straight-line distance between the two points, but by choosing a sufficiently devious route, S can be made arbitrarily large.

Example 5.10 A Path-dependent Line Integral

Here is a second example of a path-dependent line integral. Let $\mathbf{F} = A(xy\hat{\mathbf{i}} + y^2\hat{\mathbf{j}})$. The force $\mathbf{F}$ has no particular physical significance.

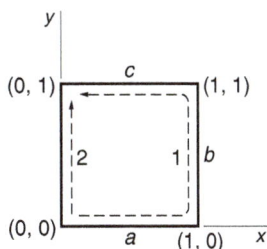

Consider the integral from (0,0) to (0,1), first along path 1 and then along path 2, as shown in the figure.

The segments of each path lie along a coordinate axis, so evaluating the integrals is simple. For path 1 we have

$$\oint_1 \mathbf{F} \cdot d\mathbf{r} = \int_a \mathbf{F} \cdot d\mathbf{r} + \int_b \mathbf{F} \cdot d\mathbf{r} + \int_c \mathbf{F} \cdot d\mathbf{r}.$$

Along segment a, $d\mathbf{r} = dx\,\hat{\mathbf{i}}$, $\mathbf{F} \cdot d\mathbf{r} = F_x\,dx = Axy\,dx$. Since $y = 0$ along the line of this integration, $\int_a \mathbf{F} \cdot d\mathbf{r} = 0$. For path b,

$$\int_b \mathbf{F} \cdot d\mathbf{r} = A \int_{x=1,y=0}^{x=1,y=1} y^2\,dy$$
$$= \frac{A}{3},$$

while for path c, where $y = 1$,

$$\int_c \mathbf{F} \cdot d\mathbf{r} = A \int_{x=1,y=1}^{x=0,y=1} xy\,dx$$
$$= A \int_1^0 x\,dx = -\frac{A}{2}.$$

Thus

$$\oint_1 \mathbf{F} \cdot d\mathbf{r} = \frac{A}{3} - \frac{A}{2}$$

$$= -\frac{A}{6}.$$

Along path 2 we have

$$\oint_2 \mathbf{F} \cdot d\mathbf{r} = A \int_{x=0,y=0}^{x=0,y=1} y^2 \, dy$$

$$= \frac{A}{3}$$

$$\neq \oint_1 \mathbf{F} \cdot d\mathbf{r}.$$

The work by the applied force is different for the two paths.

In the general case, the path of a line integral lies along some arbitrary curve and not conveniently along coordinate axes. The following general method of evaluating a line integral can be used if all else fails.

For simplicity we again consider motion in a plane. Generalization to three dimensions is straightforward.

The problem is to evaluate $\oint_a^b \mathbf{F} \cdot d\mathbf{r}$ along a specified path. The path can be characterized by an equation of the form $g(x, y) = 0$. For example, if the path is a unit circle about the origin, then all points on the path obey $x^2 + y^2 - 1 = 0$.

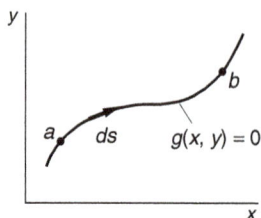

We can characterize every point on the path by a parameter s which in a practical problem could be for example distance along the path, or angle—anything just as long as each point on the path is associated with a value of s so that we can write $x = x(s), y = y(s)$. If we move along the path a short way, so that s changes by the amount ds, then the change in x is $dx = (dx/ds)ds$, and the change in y is $dy = (dy/ds)ds$. Since both x and y are determined by s, so are F_x and F_y. Hence we can write $\mathbf{F} = F_x(s)\hat{\mathbf{i}} + F_y(s)\hat{\mathbf{j}}$, and we have

$$\oint_a^b \mathbf{F} \cdot d\mathbf{r} = \int_a^b (F_x dx + F_y \, dy)$$

$$= \int_{s_a}^{s_b} \left[F_x(s)\frac{dx}{ds} + F_y(s)\frac{dy}{ds} \right] ds.$$

We have reduced the problem to the more familiar problem of evaluating a one-dimensional definite integral. The calculation is much simpler in practice than in theory. Here is an example.

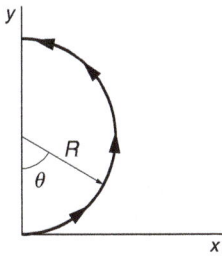

Example 5.11 Parametric Evaluation of a Line Integral

Evaluate the line integral of $\mathbf{F} = A(x^3\hat{\mathbf{i}} + xy^2\hat{\mathbf{j}})$ from $(x = 0, y = 0)$ to $(x = 0, y = 2R)$ along the semicircle shown.

The natural parameter to use here is θ, since as θ varies from 0 to π, the radius vector sweeps out the semicircle. We have

$$
\begin{aligned}
x &= R\sin\theta & y &= R(1 - \cos\theta) \\
dx &= R\cos\theta\, d\theta & dy &= R\sin\theta\, d\theta \\
F_x &= AR^3\sin^3\theta & F_y &= AR^3\sin\theta(1 - \cos\theta)^2
\end{aligned}
$$

$$
\oint \mathbf{F} \cdot d\mathbf{r} = AR^4 \int_0^\pi [(\sin^3\theta)\cos\theta + \sin\theta(1 - \cos\theta)^2 \sin\theta]\, d\theta.
$$

Evaluation of the integral is straightforward. If you are interested in carrying it through, try substituting $u = \cos\theta$.

5.4 The Conservation of Mechanical Energy

Conservative forces, where the work by the force along a path depends only on the end points, play a major role in physics. We have seen two examples of conservative forces: the uniform force (Example 5.8) and the central force (Example 5.9).

The work by a conservative force along any path from a to b is

$$
\oint_{\mathbf{r}_a}^{\mathbf{r}_b} \mathbf{F} \cdot d\mathbf{r} = \text{function of } (\mathbf{r}_b) - \text{function of } (\mathbf{r}_a)
$$

or

$$
\oint_{\mathbf{r}_a}^{\mathbf{r}_b} \mathbf{F} \cdot d\mathbf{r} = -U(\mathbf{r}_b) + U(\mathbf{r}_a), \tag{5.17}
$$

where $U(\mathbf{r})$ is a function, defined by the above expression, known as the *potential energy* function. (The reason for the sign convention will be clear in a moment.) We have not proven in general that $U(\mathbf{r})$ exists, but we have seen examples of forces where the work is indeed path-independent, so we know that U exists for at least some forces.

For a conservative force, the work–energy theorem $W_{ba} = K_b - K_a$ becomes

$$
\begin{aligned}
W_{ba} &= -U_b + U_a \\
&= K_b - K_a
\end{aligned}
$$

or, rearranging,

$$
K_a + U_a = K_b + U_b. \tag{5.18}
$$

The left-hand side of this equation, $K_a + U_a$, depends on the speed of the particle and its potential energy at $\mathbf{r}_a$, without reference to $\mathbf{r}_b$. Similarly, the right-hand side depends on the speed and potential energy at $\mathbf{r}_b$, without reference to $\mathbf{r}_a$. Because $\mathbf{r}_a$ and $\mathbf{r}_b$ are arbitrary and not specially chosen points, this can be true only if each side of the equation equals a constant. Denoting this constant by E, we have

$$K_a + U_a = K_b + U_b = E. \tag{5.19}$$

E is called the *total mechanical energy* of the particle, or, less precisely, the total energy.

We have shown that if the force is conservative, the total energy is independent of the position of the particle. In such a case, the total energy remains constant, or, in the language of physics, the energy is *conserved*. Although the conservation of mechanical energy is a derived law, which means that it has basically no new physical content, it presents such a different way of looking at a physical process compared with applying Newton's laws that we have what amounts to a totally new tool. Furthermore, although the conservation of mechanical energy follows directly from Newton's laws, it is an important key to understanding the more general law of conservation of energy, which is independent of Newton's laws and which vastly increases our understanding of nature. When we discuss this in greater detail in Sections 5.9 and 5.10, we shall see that the conservation law for mechanical energy turns out to be a special case of the more general law.

A peculiar property of energy is that the value of E is arbitrary; only *changes* in E have physical significance. This comes about because the equation

$$U_b - U_a = -\oint_a^b \mathbf{F} \cdot d\mathbf{r} \tag{5.20}$$

defines only the difference in potential energy between a and b and not the potential energy itself. We could add an arbitrary constant to U_b and the same constant to U_a and still satisfy the defining equation. However, since $E = K + U$, adding an arbitrary constant to U increases E by the same amount.

As a corollary, Eq. (5.20) implies that the work by a conservative force $\mathbf{F}$ around a closed path is zero:

$$\oint \mathbf{F} \cdot d\mathbf{r} = 0. \tag{5.21}$$

The circle on the integral sign signifies a closed path.

The following example illustrates the new perspective that energy methods bring to solving dynamical problems.

Example 5.12 Energy Solution to a Dynamical Problem

To illustrate the power of the energy method, we solve an old problem a new way using energy methods. The problem is the motion of a pendulum, which we solved using Newton's laws in Example 3.10.

The work on mass m by the gravitational force $-m\mathbf{g}$ as m moves from $y = 0$ to y is $-mgy$, and so $U(y) - U(0) = mgy$. Consequently, the total

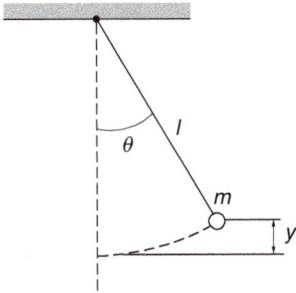

energy of the pendulum shown in the sketch is

$$E = K + U$$
$$= \tfrac{1}{2}ml^2\dot{\theta}^2 + mgy,$$

where l is the length of the pendulum and $y = l(1 - \cos\theta)$.

It is easy to evaluate E at the end of the swing because there $\theta = \theta_0$ and $\dot{\theta} = 0$. The total energy is $E = mgy = mgl(1 - \cos\theta_0)$ and the energy equation becomes

$$\tfrac{1}{2}ml^2\dot{\theta}^2 + mgl(1 - \cos\theta) = mgl(1 - \cos\theta_0).$$

Solving for $d\theta/dt$ we have

$$\frac{d\theta}{dt} = \sqrt{\frac{2g}{l}(\cos\theta - \cos\theta_0)},$$

which can be rearranged to give

$$\int \frac{d\theta}{\sqrt{\cos\theta - \cos\theta_0}} = \sqrt{\frac{2g}{l}} \int dt. \qquad (1)$$

Let us look at the solution for the case of small amplitude where we can make the small-angle approximation $\cos\theta \approx 1 - \tfrac{1}{2}\theta^2$. We obtain

$$\int \frac{d\theta}{\sqrt{\tfrac{1}{2}}\sqrt{\theta_0^2 - \theta^2}} = \sqrt{\frac{2g}{l}} \int dt$$

which we can rewrite as

$$\int \frac{d\theta/\theta_0}{\sqrt{1 - (\theta/\theta_0)^2}} = \omega \int dt$$

where we have introduced $\omega = \sqrt{g/l}$. The integral on the left has the form $\int dx/\sqrt{1 - x^2} = \arcsin x$, where $x = \theta/\theta_0$. We take the lower limits of the integrals to be $(\theta = 0, t = 0)$ and the upper limits to be (θ, t). The result is

$$\arcsin \theta/\theta_0 - 0 = \sqrt{\frac{g}{l}}(t - 0),$$

$$\theta = \theta_0 \sin\omega t.$$

Note that the energy method does not require finding a general solution for the motion and then fitting it to boundary conditions: the boundary conditions are built in. More importantly, Eq. (1) is a general equation that is not limited to the small-angle approximation. It has a mathematically exact solution in terms of functions called *elliptic integrals*, but without going into that complexity we can use Eq. (1) to find an important result: the correction to the period of a pendulum due to its finite amplitude. Such a correction would be very difficult to extract starting with the Newtonian equation of motion. The correction is worked out in Note 5.1.

5.5 Potential Energy

The concept of potential energy was introduced in the last section. Its definition is embodied in Eq. (5.20). Here are examples that illustrate potential energy for three conservative forces: the uniform force, the central force, and the spring force.

Example 5.13 Potential Energy of a Uniform Force Field

From Example 5.8, the work by a uniform force is $W_{ba} = \mathbf{F}_0 \cdot (\mathbf{r}_b - \mathbf{r}_a)$. For instance, the force on a particle of mass m due to a uniform gravitational field is $-mg\hat{\mathbf{k}}$, so if the particle moves from $\mathbf{r}_a$ to $\mathbf{r}_b$, the change in potential energy is

$$U_b - U_a = -\int_{z_a}^{z_b} (-mg)\,dz$$

$$= mg(z_b - z_a).$$

If we adopt the convention $U = 0$ at ground level where $z = 0$, then $U(h) = mgh$, where h is the height above the ground. However, a potential energy of the form $mgh + C$, where C is any constant, is just as suitable.

As an application, suppose mass m is projected upward with initial velocity $\mathbf{v}_0 = v_{0x}\hat{\mathbf{i}} + v_{0y}\hat{\mathbf{j}} + v_{0z}\hat{\mathbf{k}}$. Find the speed at height h using conservation of energy.

$$K_0 + U_0 = K(h) + U(h)$$

$$\tfrac{1}{2}mv_0^2 + 0 = \tfrac{1}{2}mv^2(h) + mgh$$

or

$$v(h) = \sqrt{v_0^2 - 2gh}.$$

Example 5.13 is trivial because motion in a uniform force field is easily found from $\mathbf{F} = m\mathbf{a}$. Nevertheless, it illustrates the ease with which the energy method solves the problem, where motion in all three directions is handled at once. In contrast, Newton's law involves three equations, one for each component of motion.

Example 5.14 Potential Energy of a Central Force

A central force, which is always conservative, has the general form $\mathbf{F} = f(r)\hat{\mathbf{r}}$, where $f(r)$ is some function of the distance to the origin. The potential energy of a particle in a central force is

$$U_b - U_a = -\int_{\mathbf{r}_a}^{\mathbf{r}_b} \mathbf{F} \cdot d\mathbf{r}$$

$$= -\int_{r_a}^{r_b} f(r)\, dr.$$

The inverse square force $f(r) = A/r^2$ is an important example of a central force. The gravitational force between two masses m_1 and m_2, $\mathbf{F} \propto (m_1 m_2/r^2)\hat{\mathbf{r}}$, is one instance, and the Coulomb electrostatic force between two charges q_1 and q_2, $\mathbf{F} \propto (q_1 q_2/r^2)\hat{\mathbf{r}}$, is another.

$$U_b - U_a = -\int_{r_a}^{r_b} \frac{A}{r^2} dr$$

$$= \frac{A}{r_b} - \frac{A}{r_a}.$$

To obtain the general potential energy function, we replace r_b by the radial variable r. Then

$$U(r) = \frac{A}{r} + \left(U_a - \frac{A}{r_a}\right)$$

$$= \frac{A}{r} + C.$$

The constant C has no physical meaning, because only changes in U are physically significant, so we are free to give C any value we like. A convenient choice in this case is $C = 0$, which corresponds to taking $U(\infty) = 0$. With this convention we have

$$U(r) = \frac{A}{r}.$$

Example 5.15 Potential Energy of the Three-dimensional Spring Force

The linear restoring force, or spring force, is among the important forces in physics. To show that the spring force is conservative, consider a spring of equilibrium length r_0 with one end attached at the origin.

If the spring is stretched to length r along direction $\hat{\mathbf{r}}$, it exerts a force

$$\mathbf{F}(r) = -k(r - r_0)\hat{\mathbf{r}}.$$

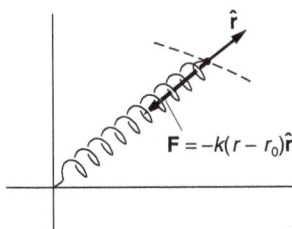

Because the force is central, it is conservative. The potential energy is given by

$$U(r) - U(a) = -\int_a^r (-k)(r - r_0)\, dr$$
$$= \tfrac{1}{2}k(r - r_0)^2\Big|_a^r .$$
$$= \tfrac{1}{2}k[(r - r_0)^2 - (r_a - r_0)^2].$$

Hence

$$U(r) = \tfrac{1}{2}k(r - r_0)^2 + C.$$

Conventionally, we choose the potential energy to be zero at equilibrium: $U(r_0) = 0$, which gives

$$U(r) = \tfrac{1}{2}k(r - r_0)^2.$$

When several conservative forces act on a particle, the potential energy is the sum of the potential energies for each force. In the following example, two conservative forces act.

Example 5.16 Bead, Hoop, and Spring

A bead of mass m slides without friction on a vertical hoop of radius R. The bead moves under the combined action of gravity and a spring attached to the bottom of the hoop. For simplicity, we assume that the equilibrium length of the spring is zero, so that the force due to the spring is $-kr$, where r is the instantaneous length of the spring, as shown. The bead is released at the top of the hoop with negligible speed. How fast is the bead moving at the bottom of the hoop?

At the top of the hoop, the gravitational potential energy of the bead is $mg(2R)$ and the potential energy due to the spring is $\tfrac{1}{2}k(2R)^2 = 2kR^2$.

Hence the initial potential energy is

$$U_i = 2mgR + 2kR^2.$$

The total potential energy at the bottom of the hoop is

$$U_f = 0.$$

Because both forces are conservative, the mechanical energy is constant and we have

$$K_i + U_i = K_f + U_f.$$

The initial kinetic energy is zero and we obtain

$$K_f = U_i - U_f$$

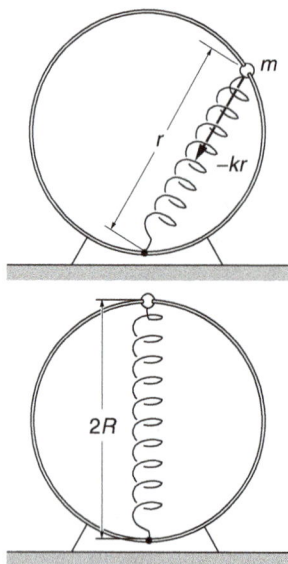

or

$$\tfrac{1}{2}mv_f^2 = 2mgR + 2kR^2.$$

Hence

$$v_f = 2\sqrt{gR + \frac{kR^2}{m}}.$$

5.6 What Potential Energy Tells Us about Force

In many physical problems, it is easier to find the potential energy than to calculate the force. The procedure for then finding the force from the potential energy turns out to be straightforward, as we shall show in this section for a one-dimensional system. The general case of three dimensions is discussed in Note 5.2.

Suppose that we have a one-dimensional system, such as a mass on a spring, where the force is $F(x)$ and the potential energy is

$$U_b - U_a = -\int_{x_a}^{x_b} F(x)\,dx.$$

Consider the change in potential energy ΔU as the particle moves from some point x to $x + \Delta x$:

$$U(x + \Delta x) - U(x) \equiv \Delta U$$
$$= -\int_x^{x+\Delta x} F(x)\,dx.$$

For Δx sufficiently small, $F(x)$ can be considered constant over the range of integration and we have

$$\Delta U \approx -F(x)[(x + \Delta x) - x]$$
$$= -F(x)\Delta x$$

or

$$F(x) \approx -\frac{\Delta U}{\Delta x}.$$

In the limit $\Delta x \to 0$ we have

$$F(x) = -\frac{dU}{dx}. \tag{5.22}$$

The result is reasonable: potential energy is the negative integral of the force so it follows that force is the negative derivative of the potential energy.

5.7 Energy Diagrams

We can often find key features of the motion of a one-dimensional system by using an *energy diagram*, in which the total energy E and the potential energy U are plotted as functions of position. The kinetic energy

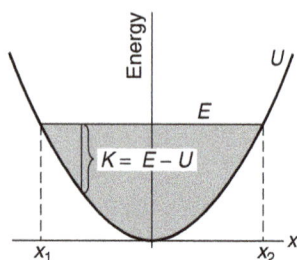

$K = E - U$ is then easily found by inspection. Because kinetic energy cannot be negative, the motion of the system is constrained to regions where $U \le E$.

Here is the energy diagram for a harmonic oscillator. The potential energy $U = kx^2/2$ is a parabola centered at the origin.

Because the total energy is constant for a conservative system, E does not vary with the position x and is represented by a horizontal straight line. Motion is limited to the shaded region where $E \ge U$; the limits of the motion, x_1 and x_2 in the sketch, are called the turning points.

Here is what the diagram tells us. The kinetic energy, $K = E - U$, is greatest at the origin and decreases as the particle flies past the origin in either direction. At a turning point $K = 0$, and the particle comes momentarily to rest. The particle then accelerates back toward the origin with increasing kinetic energy, and the cycle is repeated.

The harmonic oscillator provides a good example of bounded motion. As E increases, the turning points move farther and farther apart, but the particle is never free. If E is decreased, the amplitude of motion decreases, until finally for $E = 0$ the particle lies at rest at $x = 0$.

Quite a different behavior occurs if U does not increase indefinitely with distance. For instance, consider the case of a particle constrained to a radial line and acted on by a repulsive inverse-square law force $(A/r^2)\hat{\mathbf{r}}$. Here $U = A/r$, where A is positive.

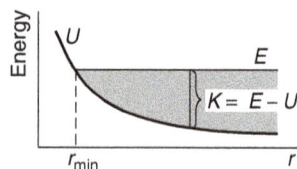

There is a distance of closest approach, r_{min}, as shown in the diagram, but the motion is not bounded for large r because U decreases with distance while the total energy remains constant. If the particle is shot toward the origin, it gradually loses kinetic energy until it comes momentarily to rest at r_{min}. The motion then reverses and the particle moves out toward infinity. The final and initial speeds at any point are identical; the collision merely reverses the velocity.

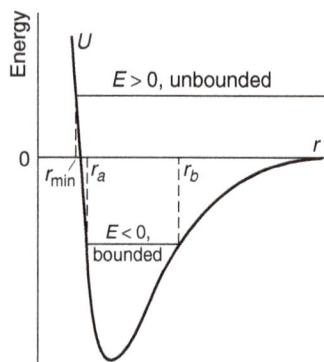

With some potentials, either bounded or unbounded motion can occur depending upon the energy. For instance, consider the interaction between two atoms. The energy diagram for a typical two-atom system is shown in the sketch.

At large separations, the atoms attract each other weakly with the van der Waals force, which varies as $1/r^7$. As the atoms approach, the electron clouds begin to overlap, producing strong forces that can be either attractive or repulsive depending on the details of the electron configuration. If the force is attractive, the potential energy decreases with decreasing r. At very short distances atoms always repel each other strongly because of the repulsion of the positively charged nuclei, and U increases rapidly.

For positive energy $E > 0$, the motion is unbounded and the atoms are free to fly apart. As the diagram indicates, the distance of closest approach, r_{min}, does not change appreciably as E is increased. The steep slope of the potential energy curve at small r means that the atoms behave much like hard spheres—the distance of closest approach r_{min} is not sensitive to the energy of collision.

The situation is quite different for $E < 0$. Then the motion is bounded for both small and large separations; the atoms never approach closer than r_a or move farther apart than r_b. A bound system of two atoms is, of course, a molecule, and the sketch represents a typical diatomic molecule energy diagram.

If two atoms collide with positive energy, they cannot form a molecule unless some means is available for losing enough energy to make E negative. In general, a third body is necessary to carry off the excess energy. Sometimes the third body is a surface, which is the reason surface catalysts are used to speed certain reactions. For instance, atomic hydrogen is stable in the gas phase even though the hydrogen molecule is tightly bound. However, if a piece of platinum is inserted in the hydrogen, the atoms immediately join to form molecules. What happens is that hydrogen atoms tightly adhere to the surface of the platinum, and if a collision occurs between two atoms on the surface, the excess energy is released to the surface, and the molecule, which is not strongly attracted to the surface, leaves. The energy delivered to the surface is so large that the platinum glows brightly. A third atom can also carry off the excess energy, but for this to happen the two atoms must collide when a third atom is nearby. This is a rare event at low pressures, but it becomes increasingly important at higher pressures. Another possibility is for the two atoms to lose energy by the emission of light, but this occurs rarely and is usually unimportant.

5.8 Non-conservative Forces

We have stressed conservative forces and potential energy in this chapter because they play an important role in physics, but in many physical processes non-conservative forces like friction are present. In this section we shall see how to extend the work–energy theorem to include non-conservative forces.

Often both conservative and non-conservative forces act on the same system. For instance, an object falling through the air experiences the conservative gravitational force and the non-conservative force of air friction. We can write the total force $\mathbf{F}$ as

$$\mathbf{F} = \mathbf{F}^c + \mathbf{F}^{nc}$$

where $\mathbf{F}^c$ and $\mathbf{F}^{nc}$ are the conservative and the non-conservative forces, respectively. Since the work–energy theorem is true whether or not the forces are conservative, the total work by $\mathbf{F}$ as the particle moves from a to b is

$$W_{ba}^{\text{total}} = \oint_a^b \mathbf{F} \cdot d\mathbf{r}$$

$$= \oint_a^b \mathbf{F}^c \cdot d\mathbf{r} + \oint_a^b \mathbf{F}^{nc} \cdot d\mathbf{r}$$

$$= -U_b + U_a + W_{ba}^{nc}.$$

Here U is the potential energy associated with the conservative force and W_{ba}^{nc} is the work by the non-conservative force. The work–energy theorem, $W_{ba}^{\text{total}} = K_b - K_a$, now has the form

$$-U_b + U_a + W_{ba}^{\text{nc}} = K_b - K_a$$

or

$$K_b + U_b - (K_a + U_a) = W_{ba}^{\text{nc}}.$$

If we define the total mechanical energy by $E = K + U$, as before, then E is no longer constant but depends on the state of the system. We have

$$E_b - E_a = W_{ba}^{\text{nc}}. \tag{5.23}$$

This result is a generalization of the statement of conservation of mechanical energy that we discussed in Section 5.4. If non-conservative forces do no work, then $E_b = E_a$ and mechanical energy is conserved. In general, the effect of non-conservative forces is to alter the mechanical energy. In particular, the work by friction is always negative and the mechanical energy decreases. Nevertheless, energy methods continue to be useful; we simply must be careful not to overlook the work by the non-conservative forces W_{ba}^{nc}. Here is an example.

Example 5.17 Block Sliding Down an Inclined Plane
A block of mass M slides down a plane of angle θ. The problem is to find the speed of the block after it has descended through height h, assuming that it starts from rest and that the coefficient of friction μ is constant.

Initially the block is at rest at height h; finally the block is moving with speed v at height 0. Hence

$$
\begin{array}{ll}
U_a = Mgh & U_b = 0 \\
K_a = 0 & K_b = \frac{1}{2}Mv^2 \\
E_a = Mgh & E_b = \frac{1}{2}Mv^2.
\end{array}
$$

The non-conservative force is $f = \mu N = \mu Mg\cos\theta$. The non-conservative work is therefore

$$
\begin{aligned}
W_{ba}^{\text{nc}} &= \int_a^b \mathbf{f} \cdot d\mathbf{r} \\
&= -fs \\
&= -\mu N s \\
&= -(\mu Mg\cos\theta)s
\end{aligned}
$$

where s is the distance the block slides. The negative sign arises because the direction of $\mathbf{f}$ is always opposite to the displacement, so that

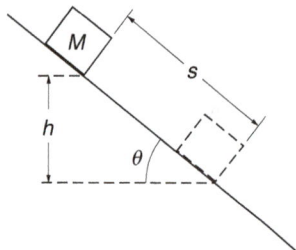

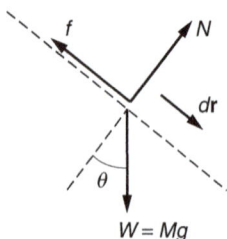

$\mathbf{f} \cdot d\mathbf{r} = -f ds$. Using $s = h/\sin\theta$, we have

$$W_{ba}^{nc} = -\mu Mg\cos\theta\frac{h}{\sin\theta}$$
$$= -\mu\cot\theta Mgh.$$

The energy equation $E_b - E_a = W_{ba}^{nc}$ becomes

$$\tfrac{1}{2}Mv^2 - Mgh = -\mu\cot\theta Mgh,$$

which gives

$$v = \sqrt{2(1 - \mu\cot\theta)gh}.$$

Since all the forces acting on the block are constant, the expression for v could easily be found by applying our results for motion under uniform acceleration; the energy method does not represent much of a shortcut here. The power of the energy method lies in its generality. For instance, suppose that the coefficient of friction varies along the surface so that the friction force is $f = \mu(x)Mg\cos\theta$. The work by friction is

$$W_{ba}^{nc} = -Mg\cos\theta\int_a^b \mu(x)\,dx,$$

and the final speed is easily found. In contrast, there is no simple way to find the speed by integrating the acceleration with respect to time.

5.9 Energy Conservation and the Ideal Gas Law

As far as we know, the basic interactions in nature, for instance the force of gravity and the forces of electric and magnetic interactions, are all conservative. This leads to a puzzle: if the basic forces are conservative, how do non-conservative forces arise? The resolution of this problem lies in the point of view we adopt in describing a physical system, and in our willingness to broaden the concept of energy.

Consider friction, the most familiar non-conservative force. When a block slides across a table mechanical energy is lost due to friction, and the block's speed decreases. Something else also happens: the block and the table become warmer. Up to now, our discussion of energy has not involved the idea of temperature; a block of mass M moving with speed v possesses kinetic energy $\tfrac{1}{2}Mv^2$ whether the block is hot or cold. Similarly, a harmonic oscillator possesses kinetic and potential energy, but these have nothing to do with temperature. Nevertheless, if we look carefully, we find that the heating of the system bears a definite relation to the mechanical energy that is dissipated.

The British physicist James Prescott Joule was the first to establish a quantitative relation between heat and energy, although others, notably Robert Mayer in Germany, had remarked on it qualitatively somewhat earlier. (Mayer showed that water could be heated just by shaking it.) In the early 1840's, Joule carried out a series of meticulous experiments on the heating of water by a paddle wheel driven by a falling weight, and showed that the loss of mechanical energy by friction is accompanied by a corresponding rise in the temperature of the water. Joule concluded that heat is a form of energy and that the sum of the mechanical energy and the heat energy of a system is conserved.

The development of kinetic theory explained Joule's idea of heat energy in terms of the atomic picture of matter. Kinetic theory relates the macroscopic properties of a gas, for instance its pressure and temperature, to a microscopic model in which the gas consists of many small particles. In the first approximation, the particles move freely except when they collide like small hard spheres. An early triumph of kinetic theory was its application to the ideal gas law. The ideal gas law, sometimes called the law of Gay-Lussac after the French physicist who established it experimentally, relates the pressure $\mathcal{P}$, volume $\mathcal{V}$, and temperature T of a given quantity of gas. The quantity of gas, N_{mol}, is measured in units of the gram-mole, where one mole has a mass in grams equal to the molecular weight of the species.

The ideal gas law is

$$\mathcal{P}\mathcal{V} = N_{mol}RT. \tag{5.24}$$

Here T is the temperature measured in kelvin (K), a temperature scale whose zero is at approximately $-273\,°C$ and for which the difference between the boiling and freezing temperatures of water is 100 K. (The degrees symbol is not used with K.) R is an empirical constant, the *universal gas constant*, which has the value $R \approx 8.314\,\mathrm{J/(mol \cdot K)}$.

In Example 4.22 we derived an expression for the pressure of a gas in terms of its mass density ρ and the mean squared velocity $\overline{v^2}$ of its particles in the form $\mathcal{P} = \frac{1}{3}\rho\overline{v^2}$. Multiplying by the volume of the gas gives

$$\mathcal{P}\mathcal{V} = \frac{1}{3}\rho\mathcal{V}\overline{v^2}.$$

The product $\rho\mathcal{V}$ is the total mass M_{tot} of the gas.

The number of particles in a mole of gas is known as *Avogadro's number*, $N_A \approx 6.022 \times 10^{23}$. If the mass of each particle is m, then for a number of moles N_{mol}, the total mass is $M = N_{mol}N_A m$ and

$$\mathcal{P}\mathcal{V} = \frac{1}{3}M\overline{v^2} = \frac{1}{3}N_{mol}N_A m\overline{v^2}. \tag{5.25}$$

Comparing Eqs. (5.24) and (5.25) we see that

$$\frac{1}{3}N_{mol}N_A m\overline{v^2} = N_{mol}RT$$

$$\frac{1}{3}m\overline{v^2} = \frac{R}{N_A}T. \tag{5.26}$$

The ratio R/N_A plays a fundamental role in statistical physics and is known as the *Boltzmann constant* $\mathrm{k} \approx 1.380 \times 10^{-23}$ J/K, named after Ludwig Boltzmann, a founder of kinetic theory. We can rewrite Eq. (5.26) as

$$\frac{1}{2}m\overline{v^2} = \frac{3}{2}kT. \tag{5.27}$$

This simple equation encompasses a great deal of physics. First, it provides a physical explanation for temperature: temperature is a measure of the mean kinetic energy of atoms in an ideal gas. Zero temperature, "absolute zero," is the temperature at which translational motion ceases. Second, it reveals that thermal energy is microscopic kinetic energy, but kinetic energy of a particular kind, involving *random* motion. Randomness was introduced in Example 4.22 when we made the assumption

$$\overline{v_x^2} = \overline{v_y^2} = \overline{v_z^2} = \frac{1}{3}\overline{v^2}.$$

Without this assumption, pressure would vary with direction, which is never seen. A consequence of this relation is that

$$\frac{1}{2}m\overline{v_x^2} = \frac{1}{2}m\overline{v_y^2} = \frac{1}{2}m\overline{v_z^2} = \frac{1}{2}kT.$$

The energy is equally divided among the three modes of translational motion. This is an example of a general theorem called the *equipartition theorem*: if the energy of a system can be written as a sum of quadratic terms such as $\frac{1}{2}mv_x^2$, $p_x^2/2m$, or $\frac{1}{2}kx^2$, then in thermal equilibrium, the average energy associated with each term is $\frac{1}{2}kT$. The equipartition theorem is derived in texts on statistical mechanics. The theorem is powerful and very useful, but under some conditions it breaks down. In fact, the first clues on the quantum nature of matter came from a failure of the equipartition theorem.

Another feature of thermal energy that distinguishes it from mechanical energy is that heat energy naturally flows from hot systems to cold systems as soon as they are put in contact, but never in the other direction. This arises because of the random nature of thermal energy and the statistical properties that govern its flow. Consequently, heat energy is a fundamentally new type of energy, but it fits into the larger framework of Newtonian mechanics.

Example 5.18 Heat Capacity of a Gas

The heat capacity of a system is the amount of energy required to raise the temperature 1 K. This is proportional to the mass of the system, so we shall consider the heat capacity of one mole (the molar heat capacity). The amount of energy required depends on whether or not the system does work as the heat is added. If a gas expands, it does work on the container, and so let us consider heat capacity where the volume is kept constant, conventionally denoted by C_V. From

Eq. (5.27) we see that the energy of one mole of ideal gas is

$$E = N_A \frac{1}{2} m \overline{v^2}$$
$$= \frac{3}{2} N_A k T$$
$$= \frac{3}{2} R T.$$

If only the three translational modes of motion contribute, the energy required to increase the temperature 1 K is

$$C_V = \frac{3}{2} R = 12.47 \text{ J/mol} \cdot \text{K}.$$

Gas data tables list the value for C_V for the monatomic noble gases helium and argon as 12.5 J/mol · K, consistent with this result.

5.10 Conservation Laws

We discussed conservation of momentum in Chapter 4 and conservation of energy in this chapter. They are two of the most fundamental conservation laws in physics, but they are different in character. Consider the momentum of an isolated system of particles. The ith particle at some instant has momentum $m_i \mathbf{v}_i$ that may, however, change due to collisions. Nevertheless, the particle's momentum does not transform into a different physical quantity, and the total momentum remains unchanged.

Energy, on the other hand, is a chameleon that can change from one form to another. A simple example is the interplay of kinetic and potential energy discussed earlier in this chapter. Conservation of mechanical energy is simply a consequence of Newton's laws and tells us nothing new. We then enlarged our concept of energy to include thermal energy. There are additional forms of energy beyond mechanical and thermal: chemical, electromagnetic, and nuclear, to name a few. When all forms are included, total energy has always been found to be conserved. It is only when one or more forms are overlooked in the accounting that the energy of a system seems not to be conserved. When talking about mechanical energy, the most common reason that energy conservation appears to fail is when some mechanical energy is transformed into heat. Nevertheless, the total energy, mechanical plus thermal, is conserved.

Conservation of energy may look like an act of desperation: whenever mechanical energy appears or disappears, we seem to invent some new kind of energy to balance the energy books. Luckily, only a few kinds of energy are known, and they have all been verified experimentally. In 1905, Einstein's "miraculous" year when he published his theory of special relativity and several other great discoveries, Einstein postulated two new forms of energy. One was associated with the photoelectric effect, in which the energy of light is carried by discrete packets of energy ("quanta") called photons. A photon of frequency v carries energy

$E = h\nu$, where h is Planck's constant. This was soon verified by measuring the kinetic energy of electrons ejected from a metal due to incident light of known frequency.

Einstein's second idea was much more startling—that mass itself is a form of energy, according to his famous relation $E = mc^2$. (We shall show his proof in Chapter 13.) Because the speed of light is so high, even a small mass is equivalent to a large amount of energy. Chemists are therefore justified in assuming that mass is conserved in chemical reactions, because the mass change is very small for chemical energies, which are in the eV range ($1 \text{ eV} \approx 1.6 \times 10^{-19}$ J). A particularly violent chemical reaction is the reaction of hydrogen gas (H_2) with fluorine gas (F_2) to give hydrogen fluoride (HF). If one mole of hydrogen (2.016 g) reacts with one mole of fluorine (37.996 g), the energy released is $\Delta E = 6.6 \times 10^5$ J, but the mass change is only $\Delta m = \Delta E/c^2 = 7.3 \times 10^{-9}$ g, smaller than the masses of the reactants by a factor of nearly 10^{10}.

Many nuclear reactions involve energies millions of times greater than the energies of chemical reactions. In the 1930s, experimenters were able to measure nuclear masses with enough accuracy to show that the energy released in a nuclear reaction agrees with the known mass difference Δm according to $\Delta E = \Delta m\, c^2$. For example, an atom of radium-226, ^{226}Ra, spontaneously emits an α-ray (a nucleus of helium-4, ^{4}He) having a kinetic energy of 4.78 MeV, leaving a residual nucleus of radon-222, ^{222}Rn:

$$^{226}\text{Ra} \rightarrow {}^{222}\text{Rn} + {}^4\text{He}.$$

(The number at the upper left of the element's symbol is the mass number, the total number of protons and neutrons in the nucleus.) The difference between the initial mass ^{226}Ra and the final masses ^{222}Rn plus ^{4}He is 8.80×10^{-30} kg and the mass energy accounts closely for the kinetic energy of the α-ray plus the small kinetic energy of the recoiling ^{222}Rn nucleus. Mass is indeed a form of energy, and can be converted to mechanical energy.

Example 5.19 Conservation Laws and the Neutrino

Some unstable nuclei emit an energetic electron, a process called β-decay. In one form of β-decay, a neutron in the unstable nucleus becomes a proton, and a negatively charged electron (a β-ray) is emitted. Note that there is no change in the net electric charge, either in β-decay or in α-decay. Conservation of charge appears to be a fundamental conservation law; no processes have ever been observed where net charge is created. When it became possible to measure the energy of β-rays experimentally, physicists were nonplussed. Unlike α-decay, where the emitted α-rays have definite energies, the β-rays were found to have a continuous spectrum of energy, from zero energy to a maximum depending on the nucleus. The graph shows experimental data. (Source: G. J. Neary, Roy. Phys. Soc. (London) A175, 71 (1940).)

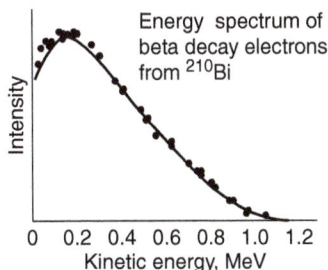

Energy spectrum of beta decay electrons from ^{210}Bi

Intensity

0   0.2 0.4 0.6 0.8 1.0 1.2
Kinetic energy, MeV

A widely held opinion at the time was that these poorly understood β-decay processes did not obey conservation of energy. Wolfgang Pauli in Germany disagreed. Pauli, an outstanding theorist in an era of outstanding physicists, remained convinced that conservation of energy was a fundamental principle of physics, and he postulated that the missing energy was being carried off by an undetected particle emitted along with the β-ray.

Several properties of the unknown particle could be inferred from the β-decay measurements.

(1) The particle is neutral because all charge is already accounted for and charge is conserved. Furthermore, a charged particle would interact strongly with matter and could be detected. Inability to detect the particle also implied that it interacts hardly at all with ordinary matter, (2) The maximum β-ray energy evidently corresponds to the electron receiving all the available energy, and must therefore be equal to the decay energy ΔE. The mass difference Δm of the reaction, not including the mass of the unknown particle, is found to account within experimental error for the observed energy according to $\Delta m = \Delta E / c^2$. Hence the unknown particle must have very small mass. The particle was initially thought to be massless, like a photon, but it has since been found to have a very small but non-zero mass, less than 10^{-5} of the electron mass.

Based on these inferred properties, the unknown particle was named the neutrino ("little neutral one"). Nuclear reactors produce abundant amounts of neutrinos, and neutrinos were first detected directly in 1956, using neutrinos from a reactor.

The Sun is a copious source of neutrinos from the nuclear reactions in its interior, producing a flux of $\approx 10^{11}$ neutrinos cm$^{-2} \cdot$ s^{-1} at the Earth's surface. The interaction of neutrinos with matter is so slight that almost all of the solar neutrinos pass directly through the Earth.

5.11 World Energy Usage

Energy is unique among the basic concepts of science in the variety of its uses and the multitude of units that are employed. Energy plays a prominent role in the public agenda, because it is inextricably linked to quality of life and to concerns about the environment. In dealing with these matters, a nodding acquaintance with energy units and usage is valuable. We summarize a few of the facts in this section. Further discussion of energy in society is given in *Physics and Energy*, Robert L. Jaffe and Washington Taylor, Cambridge University Press (2013).

Although energy takes many forms, all units of energy and power are defined by conversion factors from the primary SI units, joules (J) and watts (W), respectively. The calorie (cal) was originally defined as the energy required to raise the temperature of 1 gram of water by 1 degree C, but based on experiments initiated by James Prescott Joule, 1 calorie is now defined as 4.1868 J. The "calorie" employed in nutrition is actually a kilocalorie (kcal).

Some non-SI units are retained for historical reasons. New units are generally introduced for practical reasons to avoid using unusually large or small powers of 10. Scientists and engineers have therefore devised units appropriate to their particular fields. For example, physicists and chemists typically express the binding energy of atoms in a molecule using the unit of electron-volts, because molecular binding energies are of the order of a few electron-volts. But electron-volts would be very inconvenient for a petroleum engineer who wanted to express the vastly greater amount of energy from a barrel of oil.

Table 1 summarizes some of the more commonly used units of energy and power. Table 2 gives some statistics on global energy production, Table 3 gives energy production in the United States, and Table 4 gives energy consumption per capita in selected countries.

Example 5.20 Energy and Water Flow from Hoover Dam

Hoover Dam on the Arizona–Nevada border generates power and regulates the water flow of the Colorado River by controlling the outlet of the enormous reservoir of Lake Mead fed by the river. The water falling through the generator turbines supplies the neighboring states with a significant fraction of their electric power and the outflow is a major contributor to irrigation water. Power and irrigation both depend on the water flow. We can estimate the efficiency by which Hoover Dam generates energy by comparing the potential energy lost by the outflowing water to the actual electric energy produced.

According to the U.S. Department of the Interior, the average energy generation from Hoover Dam for 1999 through 2008 was about 4.2×10^{12} watt-hours annually. The head (height the water falls) varies between 590 feet and 420 feet, with an average of about 520 feet.

The water transfers momentum to the turbine blades at the bottom of the penstocks. Assume for simplicity that the water loses all its energy to the turbines, notwithstanding that a visitor will see water roaring out from the base of the dam. The maximum energy available is Mgh where M is the mass of water falling on the turbines, and h is the head. Given the energy and the head, we can calculate the mass M, and from this the volume of water, assuming that the energy conversion is 100% efficient. The only challenging part of the calculation is converting the various units to SI for consistency. (Section 2.8 presents a systematic method for converting units.)

1 ft = 12 inches = 2.54 cm/in × 12 in ≈ 30.5 cm = 0.305 m. The head in meters is then 520 ft × 0.305 m/ft = 159 m.

Because 1 hour = 3600 s, the annual energy generated in joules/yr (1 J = 1 W · s) is

$$E = 3600 \text{ s/hr} \times (4.2 \times 10^{12} \text{ W} \cdot \text{hr}) = 1.51 \times 10^{16} \text{J/yr}.$$

Let M be the total mass of water that flows through the dam annually, and equate its potential energy to the energy produced. We have (using 1 J = 1 kg · m^2/s^2)

$$M = \frac{E}{gh} = \frac{1.51 \times 10^{16} \text{ kg} \cdot \text{m}^2/\text{s}^2}{9.80 \text{ m/s}^2 \times 159 \text{ m}} = 9.69 \times 10^{12} \text{ kg annually}$$

for the mass of annual water flow. Taking the density of water to be 1000 kg/m^3, the total volume is

$$V = 9.69 \times 10^{12} \text{ kg} \times \frac{1 \text{ m}^3}{1000 \text{ kg/yr}} = 9.69 \times 10^9 \text{m}^3.$$

Engineers, surveyors, and others involved with major water projects in the U.S. use a volume unit called the *acre-foot* rather than the SI unit m^3. An acre is an area of 4047 m^2, so that

$$1 \text{ acre-foot} \approx 4047 \text{ m}^2 \times 0.305 \text{ m} = 1234 \text{ m}^3.$$

The annual volume in acre-feet required to generate the energy would be

$$V = 9.69 \times 10^9 \text{ m}^3/\text{yr} \times \frac{1 \text{ acre-foot}}{1234 \text{ m}^3} = 7.0 \times 10^6 \text{ acre-foot/yr}.$$

According to the Colorado River Compact of 1922, the flow to the states below Lake Mead was to average 7.5×10^6 acre-foot per year. If these figures are accurate, then the efficiency of the power generation is estimated to be

$$\text{efficiency} = \frac{\text{energy produced}}{\text{energy available}} = \frac{7.0 \times 10^6 \text{ acre-foot/yr}}{7.5 \times 10^6 \text{ acre-foot/yr}} = 93\%.$$

Our estimate is not precise, because we do not know the actual conditions of the river flow, dam head, and energy production for a specific period. Also, we have neglected the energy of the outflow water. Nevertheless, this analysis suggests that the efficiency of hydroelectric power generation is impressively high.

Table 1. Energy, power, and related units[*]
Entries with "=" signs are exact definitions.

Name	Symbol	SI value	Comment
joule	J	–	SI unit of energy
watt	W	$= 1$ J/s	SI unit of power
erg	erg	$= 10^{-7}$ J	cgs unit of energy
electron-volt	eV	$\approx 1.60 \times 10^{-19}$ J	widely used in physics[1]
photon energy	$h\nu$	–	energy of light quantum[2]
kelvin	K	–	informal energy unit[3]
calorie	cal	$= 4.1868$ J	older unit of heat[4]
Calorie	Cal, kcal	$= 4186.8$ J	used in nutrition and physiology[5]
solar constant	–	$\approx 1.368 \times 10^3$ W/m^2	mean power/area from the Sun[6]
kilowatt-hour	kWh	$= 3.6 \times 10^6$ J	domestic unit of energy
horsepower	hp	$= 746$ W	engineering unit of power[7]
British thermal unit	Btu	$\approx 1.06 \times 10^3$ J	domestic unit of heat energy[8]
ton of oil equivalent	toe	$\approx 4.19 \times 10^{10}$ J	industrial energy unit[9]
kg of oil equivalent	kgoe	$\approx 4.19 \times 10^7$ J	industrial energy unit
therm	tm	$\approx 1.06 \times 10^8$ J	industrial energy unit[10]
quadrillion Btu	quad	$\approx 1.06 \times 10^{18}$ J	global energy unit
terawatt-years	TWyr	$\approx 3.15 \times 10^{19}$ J	global energy unit

[*] *Sources: Guide for the Use of the International System of Units (SI)*, NIST, US Department of Commerce, and Graham Woan, *The Cambridge Handbook of Physics Formulas*, Cambridge University Press (2003).

[1] The ionization energies of atoms and the reaction energies of molecules are typically in the eV range. Nuclear and particle physics involve phenomena in the MeV, GeV, and TeV range.

[2] $h \approx 6.63 \times 10^{-34}$ J · s is Planck's constant, and ν is the frequency of the light. The median energy of photons from the Sun is about 2.5 J.

[3] The mean thermal energy of a system in thermal equilibrium is characterized by the energy kT, where $k \approx 1.38 \times 10^{-23}$ J/K is Boltzmann's constant and T is the absolute temperature in kelvin. Informal usage such as "an energy of 5.0 nanokelvin" is jargon but clear from the context.

[4] Before the relation between heat and energy was understood, the calorie was defined as the heat required to raise the temperature of 1 gram of water 1 degree C. James Prescott Joule measured "the mechanical equivalent of heat" in the 1840s. Today the calorie is defined as 4.1868 J.

[5] The calorie is inconveniently small for some purposes, and the kilocalorie (the "large calorie") is more commonly used in some fields.

[6] At a mean radius of the Earth's orbit (the mean semi-major axis), and at the top of the Earth's atmosphere.

[7] Horsepower came originally from an estimate of the average power a horse could produce. Horsepower has several slightly varying definitions; the given value is horsepower (electric), which is defined to be 746 W.

[8] The Btu was originally the energy required to heat 1 pound of water 1 degree Fahrenheit.

[9] An estimate of the energy released by burning 1000 kg of crude oil.

[10] 10^5 btu, approximately the energy of burning 100 cubic feet of natural gas.

Table 2. Global energy supply (2008)[**]

Total supply, 12,267 Mtoe ($\approx$ 485 quad)

Source	% of total
oil	33.2
coal/peat	27.0
natural gas	21.1
combustibles and waste	10.0
nuclear	5.8
hydro	2.2
other	0.7

[**]*Source: 2010 Key World Energy Statistics,* International Energy Agency, Paris.

Table 3. U.S. energy supply (2009)[†]

Total use 94.6 quad

Source	% of total
petroleum	35.3
natural gas	23.4
coal	19.7
renewable energy	7.7
nuclear electric power	13.9

[†]*Source: U.S. Energy Information Administration/Annual Energy Review 2009.*

Table 4. Per capita energy consumption of selected nations[††]

Units: kgoe per person

Country	Energy per capita	Country	Energy per capita	Country	Energy per capita
Albania	767	Argentina	1058	Australia	5898
Austria	4125	Belgium	5892	Benin	306
Brazil	1124	Bulgaria	2592	Canada	8473
China	1316	Congo	300	Czech Rep.	4419
Denmark	3634	Egypt	828	El Salvador	673
Finland	6555	France	4397	Germany	4187
Iceland	12209	India	491	Indonesia	814
Jordan	1296	Kazakhstan	3462	Kuwait	11102
Mexico	1701	Nepal	338	New Zealand	4218
Norway	7153	Pakistan	490	Poland	2429
Qatar	19456	Russian Fed.	4519	Saudi Arabia	6068
United Kingdom	3895	United States	7886	Yemen	321

[††]*Source:* International Energy Administration, Statistics Division, 2007 Energy Balances of OECD Countries (2008 edition) and Energy Balances of Non-OECD countries (2007 edition).

Note 5.1 Correction to the Period of a Pendulum

To first order, a pendulum displays simple harmonic motion and its period does not depend on the amplitude of its swing. However, the motion of a pendulum is not exactly simple harmonic motion. In this Note we calculate the correction to the period due to finite amplitude.

Our starting point is the equation for the motion of a pendulum derived in Example 5.12:

$$\int \frac{d\theta}{\sqrt{\cos\theta - \cos\theta_0}} = \sqrt{\frac{2g}{l}} \int dt. \tag{1}$$

This equation is exact. To obtain a more accurate solution for the period than given by the small-angle approximation, it is helpful to use the identity $\cos\theta = 1 - 2\sin^2(\theta/2)$. This gives

$$\cos\theta - \cos\theta_0 = 2[\sin^2(\theta_0/2) - \sin^2(\theta/2)]. \tag{2}$$

Introducing Eq. (2) in Eq. (1) gives

$$\int \frac{d\theta}{\sqrt{2}\sqrt{\sin^2(\theta_0/2) - \sin^2(\theta/2)}} = \sqrt{\frac{2g}{l}} \int dt. \tag{3}$$

Now let us change variables as follows:

$$\sin u = \frac{\sin(\theta/2)}{\sin(\theta_0/2)}. \tag{4}$$

The motivation for this is that although θ is periodic, as the pendulum swings through a cycle, θ varies between $-\theta_0$ and θ_0. On the other hand, u varies between $-\pi$ and $+\pi$. If we let

$$K = \sin\frac{\theta_0}{2},$$

then

$$\sin\frac{\theta}{2} = K\sin u$$

and

$$d\theta = \left(\sqrt{\frac{1 - \sin^2 u}{1 - K^2\sin^2 u}}\right) 2K\,du. \tag{5}$$

Substituting Eqs. (4) and (5) in Eq. (3) gives

$$\int \frac{du}{\sqrt{1 - K^2\sin^2 u}} = \sqrt{\frac{g}{l}} \int dt.$$

Let us take the integral over one period. The limits on u are 0 and 2π, while t ranges from 0 to T. We have

$$\int_0^{2\pi} \frac{du}{\sqrt{1 - K^2\sin^2 u}} = \sqrt{\frac{g}{l}}\,T. \tag{6}$$

The integral on the left is an *elliptic integral*: specifically, it is a complete elliptic integral of the first kind. Values for this function are available from computed tables. However, for our purposes it is more convenient to expand the integrand:

$$\frac{1}{\sqrt{(1 - K^2 \sin^2 u)}} = 1 + \tfrac{1}{2}K^2 \sin^2 u + \cdots$$

and

$$T = \sqrt{\frac{l}{g}} \int_0^{2\pi} du(1 + \tfrac{1}{2}K^2 \sin^2 u + \cdots)$$

$$= \sqrt{\frac{l}{g}}\left(2\pi + \frac{2\pi}{4}K^2 + \cdots\right)$$

$$= 2\pi\sqrt{\frac{l}{g}}\left(1 + \frac{1}{4}\sin^2\frac{\theta_0}{2} + \cdots\right).$$

If $\theta_0 \ll 1$, then $\sin^2(\theta_0/2) \approx \theta_0^2/4$, and we have

$$T = 2\pi\sqrt{\frac{l}{g}}(1 + \tfrac{1}{16}\theta_0^2 + \cdots). \tag{7}$$

The fractional change in period due to finite amplitude θ_0 is

$$\frac{\Delta T}{T} = \frac{T(\theta_0) - T(\theta_0 = 0)}{T} = \frac{1}{16}\theta_0^2.$$

For an amplitude of 0.1 rad, about 6°, the period is increased by about 1 part in 10^4, slowing a clock by roughly a minute a day. For larger amplitudes, higher order terms in Eq. (7) can be introduced, but at that point it is better to go to the exact solution. Note that as $\theta_0 \to \pi$, $T \to \infty$.

Note 5.2 Force, Potential Energy, and the Vector Operator ∇

We have shown that in the case of one dimension, force and potential are related by the integral relation

$$\int_a^b \mathbf{F} \cdot d\mathbf{r} = -[U(b) - U(a)] \tag{1}$$

and by the differential relation

$$F_x = -\frac{dU}{dx}.$$

In this Note we shall extend the differential relation to the general case of more than one independent variable.

Working in three-dimensional Cartesian coordinates, Eq. (1) becomes

$$F_x\Delta x + F_y\Delta y + F_z\Delta z \approx -\Delta U(x, y, z) \tag{2}$$

for small path increments $\Delta x, \Delta y, \Delta z$. Now suppose that $y = y_0$ and $z = z_0$, where y_0 and z_0 are constants. It follows that $\Delta y = 0$ and $\Delta z = 0$, so that Eq. (2) becomes

$$F_x\Delta x \approx -\Delta U(x, y_0, z_0)$$

or

$$F_x \approx -\frac{\Delta U(x, y_0, z_0)}{\Delta x}. \tag{3}$$

Equation (3) looks like a derivative (before we take the limit), but here U is a function of several independent variables, only one of which is allowed to vary.

Equation (3) tells us how fast U changes when only one of the independent variables, here x, varies. This special type of derivative is called a *partial derivative*, and is denoted by the symbol ∂ instead of d when we take the limit $\Delta x \to 0$:

$$\begin{aligned} F_x &= -\lim_{\Delta x \to 0} \frac{\Delta U}{\Delta x} \\ &= -\frac{\partial U}{\partial x}. \end{aligned} \tag{4}$$

Because the partial derivative in Eq. (4) is with respect to x, this tells us that we need to hold y and z constant when evaluating the derivative.

By the symmetry of Cartesian coordinates, we can write

$$F_x \,\hat{\mathbf{i}} + F_y \,\hat{\mathbf{j}} + F_z \,\hat{\mathbf{k}} = -\left(\hat{\mathbf{i}}\, \frac{\partial U}{\partial x} + \hat{\mathbf{j}}\, \frac{\partial U}{\partial y} + \hat{\mathbf{k}}\, \frac{\partial U}{\partial z} \right). \tag{5}$$

As a simple example, consider the potential energy $U = mgz$ of a mass m in a downward uniform gravitational field, where z is the height above the ground. Then $F_x = 0$, $F_y = 0$, and $F_z = -(\partial U / \partial z)\, \hat{\mathbf{k}} = -mg\, \hat{\mathbf{k}}$.

∇ and the Gradient

The form $\left(\hat{\mathbf{i}}\, \frac{\partial}{\partial x} + \hat{\mathbf{j}}\, \frac{\partial}{\partial y} + \hat{\mathbf{k}}\, \frac{\partial}{\partial z} \right)$ is called a *vector operator*, because it has components like a vector and its partial derivatives operate on a quantity placed to its right. When operating on a scalar function such as potential energy, it is also called the *gradient operator*. To simplify the notation, we write

$$\nabla \equiv \left(\hat{\mathbf{i}}\, \frac{\partial}{\partial x} + \hat{\mathbf{j}}\, \frac{\partial}{\partial y} + \hat{\mathbf{k}}\, \frac{\partial}{\partial z} \right)$$

where ∇ is called "del" or sometimes "nabla" (after an ancient Hebrew harp of similar shape).

With this notation, the relation between force and potential energy can be written

$$\mathbf{F} = -\nabla U. \tag{6}$$

When ∇ operates on a scalar to give a vector, as in Eq. (6), the combination ∇U is called the *gradient* of U, sometimes written $\overrightarrow{\text{grad}}\, U$.

To see where the name gradient comes from, use Eq. (6) in Eq. (1) and integrate from $a = (x_1, y_1, z_1)$ to $b = (x_2, y_2, z_2)$:

$$\int_a^b \nabla U \cdot d\mathbf{r} = U(b) - U(a).$$

This result does not make use of any specific properties of U, so it is a general property of the gradient and holds for any differentiable function, say $h(x, y, z)$, and for any displacement, say $d\mathbf{s}$:

$$\int_a^b \boldsymbol{\nabla} h \cdot d\mathbf{s} = h(b) - h(a).$$

The gradient tells how much a function changes due to a given displacement.

Contour Lines and the Gradient

The equation $U(x, y, z) = \text{constant} = C$ defines for each value of C a surface known as a *constant energy* surface. A particle constrained to move on such a surface has constant potential energy. For example, the gravitational potential energy of a particle m at distance $r = \sqrt{x^2 + y^2 + z^2}$ from particle M fixed at the origin is $U = -GMm/r$, so the surfaces of constant energy are given by

$$-\frac{GMm}{r} = C$$

or

$$r = -\frac{GMm}{C}.$$

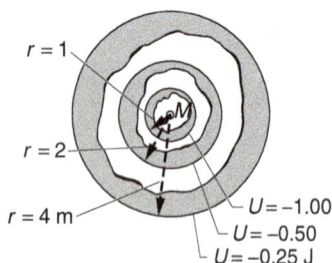

The constant energy surfaces are spheres centered on M, as shown in the drawing. (We have taken $GMm = 1$ N·m^2 for convenience.) Constant energy surfaces are usually difficult to draw, and for this reason it is generally easier to visualize U by considering the lines of intersection of the constant energy surfaces with a plane.

These lines are sometimes referred to as constant energy lines or, more simply, contour lines. The contour lines of a function are analogous to the contours of constant altitude shown on the topographic map of a hilly countryside.

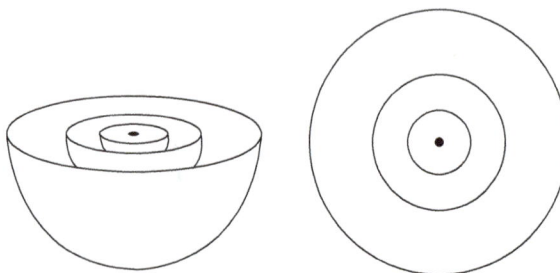

Consider now

$$\int_a^b \boldsymbol{\nabla} U \cdot d\mathbf{s} = U(b) - U(a).$$

If we take $d\mathbf{s}$ to be an arbitrary displacement tangentially along one of the circular contour lines, then $U(b) - U(a) = 0$ because the potential is constant on a contour line. The dot product on the left-hand side is accordingly 0, and we conclude that the gradient must be perpendicular to $d\mathbf{s}$.

This is a fundamental result; the gradient vector of a function is always perpendicular to the contour lines of the function.

A second fundamental result is that if we make a displacement from one contour line toward another, the function changes most rapidly along the direction of the gradient, because if the displacement $d\mathbf{s}$ is parallel to ∇U, the dot product is as large as possible. In terms of a hilly countryside, the gradient is the line of steepest descent down a hill—hence the origin of the name gradient.

Divergence

We have seen how ∇ operates on a scalar to give the gradient vector. We now ask, what happens if ∇ operates on a vector? There are two ways to do this: with the dot product to give a scalar, and with the cross product to give a vector. Both methods have important physical applications, especially in electromagnetism, but we leave proofs and most details to a more thorough study of mathematical physics.

The dot product of ∇ and a vector $\mathbf{F}$ is a scalar called the *divergence* of $\mathbf{F}$:

$$\nabla \cdot \mathbf{F} = \text{divergence of } \mathbf{F} = div\ \mathbf{F}.$$

In Cartesian coordinates, the divergence of a vector $\mathbf{F}$ is

$$\nabla \cdot \mathbf{F} = \frac{\partial F_x}{\partial x} + \frac{\partial F_y}{\partial y} + \frac{\partial F_z}{\partial z}.$$

For a more physical interpretation of the divergence, consider a positive electric charge Q fixed at the origin. By moving a small positive charge q to various points in space around Q, we can map out the magnitude and direction of the electric force $\mathbf{F}$ on q, an operation that defines the electric field $\mathbf{E} = \mathbf{F}/q$. The sketch shows the lines of the electric field. They radiate outward, and give a sense of "divergence."

In contrast, the lines of a uniform force field give no impression of divergence. A uniform force field has zero divergence, a self-evident result, because the partial derivatives of the components are all identically zero.

We now show that for the electric field $\mathbf{E} = (kQ/r^2)\hat{\mathbf{r}}$ produced by Q, the divergence is not identically zero. To avoid the mathematical complexities of dealing with a point charge of zero radius, suppose that Q is actually a ball of charge having radius a and uniform charge density ρ,

$$\rho = \frac{Q}{(4/3)\pi a^3}.$$

Because the electric field and the gravitational field are both inverse-square central forces, we can take over results from Section 3.3.1 wholesale, for example that if $r \leq a$, only the charge within r contributes to the

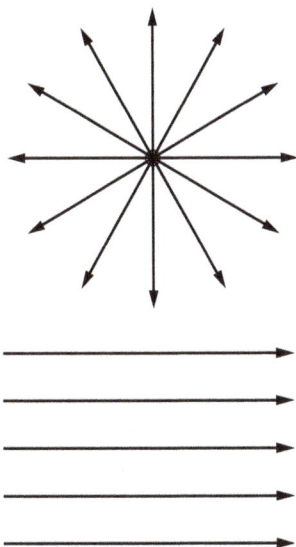

field. The field inside the ball is easily found to be

$$\mathbf{E} = \frac{kQr}{a^3}\hat{\mathbf{r}}$$

$$= \frac{kQ}{a^3}(x\hat{\mathbf{i}} + y\hat{\mathbf{j}} + z\hat{\mathbf{k}}).$$

It follows that $\mathbf{\nabla} \cdot \mathbf{E} = 3kQ/a^3 \neq 0$ inside the ball. Outside the ball, $\mathbf{\nabla} \cdot \mathbf{E} = 0$, but the important feature is that the divergence is not identically zero everywhere.

If we integrate $\mathbf{\nabla} \cdot \mathbf{E}$ over any volume V for $r \geq a$, we find

$$\int_0^r \mathbf{\nabla} \cdot \mathbf{E} dV = \int_0^r \frac{3kQ}{a^3} dV$$

$$= \frac{3kQ}{a^3} \int_0^a dV$$

$$= 4\pi kQ.$$

The result, which is independent of a and therefore also holds for a point charge, shows that the volume integral of the divergence tells us about the source of the field, in this case the charge Q.

Curl

The remaining operation with $\mathbf{\nabla}$ is to take the cross product with some vector $\mathbf{F}$ to give a new vector called the *curl* of $\mathbf{F}$:

$$\mathbf{\nabla} \times \mathbf{F} = \text{curl of } \mathbf{F} = \vec{curl}\,\mathbf{F}.$$

We can calculate the components of $\vec{curl}\,\mathbf{F}$ in Cartesian coordinates using the determinantal form

$$\begin{vmatrix} \hat{\mathbf{i}} & \hat{\mathbf{j}} & \hat{\mathbf{k}} \\ \frac{\partial}{\partial x} & \frac{\partial}{\partial y} & \frac{\partial}{\partial z} \\ F_x & F_y & F_z \end{vmatrix}$$

to give

$$\mathbf{\nabla} \times \mathbf{F} = \hat{\mathbf{i}}\left(\frac{\partial F_z}{\partial y} - \frac{\partial F_y}{\partial z}\right) + \hat{\mathbf{j}}\left(\frac{\partial F_x}{\partial z} - \frac{\partial F_z}{\partial x}\right) + \hat{\mathbf{k}}\left(\frac{\partial F_y}{\partial x} - \frac{\partial F_x}{\partial y}\right).$$

The curl of a force $\mathbf{F}$ has an important application in mechanics if we know the force as a function of position. If $\mathbf{F}$ is conservative $\mathbf{F} = \mathbf{F}^c$ then its curl is 0 everywhere; if $\mathbf{F}$ is non-conservative $\mathbf{F} = \mathbf{F}^{nc}$, its curl is not zero everywhere.

How the Curl Got Its Name

The curl was invented to help describe the properties of moving fluids. To see how the curl is connected with "curliness" or rotation, consider an

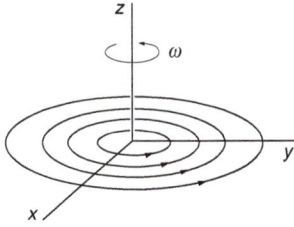

idealized whirlpool turning with constant angular velocity ω about the z axis.

The velocity of the fluid at $\mathbf{r}$ is $\mathbf{v} = r\omega\hat{\boldsymbol{\theta}}$, where $\hat{\boldsymbol{\theta}}$ is the unit vector in the tangential direction. In Cartesian coordinates,

$$\mathbf{v} = r\omega(-\sin\omega t\,\hat{\mathbf{i}} + \cos\omega t\,\hat{\mathbf{j}})$$

$$= r\omega\left(-\frac{y}{r}\hat{\mathbf{i}} + \frac{x}{r}\hat{\mathbf{j}}\right)$$

$$= -\omega y\hat{\mathbf{i}} + \omega x\hat{\mathbf{j}}.$$

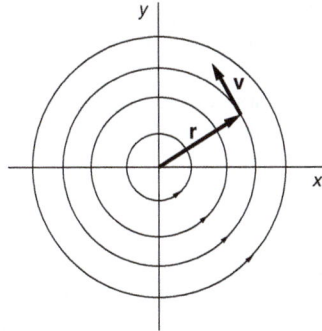

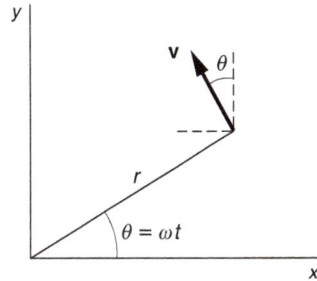

The curl of $\mathbf{v}$ is

$$\boldsymbol{\nabla}\times\mathbf{v} = \begin{vmatrix} \hat{\mathbf{i}} & \hat{\mathbf{j}} & \hat{\mathbf{k}} \\ \frac{\partial}{\partial x} & \frac{\partial}{\partial y} & \frac{\partial}{\partial z} \\ -\omega y & \omega x & 0 \end{vmatrix}$$

$$= \hat{\mathbf{k}}\left[\frac{\partial}{\partial x}(\omega x) + \frac{\partial}{\partial y}(\omega y)\right]$$

$$= 2\omega\hat{\mathbf{k}}$$

$$\neq 0.$$

If a paddle wheel is placed in the liquid, it will start to rotate. The rotation will be a maximum when the axis of the wheel points along the z axis parallel to $\boldsymbol{\nabla}\times\mathbf{v}$. In Europe, curl is often called "rot" (for "rotation").

Problems

For problems marked *, refer to page 521 for a hint, clue, or answer.

5.1 *Loop-the-loop**

A small block of mass m starts from rest and slides along a frictionless loop-the-loop as shown in the sketch on the next page. What should be the initial height z, so that m pushes against the top of the track (at a) with a force equal to its weight?

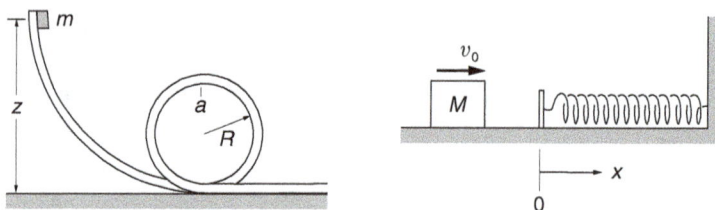

5.2 Block, spring, and friction

A block of mass M slides along a horizontal table with speed v_0. At $x = 0$ it hits a spring with spring constant k and begins to experience a friction force, as indicated in the right-hand sketch. The coefficient of friction is variable and is given by $\mu = bx$, where b is a constant. Find the distance l the block travels before coming to a permanent stop.

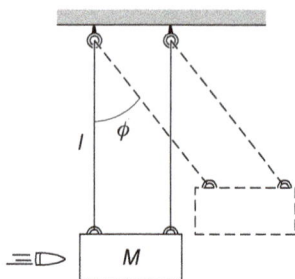

5.3 Ballistic pendulum*

A simple way to measure the speed of a bullet is with a *ballistic pendulum*. As illustrated, this consists of a wooden block of mass M into which the bullet is shot. The block is suspended from cables of length l, and the impact of the bullet causes it to swing through a maximum angle ϕ, as shown. The initial speed of the bullet is v, and its mass is m.

(*a*) How fast is the block moving immediately after the bullet comes to rest? (Assume that this happens quickly.)

(*b*) Show how to find the velocity of the bullet by measuring m, M, l, and ϕ.

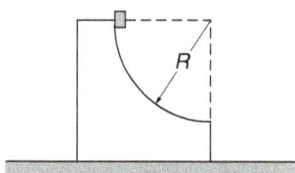

5.4 Sliding on a circular path*

A small cube of mass m slides down a circular path of radius R cut into a large block of mass M, as shown. M rests on a table, and both blocks move without friction. The blocks are initially at rest, and m starts from the top of the path.

Find the velocity v of the cube as it leaves the block.

5.5 Work on a whirling mass

Mass m whirls on a frictionless table, held to circular motion by a string which passes through a hole in the table. The string is pulled so that the radius of the circle changes from r_i to r_f.

(*a*) Show that the quantity $L = mr^2\dot\theta$ remains constant.

(*b*) Show that the work in pulling the string equals the increase in kinetic energy of the mass.

5.6 Block sliding on a sphere*

A small block slides from rest from the top of a frictionless sphere of radius R, as shown on the next page. How far below the top x does it lose contact with the sphere? The sphere does not move.

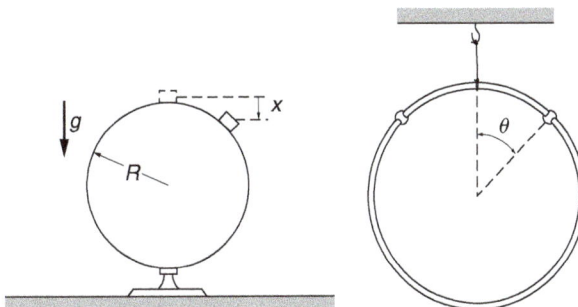

5.7 *Beads on hanging ring**
A ring of mass M hangs from a thread, and two beads of mass m slide on it without friction, as shown. The beads are released simultaneously from the top of the ring and slide down opposite sides. Show that the ring will start to rise if $m > 3M/2$, and find the angle at which this occurs.

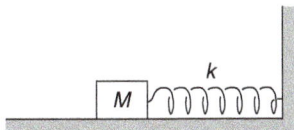

5.8 *Damped oscillation**
The block shown in the drawing is acted on by a spring with spring constant k and a weak friction force of constant magnitude f. The block is pulled distance x_0 from equilibrium and released. It oscillates many times and eventually comes to rest.

(*a*) Show that the decrease of amplitude is the same for each cycle of oscillation.

(*b*) Find the number of cycles n the mass oscillates before coming to rest.

5.9 *Oscillating block*
A block of mass M on a horizontal frictionless table is connected to a spring (spring constant k). The block is set in motion so that it oscillates about its equilibrium point with a certain amplitude A_0. The period of motion is $T_0 = 2\pi \sqrt{M/k}$.

(*a*) A lump of sticky putty of mass m is dropped onto the block. The putty sticks without bouncing. The putty hits M at the instant when the velocity of M is zero. Find

(1) The new period.

(2) The new amplitude.

(3) The change in the mechanical energy of the system.

(*b*) Repeat part (*a*), but this time assume that the sticky putty hits M at the instant when M has its maximum velocity.

5.10 *Falling chain**
A chain of total mass M and length l is suspended vertically with its lowest end touching a scale. The chain is released and falls onto the scale.

What is the reading of the scale when a length of chain, x, has fallen? (Neglect the size of individual links.)

5.11 *Dropped soldiers*

It is told that during World War II the Russians, lacking sufficient parachutes for airborne operations, occasionally dropped soldiers inside bales of hay onto snow.

The human body can survive an average pressure on impact of 30 lb/in^2. Suppose that the lead plane drops a dummy bale equal in weight to a loaded one from an altitude of 100 ft, and that the pilot observes that it sinks about 2 ft into the snow. If the weight of an average soldier is 180 lb and his effective area is 5 ft^2, is it safe to drop the men?

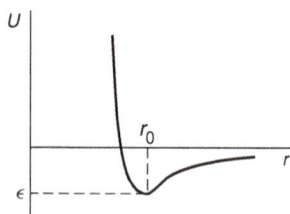

5.12 *Lennard-Jones potential*

A commonly used potential energy function to describe the interaction between two atoms is the Lennard-Jones 6-12 potential given by

$$U = \epsilon\left[\left(\frac{r_0}{r}\right)^{12} - 2\left(\frac{r_0}{r}\right)^6\right].$$

(a) Find the position of the potential minimum and its value.

(b) Near the minimum the atoms execute simple harmonic motion. Find the frequency of oscillation.

5.13 *Bead and gravitating masses*

A bead of mass m slides without friction on a smooth rod along the x axis. The rod is equidistant between two spheres of mass M. The spheres are located at $x = 0, y = \pm a$ as shown, and attract the bead gravitationally.

(a) Find the potential energy of the bead.

(b) The bead is released at $x = 3a$ with velocity v_i toward the origin. Find the speed as it passes the origin.

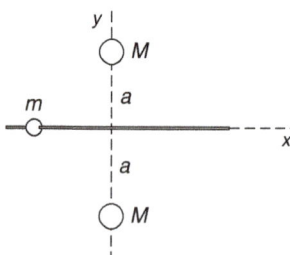

5.14 *Particle and two forces*

A particle of mass m moves in one dimension along the positive x axis. It is acted on by a constant force directed toward the origin with magnitude B, and an inverse-square law repulsive force with magnitude A/x^2.

(a) Find the potential energy function $U(x)$.

(b) Sketch the energy diagram for the system when the maximum kinetic energy is $K_0 = \frac{1}{2}mv_0^2$.

(c) Find the equilibrium position, x_0.

5.15 *Sportscar power*

A 1800-lb sportscar accelerates to 60 mi/h in 4 s. What is the average power that the engine delivers to the car's motion during this period? For consistency, we are using the definition 1 hp = 746 W.

5.16 *Snowmobile and hill*

A snowmobile climbs a hill at 15 mi/hr. The hill has a grade of 1 ft rise for every 40 ft. The resistive force due to the snow is

5% of the vehicle's weight. How fast will the snowmobile move downhill, assuming its engine delivers the same power?

5.17 *Leaper**

A 55-kg athlete leaps into the air from a crouching position. Her center of mass rises 60 cm as her feet leave the ground and then it continues another 80 cm to the top of the leap. What is the average power she develops, assuming the force on the ground is constant?

5.18 *Sand and conveyor belt*

Sand runs from a hopper at constant rate dm/dt onto a horizontal conveyor belt driven at constant speed V by a motor.

(a) Find the power needed to drive the belt.

(b) Compare the answer to (a) with the rate of change of kinetic energy of the sand. Can you account for the difference?

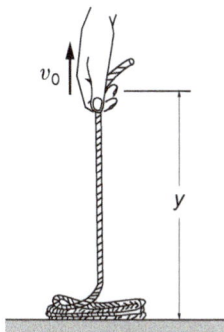

5.19 *Coil of rope*

A uniform rope of mass density λ per unit length is coiled on a smooth horizontal table. One end is pulled straight up with constant speed v_0, as shown.

(a) Find the force exerted on the end of the rope as a function of height y.

(b) Compare the power delivered to the rope with the rate of change of the rope's total mechanical energy.

6 TOPICS IN DYNAMICS

6.1 Introduction

This chapter illustrates applications of Newtonian mechanics and the conservation laws for momentum and energy that we have already studied; no new concepts are introduced. We shall apply these ideas to analyze some phenomena that appear again and again in the broad landscape of physics: small oscillations, stable and unstable motion, normal modes of vibration in bound systems, and the general properties of collisions that follow from the conservation laws.

6.2 Small Oscillations in a Bound System

The interatomic potential we discussed in Section 5.7 illustrates a universal feature of bound systems: because the potential energy has a minimum at equilibrium, nearly every bound system oscillates like a harmonic oscillator about its equilibrium position if it is slightly perturbed. This is suggested by the appearance of the energy diagram near the minimum—U has a nearly parabolic shape much like a harmonic oscillator potential. If the total energy is sufficiently low for the motion to be restricted to the region where the curve is closely parabolic, as illustrated in the sketch, the system must behave like a harmonic oscillator, as we shall now prove.

As we discussed in Note 1.2, any "well-behaved" differentiable function can be expanded in a Taylor's series about a given point. If we expand $U(r)$ about the position of the potential minimum r_0, then

$$U(r) = U(r_0) + (r - r_0) \left.\frac{dU}{dr}\right|_{r_0} + \frac{1}{2}(r - r_0)^2 \left.\frac{d^2U}{dr^2}\right|_{r_0} + \cdots .$$

Because U has its minimum at r_0, its first derivative is $dU/dr = 0$ at r_0. Furthermore, for sufficiently small displacements, we can neglect terms beyond the third in the power series. In this case

$$U(r) \approx U(r_0) + \frac{1}{2}(r - r_0)^2 \left.\frac{d^2U}{dr^2}\right|_{r_0} .$$

This is the potential energy of a harmonic oscillator,

$$U(x) = \text{constant} + \frac{1}{2}k(x - x_0)^2.$$

Comparing these two equations, we can identify the effective spring constant in the bound system as

$$k = \left.\frac{d^2U}{dr^2}\right|_{r_0} . \tag{6.1}$$

Because bound systems have a potential energy minimum at equilibrium, we naturally expect that they will behave like harmonic oscillators for small displacements (except for anomalous cases where the second derivative vanishes at equilibrium). For this reason, the harmonic oscillator approximation is relevant to phenomena ranging from molecular vibrations to oscillations in the shape of the Earth.

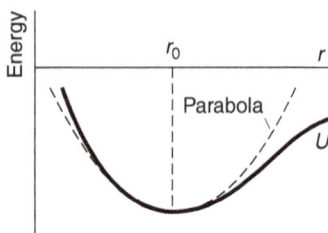

Energy

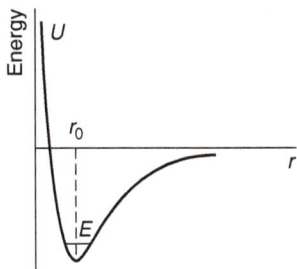

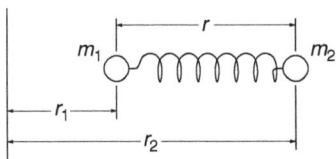

Example 6.1 Molecular Vibrations

Suppose that two atoms, masses m_1 and m_2, are bound in a diatomic molecule with energy so low that their separation is always close to the equilibrium value r_0.

With the parabola approximation, the molecule can be modeled as shown, with m_1 and m_2 connected by a spring of equilibrium length r_0 and spring constant k.

The effective spring constant from Eq. (6.1) is $k = (d^2U/dr^2)|_{r_0}$. How can we find the vibration frequency of the molecule?

The equations of motion of the two masses are

$$m_1\ddot{r}_1 = k(r - r_0)$$
$$m_2\ddot{r}_2 = -k(r - r_0),$$

where $r = r_2 - r_1$ is the instantaneous separation of the atoms. To find the equation of motion for r, divide the first equation by m_1 and the second by m_2, and subtract. The result is

$$\ddot{r}_2 - \ddot{r}_1 = \ddot{r} = -k\left(\frac{1}{m_1} + \frac{1}{m_2}\right)(r - r_0)$$

or

$$\ddot{r} = -\frac{k}{\mu}(r - r_0),$$

where $\mu = m_1 m_2/(m_1 + m_2)$ has the dimension of mass and is called the *reduced mass*.

For a harmonic oscillator with equation of motion $\ddot{x} = -(k/m)(x - x_0)$ the frequency of oscillation is $\omega = \sqrt{k/m}$, so by analogy, the vibrational frequency of the molecule is

$$\omega = \sqrt{\frac{k}{\mu}}$$

$$= \sqrt{\frac{d^2U}{dr^2}\bigg|_{r_0} \frac{1}{\mu}}.$$

This vibrational motion, characteristic of all molecules, can be identified by the light the molecule radiates or absorbs. The vibrational frequencies typically lie in the near-infrared ($< 5 \times 10^{14}$ Hz), and by measuring the frequency we can find the value of d^2U/dr^2 at the potential energy minimum. For the hydrogen chloride (HCl) molecule, the effective spring constant turns out to be 4.8×10^5 dynes/cm = 480 N/m

(≈ 3 lb/in). For the nitric oxide (NO) molecule, $k \approx 1550$ N/m. Not surprisingly, the energy to separate the atoms is about three times stronger in NO than in HCl.

For large amplitudes the higher order terms in the Taylor's series start to play a measurable role, and lead to slight departures of the oscillator from its ideal harmonic behavior. Observing these slight "anharmonicities" gives further details regarding the shape of the potential energy curve.

Example 6.2 Lennard-Jones potential

The Lennard-Jones 6-12 potential, given by

$$U = \epsilon \left[\left(\frac{r_0}{r} \right)^{12} - 2 \left(\frac{r_0}{r} \right)^6 \right], \tag{1}$$

is a commonly used potential energy function to describe the interaction between two atoms. The figure shows U for the chlorine diatomic molecule Cl_2, where $r_0 = 2.98\,\text{Å} = 2.98 \times 10^{-10}$ m and $\epsilon = 2.48\,\text{eV} = 3.97 \times 10^{-19}$ J.

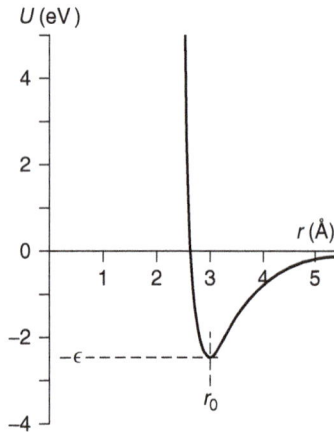

The term $(r_0/r)^{12}$ rises steeply for $r < r_0$, which models the strong "hard sphere" repulsion between two atoms at close separation. The term $(r_0/r)^6$ decreases slowly for $r > r_0$ to model the long attractive tail between two atoms at larger separations. The two terms together produce a potential capable of binding atoms, as indicated in the sketch.

We shall calculate the frequency of small oscillations about equilibrium for two identical atoms of mass m bound by the Lennard-Jones potential.

From Eq. (1) the first derivative of U with respect to r is

$$\frac{dU}{dr} = \left(\frac{\epsilon}{r_0} \right) \left[-12 \left(\frac{r_0}{r} \right)^{13} + 12 \left(\frac{r_0}{r} \right)^7 \right]. \tag{2}$$

From Eq. (2), $dU/dr = 0$ at $r = r_0$, verifying that the equilibrium radius is r_0. Substituting $r = r_0$ in Eq. (1), the depth of the potential well at equilibrium is $U(r_0) = -\epsilon$.

The second derivative of U with respect to r is

$$\frac{d^2U}{dr^2} = \left(\frac{\epsilon}{r_0^2} \right) \left[(12)(13) \left(\frac{r_0}{r} \right)^{14} - (12)(7) \left(\frac{r_0}{r} \right)^8 \right].$$

According to Eq. (6.1), the effective spring constant is then

$$k = \frac{d^2U}{dr^2} \bigg|_{r_0}$$
$$= \frac{72\epsilon}{r_0^2}.$$

The reduced mass is $\mu = m^2/2m = m/2$ and the vibrational frequency ω is therefore

$$\omega = \sqrt{k/\mu}$$
$$= 12\sqrt{\epsilon/r_0^2 m}.$$

Let's apply this result to a real molecule, for example the chlorine diatomic molecule Cl_2, for which $m = 5.89 \times 10^{-26}$ kg, and for which the calculated values of r_0 and ϵ are $r_0 = 2.98 \times 10^{-10}$ m and $\epsilon = 3.97 \times 10^{-19}$ J. We find $\omega = 1.05 \times 10^{14}$ rad/s, in excellent agreement with the experimentally measured vibrational frequency 1.05×10^{14} rad/s.

6.2.1 Quadratic Energy Forms

In many problems it is natural to write the energies using variables other than linear displacement. The energies in some regimes of motion often have the quadratic forms

$$U = \tfrac{1}{2}Aq^2 + \text{constant}$$
$$K = \tfrac{1}{2}B\dot{q}^2, \tag{6.2}$$

where q represents a variable appropriate to the problem. For the elementary case of a mass on a spring we have

$$U = \tfrac{1}{2}kx^2$$
$$K = \tfrac{1}{2}m\dot{x}^2$$

and

$$\omega = \sqrt{\frac{k}{m}}.$$

By analogy with a mass on a spring, the angular frequency of the system described by Eq. (6.2) is

$$\omega = \sqrt{\frac{A}{B}}.$$

To show explicitly that any system whose energy has the form of Eq. (6.2) oscillates harmonically with a frequency $\sqrt{A/B}$, note that the total energy of the system is

$$E = K + U$$
$$= \tfrac{1}{2}B\dot{q}^2 + \tfrac{1}{2}Aq^2 + \text{constant}.$$

Because the system is conservative, E is constant. Differentiating the energy equation with respect to time gives

$$\frac{dE}{dt} = B\dot{q}\ddot{q} + Aq\dot{q}$$
$$= 0$$

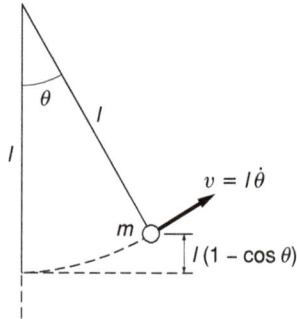

or

$$\ddot{q} + \frac{A}{B}q = 0.$$

Hence q undergoes harmonic motion with angular frequency $\sqrt{A/B}$.

Once we have identified the kinetic and potential energies of a bound system, we can find the angular frequency of small oscillations by inspection. For instance, the energies of a pendulum are

$$U = mgl(1 - \cos\theta) \approx \tfrac{1}{2}mgl\theta^2$$
$$K = \tfrac{1}{2}ml^2\dot{\theta}^2$$

so that

$$\omega = \sqrt{g/l}.$$

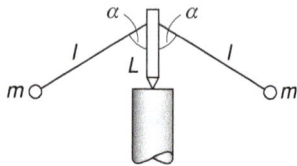

Example 6.3 Small Oscillations of a Teeter Toy

The teeter toy consists of two identical weights hanging on drooping arms from a peg, as shown.

In this example we find the period of oscillation of the teeter toy when it is rocking from side to side. For simplicity, we shall consider only rocking motion in the vertical plane.

Let us evaluate the potential energy when the teeter toy is cocked at angle θ, as shown in the sketch.

If we take the zero of gravitational potential at the pivot, we have

$$U(\theta) = mg[L\cos\theta - l\cos(\alpha + \theta)] + mg[L\cos\theta - l\cos(\alpha - \theta)].$$

Using the identity $\cos(\alpha \pm \theta) = \cos\alpha\cos\theta \mp \sin\alpha\sin\theta$, we can rewrite $U(\theta)$ as

$$U(\theta) = 2mg\cos\theta(L - l\cos\alpha)$$
$$= -A\cos\theta,$$

where $A = 2mg(l\cos\alpha - L) = $ constant. Using the small-angle approximation (or alternatively, expanding $U(\theta)$ in a Taylor's series about $\theta = 0$), we have

$$U(\theta) = -A\left(1 - \frac{\theta^2}{2} + \cdots\right)$$
$$\approx -A + \tfrac{1}{2}A\theta^2.$$

To find the kinetic energy, let s be the distance of each mass from the pivot, as shown.

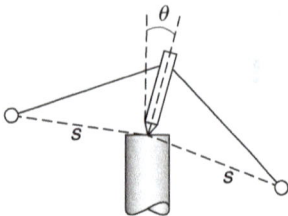

If the toy rocks with angular speed $\dot{\theta}$, the speed of each mass is $s\dot{\theta}$, and the total kinetic energy is

$$K = \tfrac{1}{2}(2m)s^2\dot{\theta}^2$$
$$= \tfrac{1}{2}B\dot{\theta}^2,$$

where $B = 2ms^2$.

Hence the frequency of oscillation is

$$\omega = \sqrt{\frac{A}{B}}$$
$$= \sqrt{\frac{g(l\cos\alpha - L)}{s^2}}.$$

6.3 Stability

The result $F = -dU/dx$ that we derived in Section 5.6 makes it possible to calculate the force from a known potential energy function. In addition, the result is helpful for visualizing the stability of a system from a diagram of the potential energy.

Consider the case of a harmonic oscillator, where the potential energy $U = kx^2/2$ is described by a parabola.

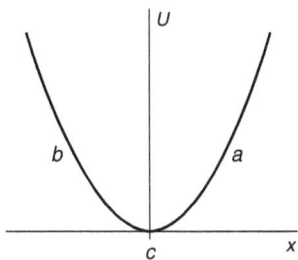

At point a, $dU/dx > 0$ and so the force is negative. At point b, $dU/dx < 0$ and the force is positive. At c, $dU/dx = 0$ and the force is zero. The force is directed toward the origin no matter which way the particle is displaced, and the force vanishes only when the particle is at the origin. The minimum of the potential energy curve coincides with the equilibrium position of the system. Evidently this is a stable equilibrium, because any displacement of the system produces a force that tends to push the particle back toward its resting point.

Whenever $dU/dx = 0$, a system is in equilibrium. If this occurs at a maximum of U, where $d^2U/dx^2 < 0$, the equilibrium is not stable because a positive displacement would create a positive force, which would tend to increase the displacement, and a negative displacement would in this case produce a negative force, again causing the displacement to become larger. Equilibrium is stable if $d^2U/dx^2 > 0$.

A pendulum consisting of a mass m supported by a rigid rod of length l and negligible mass offers a good illustration of stability. If we take the potential energy to be zero at the bottom of its swing, we see that

$$U(\theta) = mgz$$
$$= mgl(1 - \cos\theta).$$

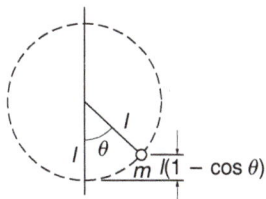

The pendulum is in equilibrium for $\theta = 0$ and $\theta = \pi$. A pendulum will quite happily hang downward for as long as you please but it will not hang vertically upward for long. $dU/d\theta = 0$ at $\theta = \pi$, but U has a maximum there (at $\theta = \pi$, $d^2U/d\theta^2 = -mgl < 0$) and the equilibrium is not stable.

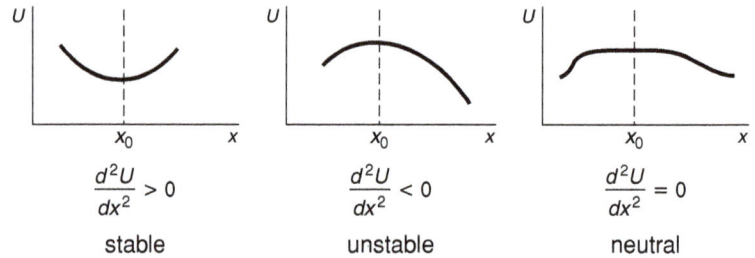

$\dfrac{d^2U}{dx^2} > 0$	$\dfrac{d^2U}{dx^2} < 0$	$\dfrac{d^2U}{dx^2} = 0$
stable	unstable	neutral

Example 6.4 Stability of the Teeter Toy

The teeter toy, which we looked at in Example 6.3, is unexpectedly stable—the toy can be spun or rocked with little danger of toppling over. We can see why this is so by looking at its potential energy. For simplicity, we consider only rocking motion in the vertical plane and assume that all the mass is in the weights.

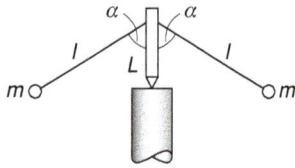

Let us evaluate the potential energy when the teeter toy is cocked at angle θ, as shown in the sketch. If we take the zero of gravitational potential at the pivot, we have

$$U(\theta) = mg[L\cos\theta - l\cos(\alpha + \theta)] + mg[L\cos\theta - l\cos(\alpha - \theta)].$$

Using the identity $\cos(\alpha \pm \theta) = \cos\alpha\cos\theta \mp \sin\alpha\sin\theta$, we can rewrite $U(\theta)$ as

$$U(\theta) = 2mg\cos\theta(L - l\cos\alpha).$$

Equilibrium occurs when

$$\frac{dU}{d\theta} = -2mg\,\sin\theta(L - l\,\cos\alpha) = 0.$$

The solution is $\theta = 0$, as we expect from symmetry. (We reject the solution $\theta = \pi$ on the grounds that θ must be limited to less than $\pi/2$.)

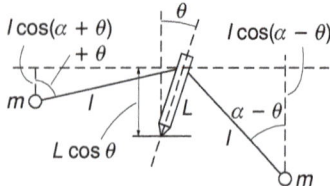

To investigate the stability of the equilibrium position, we must examine the second derivative of the potential energy. We have

$$\frac{d^2U}{d\theta^2} = -2mg\cos\theta(L - l\cos\alpha).$$

At equilibrium,

$$\left.\frac{d^2U}{d\theta^2}\right|_{\theta=0} = -2mg(L - l\cos\alpha).$$

For stability, we require the second derivative to be positive, so that $L - l\cos\alpha < 0$, or $L < l\cos\alpha$. The conclusion is that for the teeter toy to be stable, the weights must hang below the pivot, as can be seen from the sketch.

In Example 6.3, we found that the oscillation frequency of the teeter toy is

$$\omega = \sqrt{\frac{g(l\cos\alpha - L)}{s^2}} \tag{1}$$

and in this example we showed that $(l\cos\alpha - L) > 0$ for stability. Equation (1) shows that as $(l\cos\alpha - L) \to 0$, then $\omega \to 0$ and the period of oscillation becomes very long. In the limit $(l\cos\alpha - L) = 0$, the system is in neutral equilibrium. If $(l\cos\alpha - L) < 0$, the system is unstable and the teeter toy topples, with θ varying exponentially instead of harmonically.

As the previous example shows, a low frequency of oscillation is associated with a system operating near the threshold of stability. This is a general property of stable systems because a low frequency of oscillation corresponds to a weak restoring force. When a ship rolled by a wave oscillates about equilibrium, the period of the roll should be long for comfort. The hull is therefore designed so that its center of gravity is as high as possible consistent with stability. Lowering the center of gravity makes the system "stiffer." The roll becomes quicker and less comfortable, but the ship becomes intrinsically more stable.

6.4 Normal Modes

A single harmonic oscillator is among the simplest systems in physics. The next step up in complexity is a system of two harmonic oscillators. This situation is not particularly interesting if the oscillators are isolated but if they interact even slightly the system is transformed and acquires unique properties.

To illustrate what happens, let's consider a system of two identical pendulums of length l and mass m, described by angles θ_1 and θ_2. In the small-angle approximation, $\sin\theta \approx \theta$, the equations of motion are

$$ml\ddot{\theta}_1 = -mg\theta_1, \tag{6.3a}$$
$$ml\ddot{\theta}_2 = -mg\theta_2. \tag{6.3b}$$

Taking $\omega_0^2 = g/l$, these equations take the standard form $\ddot{\theta} + \omega_0^2\theta = 0$ with solutions $\theta(t) = A\sin(\omega_0 t + \phi)$. The constants A and ϕ can be chosen to satisfy the initial conditions.

To create an interaction between the pendulums we couple them with a weak spring of constant k. The pivots of the pendulums are chosen to be separated by the length of the spring, so when the pendulums hang

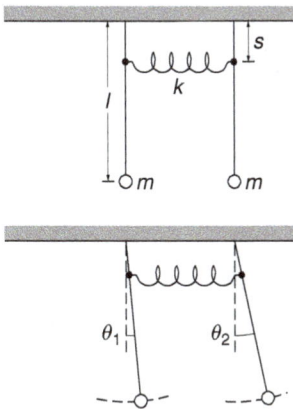

straight down the spring is unstretched, as shown. The spring is attached at a short distance s from the pivot points, so that when the pendulums move the spring is stretched by distance $s(\theta_1 - \theta_2)$. The equations of motion are

$$ml\ddot{\theta}_1 = -mg\theta_1 - ks(\theta_1 - \theta_2), \tag{6.4a}$$

$$ml\ddot{\theta}_2 = -mg\theta_2 + ks(\theta_1 - \theta_2). \tag{6.4b}$$

Dividing these equations by ml, and letting $\omega_0^2 = g/l$, $\kappa^2 = ks/ml$, and introducing $\Omega^2 = \omega_0^2 + \kappa^2$, Eqs. (6.4) become

$$\ddot{\theta}_1 + \Omega^2\theta_1 = \kappa^2\theta_2, \tag{6.5a}$$

$$\ddot{\theta}_2 + \Omega^2\theta_2 = \kappa^2\theta_1. \tag{6.5b}$$

It is reasonable to guess that the motions are periodic and try solutions of the form

$$\theta_1(t) = A_1 \sin(\omega t + \phi), \tag{6.6a}$$

$$\theta_2(t) = A_2 \sin(\omega t + \phi). \tag{6.6b}$$

θ_1 and θ_2 have been given the same time-dependence because if they differed, Eqs. (6.5) could not be satisfied at all times. Whether a value of ω can be found that satisfies Eqs. (6.5) remains to be seen. Substituting Eqs. (6.6) in Eqs. (6.5) gives

$$(\Omega^2 - \omega^2)A = \kappa^2 B, \tag{6.7a}$$

$$(\Omega^2 - \omega^2)B = \kappa^2 A, \tag{6.7b}$$

and substituting Eq. (6.7b) in Eq. (6.7a) leads to

$$(\Omega^2 - \omega^2)^2 = \kappa^4 \quad \text{or} \tag{6.8a}$$

$$\Omega^2 - \omega^2 = \pm\kappa^2. \tag{6.8b}$$

Hence we have two possibilities for ω. Recalling that $\Omega^2 = \omega_0^2 + \kappa^2$, we have

$$\omega_0^2 + \kappa^2 - \omega^2 = \kappa^2, \tag{6.9a}$$

$$\omega_0^2 + \kappa^2 - \omega^2 = -\kappa^2. \tag{6.9b}$$

The solutions for ω^2 are

$$\omega_+^2 = \omega_0^2 + 2\kappa^2 \qquad\qquad A = -B, \tag{6.10a}$$

$$\omega_-^2 = \omega_0^2 \qquad\qquad A = B. \tag{6.10b}$$

Evidently there are two ways for the system to display simple periodic motion. At frequency ω_+, the pendulums swing in opposite directions with the same amplitude. The frequency is larger than the free pendulum frequency ω_0 because the pendulums pull against each other as they swing apart, or push as they swing together, increasing the restoring force. At frequency ω_-, the pendulums swing together with the same amplitude. The spring is never stretched and the frequency is the same as for a free pendulum.

These two motions are known as *normal modes* of the system. A normal mode of a coupled oscillator system is a superposition of amplitudes of the oscillators in which all the motions are at a single frequency. In a sense, a normal mode is a new oscillator with a unique frequency. Written out in full, the modes are

Mode 1, frequency $\omega_+ = \sqrt{\omega_0^2 + 2\kappa^2 x}$, $\theta_1 = -\theta_2$
$$\theta_1(t) = A_+ \sin(\omega_+ t + \phi_+)$$
$$\theta_2(t) = -A_+ \sin(\omega_+ t + \phi_+)$$

Mode 2, frequency $\omega_- = \omega_0$, $\theta_1 = \theta_2$
$$\theta_1(t) = A_- \sin(\omega_- t + \phi_-)$$
$$\theta_2(t) = A_- \sin(\omega_- t + \phi_-)$$

The equations of motion for the coupled oscillators are linear in the displacements and so solutions to the equations of motion can be added to yield new solutions. Any linear combination of the normal modes can be added, so the most general possible motion of the system is

$$\theta_1(t) = A_+ \sin \omega_+(t + \phi_+) + A_- \sin \omega_-(t + \phi_-) \tag{6.11a}$$

$$\theta_2(t) = -A_+ \sin \omega_-(t + \phi_+) + A_- \sin \omega_-(t + \phi_-). \tag{6.11b}$$

We are free to choose the amplitudes A_+ and A_-. If $A_- = 0$ the pendulums execute simple harmonic motion at the frequency ω_+. The most general motion of each pendulum has components at both of the system's normal mode frequencies.

In the general case of coupled oscillators, a normal mode analysis yields the frequencies and relative amplitudes of all the normal modes. The actual motion depends on the initial conditions. If energy is put into one normal mode, it remains in that mode with the individual oscillators vibrating with their own particular relative amplitude. If several modes are excited, the excitations can interfere with each other, causing the energy to move among the individual oscillators as the following example shows.

Example 6.5 Energy Transfer Between Coupled Oscillators

Consider again the two-pendulum and spring system we have just discussed. Suppose that one of the pendulums, called pendulum 1, is displaced to angle θ_0 and released at $t = 0$. Initially, all of the energy of the system is in pendulum 1. The problem is to see what happens to the energy as time goes by.

The initial conditions of the system are $\theta_1(0) = \theta_0, \theta_2(0) = 0$. Because the pendulums are initially at rest, the phase angles ϕ in Eqs. (6.11) are

$\pi/2$. The solutions are of the form $\cos \omega t$, and Eqs. (6.11) become

$$\theta_1(t) = A_+ \cos(\omega_+ t) + A_- \cos(\omega_- t)$$

$$\theta_2(t) = -A_+ \cos(\omega_+ t) + A_- \cos(\omega_- t).$$

To satisfy the condition $\theta_2(0) = 0$, we take $A_+ = A_- = \theta_0/2$. Then

$$\theta_1(t) = \frac{\theta_0}{2}[\cos(\omega_+ t) + \cos(\omega_- t)] \tag{1a}$$

$$\theta_2(t) = \frac{\theta_0}{2}[-\cos(\omega_+ t) + \cos(\omega_- t)]. \tag{1b}$$

We can write this in a more symmetric form by introducing the average frequency $\bar{\omega} = (\omega_+ + \omega_-)/2$, and the difference frequency $\delta = \omega_+ - \omega_-$, so that $\omega_+ = \bar{\omega} + \delta$, $\omega_- = \bar{\omega} - \delta$. Using the identity $\cos(A + B) = \cos A \cos B - \sin A \sin B$, we can write Eqs. (6.13) as

$$\theta_1(t) = \theta_0 \cos \delta t \cos \bar{\omega} t$$

$$\theta_2(t) = \theta_0 \sin \delta t \cos \bar{\omega} t.$$

The system oscillates at average frequency $\bar{\omega}$, but the amplitude of oscillation slowly swings back and forth between the two pendulums at frequency δ. Initially the energy is in pendulum 1, but when $\delta t = \pi/x$, the energy has transferred to pendulum 2. The energy eventually returns to pendulum 1. As the coupling is decreased, δ grows smaller and it takes longer for the energy to transfer. Nonetheless, it will get there eventually. In the mechanical design of complex systems such as airplanes, care needs to be taken to avoid accidental resonances in the system. Even if the coupling between the oscillators is weak, energy transfer can be large with results that could be calamitous.

Example 6.6 Normal Modes of a Diatomic Molecule

In this example, we shall again analyze the diatomic molecule model presented in Example 6.1, but from the point of view of normal modes, a description that can be extended to polyatomic molecules modeled as several masses connected by springs. The diatomic molecule is modeled as two masses m_1 and m_2 connected by a spring of constant k. We consider only motion along the x axis. with coordinates x_1, x_2. The equations of motion are

$$m_1 \ddot{x}_1 = k[(x_2 - x_{20}) - (x_1 - x_{10})]$$

$$m_2 \ddot{x}_2 = -k[(x_2 - x_{20}) - (x_1 - x_{10})],$$

where x_{10}, x_{20} are the equilibrium coordinates of the masses. Note that the forces are equal and opposite, as required by Newton's third law.

These equations can be simplified by changing the dependent variable to eliminate the equilibrium coordinates. Let $x_i' = x_i - x_{i0}$, where

$i = (1, 2)$. The equations of motion then become

$$m_1 \ddot{x}'_1 = k(x'_2 - x'_1) \tag{1a}$$
$$m_2 \ddot{x}'_2 = -k(x'_2 - x'_1). \tag{1b}$$

To illustrate a general approach to molecular vibrations, suppose we have a polyatomic molecule model with N masses and several springs coupling them. We now look for special solutions of the form

$$x'_i = a_i \sin(\omega t + \phi) \qquad i = 1, \ldots, N$$

where a_i is the vibration amplitude of the ith mass. Note that in the special solution we are looking for, each mass vibrates with the same angular frequency ω. The phase factor ϕ is also the same for each mass. We justify the existence of such a solution by arguing that if the masses were vibrating with different frequencies, it would not be possible to conserve linear momentum for an isolated molecule.

These special solutions are called the *normal modes* of the system, analogous to the normal modes of the two coupled pendulums discussed in Section 6.4. If there are N' equations of motion, there are N' normal modes, each with its own special frequency. There is always a normal mode with zero frequency, corresponding to the trivial case of a stationary system with no vibrations, leaving $N' - 1$ non-trivial frequencies.

To show an example of a normal mode, and to keep the algebra simple, we go back to the diatomic molecule model with masses m_1 and m_2. There are two equations of motion, so we expect one non-trivial vibration frequency. Using

$$x'_i = a_i \sin(\omega t + \phi)$$
$$\ddot{x}'_i = -\omega^2 a_i \sin(\omega t + \phi)$$

in Eqs. (1a) and (1b) gives

$$(k - \omega^2 m_1)a_1 = ka_2 \tag{2a}$$
$$(k - \omega^2 m_2)a_2 = ka_1. \tag{2b}$$

The factor $\sin(\omega t + \phi)$ occurs in every term and cancels throughout. Solving Eq. (2a) for a_2 and substituting in Eq. (2b) gives the following equation for the amplitude a_1, after some simplification:

$$\omega^2[\omega^2 m_1 m_2 - k(m_1 + m_2)]a_1 = 0. \tag{3}$$

There are several possible solutions for Eq. (3). One solution is $a_1 = 0$; it follows from Eq. (2a) that $a_2 = 0$ also. This is a trivial solution for a stationary non-vibrating system. Similarly, the solution $\omega^2 = 0$

also corresponds to a non-vibrating system. The interesting non-trivial solution is

$$\omega^2 = \frac{k(m_1 + m_2)}{m_1 m_2}$$

$$= \frac{k}{\mu},$$

as we expect.

Now that we have the normal mode frequency, we can solve Eq. (2a) for the relative amplitudes a_2/a_1:

$$\frac{a_2}{a_1} = -\frac{m_1}{m_2}.$$

The two masses move in opposite directions, with amplitudes that ensure conservation of linear momentum. The actual amplitudes depend on the initial conditions.

Treating a polyatomic molecule model follows the same lines, with more complicated algebra, as we shall see in the following example. Nonetheless, the final result is again the normal mode frequencies ω_i in terms of the masses and spring constants. The important point is that the ω_i are the only possible non-trivial vibrational frequencies of the system, and any possible motion is therefore a linear combination of the normal modes

$$\sum_{i=1}^{N'-1} A_i \sin(\omega_i t + \phi_i).$$

The amplitudes A_i and the phase angles ϕ_i depend on the initial conditions. It is easy to show that the total energy in a normal mode is proportional to A_i^2.

Example 6.7 Linear Vibrations of Carbon Dioxide

CO_2 is a linear molecule consisting of a central carbon atom with an oxygen atom bound on either side. Restricting ourselves to linear vibrations, a mass-and-spring model has three masses m, M, and m and two springs in line, as shown. Both springs have the same spring constant k representing the carbon–oxygen bonds.

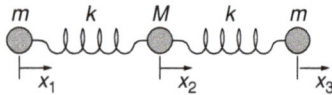

Denoting the departures of the atoms from their equilibrium positions by x_1, x_2, x_3, the equations of motion are

$$m\ddot{x}_1 = -k(x_1 - x_2) \tag{1a}$$

$$M\ddot{x}_2 = -k(x_2 - x_1) - k(x_2 - x_3) \tag{1b}$$

$$m\ddot{x}_3 = -k(x_3 - x_2). \tag{1c}$$

The notation is simplified by defining frequencies $\omega_0 = \sqrt{k/m}$ and $\Omega_0 = \sqrt{k/M}$. Equations (6.17) become

$$\ddot{x}_1 + \omega_0^2(x_1 - x_2)) = 0$$
$$\ddot{x}_2 + \Omega_0^2(2x_2 - x_1 - x_2) = 0$$
$$\ddot{x}_3 + \omega_0^2(x_3 - x_2) = 0.$$

If we search for solutions of the form $x = a\sin(\omega t + \phi)$, we obtain

$$\omega^2 a_1 - \omega_0^2(a_1 - a_2) = 0 \tag{2a}$$
$$\omega^2 a_2 - \Omega_0^2(2a_2 - a_1 - a_3) = 0 \tag{2b}$$
$$\omega^2 a_3 - \omega_0^2(a_3 - a_2) = 0. \tag{2c}$$

There is a general method for finding the values of ω^2 that satisfy these equations, in fact any number of such coupled equations, but we can find the solutions by straightforward algebra. (The general method involves finding the roots of the determinant of the coefficients, a well-known result from linear algebra for the solution of homogeneous equations.)

If $a_2 \neq 0$, then Eqs. (2a) and (2c) give $a_1 = a_3 = [\omega_0^2/(\omega^2 - \omega_0^2)]a_2$. Substituting these in Eqs. (2b) gives

$$\omega^2(\omega^2 - \omega_0^2) - 2\Omega_0^2) = 0. \tag{3}$$

$\omega = 0$ is a solution but not a normal mode; if $\ddot{x} = 0$, $x = x_0 + vt$. This solution describes center of mass motion for a free molecule. The second root of Eq. (3) is $\omega = \sqrt{\omega_0^2 + 2\Omega_0^2}$. This is an antisymmetric mode in which the two oxygen atoms move in one direction while the carbon atom moves in the opposite direction, as shown in (a). From Eq. (2a) the relative amplitude of the carbon and oxygen motions is $a_1/a_2 = -\omega_0^2/2\Omega_0^2 = -M/2m$, as one expects in order to conserve total momentum.

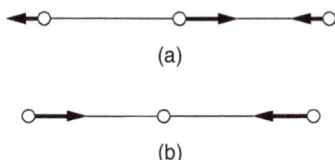

The remaining solution is for the symmetric case: $a_2 = 0$, $a_1 = -a_3$, and $\omega = \omega_o$. Here the carbon atom remains at rest while the oxygen atoms move in opposite directions at frequency ω, as shown in (b).

(a)

(b)

6.5 Collisions and Conservation Laws

In the early 1900s, Ernest Rutherford and his colleagues at the University of Manchester in England carried out experiments that set physics on a path that persists to this day. They bombarded thin metal foils with energetic α-rays (nuclei of ^{4}He) emitted by radioactive elements. Most of the α-rays were deflected ("scattered") through relatively small angles, but the experimenters were amazed to find that some of the α-rays scattered through very large angles, bouncing back from the foil. Ruther-ford said that it was as incredible as if an artillery shell bounced back from a sheet of paper. Interpretation of these experiments led to our model of the atom: a small dense nucleus containing most of the mass, surrounded by a cloud of electrons. Ever since this ground-breaking work physicists have employed collision and scattering experiments to investigate the properties of particles, for example to map out the forces particles exert or how their charge or mass is distributed. The Large Hadron Collider, a circular proton accelerator 8500 m in diameter lo-cated near Geneva on the French–Swiss border, was built to collide en-ergetic protons with matter to stimulate reactions that might reveal theo-retically predicted particles—or possibly some new physics.

In this section, we shall use conservation laws to derive general results that must hold for any collision, regardless of the nature of the forces in-volved. We shall use only non-relativistic ("classical") mechanics here, leaving the relativistic treatment of collisions until Chapter 13. Rela-tivistic effects typically depend on the factor $\sqrt{1 - v^2/c^2}$, where v is the speed of a particle and c is the speed of light. Classical mechanics there-fore holds to good accuracy if $v^2/c^2 \ll 1$. Another way of expressing this criterion is the ratio of the classical kinetic energy of a particle to its *rest mass* energy $= m_0 c^2$:

$$\frac{2 \times \text{classical kinetic energy}}{\text{rest mass energy}} = \frac{m_0 v^2}{m_0 c^2}$$
$$= \frac{v^2}{c^2}$$
$$\ll 1$$

for classical mechanics to hold accurately.

The rest mass m_0 of a particle is its mass when it is at rest. Rest mass is commonly expressed in energy units using $E = m_0 c^2$; for an electron m_0 is 0.51 MeV, and for a proton it is 938 MeV. The Large Hadron Collider can accelerate protons to energies as high as 14 TeV. Collisions in this energy range are strongly relativistic, but conservation laws (appropriate to relativistic mechanics) still apply.

6.5.1 Stages of a Collision

The drawings show three stages during the collision of two particles.

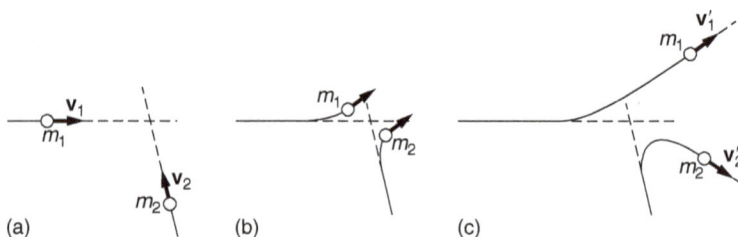

(a)   (b)   (c)

In (*a*), long before the collision, each particle is effectively free, since the interaction forces are generally important only at small separations. As the particles approach, (*b*), the momentum and energy of each particle change due to the interaction forces. Finally, long after the collision, (*c*), the particles are again free and move along straight lines with new directions and velocities. Experimentally, we usually know the initial velocities $\mathbf{v}_1$ and $\mathbf{v}_2$; often one particle is initially at rest in a target and is bombarded by particles of known energy. The experiment might consist of measuring the final velocities $\mathbf{v}'_1$ and $\mathbf{v}'_2$ with suitable particle detectors.

Assuming that external forces are negligible, the total momentum is conserved and we have

$$\mathbf{P}_i = \mathbf{P}_f.$$

For a two-body collision, this becomes

$$m_1\mathbf{v}_1 + m_2\mathbf{v}_2 = m_1\mathbf{v}'_1 + m_2\mathbf{v}'_2. \tag{6.20}$$

Equation (6.20) is equivalent to three scalar equations. However, the components of $\mathbf{v}'_1$ and $\mathbf{v}'_2$ comprise six unknowns, so a complete solution requires three more equations. The energy equation provides one additional relation between the velocities, as we now show.

6.5.2 Elastic and Inelastic Collisions

Consider a collision on a linear air track between two riders of equal mass M fitted with coil springs.

Suppose that initially rider 1 has speed v and rider 2 is at rest. After the collision, 1 is at rest and 2 moves to the right with speed v. It is clear that momentum has been conserved and that the total kinetic energy of the two bodies, $Mv^2/2$, is the same before and after the collision. A collision in which the total kinetic energy is unchanged is called an *elastic* collision. A collision is elastic if the interaction forces are conservative, like the spring force in this example.

As a second experiment, take the same two riders and replace the springs with lumps of sticky putty. Let 2 be initially at rest, as before.

After the collision, the riders stick together and move off with speed v'. By conservation of momentum, $Mv = 2Mv'$, so that $v' = v/2$. The initial kinetic energy of the system is $Mv^2/2$, but the final kinetic energy is $(2M)v'^2/2 = Mv^2/4$. In this collision, the kinetic energy is only half as much after the collision as before. The kinetic energy has changed because the interaction forces were non-conservative; part of the initial kinetic energy was transformed to random heat energy in the putty during the collision. A collision in which the total kinetic energy is not conserved is called an *inelastic* collision.

Although the total energy of any system is always conserved, part of the kinetic energy in a collision may be converted to some other form. To take this into account, we write the conservation of energy equation for collisions as

$$K_i = K_f + Q, \tag{6.21}$$

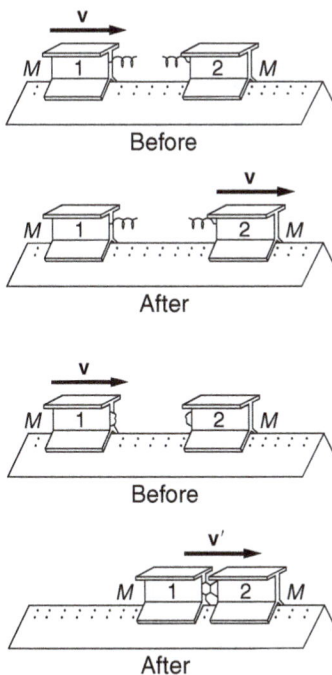

Before

After

Before

After

where $Q = K_i - K_f$ is the amount of kinetic energy converted to another form. For a two-body collision, Eq. (6.21) becomes

$$\tfrac{1}{2}m_1v_1{}^2 + \tfrac{1}{2}m_2v_2{}^2 = \tfrac{1}{2}m_1v_1'{}^2 + \tfrac{1}{2}m_2v_2'{}^2 + Q. \tag{6.22}$$

In most collisions on the everyday scale, kinetic energy is lost and Q is positive. However, Q can be negative if internal energy of the system is converted to kinetic energy in the collision. Such collisions are some-times called *superelastic*—the kinetic energy of the outgoing particles is greater than for the incoming particles. They are important in atomic and nuclear physics; in many nuclear reactions, the total mass of the products is less than the total mass of the reactants. According to the special the-ory of relativity, discussed in Chapter 13, $Q = (M_{\text{products}} - M_{\text{reactants}})c^2 = \Delta mc^2 < 0$. Superelastic collisions are rarely encountered in the everyday world, but one example would be the collision of two cocked mouse-traps, releasing the energy stored in the springs.

6.5.3 Collisions in One Dimension

If we have a two-body collision in which the particles are constrained to move along a straight line, the conservation laws Eqs. (6.11) and (6.22) completely determine the final velocities, regardless of the nature of the interaction forces. With the velocities shown in the sketch, the conserva-tion laws give

Momentum:

Before

After

$$m_1v_1 + m_2v_2 = m_1v_1' + m_2v_2'.$$

Energy:

$$\tfrac{1}{2}m_1v_1{}^2 + \tfrac{1}{2}m_2v_2{}^2 = \tfrac{1}{2}m_1v_1'{}^2 + \tfrac{1}{2}m_2v_2'{}^2 + Q.$$

These equations can be solved for v_1' and v_2' in terms of m_1, m_2, v_1, v_2, and Q. The following example illustrates the process.

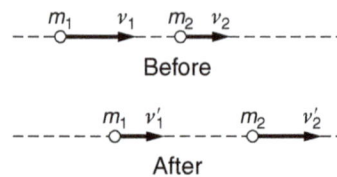

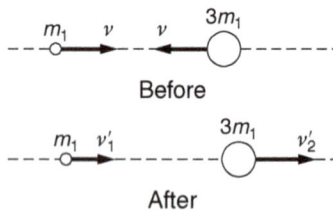

Example 6.8 Elastic Collision of Two Balls

Consider the one-dimensional elastic collision of two balls of masses m_1 and m_2, with $m_2 = 3m_1$. Suppose that the balls have equal and opposite velocities **v** before the collision; the problem is to find the final velocities.

Before

After

The conservation laws yield

$$m_1v - 3m_1v = m_1v_1' + 3m_1v_2' \tag{1}$$

$$\tfrac{1}{2}m_1v^2 + \tfrac{1}{2}(3m_1)v^2 = \tfrac{1}{2}m_1v_1'{}^2 + \tfrac{1}{2}(3m_1)v_2'{}^2. \tag{2}$$

We can eliminate v_1' using Eq. (1):

$$v_1' = -2v - 3v_2'. \tag{3}$$

Inserting this in Eq. (2) gives

$$4v^2 = (-2v - 3v_2')^2 + 3v_2'^2$$
$$= 4v^2 + 12vv_2' + 12v_2'^2$$

or

$$0 = 12vv_2' + 12v_2'^2. \tag{4}$$

Equation (4) has two solutions: $v_2' = -v$ and $v_2' = 0$. The corresponding values of v_1' can be found from Eq. (3).

Solution 1:

$$v_1' = v$$
$$v_2' = -v.$$

Solution 2:

$$v_1' = -2v$$
$$v_2' = 0.$$

We recognize that solution 1 simply restates the initial conditions: the particles simply miss each other. We always obtain such a trivial "solution" in this type of problem because the initial velocities evidently satisfy the conservation law equations.

Solution 2 is the non-trivial one. It shows that after the collision, m_1 is moving to the left with twice its original speed and the heavier ball is at rest.

6.5.4 Collisions and Center of Mass Coordinates

It is frequently simpler to treat three-dimensional collision problems in the center of mass (C) coordinate system than in the laboratory (L) system.

Consider two particles of masses m_1 and m_2 with velocities $\mathbf{v}_1$ and $\mathbf{v}_2$, respectively. The center of mass velocity V is

$$\mathbf{V} = \frac{m_1 \mathbf{v_1} + m_2 \mathbf{v_2}}{m_1 + m_2}.$$

As shown in the velocity diagram, $\mathbf{V}$ lies on the line joining $\mathbf{v}_1$ and $\mathbf{v}_2$.

The velocities in the C system are

$$\mathbf{v}_{1c} = \mathbf{v}_1 - \mathbf{V}$$
$$= \frac{m_2}{m_1 + m_2}(\mathbf{v}_1 - \mathbf{v}_2),$$

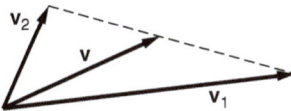

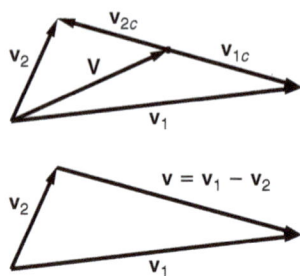

and

$$\mathbf{v}_{2c} = \mathbf{v}_2 - \mathbf{V}$$
$$= \frac{-m_1}{m_1 + m_2}(\mathbf{v}_1 - \mathbf{v}_2).$$

$\mathbf{v}_{1c}$ and $\mathbf{v}_{2c}$ lie back to back along the relative velocity vector $\mathbf{v} = \mathbf{v}_1 - \mathbf{v}_2$. The momenta in the C system are

$$\mathbf{p}_{1c} = m_1\mathbf{v}_{1c}$$
$$= \frac{m_1 m_2}{m_1 + m_2}(\mathbf{v}_1 - \mathbf{v}_2)$$
$$= \mu\mathbf{v}$$
$$\mathbf{p}_{2c} = m_2\mathbf{v}_{2c}$$
$$= \frac{-m_1 m_2}{m_1 + m_2}(\mathbf{v}_1 - \mathbf{v}_2)$$
$$= -\mu\mathbf{v}.$$

Here $\mu = m_1 m_2/(m_1 + m_2)$ is the reduced mass of the system, the natural unit of mass in a two-particle system. The total momentum in the C system is zero, as we expect.

The total momentum in the L system is

$$m_1\mathbf{v}_1 + m_2\mathbf{v}_2 = (m_1 + m_2)\mathbf{V}$$

and since total momentum is conserved in any collision, $\mathbf{V}$ is constant. We can use this result to help visualize the velocity vectors before and after the collision.

Sketch (a) shows the trajectories and velocities of two colliding particles before and after the collision. Sketch (b) shows the initial velocities in the L and C systems. All the vectors lie in the same plane, and $\mathbf{v}_{1c}$ and $\mathbf{v}_{2c}$ must be back to back since the total momentum in the C system is zero. After the collision, sketch (c), the velocities in the C system are again back to back. Sketch (c) also shows the final velocities in the lab system. Note that the plane of sketch (c) is not necessarily the plane of sketch (b).

Evidently the geometrical relation between initial and final velocities in the L system is quite complicated. Fortunately, the situation in the C system is much simpler. The initial and final velocities in the C system determine a plane known as the plane of scattering. For clarity, the initial and final velocities are shown in the same plane.

The interaction force must be known in order to calculate Θ. Conversely, by measuring the deflection we can learn about the interaction force. However, here we shall simply assume that the interaction has caused some deflection in the C system.

An important simplification occurs if the collision is elastic. Conservation of energy applied to the C system gives, for elastic collisions,

$$\tfrac{1}{2}m_1 v_{1c}^2 + \tfrac{1}{2}m_2 v_{2c}^2 = \tfrac{1}{2}m_1 v_{1c}'^2 + \tfrac{1}{2}m_2 v_{2c}'^2.$$

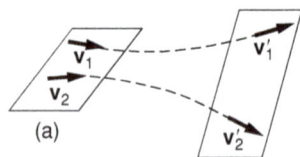

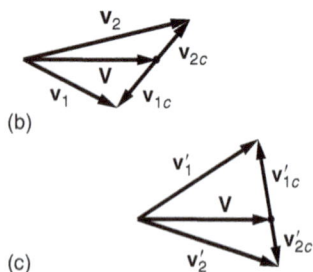

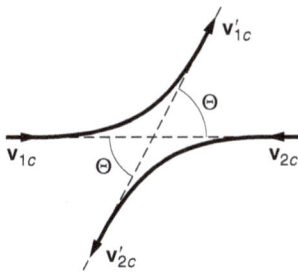

Total momentum is zero in the C system. We therefore have

$$m_1 v_{1c} - m_2 v_{2c} = 0$$
$$m_1 v'_{1c} - m_2 v'_{2c} = 0.$$

Using momentum conservation to eliminate v_{2c} and v'_{2c} from the energy equation gives

$$\frac{1}{2}\left(m_1 + \frac{m_1^2}{m_2}\right)v_{1c}^2 = \frac{1}{2}\left(m_1 + \frac{m_1^2}{m_2}\right)v'^2_{1c}$$

or

$$v_{1c} = v'_{1c}.$$

Similarly,

$$v_{2c} = v'_{2c}.$$

In an elastic collision, the speed of each particle in the C system is the same before and after the collision; the velocity vectors simply rotate in the scattering plane. In many experiments, one of the particles, say m_2, is initially at rest in the laboratory.

From conditions before the collision we have

$$\mathbf{V} = \frac{m_1}{m_1 + m_2}\mathbf{v}_1$$
$$\mathbf{v}_{1c} = \mathbf{v}_1 - \mathbf{V}$$
$$= \frac{m_2}{m_1 + m_2}\mathbf{v}_1$$
$$\mathbf{v}_{2c} = -\mathbf{V}$$
$$= -\frac{m_1}{m_1 + m_2}\mathbf{v}_1.$$

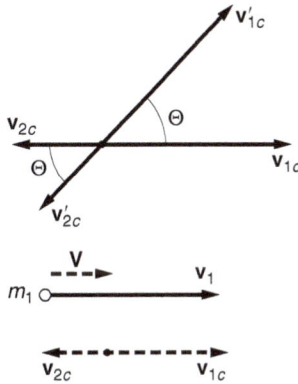

The sketches show the trajectories after the collision in the C and L systems. $\mathbf{v}'_1$ makes angle θ_1 in L, and $\mathbf{v}'_2$ makes angle θ_2 in L. Because these angles are in L, they are in principle measurable in the lab. The velocity diagrams can be used to relate θ_1 and θ_2 to the scattering angle Θ, as we shall see in the following example.

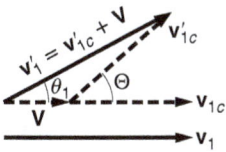

Example 6.9 Limitations on Laboratory Scattering Angle

Consider the elastic scattering of a particle of mass m_1 and velocity $\mathbf{v}_1$ from a second particle of mass m_2 at rest.

The scattering angle Θ in the C system is unrestricted, but the conservation laws impose limitations on the laboratory angles, as we now show.

The center of mass velocity has magnitude

$$V = \frac{m_1 v_1}{m_1 + m_2} \tag{1}$$

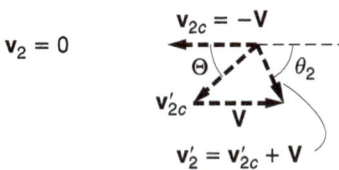

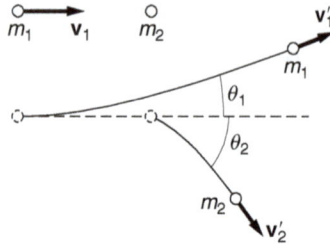

and is parallel to $\mathbf{v}_1$. The initial velocities in the C system are

$$\mathbf{v}_{1c} = \frac{m_2}{m_1 + m_2}\mathbf{v}_1 \tag{2}$$

$$\mathbf{v}_{2c} = -\frac{m_1}{m_1 + m_2}\mathbf{v}_1.$$

Suppose m_1 is scattered through angle Θ in the C system.

From the velocity diagram we see that the laboratory scattering angle of the incident particle is given by

$$\tan\theta_1 = \frac{v'_{1c}\sin\Theta}{V + v'_{1c}\cos\Theta}.$$

Since the scattering is elastic, $v'_{1c} = v_{1c}$. Hence

$$\tan\theta_1 = \frac{v_{1c}\sin\Theta}{V + v_{1c}\cos\Theta}$$

$$= \frac{\sin\Theta}{(V/v_{1c}) + \cos\Theta}.$$

From Eqs. (1) and (2), $V/v_{1c} = m_1/m_2$. Therefore

$$\tan\theta_1 = \frac{\sin\Theta}{(m_1/m_2) + \cos\Theta}. \tag{3}$$

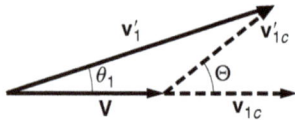

The scattering angle Θ depends on the details of the interaction, but in general it can assume any value. If $m_1 < m_2$, it follows from Eq. (3) or the geometric construction in sketch (a) that θ_1 is unrestricted. However, the situation is quite different if $m_1 > m_2$. In this case θ_1 is never greater than a certain angle $\theta_{1,\max}$. As sketch (b) shows, the maximum value of θ_1 occurs when $\mathbf{v}'_1$ is perpendicular to $\mathbf{v}'_{1c}$. In this case $\sin\theta_{1,\max} = v_{1c}/V = m_2/m_1$. If $m_1 \gg m_2$, $\theta_{1,\max} \approx m_2/m_1$ and the maximum scattering angle in L approaches zero.

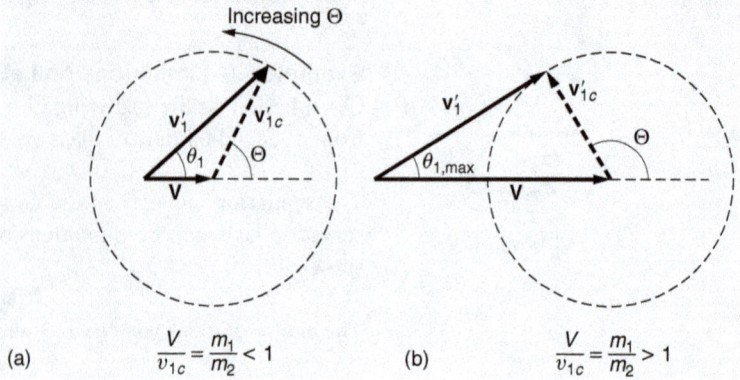

(a) $\dfrac{V}{v_{1c}} = \dfrac{m_1}{m_2} < 1$ (b) $\dfrac{V}{v_{1c}} = \dfrac{m_1}{m_2} > 1$

As a homely illustration, if $m_1/m_2 < 1$ as in (a), this is like a stream of marbles bouncing off a bowling ball; the marbles scatter in all directions. On the other hand, if a moving bowling ball scatters off marbles, then $m_1/m_2 \gg 1$ and the bowling ball is hardly deflected at all, continuing much along its original path.

A special case of Eq. (3) is $m_1/m_2 = 1$. Then

$$\tan \theta_1 = \frac{\sin \Theta}{1 + \cos \Theta}$$
$$= \tan(\Theta/2)$$

so that

$$\theta_1 = \Theta/2.$$

Problems
For problems marked *, refer to page 522 for a hint, clue, or answer.

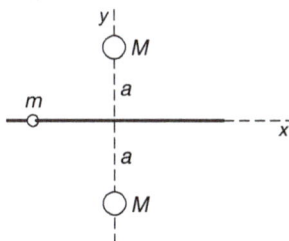

6.1 *Oscillation of bead with gravitating masses*

A bead of mass m slides without friction on a smooth rod along the x axis. The rod is equidistant between two spheres of mass M. The spheres are located at $x = 0, y = \pm a$ as shown, and attract the bead gravitationally.

Find the frequency of small oscillations of the bead about the origin.

6.2 *Oscillation of a particle with two forces*

A particle of mass m moves in one dimension along the positive x axis. It is acted on by a constant force directed toward the origin with magnitude B, and an inverse-square law repulsive force with magnitude A/x^2.

What is the frequency of small oscillations about the equilibrium point x_0?

6.3 *Normal modes and symmetry*

Four identical masses m are joined by three identical springs, of spring constant k, and are constrained to move on a line, as shown.

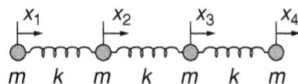

There is a high degree of symmetry in this problem, so that one can guess the normal mode motions by inspection, without a lengthy calculation. Once the relative amplitudes of the normal mode motions are known, the normal mode vibrational frequencies follow directly.

Because of the symmetry, the normal mode amplitudes must obey $x_1 = \pm x_4$ and $x_2 = \pm x_3$. Another condition is that the center of mass must remain at rest. The possibilities are:

$(x_4 = x_1)$ and $(x_3 = x_2)$

$(x_4 = -x_1)$ and $(x_3 = -x_2)$.

The normal mode equations lead to three possible non-trivial vibrational frequencies and three corresponding normal modes. Find the normal mode frequencies. It is convenient to use the dimensionless parameter $\beta = \omega^2/\omega_0^2$, where ω is a frequency to be found and $\omega_0 \equiv \sqrt{k/m}$.

6.4　*Bouncing ball**

A ball falls to the floor and bounces inelastically, eventually coming to rest. The speed immediately after a collision is e times the speed immediately before (e is the *coefficient of restitution* where $e < 1$). If the speed is v_0 just before the first bounce at $t = 0$, find the additional time needed to come to rest.

6.5　*Marble and superball*

A small ball of mass m is placed on top of a "superball" of mass M, and the two balls are dropped to the floor from height h. How high does the small ball rise after the collision? Assume that collisions with the superball are elastic, and that $m \ll M$. To help visualize the problem, assume that the balls are slightly separated when the superball hits the floor. (If you are surprised by the result, try demonstrating the problem with a marble and a superball.)

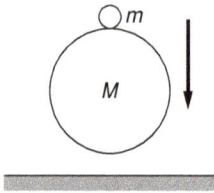

6.6　*Three car collision*

Consider three identical cars A, B, and C. Cars B and C are at rest with their brakes off. Car A, with kinetic energy E_0, plows into B at high speed, pushing B into C. If the collisions are completely inelastic, find the kinetic energy of the three-car pile-up after the collisions.

6.7　*Proton collision*

A proton makes a head-on collision with an unknown particle at rest. The proton rebounds straight back with $\frac{4}{9}$ of its initial kinetic energy.

Find the ratio of the mass of the unknown particle to the mass of the proton, assuming that the collision is elastic.

6.8　*Collision of m and M*

A particle of mass m and initial velocity v_0 collides elastically with a particle of unknown mass M coming from the opposite direction as shown in the left-hand sketch on the next page. After the collision, m has velocity $v_0/2$ at right angles to the incident direction, and M moves off in the direction shown in the left-hand sketch on the following page. Find the ratio M/m.

6.9 *Collision of m and 2m*

Particle A of mass m has initial velocity v_0. After colliding with particle B of mass $2m$ initially at rest, the particles follow the paths shown in the right-hand sketch. Find θ.

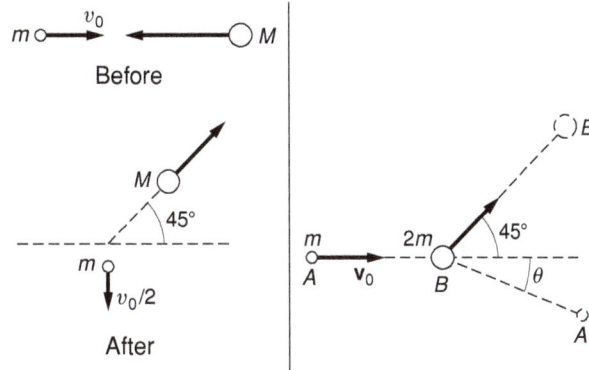

6.10 *Nuclear reaction in the L system*

In the L system, a particle of mass m_1 with kinetic energy E_1 strikes a particle of mass m_2 initially at rest. A nuclear reaction occurs, with the release of a particle of mass m_3 at angle θ and energy E_3, and a particle of mass m_4 at angle ϕ and energy E_4. (Angles are measured in L from the incident line.) ϕ and E_4 are not measured.

Find an expression for the energy Q released in the reaction in terms of the masses, energies E_1 and E_3, and the angle θ.

6.11 *Uranium fission*

In a nuclear reactor powerplant, a very slow ("thermal") neutron has high probability of reacting with the uranium-235 in the fuel rods. The ^{235}U then fissions asymmetrically into a light fragment (most likely strontium ^{97}Sr) and a heavy fragment (most likely xenon ^{138}Xe), releasing energy 170 MeV.

The fission also produces a few fast neutrons. These neutrons are slowed to thermal speeds by collisions as they pass through a moderator, possibly helium, graphite, or even ordinary water. Once slowed, they can induce additional fission events, so that the process becomes a self-sustaining chain reaction.

(*a*) What are the energies of the two fragments (in MeV) immediately after fission? Neglect energy carried off by the fast neutrons.

(*b*) A 1 keV fast neutron (relative mass 1) in a moderator collides elastically with a helium atom ^{4}He (relative mass 4) at rest. What is the maximum amount of energy the neutron can lose?

6.12 *Hydrogen fusion*

The Sun generates most of its energy from a chain of nuclear reactions, beginning with the fusion of the nuclei of two atoms of ordinary hydrogen ^{1}H. The process in the Sun requires high densities, about 160 times the density of water on the Earth, and extremely high temperatures, about 1.5×10^7 K. These conditions enable the nuclei to approach close enough against the repulsive Coulomb force to allow the nuclei to come within the range of the strong nuclear force.

Efforts on the Earth to generate energy from fusion have centered mainly on the reaction between the two heavier isotopes of hydrogen, deuterium ^{2}D (one proton, one neutron) and tritium ^{3}T (one proton, two neutrons):

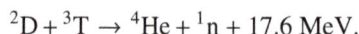

$$^2\text{D} + {}^3\text{T} \rightarrow {}^4\text{He} + {}^1\text{n} + 17.6 \text{ MeV}.$$

The products are ^{4}He (relative mass 4) and a neutron (relative mass 1).

What are the energies of the products, in MeV? The deuterium and tritium are essentially at rest.

6.13 *Nuclear reaction of α-rays with lithium**

A thin target of lithium is bombarded by helium nuclei (α-rays) of energy E_0. The lithium nuclei are initially at rest in the target but are essentially unbound. When an α-ray enters a lithium nucleus, a nuclear reaction can occur in which the compound nucleus splits apart into a boron nucleus and a neutron. The collision is inelastic, and the final kinetic energy is less than E_0 by 2.8 MeV. The relative masses of the particles are: helium, mass 4; lithium, mass 7; boron, mass 10; neutron, mass 1. The reaction can be symbolized

$$^7\text{Li} + {}^4\text{He} \rightarrow {}^{10}\text{B} + {}^1\text{n} - 2.8 \text{ MeV}.$$

(*a*) What is $E_{0,\text{threshold}}$, the minimum value of E_0 for which neutrons can be produced?

(*b*) Show that if the incident energy falls in the range $E_{0,\text{threshold}} < E_0 < E_{0,\text{threshold}} + 0.27$ MeV, the neutrons ejected in the forward direction do not all have the same energy but must have either one or the other of two possible energies. (You can understand the origin of the two groups by looking at the reaction in the center of mass system.)

6.14 *Superball bouncing between walls**

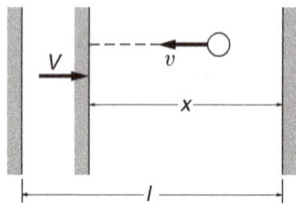

A "superball" of mass m bounces back and forth with speed v between two parallel walls, as shown. The walls are initially separated by distance l. Gravity is neglected and the collisions are perfectly elastic.

(*a*) Find the time-average force F on each wall.

(*b*) If one surface is slowly moved toward the other with speed $V \ll v$, the bounce rate will increase due to the shorter distance between collisions, and because the ball's speed increases when it bounces from the moving surface. Show that

$$\frac{dv}{dt} \cong \frac{vV}{x}; \frac{dv}{dx} \cong -\frac{v}{x};$$

and find $v(x)$.

(*c*) Find the average force at distance x.

6.15 *Center of mass energy*

Show that the energy of two non-interacting particles with masses M_a and M_b can be written $E = E_0 + E'$ where $E_0 = \frac{1}{2}MV^2$ is the energy of c. of m. motion, $E' = \frac{1}{2}\mu V_r^2$ is the energy in the C system $M = M_a + M_b$, **V** is the velocity of the c. of m. in the C system, and **V**$_r$ is the relative velocity of the particles.

6.16 *Converting between C and L systems**

A particle of mass m and velocity v_0 collides elastically with a particle of mass M initially at rest and is scattered through angle Θ in the center of mass C system.

(*a*) Find the final velocity of m in the laboratory L system.

(*b*) Find the fractional loss of kinetic energy of m.

6.17 *Colliding balls*

Two balls, of mass m and mass $2m$, approach from perpendicular directions with identical speeds v and collide. After the collision, the more massive ball moves with the same speed v but downward, perpendicular to its original direction. The less massive ball moves with speed U at an angle θ with respect to the horizontal. Assume that no external forces act during the collision.

(*a*) Calculate the final speed U of the less massive ball and the angle θ.

(*b*) Determine how much kinetic energy is lost or gained by the two balls during the collision. Is this collision elastic, inelastic, or superelastic?

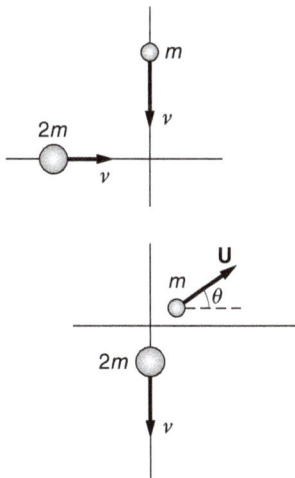

7 ANGULAR MOMENTUM AND FIXED AXIS ROTATION

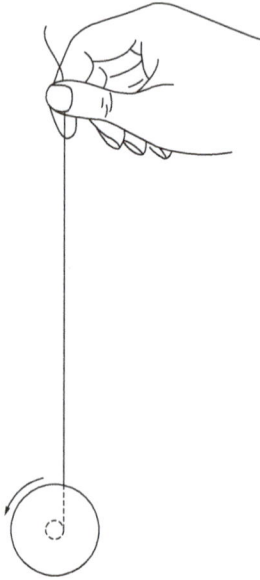

7.1 Introduction

Our discussion of the principles of mechanics has so far neglected an important issue—the rotational motion of solid bodies. For example, consider the common yo-yo running up and down its string as the spool winds and unwinds. In principle we can predict the motion because each particle of the yo-yo moves according to Newton's laws. The motion is simple but attempting to analyze it on a particle-by-particle basis would quickly prove to be hopeless. To treat the rotational motion of extended bodies as a whole, we need to develop a simple method, which is the goal of this chapter.

In attacking the problem of translational motion, we introduced the concepts of force, linear momentum, and center of mass. In this chapter we introduce precisely analogous concepts for rotational motion: torque, angular momentum, and moment of inertia.

Our goal, of course, is much more ambitious than merely to understand yo-yo motion; our goal is to find a general way to analyze the motion of rigid bodies under any combination of applied forces. Happily, as we shall demonstrate, this problem can be divided into two much simpler problems—finding the center of mass motion, a problem that we have already solved, and finding the rotational motion around the center of mass, the task at hand. The justification for this separation is a theorem of rigid body motion, called Chasles' theorem, which asserts that any displacement of a rigid body can be decomposed into two independent motions: a translation of the center of mass and a rotation around the center of mass. A few minutes spent playing with a rigid body such as a book or a chair should convince you that such a separation is plausible. Note that the theorem does not say that this is the *only* way to represent a general displacement—merely that it is one possible way of doing so. A formal proof of Chasles' theorem is presented in Note 7.1 at the end of the chapter, but the proof is not essential at this point in our discussion. What is important is being able to visualize any displacement as the combination of a single translation and a single rotation.

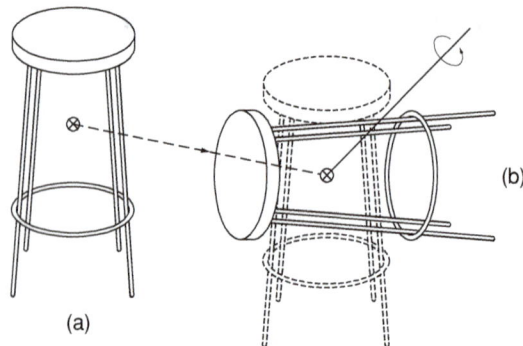

With reference to the sketch, to bring the body from position (a) to some new position (b), first translate it so that the center of mass coincides with the new position of the center of mass, and then rotate it around the appropriate axis through the center of mass until the body is in the desired position.

Leaving aside extended bodies for the moment, we start in the great tradition of physics by considering the simplest possible system—a particle. Because a particle has no size, its orientation in space is of no consequence, and it might seem that we need be concerned only with translational motion. Nevertheless, particle motion is useful for introducing the concepts of *angular momentum* and *torque*. We shall then move progressively to more complex systems, culminating, in Chapter 8, with a treatment of the general motion of a rigid body.

7.2 Angular Momentum of a Particle

Here is the formal definition of the angular momentum $\mathbf{L}$ of a particle that has momentum $\mathbf{p}$ and is at position $\mathbf{r}$ with respect to a given coordinate system:

$$\mathbf{L} = \mathbf{r} \times \mathbf{p}. \tag{7.1}$$

The unit of angular momentum is $\text{kg} \cdot \text{m}^2/\text{s}$ in the SI system or $\text{g} \cdot \text{cm}^2/\text{s}$ in CGS. There are no special names for these units.

Two aspects of angular momentum deserve comment. First, $\mathbf{L}$ explicitly involves the position vector $\mathbf{r}$. The value of $\mathbf{L}$ therefore depends not only on the motion of the particle, but also on its location with respect to the origin of a particular coordinate system. This is in contrast to the situation for linear momentum $\mathbf{p}$, which is independent of the coordinate system. Consequently one cannot meaningfully speak of the angular momentum of a particle alone; one must always identify the coordinate system.

The second unusual aspect of angular momentum is that it is the first physical quantity that we have encountered that involves the cross product. You may recall from Chapter 1 that $\mathbf{r} \times \mathbf{p}$ is a vector whose magnitude is $|\mathbf{r}||\mathbf{p}| \sin \alpha$ where α is the angle between $\mathbf{r}$ and $\mathbf{p}$.

Probably the least intuitive aspect of angular momentum is its direction. The vectors $\mathbf{r}$ and $\mathbf{p}$ determine a plane (sometimes known as the plane of motion), and by the properties of the cross product, $\mathbf{L}$ is perpendicular to this plane. Although there is nothing particularly "natural" about this definition of angular momentum, we will see that $\mathbf{L}$ so defined obeys a simple but important dynamical equation.

The diagram shows the trajectory and instantaneous position and momentum of a particle. $\mathbf{L} = \mathbf{r} \times \mathbf{p}$ is perpendicular to the plane of $\mathbf{r}$ and $\mathbf{p}$, and points in the direction dictated by the right-hand rule for vector multiplication. (Point the index finger of your right hand along $\mathbf{r}$ and orient your hand so that you can bend the other fingers toward $\mathbf{p}$. Your thumb then points in the direction of $\mathbf{L}$.)

If $\mathbf{r}$ and $\mathbf{p}$ lie in the x–y plane, then $\mathbf{L}$ lies along the z axis. $\mathbf{L}$ points in the positive z direction if the "sense of rotation" as the point moves with respect to the origin is counterclockwise, and in the negative z direction

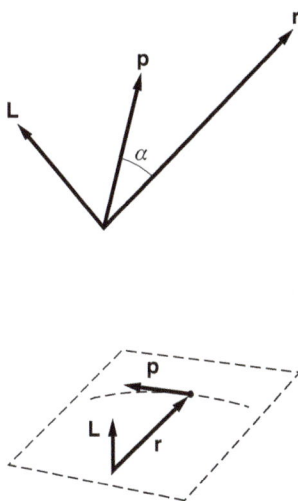

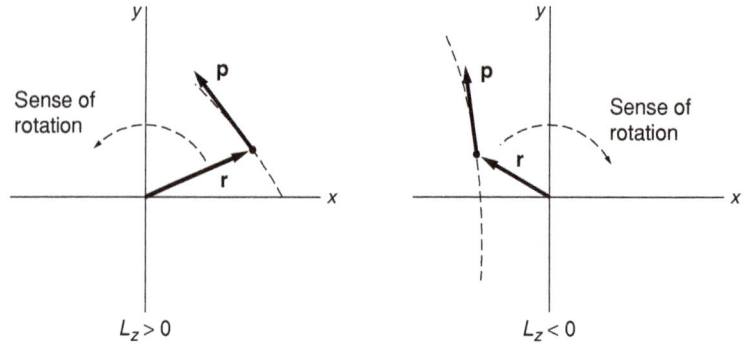

$L_z > 0$ $L_z < 0$

if the sense of rotation is clockwise. Note that the sense of rotation is well defined no matter whether the trajectory is curved or straight.

The distinction between positive and negative rotation breaks down if the trajectory aims at the origin, but in this case $\mathbf{r}$ and $\mathbf{p}$ are parallel and $\mathbf{L} = 0$.

It is helpful to be able to visualize angular momentum geometrically as well as being able to calculate it algebraically. The results, of course, agree. To be concrete we shall take the motion to lie in the x–y plane so that $\mathbf{L}$ lies along the z axis.

To understand $\mathbf{L}$ geometrically, let us decompose $\mathbf{r}$ into a component $r_\perp$ that is perpendicular to the trajectory and $r_\parallel$ that is parallel. The length of $r_\perp$ is

$$r_\perp = r \sin(\pi - \phi) = r \sin\phi.$$
$$L_z = (\mathbf{r} \times \mathbf{p})_z = rp \, \sin\phi = r_\perp p.$$

Alternatively, we can decompose $\mathbf{p}$ into a component $p_\parallel$ parallel to $\mathbf{r}$ and a component $p_\perp$ perpendicular to $\mathbf{r}$. Then

$$L_z = (\mathbf{r} \times \mathbf{p})_z = rp \, \sin\phi = rp_\perp.$$

Another way to find L_z is to calculate $\mathbf{r} \times \mathbf{p}$ algebraically. For motion in the x–y plane, $\mathbf{r} = (x, y, 0)$ and $\mathbf{p} = m(v_x, v_y, 0)$. The cross product (written in the determinantal form described in Section 1.6) is

$$\mathbf{L} = \mathbf{r} \times \mathbf{p}$$

$$= m \begin{vmatrix} \hat{\mathbf{i}} & \hat{\mathbf{j}} & \hat{\mathbf{k}} \\ x & y & 0 \\ v_x & v_y & 0 \end{vmatrix}$$

$$= m(xv_y - yv_x)\hat{\mathbf{k}}.$$

This algebraic result has a direct geometrical interpretation. The motion is in the x–y plane; we shall look first at the motion in the y direction, and then in the x direction. As the drawing on the next page shows, the motion p_y in (a) represents a counterclockwise rotation and contributes angular momentum $+mxv_y$. The motion p_x in (b) represents a clockwise

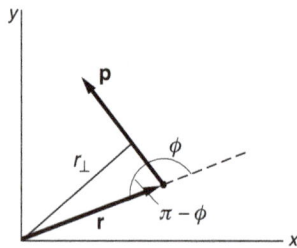

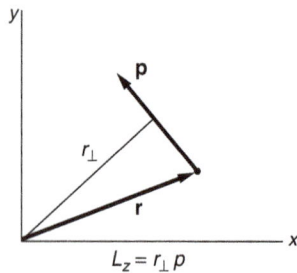

$L_z = r_\perp p$

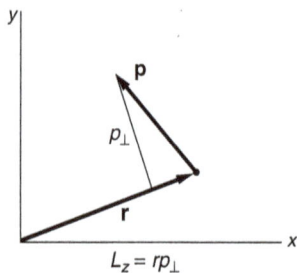

$L_z = rp_\perp$

(a) $L_z = xp_y$ (b) $L_z = yp_x$ (c) $L_z = xp_y - yp_x$

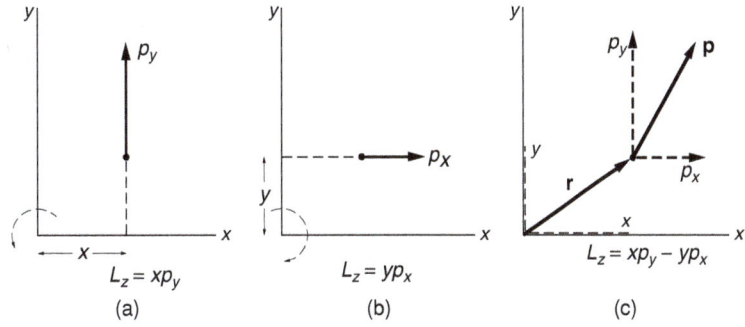

rotation and creates angular momentum $-myv_x$ around the origin. The sum of these is $m(xv_y - yv_x)$, as we found before.

We have illustrated angular momentum using motion in the x–y plane where the angular momentum lies along the z axis. There is no difficulty applying these methods to the general case where **L** has components along all three axes.

Example 7.1 Angular Momentum of a Sliding Block 1

A block of mass m and negligible dimensions slides freely in the x direction with velocity $\mathbf{v} = v\hat{\mathbf{i}}$, as shown in the sketch. What is its angular momentum $\mathbf{L}_A$ around origin A and its angular momentum $\mathbf{L}_B$ around origin B?

As shown in the drawing the vector from origin A to the block is $\mathbf{r}_A = x\hat{\mathbf{i}}$. Since $\mathbf{r}_A$ is parallel to $\mathbf{v}$, their cross product is zero:

$$\mathbf{L}_A = m\mathbf{r}_A \times \mathbf{v}$$
$$= 0.$$

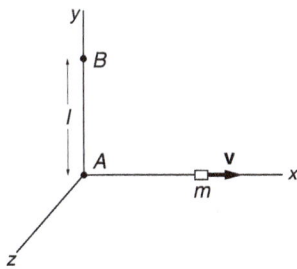

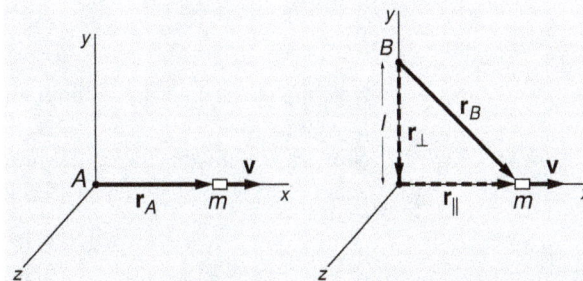

Taking origin B, we can resolve $\mathbf{r}_B$ into a component $\mathbf{r}_\parallel$ parallel to $\mathbf{v}$ and a component $\mathbf{r}_\perp$ perpendicular to $\mathbf{v}$. Then

$$\mathbf{L}_B = m\mathbf{r}_B \times \mathbf{v} = m(\mathbf{r}_\parallel + \mathbf{r}_\perp) \times \mathbf{v}.$$

With $\mathbf{r}_\parallel \times \mathbf{v} = 0$ and $|\mathbf{r}_\perp \times \mathbf{v}| = lv\hat{\mathbf{k}}$ we have

$$\mathbf{L}_B = mlv\hat{\mathbf{k}}.$$

$\mathbf{L}_B$ lies in the positive z direction because the sense of rotation is counterclockwise around the z axis.

To calculate $\mathbf{L}_B$ formally we can write $\mathbf{r}_B = x\hat{\mathbf{i}} - l\hat{\mathbf{j}}$ and evaluate $\mathbf{r}_B \times \mathbf{v}$ using the determinantal form

$$\mathbf{L}_B = m\mathbf{r}_B \times \mathbf{v}$$

$$= m \begin{vmatrix} \hat{\mathbf{i}} & \hat{\mathbf{j}} & \hat{\mathbf{k}} \\ x & -l & 0 \\ v & 0 & 0 \end{vmatrix}$$

$$= mlv\hat{\mathbf{k}}$$

as before.

The following example emphasizes again how $\mathbf{L}$ depends on the choice of origin.

Example 7.2 Angular Momentum of the Conical Pendulum

Let us return to the conical pendulum, which we encountered in Example 2.10. Assume that the pendulum is in steady circular motion with constant angular speed ω.

We begin by evaluating $\mathbf{L}_A$, the angular momentum around origin A. From the sketch we see that $\mathbf{L}_A$ lies in the positive z direction. $\mathbf{L}_A$ has magnitude $|\mathbf{r}_\perp||\mathbf{p}| = |\mathbf{r}||\mathbf{p}| = rp$, where r is the radius of the circular motion. Since

$$|\mathbf{p}| = Mv = Mr\omega$$

we have

$$\mathbf{L}_A = Mr^2\omega\hat{\mathbf{k}}.$$

Note that $\mathbf{L}_A$ is constant, in both magnitude and direction.

Now let us evaluate the angular momentum around the origin B located at the pivot. The magnitude of $\mathbf{L}_B$ is

$$|\mathbf{L}_B| = |\mathbf{r}' \times \mathbf{p}|$$

$$= |\mathbf{r}'||\mathbf{p}| = l|\mathbf{p}|$$

$$= Mlr\omega,$$

where $|\mathbf{r}'| = l$, the length of the string. It is again apparent that $\mathbf{L}$ depends on the origin we choose.

Unlike $\mathbf{L}_A$, the direction of $\mathbf{L}_B$ is not constant. $\mathbf{L}_B$ is perpendicular to both $\mathbf{r}'$ and $\mathbf{p}$. The sketches show $\mathbf{L}_B$ at different times. Two sketches

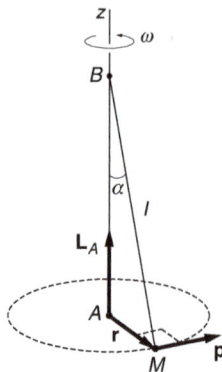

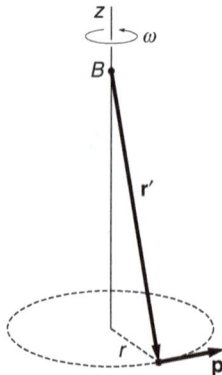

are given, to emphasize that only the magnitude and direction of **L** are important, not the position at which we choose to draw it. The magnitude of $\mathbf{L}_B$ is constant but its *direction* is obviously not constant; as the bob swings around, $\mathbf{L}_B$ sweeps out the shaded cone shown in the sketch at the right. The z component of $\mathbf{L}_B$ is constant but the horizontal component travels around the circle with the bob. We shall see the dynamical consequences of this in Example 7.9.

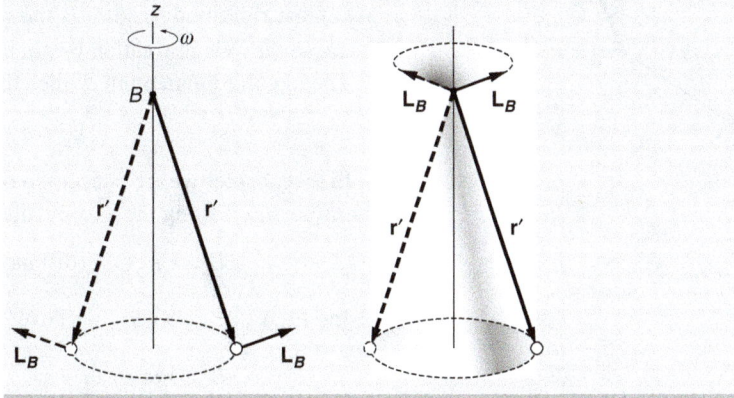

7.3 Fixed Axis Rotation

The most prominent application of angular momentum in Newtonian mechanics is to the analysis of the motion of rigid bodies. In general, rigid body motion involves free rotation around any axis—for instance, the motion of a stick flung into the air, spinning and tumbling. Analysis of the general case involves mathematical complexities that we are going to take up in Chapter 8. In this chapter we restrict ourselves to the special, but important, case of rotation around a fixed axis. By fixed axis we mean that the *direction* of the axis of rotation is always along the same line, but the axis itself may translate. For example, a car wheel attached to an axle undergoes fixed axis rotation as long as the car drives straight ahead. If the car turns, the wheel must rotate around a vertical axis while simultaneously spinning on the axle; the motion is no longer fixed axis rotation. If the wheel flies off the axle and wobbles down the road, the motion is definitely not rotation around a fixed axis.

We can choose the axis of rotation to be in the z direction without loss of generality. The rotating object can be a wheel or a stick, or anything we choose, the only restriction being that it is rigid—which is to say that its shape does not change as it rotates.

When a rigid body rotates around an axis, every particle in the body remains at a fixed distance from the axis. If we choose a coordinate system with its origin on the axis, then for every particle in the body, $|\mathbf{r}| = $ constant. The only way that $\mathbf{r}$ can change while $|\mathbf{r}|$ remains constant is for the velocity to be perpendicular to $\mathbf{r}$. In this chapter and

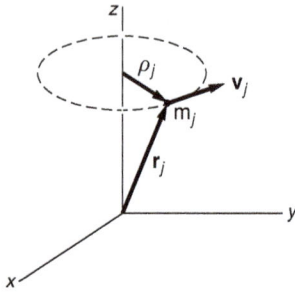

the next we shall use the symbol ρ to denote perpendicular distance to the axis of rotation. In contrast, r stands for the distance to the origin: $r = \sqrt{x^2 + y^2 + z^2}$.

Consider a body rotating around the z axis, so that

$$|\mathbf{v}_j| = |\dot{\mathbf{r}}_j|$$
$$= \omega \rho_j \tag{7.2}$$

where ρ_j is the perpendicular distance from the axis of rotation to particle m_j of the rigid body and ω is the rate of rotation (the angular speed). Because the axis of rotation lies in the z direction, $\rho_j = \sqrt{x_j^2 + y_j^2}$.

The angular momentum of the jth particle, $\mathbf{L}_j$, is

$$\mathbf{L}_j = \mathbf{r}_j \times m_j \mathbf{v}_j.$$

In this chapter we are concerned only with L_z, the component of angular momentum along the axis of rotation. Since $\mathbf{v}_j$ lies in the x–y plane,

$$L_{j,z} = m_j v_j \times (\text{distance to } z \text{ axis}) = m_j v_j \rho_j.$$

Using Eq. (7.2), $v_j = \omega \rho_j$, we have

$$L_{j,z} = m_j \rho_j^2 \omega.$$

The z component of the total angular momentum of the body L_z is the sum of the individual z components:

$$L_z = \sum_j L_{j,z}$$

$$= \sum_j m_j \rho_j^2 \omega, \tag{7.3}$$

where the sum is over all particles of the body. We have taken ω to be constant throughout, because the body is rigid.

7.3.1 Moment of Inertia

Equation (7.3) can be written

$$L_z = I\omega, \tag{7.4}$$

where

$$I \equiv \sum_j m_j \rho_j^2. \tag{7.5}$$

I is a geometrical quantity called the *moment of inertia*. I depends on the distribution of mass in the body with respect to the axis of rotation. (We shall give a more general definition for I in Chapter 8 when we talk about unrestricted rigid body motion.)

Equation (7.4) reveals a close analogy between angular momentum around an axis and linear momentum along an axis $P = Mv$. The moment of inertia plays the same role in rotational motion that mass plays in linear momentum. This is but one of many analogies between rotational motion and linear motion.

For continuously distributed matter we can replace the sum over particles in Eq. (7.5) by an integral over differential mass elements. In this case

$$\sum_j m_j \rho_j^{\,2} \rightarrow \int \rho^2 dm,$$

and

$$I = \int \rho^2 dm$$
$$= \int (x^2 + y^2) dm.$$

To evaluate such an integral we generally replace the mass element dm by $dm = w dV$, the product of the density (mass per unit volume) w at the position of dm and the volume dV occupied by dm. (Often ρ is used to denote density, but that would cause confusion here.) We can write

$$I = \int \rho^2 dm$$
$$= \int (x^2 + y^2) w\, dV.$$

Before proceeding to analyze the physics of fixed axis rotation it is useful to calculate moments of inertia of some simple objects having high degrees of symmetry, where calculation of the moment of inertia is straightforward.

Example 7.3 Moments of Inertia of Some Simple Objects
(a) Uniform thin ring of mass M and radius R, around the axis of symmetry of the ring.
The moment of inertia around the axis is given by $I = \int \rho^2 dm$. Since the ring is thin, $dm = \lambda ds$, where $\lambda = M/2\pi R$ is the mass per unit length of the ring. All points on the ring are distance R from the axis so that $\rho = R$, and we have

$$I_{\text{ring}} = \int_0^{2\pi R} R^2 \lambda\, ds$$
$$= R^2 \left(\frac{M}{2\pi R}\right) s \Big|_0^{2\pi R}$$
$$= MR^2.$$

(b) Uniform thin disk of mass M and radius R, around the axis of symmetry of the disk.
We can subdivide the disk into a series of thin rings with radius ρ, width $d\rho$, and moment of inertia dI. Then $I = \int dI$.

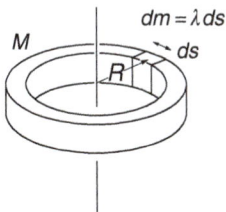

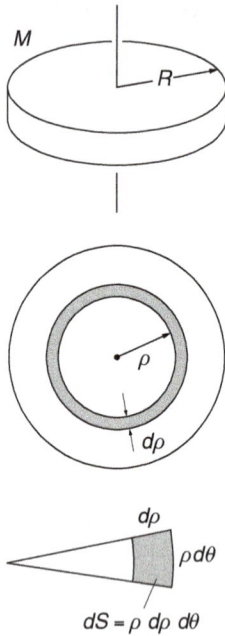

The area of one of the thin rings is $dA = 2\pi\rho\, d\rho$, and its mass is

$$dm = M\frac{dA}{A} = \frac{M2\pi\rho\, d\rho}{\pi R^2}$$

$$= \frac{2M\rho\, d\rho}{R^2}.$$

$$dI = \rho^2\, dm = \frac{2M\rho^3\, d\rho}{R^2}$$

$$I = \int_0^R \frac{2M\rho^3\, d\rho}{R^2}$$

$$= \frac{1}{2}MR^2.$$

Let us also solve this problem by double integration to illustrate the most general approach.

$$I = \int \rho^2\, dm$$

$$= \int \rho^2\sigma\, dS,$$

where σ is the mass per unit area and dS is a differential element of area. For the uniform disk, $\sigma = M/\pi R^2$. Polar coordinates are the obvious choice for this calculation. In plane polar coordinates, $dS = \rho\, d\rho\, d\theta$. Then

$$I_{\text{disk}} = \int \rho^2\, \sigma dS$$

$$= \left(\frac{M}{\pi R^2}\right)\int \rho^2\, dS$$

$$= \left(\frac{M}{\pi R^2}\right)\int_0^R \int_0^{2\pi} \rho^2\rho\, d\rho\, d\theta$$

$$= \left(\frac{2M}{R^2}\right)\int_0^R \rho^3\, d\rho$$

$$= \frac{1}{2}MR^2,$$

as before.

(c) Uniform thin stick of mass M and length L, around a perpendicular axis through its midpoint.

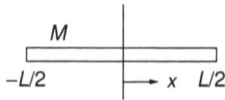

$$I_{o,\text{rod}} = \int_{-L/2}^{+L/2} x^2\, dm$$

$$= \frac{M}{L}\int_{-L/2}^{+L/2} x^2\, dx$$

$$= \frac{M}{L}\frac{1}{3}x^3\, \Big|_{-L/2}^{+L/2}$$

$$= \frac{1}{12}ML^2.$$

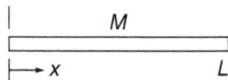

(d) Uniform thin stick around a perpendicular axis at its end.

$$I_{\text{rod}} = \frac{M}{L} \int_0^L x^2 dx$$

$$= \frac{1}{3} ML^2.$$

(e) Uniform sphere of mass M and radius R, around an axis through its center.
We quote this result but leave the proof for a problem.

$$I_{\text{sphere}} = \frac{2}{5} MR^2.$$

7.3.2 The Parallel Axis Theorem

This handy theorem tells us I, the moment of inertia around any axis, provided that we know I_0, the moment of inertia around an axis through the center of mass parallel to the first. If the mass of the body is M and the distance between the axes is l, the theorem states that

$$I = I_0 + Ml^2.$$

To prove this, consider the moment of inertia of the body around an axis that we choose to lie in the z direction. The perpendicular vector from the z axis to particle j is

$$\boldsymbol{\rho}_j = x_j \hat{\mathbf{i}} + y_j \hat{\mathbf{j}},$$

and

$$I = \sum_j m_j \rho_j{}^2.$$

If the center of mass is at $\mathbf{R} = X\hat{\mathbf{i}} + Y\hat{\mathbf{j}} + Z\hat{\mathbf{k}}$, the vector perpendicular from the z axis to the center of mass is

$$\mathbf{R}_\perp = X\hat{\mathbf{i}} + Y\hat{\mathbf{j}}.$$

If the vector from the axis through the center of mass to particle j is $\boldsymbol{\rho}'_j$, then the moment of inertia around the center of mass is

$$I_0 = \Sigma m_j \rho'^2_j.$$

From the diagram we see that

$$\boldsymbol{\rho}_j = \boldsymbol{\rho}'_j + \mathbf{R}_\perp,$$

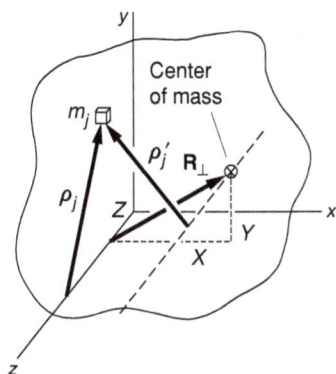

so that

$$I = \Sigma m_j \rho_j{}^2$$

$$= \Sigma m_j (\boldsymbol{\rho}'_j + \mathbf{R}_\perp)^2$$

$$= \Sigma m_j (\rho'^2_j + 2\boldsymbol{\rho}'_j \cdot \mathbf{R}_\perp + R_\perp{}^2).$$

The middle term vanishes by definition of the center of mass:

$$\Sigma m_j \boldsymbol{\rho}'_j = \Sigma m_j(\boldsymbol{\rho}_j - \mathbf{R}_\perp) = M(\mathbf{R}_\perp - \mathbf{R}_\perp)$$
$$= 0.$$

If we designate the magnitude of $\mathbf{R}_\perp$ by l, then

$$I = I_0 + Ml^2.$$

For example, in Example 7.3(c) we showed that the moment of inertia of a stick around its center of mass (the midpoint) is $ML^2/12$. The moment of inertia around its end, which is $L/2$ away from the center of mass, is therefore

$$I_a = \frac{1}{12}ML^2 + M\left(\frac{L}{2}\right)^2$$
$$= \frac{1}{3}ML^2,$$

the result we found in Example 7.3(d).

Similarly, the moment of inertia of a disk around an axis at the rim, perpendicular to the plane of the disk, is

$$I_a = \frac{1}{2}MR^2 + MR^2 = \frac{3}{2}MR^2.$$

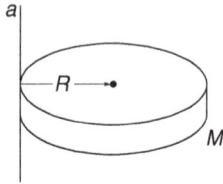

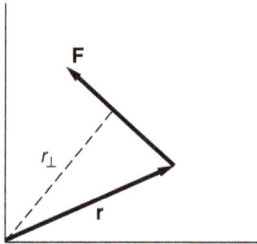

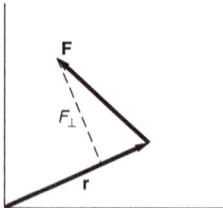

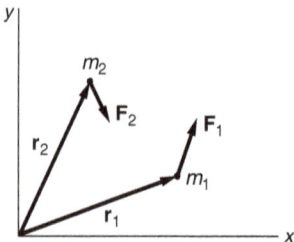

7.4 Torque

To continue our development of the dynamics of rotational motion we introduce a new quantity, *torque* $\boldsymbol{\tau}$, which plays a role in rotational motion analogous to the role of force in linear motion. The torque due to force $\mathbf{F}$ that acts on a particle at position $\mathbf{r}$ is defined to be

$$\boldsymbol{\tau} = \mathbf{r} \times \mathbf{F}. \tag{7.6}$$

In Section 7.2 we discussed several ways to evaluate angular momentum, $\mathbf{r} \times \mathbf{p}$. The methods we developed for calculating the cross product can also be applied to torque $\mathbf{r} \times \mathbf{F}$. For example, we have

$$|\boldsymbol{\tau}| = |\mathbf{r}_\perp||\mathbf{F}|$$

$$|\boldsymbol{\tau}| = |\mathbf{r}||\mathbf{F}_\perp|$$

$$\boldsymbol{\tau} = \begin{vmatrix} \hat{\mathbf{i}} & \hat{\mathbf{j}} & \hat{\mathbf{k}} \\ x & y & z \\ F_x & F_y & F_z \end{vmatrix}.$$

We can also associate a "sense of rotation" using $\mathbf{r}$ and $\mathbf{F}$. Assume that all the vectors shown in the sketch are in the x–y plane. The torque on m_1 due to $\mathbf{F}_1$ is along the positive z axis (out of the paper) and the torque on m_2 due to $\mathbf{F}_2$ is along the negative z axis (into the paper).

It is important to appreciate that torque and force are inherently different quantities. For one thing, torque depends on the origin we choose but force does not. For another, we see from the definition $\boldsymbol{\tau} = \mathbf{r} \times \mathbf{F}$ that $\boldsymbol{\tau}$ and $\mathbf{F}$ are always mutually perpendicular. There can be a torque on a system with zero net force, and there can be force with zero net torque. In general, there may be both torque and force. These three cases are illustrated in the sketches. (The torques are evaluated around the centers of the disks.)

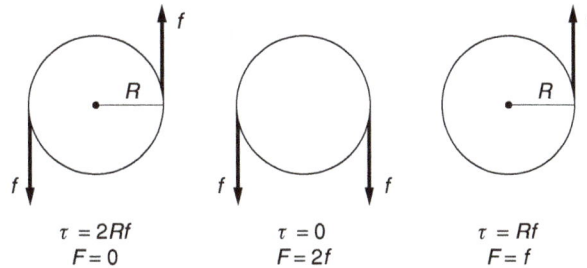

$$\tau = 2Rf \qquad\qquad \tau = 0 \qquad\qquad \tau = Rf$$
$$F = 0 \qquad\qquad\quad F = 2f \qquad\qquad F = f$$

Example 7.4 Torque due to Gravity

We often encounter systems in which there is a torque exerted by gravity. Examples include a pendulum, a child's top, and a falling chimney. In the usual case of a uniform gravitational field, the torque on a body around any point is $\mathbf{R} \times \mathbf{W}$, where $\mathbf{R}$ is a vector from the point to the center of mass and $\mathbf{W}$ is the weight. Here is the proof.

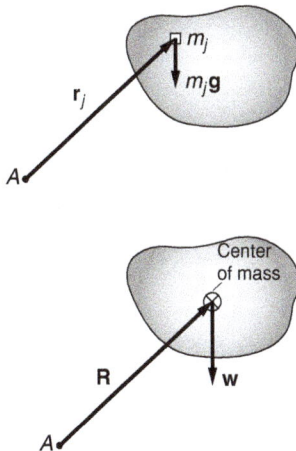

The problem is to find the torque on a body of mass M around origin A when the applied force is due to a uniform gravitational field $\mathbf{g}$. We can regard the body as a collection of particles. The torque $\boldsymbol{\tau}_j$ on the jth particle is

$$\boldsymbol{\tau}_j = \mathbf{r}_j \times m_j \mathbf{g},$$

where $\mathbf{r}_j$ is the position vector of the jth particle from origin A, and m_j is its mass. The total torque is

$$
\begin{aligned}
\boldsymbol{\tau} &= \sum_j \boldsymbol{\tau}_j \\
&= \sum_j \mathbf{r}_j \times m_j \mathbf{g} \\
&= \left(\sum_j m_j \mathbf{r}_j \right) \times \mathbf{g}.
\end{aligned}
$$

By definition of center of mass,

$$\sum m_j \mathbf{r}_j = M\mathbf{R},$$

where $\mathbf{R}$ is the position vector of the center of mass. Hence

$$\tau = M\mathbf{R} \times \mathbf{g} = \mathbf{R} \times M\mathbf{g}$$
$$= \mathbf{R} \times \mathbf{W}.$$

Example 7.5 Torque and Force in Equilibrium

For a system to be in equilibrium, the total force and the total torque must vanish. In calculating the torque, one is free to choose the origin because torque must vanish around every point for a body in equilibrium. The most convenient origin is generally a point where several forces act, since then the torques due to these forces all vanish.

A uniform rod of length $\pi R/2$ is bent in the shape of a quadrant of radius R. The rod has one end on the ground and the other leaning against a frictionless wall, as shown. The problem is to calculate the force against the wall, which equals the force component A of the wall on the quadrant.

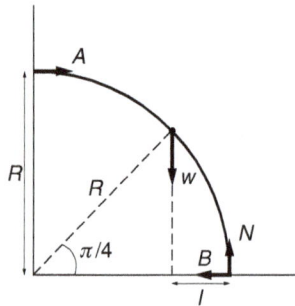

The center of mass is halfway along the rod at $\pi R/4$. For translational equilibrium, we have in the vertical direction $N = W$, where W is the weight, acting at the center of mass. The frictionless wall does not exert a vertical force. In the horizontal direction, $A = B$. For rotational equilibrium, let us evaluate the torque about the point where the quadrant rests on the ground. The torque is $\tau = Wl - AR$, and is 0 at equilibrium. (Counterclockwise torque is positive out of the plane of the paper.) From the geometric construction, we see that $l = R - R/\sqrt{2} \approx 0.293R$. Consequently, the force against the wall is $A \approx 0.293W$.

The same result is obtained if torques are taken about the point where the quadrant rests on the wall, because the torque about any point must be 0 for a body in equilibrium.

7.5 Torque and Angular Momentum

Torque is important because it determines the rate of change of angular momentum:

$$\frac{d\mathbf{L}}{dt} = \frac{d}{dt}(\mathbf{r} \times \mathbf{p})$$
$$= \left(\frac{d\mathbf{r}}{dt} \times \mathbf{p}\right) + \left(\mathbf{r} \times \frac{d\mathbf{p}}{dt}\right).$$

The first bracketed term vanishes: $(d\mathbf{r}/dt) \times \mathbf{p} = \mathbf{v} \times m\mathbf{v} = 0$, because the cross product of parallel vectors is zero. Also, $d\mathbf{p}/dt = \mathbf{F}$, by Newton's second law. Hence the second bracketed term is $\mathbf{r} \times \mathbf{F} = \boldsymbol{\tau}$, and we have

$$\boldsymbol{\tau} = \frac{d\mathbf{L}}{dt}. \tag{7.7}$$

7.5.1 Conservation of Angular Momentum

Equation (7.7) shows that if the torque is zero, $\mathbf{L}$ is constant and the angular momentum is conserved. As you have already seen from our discussion of linear momentum and energy, conservation laws are powerful tools. Because we have considered the angular momentum of a single particle, the conservation law for angular momentum has not been presented in much generality. In fact, Eq. (7.7) follows directly from Newton's second law. Only when we talk about extended systems does angular momentum assume its proper role as a new physical concept. Nevertheless, even in the present context, considerations of angular momentum lead to some surprising simplifications, as the next two examples show.

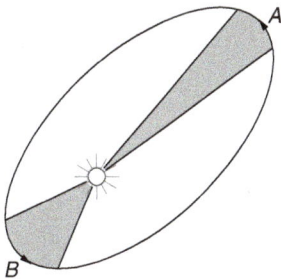

Example 7.6 Central Force Motion and the Law of Equal Areas

In 1609 the mathematician and astronomer Johannes Kepler announced his first two laws of planetary motion. The first is that the orbits of the planets are not circles but ellipses. The second is the law of equal areas: the area swept out by the radius vector from the Sun to a planet in a given time is the same for any location of the planet in its orbit.

In the sketch (not to scale), the areas swept out by the Earth during a month at two different seasons are shown shaded. The shorter radius vector at B is compensated by the greater speed of the Earth when it is nearer the Sun. We shall now show that the law of equal areas follows directly from considerations of angular momentum, and that it holds not only for motion under the gravitational force but also for motion under any central force.

Consider a particle moving under a central force $\mathbf{F}(\mathbf{r}) = f(r)\hat{\mathbf{r}}$, where $f(r)$ has any dependence on r we care to choose. The torque on the particle around the origin is $\boldsymbol{\tau} = \mathbf{r} \times \mathbf{F}(\mathbf{r}) = \mathbf{r} \times f(r)\hat{\mathbf{r}} = 0$. The angular momentum of the particle $\mathbf{L} = \mathbf{r} \times \mathbf{p}$ is therefore constant in both magnitude and direction. An immediate consequence is that the motion is confined to a plane; otherwise the direction of $\mathbf{L}$ would change with time.

We shall now prove that the rate at which area is swept out is constant, a result that leads directly to Kepler's law of equal areas.

Consider the position of the particle at t and $t + \Delta t$, when its polar coordinates are (r, θ) and $(r + \Delta r, \theta + \Delta \theta)$, respectively. The area swept out is shown shaded in the drawing.

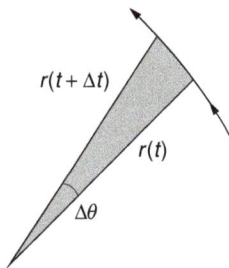

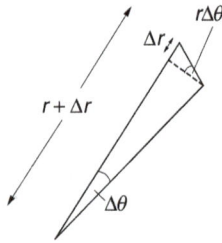

For small values of $\Delta\theta$, the area ΔA is approximately equal to the area of a triangle with base $r + \Delta r$ and altitude $r\Delta\theta$, as shown

$$\Delta A \approx \tfrac{1}{2}(r + \Delta r)(r\,\Delta\theta)$$
$$= \tfrac{1}{2}r^2\Delta\theta + \tfrac{1}{2}r\,\Delta r\,\Delta\theta.$$

The rate at which area is swept out is

$$\frac{dA}{dt} = \lim_{\Delta t \to 0} \frac{\Delta A}{\Delta t}$$
$$= \lim_{\Delta t \to 0} \frac{1}{2}\left(r^2\frac{\Delta\theta}{\Delta t} + r\frac{\Delta\theta\Delta r}{\Delta t}\right)$$
$$= \frac{1}{2}r^2\frac{d\theta}{dt}.$$

The small triangle with sides $r\Delta\theta$ and Δr is of second order and makes no contribution in the limit.

In polar coordinates the velocity of the particle is $\mathbf{v} = \dot{r}\hat{\mathbf{r}} + r\dot{\theta}\hat{\boldsymbol{\theta}}$. Its angular momentum is

$$\mathbf{L} = (\mathbf{r} \times m\mathbf{v}) = r\hat{\mathbf{r}} \times m(\dot{r}\hat{\mathbf{r}} + r\dot{\theta}\hat{\boldsymbol{\theta}}) = mr^2\dot{\theta}\hat{\mathbf{k}},$$

using $\hat{\mathbf{r}} \times \hat{\boldsymbol{\theta}} = \hat{\mathbf{k}}$. Hence

$$\frac{dA}{dt} = \frac{1}{2}r^2\dot{\theta}$$
$$= \frac{L_z}{2m}.$$

Because L_z is constant for any central force, it follows that dA/dt is constant also.

Here is another way to prove the law of equal areas, based on the vanishing of the Coriolis acceleration. For a central force, $F_\theta = 0$, so that $a_\theta = 0$. It follows that $ra_\theta = 0$, but $ra_\theta = r(2\dot{r}\dot{\theta} + r\ddot{\theta}) = (d/dt)(r^2\dot{\theta}) = 2(d/dt)(dA/dt)$. Hence $dA/dt = $ constant.

Example 7.7   Capture Cross-section of a Planet

How accurately must you aim the trajectory of an unpowered spacecraft to hit a far-off planet? Seen through a telescope, the planet has the shape of a disk. The area of the disk is πR^2, where R is the planet's radius. If gravity played no role, assuring a hit would require that we aim the spacecraft to hit this area. Fortunately, the situation is more favorable than this. Gravity tends to attract the spacecraft toward the planet, so that some trajectories that are aimed outside the planetary disk nevertheless end in a hit. Consequently, the effective area for a hit A_e is greater than the geometrical area $A_g = \pi R^2$. Our problem is to find A_e.

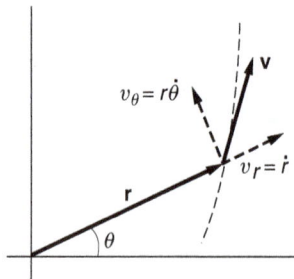

We shall neglect effects of the Sun and other planets here, although they would obviously have to be taken into account for a real space mission.

One approach to the problem would be to calculate the precise orbit of the spacecraft in the gravitational field of the planet. The procedure for this is described in Chapter 10, but such a calculation is not required, because conservation of energy and conservation of angular momentum yield the answer in a few short steps.

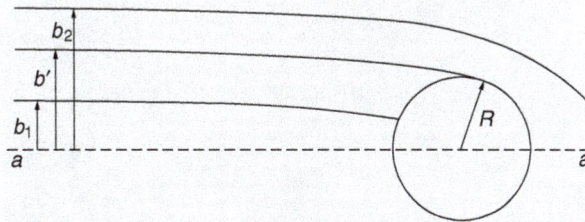

The sketch shows several possible trajectories of the spacecraft. The distance between the launch point and the target planet is assumed to be extremely large compared to R, so that the different trajectories are effectively parallel before the gravitational force of the planet becomes important. The line aa is parallel to the initial trajectories and passes through the center of the planet. The distance b between the initial trajectory and line aa is called the *impact parameter* of the trajectory. The largest value of b for which the trajectory hits the planet is indicated by b' in the sketch. The area through which the trajectory must pass to assure a hit is $A_e = \pi(b')^2$. If there were no attraction, the trajectories would be straight lines. In this case, $b' = R$ and $A_e = \pi R^2 = A_g$.

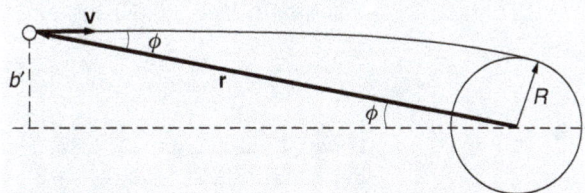

To find b', we note that the energy of the spacecraft and its angular momentum around the center of the planet are conserved. (Do you understand why angular momentum of the spacecraft is conserved while linear momentum is not?)

The kinetic energy is $K = \frac{1}{2}mv^2$, and the potential energy is $U = -mMG/r$. The total energy $E = K + U$ is

$$E = \frac{1}{2}mv^2 - \frac{mMG}{r}.$$

Initially, $r \to \infty$, and

$$L_i = -mb'v_0,$$

$$E_i = \frac{1}{2}mv_0^2.$$

A collision first occurs when this distance of closest approach is the radius of the planet. At the point of closest approach, **r** and **v** are perpendicular. If $v(R)$ is the speed at this point,

$$L_c = -mRv(R)$$

$$E_c = \frac{1}{2}mv(R)^2 - \frac{mMG}{R}.$$

Because L and E are conserved, $L_i = L_c$ and $E_i = E_c$. Hence

$$-mb'v_0 = -mRv(R) \tag{1}$$

$$\frac{1}{2}mv_0^2 = \frac{1}{2}mv(R)^2 - \frac{mMG}{R}. \tag{2}$$

Equation (1) gives $v(R) = v_0 b'/R$, and by substituting this in Eq. (2) we obtain

$$(b')^2 = R^2\left(1 + \frac{mMG/R}{mv_0^2/2}\right).$$

The effective area is

$$A_e = \pi(b')^2$$

$$= \pi R^2\left(1 + \frac{mMG/R}{mv_0^2/2}\right).$$

As we expect, the effective area is greater than the geometrical area. Since $mMG/R = -U(R)$, and $mv_0^2/2 = E$, we have

$$A_e = A_g\left(1 - \frac{U(R)}{E}\right).$$

If we "turn off" gravity $U(R) \to 0$, then $A_e \to A_g$, as we require. Furthermore, as $E \to 0, A_e \to \infty$, which means that it is impossible to miss the planet, provided that the spacecraft starts from rest. For $E = 0$, the spacecraft inevitably falls into the planet.

The gravitational force is attractive, so that $A_e > A_g$ always. The same form of force law governs the interaction of electric charges, but electric forces can be repulsive or attractive. For the repulsive force, $A_e < A_g$.

If there is a torque on a system the angular momentum must change according to $\boldsymbol{\tau} = d\mathbf{L}/dt$, as the following examples illustrate.

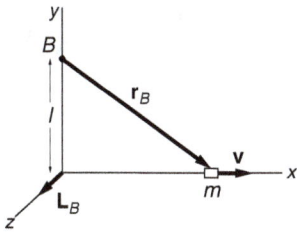

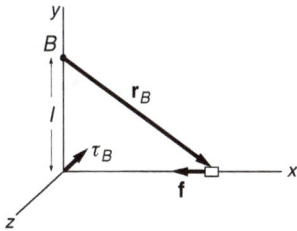

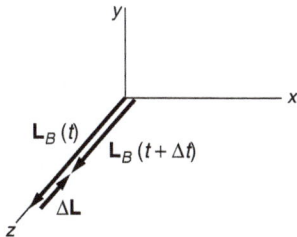

Example 7.8 Angular Momentum of a Sliding Block 2

For a simple illustration of the relation $\tau = d\mathbf{L}/dt$, consider a small block of mass m sliding along the x axis with velocity $\mathbf{v} = v\hat{\mathbf{i}}$. The angular momentum of the block around origin B is

$$\mathbf{L}_B = m\mathbf{r}_B \times \mathbf{v} \tag{1}$$

$$= mlv\hat{\mathbf{k}},$$

as we discussed in Example 7.1. If the block is sliding freely, $\mathbf{v}$ does not change, and $\mathbf{L}_B$ is therefore constant, as we expect, because there is no torque on the block.

Suppose now that the block slows because of a friction force $\mathbf{f} = -f\hat{\mathbf{i}}$. The torque on the block around origin B is

$$\tau_B = \mathbf{r}_B \times \mathbf{f}$$

$$= -lf\hat{\mathbf{k}}.$$

We see from Eq. (1) that as the block slows, $\mathbf{L}_B$ remains along the positive z direction but its magnitude decreases. Therefore, the change $\Delta\mathbf{L}_B$ in $\mathbf{L}_B$ points in the negative z direction, as shown in the sketch. The direction of $\Delta\mathbf{L}_B$ is the same as the direction of τ_B. From the fundamental relation $\tau = d\mathbf{L}/dt$, the vectors τ and $\Delta\mathbf{L}$ are always parallel.

From Eq. (1),

$$\Delta\mathbf{L}_B = ml\,\Delta v\hat{\mathbf{k}}, \tag{2}$$

where $\Delta v < 0$. Dividing Eq. (2) by Δt and taking the limit $\Delta t \to 0$, we have

$$\frac{d\mathbf{L}_B}{dt} = ml\frac{dv}{dt}\hat{\mathbf{k}}. \tag{3}$$

By Newton's second law, $m\,dv/dt = -f$ and Eq. (3) becomes

$$\frac{d\mathbf{L}_B}{dt} = -lf\hat{\mathbf{k}}$$

$$= \tau_B,$$

as we expect.

It is important to keep in mind that because τ and $\mathbf{L}$ depend on the choice of origin, the same origin must be used for both when applying the relation $\tau = d\mathbf{L}/dt$, as we were careful to do in this problem.

The angular momentum of the block in this example changed only in magnitude and not in direction, because τ and $\mathbf{L}$ happened to be along the same line. In the next example we return to the conical pendulum to study a case in which the angular momentum is constant in magnitude but changes direction due to an applied torque.

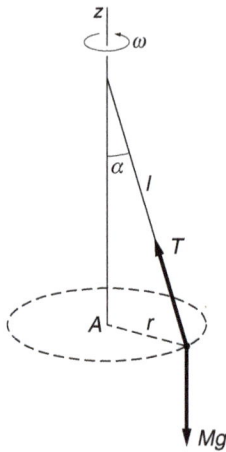

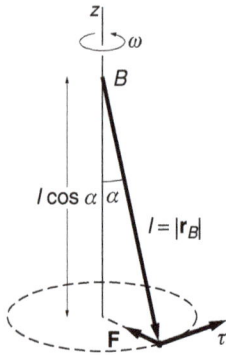

Example 7.9 Dynamics of the Conical Pendulum

In Example 7.2, we calculated the angular momentum of a conical pendulum around two different origins. For an origin at the center of the circular plane of motion, $\mathbf{L}$ is constant. In contrast, for an origin on the axis at the pivot point, $\mathbf{L}$ sweeps through space as the pendulum rotates. Nevertheless, as we shall show, the relation $\boldsymbol{\tau} = d\mathbf{L}/dt$ is satisfied for both origins.

The sketch illustrates the forces on the bob, where T is the tension in the string. For uniform circular motion there is no vertical acceleration, and consequently

$$T \cos \alpha - Mg = 0. \tag{1}$$

As the bob moves along its circular path, it accelerates toward the axis. Hence, the net force on $\mathbf{F}$ on the bob is radially inward: $\mathbf{F} = -T \sin \alpha \hat{\mathbf{r}}$. The torque on M around origin A is

$$\boldsymbol{\tau}_A = \mathbf{r}_A \times \mathbf{F}$$
$$= 0,$$

since $\mathbf{r}_A$ and $\mathbf{F}$ are both in the $\hat{\mathbf{r}}$ direction. Hence

$$\frac{d\mathbf{L}_A}{dt} = 0$$

and we have the result

$$\mathbf{L}_A = \text{constant}$$

as we already know from Example 7.2.

The situation is entirely different if we take the origin at B. The torque $\boldsymbol{\tau}_B$ is

$$\boldsymbol{\tau}_B = \mathbf{r}_B \times \mathbf{F}.$$

Hence

$$|\boldsymbol{\tau}_B| = lF \cos \alpha = lT \cos \alpha \, \sin \alpha$$
$$= Mgl \sin \alpha,$$

where we have used $T \cos \alpha = Mg$ from Eq. (1). The drawing shows that the direction of $\boldsymbol{\tau}_B$ is tangential to the line of motion of M:

$$\boldsymbol{\tau}_B = Mgl \sin \alpha \hat{\boldsymbol{\theta}}, \tag{2}$$

where $\hat{\boldsymbol{\theta}}$ is the unit tangential vector in the plane of motion.

Our problem is to show that the relation

$$\boldsymbol{\tau}_B = \frac{d\mathbf{L}_B}{dt} \tag{3}$$

is satisfied.

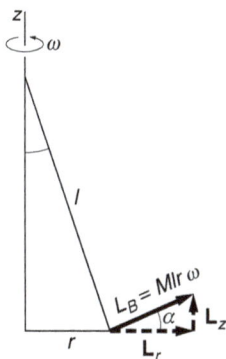

From Example 7.2, we know that $\mathbf{L}_B$ has constant magnitude $Mlr\omega$. As the diagram shows, $\mathbf{L}_B$ has a vertical component $L_z = Mlr\omega \sin \alpha$ and a horizontal radial component $L_r = Mlr\omega \cos \alpha$. Writing $\mathbf{L}_B = \mathbf{L}_z + \mathbf{L}_r$, we see that $\mathbf{L}_z$ is constant, as we expect because $\boldsymbol{\tau}_B$ has no vertical component. $\mathbf{L}_r$, however, is not constant; it changes direction as the bob swings around while the magnitude of $\mathbf{L}_r$ remains constant. We encountered such a situation in Section 1.8, where we showed that the only way a vector $\mathbf{A}$ of constant magnitude can change in time is to rotate, and that if its instantaneous rate of rotation is $d\theta/dt$, then $|d\mathbf{A}/dt| = A d\theta/dt$. We can employ this relation directly to obtain

$$\left| \frac{d\mathbf{L}_r}{dt} \right| = L_r \omega.$$

Because we shall invoke this result frequently, let us take a moment to rederive it geometrically.

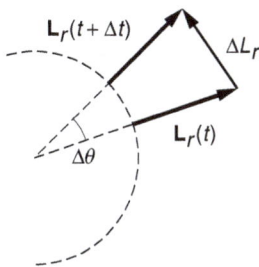

The vector diagrams show $\mathbf{L}_r$ at a time t and at $t + \Delta t$. During the interval Δt, the bob swings through angle $\Delta\theta = \omega \, \Delta t$, and $\mathbf{L}_r$ rotates through the same angle. The magnitude of the vector difference $\Delta\mathbf{L}_r = \mathbf{L}_r(t + \Delta t) - \mathbf{L}_r(t)$ is given approximately by

$$|\Delta\mathbf{L}_r| \approx L_r \Delta\theta.$$

In the limit $\Delta t \to 0$, we have

$$\frac{dL_r}{dt} = L_r \frac{d\theta}{dt}$$
$$= L_r \omega.$$

Since $L_r = Mlr\omega \cos \alpha$, we have

$$\frac{dL_r}{dt} = Mlr\omega^2 \cos \alpha.$$

$Mr\omega^2$ is the radial force, $T \sin \alpha$ so that $Mlr\omega^2 \cos \alpha = Tl \sin \alpha \cos \alpha$. Because $T \cos \alpha = Mg$, we have

$$\frac{dL_r}{dt} = Mgl \sin \alpha,$$

which agrees with the magnitude of $\boldsymbol{\tau}_B$ from Eq. (2). Furthermore, as the vector drawings indicate, $d\mathbf{L}_r/dt$ lies in the tangential direction, parallel to $\boldsymbol{\tau}_B$, as we expect.

Another way to calculate $d\mathbf{L}_B/dt$ is to write $\mathbf{L}_B$ in vector form and then differentiate:

$$\mathbf{L}_B = (Mlr\omega \sin \alpha)\hat{\mathbf{k}} + (Mlr\omega \cos \alpha)\hat{\mathbf{r}}.$$
$$\frac{d\mathbf{L}_B}{dt} = Mlr\omega \cos \alpha \frac{d\hat{\mathbf{r}}}{dt}$$
$$= Mlr\omega^2 \cos \alpha \hat{\boldsymbol{\theta}},$$

where we have used $d\hat{\mathbf{r}}/dt = \omega \hat{\boldsymbol{\theta}}$.

It is important to be able to visualize angular momentum as a vector that can rotate in space. This type of reasoning occurs often in analyzing the motion of rigid bodies; we shall find it particularly helpful in understanding gyroscope motion in Chapter 8.

7.6 Dynamics of Fixed Axis Rotation

In Chapter 4 we showed that the translational motion of a system of particles is simple to describe if we distinguish between external forces and internal forces acting on the particles. The internal forces cancel by Newton's third law, and the momentum changes only because of external forces. This leads to the law of conservation of momentum: the momentum of an isolated system is constant.

In describing rotational motion it is tempting to follow the same procedure by distinguishing between external and internal *torques*. Unfortunately, there is no way to prove from Newton's laws that the internal torques add to zero. Nevertheless, it is an experimental fact that internal torques always cancel because the angular momentum of an isolated system has never been observed to change. We shall discuss this more fully in Chapter 8, and for the remainder of this chapter we shall simply assume that only external torques change the angular momentum of a rigid body.

In this section we consider what might be called "fixed, fixed-axis" rotation, that is, rotation around an axis that is rigidly mounted and cannot translate. Examples could be the motion of a gate on its hinges or the turning of a water wheel. Motion like this, for which there is an axis of rotation at rest, is referred to as *pure rotation*. Pure rotation is important because it is simple and because it is frequently encountered.

Consider a body rotating with angular speed ω around the z axis. From Eq. (7.4) the z component of angular momentum is

$$L_z = I\omega.$$

Since $\boldsymbol{\tau} = d\mathbf{L}/dt$, where $\boldsymbol{\tau}$ is the external torque, we have

$$\tau_z = \frac{d}{dt}(I\omega)$$
$$= I\frac{d\omega}{dt}$$
$$= I\alpha,$$

where $\alpha = d\omega/dt$ is called the *angular acceleration*. We are concerned with rotation around the z axis only and so we can drop the subscript z and write

$$\tau = I\alpha. \qquad (7.8)$$

As a simple illustration, recall that in Example 7.4 we showed that the torque on a body in a uniform gravitational field is $\mathbf{R} \times \mathbf{W}$, where $\mathbf{R}$ is the vector from an origin to the center of mass and $\mathbf{W}$ is the weight. It follows that to balance an object ($\alpha = 0$) the pivot point must be at the center of mass $\mathbf{R} = 0$.

Equation (7.8) is reminiscent of $\mathbf{F} = m\mathbf{a}$. There is a close analogy between linear and rotational motion, with moment of inertia analogous to mass, torque analogous to force, and angular acceleration analogous to linear acceleration. We can develop the analogy further by evaluating the kinetic energy of a body undergoing pure rotation:

$$K = \sum_j \tfrac{1}{2} m_j v_j^2$$
$$= \sum_j \tfrac{1}{2} m_j \rho_j^2 \omega^2$$
$$= \tfrac{1}{2} I \omega^2,$$

where we have used $v_j = \rho_j \omega$ and $I = \Sigma m_j \rho_j^2$. This is clearly analogous to the kinetic energy due to the translational motion of a body: $K = \tfrac{1}{2} M V^2$, where V is the speed of the center of mass.

The method of handling problems involving rotation under applied torques is a straightforward extension of the familiar procedure for treating translational motion under applied forces, as the following example illustrates.

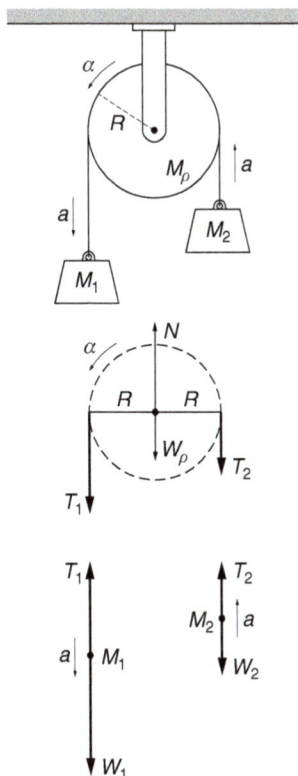

Example 7.10 Atwood's Machine with a Massive Pulley

The problem is to find the acceleration a for the arrangement shown in the sketch. The rope does not slip on the pulley and the effect of the massive pulley is to be included.

Force diagrams for the three masses are shown in the sketch, including the points of application of the forces on the pulley, which is necessary to do whenever we need to calculate torques. The pulley evidently undergoes pure rotation around its axle, so we take the axis of rotation to be the axle.

The equations of motion are

$$
\begin{array}{ll}
W_1 - T_1 = M_1 a & \text{mass}_1 \\
T_2 - W_2 = M_2 a & \text{mass}_2 \\
\tau = T_1 R - T_2 R = I\alpha & \text{net torque on pulley} \\
N - T_1 - T_2 - W_p = 0 & \text{net vertical force on pulley.}
\end{array}
$$

Note that in the torque equation, α must be positive counterclockwise to correspond to our convention that torque out of the paper is positive.

N is the force on the axle, and the equation for the net vertical force on the pulley simply assures that the pulley does not fall. Since we don't need to know N, it does not contribute to the solution.

There is a constraint relating a and α, assuming that the rope does not slip. The velocity of the rope is the velocity of a point on the surface of the wheel, $v = \omega R$, from which it follows that $a = \alpha R$.

We can now eliminate T_1, T_2, and α:

$$W_1 - W_2 - (T_1 - T_2) = (M_1 + M_2)a$$

$$T_1 - T_2 = \frac{I\alpha}{R} = \frac{Ia}{R^2}$$

$$W_1 - W_2 - \frac{Ia}{R^2} = (M_1 + M_2)a.$$

If the pulley is a simple uniform disk, we have

$$I = (M_p/2)R^2$$

and it follows that

$$a = \frac{(M_1 - M_2)g}{M_1 + M_2 + M_p/2}.$$

The pulley increases the total inertial mass of the system, but in comparison with the hanging weights, the effective mass of the pulley is only one-half its real mass.

7.7 Pendulum Motion and Fixed Axis Rotation

In Example 3.10 we analyzed the motion of a simple pendulum—a mass M hanging from a string of length l in a gravitational field g. For small-amplitude motion, we found that the pendulum executes simple harmonic motion with frequency $\omega = \sqrt{g/l}$. That solution, however, cannot be extended to a real pendulum—called a *physical pendulum*—in which the mass is not a particle but an extended structure and the support is not a massless string but a rod or some comparable structure. With the tools we have developed, we can now solve the more general problem.

We begin by describing the equation of motion of the simple pendulum using the formalism of fixed axis rotation.

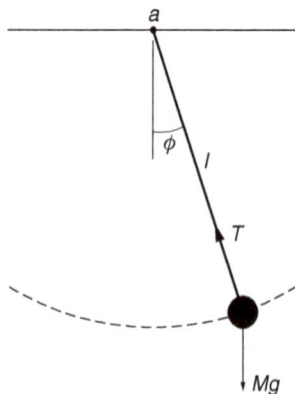

7.7.1 The Simple Pendulum

We analyzed the simple pendulum in Example 3.10 by a straightforward application of $\mathbf{F} = M\mathbf{a}$. The solution was correct, but hardly elegant. The bob on a pendulum moves on a circular arc, a case of fixed axis rotation, so the problem practically cries out to be analyzed using the language of angular momentum.

Consider the angular momentum and torque around the pivot point a. The moment of inertia of the mass is $I_a = Ml^2$. The tension in the string is a radial force and consequently exerts no torque around a; the torque arises exclusively from the weight $W = Mg$ of the bob. The torque is $\tau_a = -Wr_\perp = -Wl\sin\phi$. (The minus sign is because the torque is clockwise around a.) Using the small-angle approximation $\sin\phi \approx \phi$, the equation of motion is

$$\tau_a = I_a\alpha = I_a\ddot{\phi},$$

$$-Wl\phi = I_a\ddot{\phi},$$

$$I_a\ddot{\phi} = -Wl\phi,$$

$$\ddot{\phi} + \frac{Wl}{I_a}\phi = 0,$$

or, using $I_0 = Ml^2$,

$$\ddot{\phi} + \frac{g}{l}\phi = 0.$$

This is the equation of motion for a simple harmonic oscillator whose solution was described in Example 3.10 and proved in Example 5.2

$$\phi = A\sin\omega t + B\cos\omega t.$$

The frequency of the oscillatory motion is

$$\omega = \sqrt{\frac{Wl}{I_a}} = \sqrt{\frac{Mgl}{Ml^2}} = \sqrt{\frac{g}{l}}. \tag{7.9}$$

Caution: The symbol ω here stands for the *angular frequency* of the oscillator, a convention widely used in physics. However, we have used the same symbol to represent *angular speed*, which is also consistent with usage in physics. In this problem, the angular speed of the pendulum is $\dot{\phi}$. It is easy to confuse the two usages, particularly because they both have the same physical dimension $[T]^{-1}$, and are both measured in units of radians per second.

For the pendulum to obey the equation of motion of a harmonic oscillator we had to make the small-angle approximation: $\sin\phi \approx \phi$. Few problems in physics have exact solutions, and approximations like the small-angle approximation are often required. Because quantitative prediction is the essence of physics, it is important to determine the accuracy of an approximation wherever possible.

Note 5.1 investigates the accuracy of the small-angle approximation. The most important effect is that the finite amplitude of the swing causes a small increase in the period of the motion. For small amplitudes ϕ_0, the period T of the pendulum is

$$T \approx T_0(1 + \phi_0^2/16),$$

where $T_0 = 2\pi\sqrt{l/g}$.

7.7.2 The Physical Pendulum
Now let us turn to the physical pendulum, a real pendulum such as the one in the sketch.

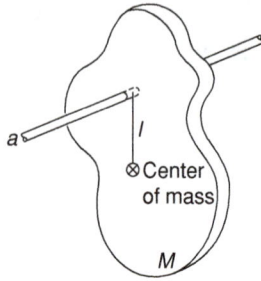

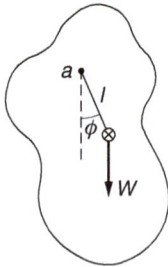

The swinging object can have any shape. Its mass is M, and its center of mass is at distance l from the pivot. The moment of inertia around the pivot, I_a, is no longer Ml^2 but a quantity that depends on the particular shape. Other than this, the analysis is identical to that for the simple pendulum that we discussed in Section 7.7.1. The pendulum executes simple harmonic motion with frequency $\omega = \sqrt{Mgl/I_a}$.

We can write this result in a simpler form if we introduce the idea of *radius of gyration*. If the moment of inertia of an object about its center of mass is I_0, the radius of gyration k is defined as

$$k \equiv \sqrt{\frac{I_0}{M}}. \tag{7.10}$$

A ring of radius R has $k_{\text{ring}} = R$; for a disk, $k_{\text{disk}} = \sqrt{1/2}R$; and for a solid sphere, $k_{\text{sphere}} = \sqrt{2/5}R$.

By the parallel axis theorem we have

$$I_a = I_0 + Ml^2$$
$$= M(k^2 + l^2),$$

so that

$$\omega = \sqrt{\frac{gl}{k^2 + l^2}}.$$

The simple pendulum corresponds to $k = 0$, in which case we obtain $\omega = \sqrt{g/l}$, as expected.

Example 7.11 Kater's Pendulum

Between the sixteenth and twentieth centuries, the most accurate measurements of g were obtained from experiments with pendulums. The method is attractive because the only quantities needed are the pendulum's dimensions and the pendulum's period, which can be determined to great accuracy by counting many swings. For precise measurements, the limiting feature turned out to be the accuracy with which the center of mass of the pendulum and its radius of gyration can be determined. A clever invention, named after the nineteenth-century English physicist, surveyor, and inventor Henry Kater, overcame this difficulty.

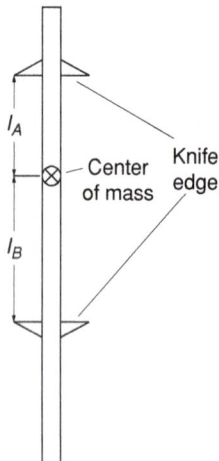

Kater's pendulum has two knife edges and the pendulum can be suspended from either. If the knife edges are distances l_A and l_B from the center of mass, then the period for small oscillations from each of these is, respectively,

$$T_A = 2\pi\sqrt{\frac{k^2 + l_A{}^2}{gl_A}}$$

$$T_B = 2\pi\sqrt{\frac{k^2 + l_B{}^2}{gl_B}}.$$

In operation, l_A or l_B is adjusted until the periods are identical: $T_A = T_B = T$. This can be done with great precision. We can then eliminate T and solve for k^2:

$$k^2 = \frac{l_A l_B^2 - l_B l_A^2}{l_B - l_A}$$

$$= l_A l_B.$$

Then

$$T = 2\pi \sqrt{\frac{l_A l_B + (l_A)^2}{g l_A}}$$

$$= 2\pi \sqrt{\frac{l_A + l_B}{g}}$$

or

$$g = 4\pi^2 \left(\frac{l_A + l_B}{T^2} \right).$$

The beauty of Kater's invention is that the only geometrical quantity needed is $l_A + l_B$, the distance between the knife edges, which can be measured to high accuracy. The position of the center of mass need not be known.

Example 7.12 Crossing Gate

A simple railroad crossing gate consists of a long narrow plank of mass M and length $2L$ pivoted at one end. While the gate is open, the gate is stored slightly off vertical, so that when the closing signal comes, the gate rotates downward. It is stopped by a support rod that keeps the gate horizontal, as shown in the sketch.

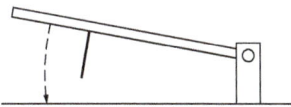

Where should the support rod be placed to minimize wear and tear on the pivot? The force $\mathbf{F}_s$ on the gate due to the support rod and the force $\mathbf{F}_p$ due to the pivot can each be resolved into a component perpendicular to the gate and a component along the gate. The components along the gate provide the radial centripetal acceleration of the rotating gate, but because we will focus on the short time of impact, these components are small compared to the large impact forces and can be neglected.

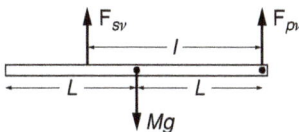

The sketch shows the vertical forces acting on the gate at the moment of impact: the component F_{sv} due to the support, the vertical component F_{pv} due to the pivot, and the weight force. We shall see that F_{pv} can be made to vanish, minimizing the force on the pivot.

Taking torques around the pivot,

$$\tau = MgL - F_{sv}l$$
$$= I_p\ddot{\theta}$$

where I_p is the moment of inertia around the pivot, L is the distance from the pivot to the center of mass, and l is the distance from the pivot to the support rod. Integrating over the short time of the collision from t to $t + \Delta t$ gives

$$I_p\dot{\theta} \approx (MgL - F_{sv}l)\Delta t$$
$$\approx -F_{sv}l\Delta t. \tag{1}$$

The torque due to the weight force can be neglected compared to the torque due to the much larger impact force.

Applying Newton's second law to the motion of the center of mass,

$$M\frac{dV}{dt} = -(F_{sv} + F_{pv}) + Mg$$

and integrating

$$MV = ML\dot{\theta} = -(F_{sv} + F_{pv}), \tag{2}$$

where the impulse due to the weight force can be neglected over the short collision time.

Solving Eq. (1) for $F_{sv}\Delta t = -I_p\dot{\theta}/l$ and using this in Eq. (2) gives

$$F_{pv}\Delta t = (I_p/l - ML)\dot{\theta}. \tag{3}$$

Equation (3) shows that if we make $I_p/l - ML = 0$, the impulse due to F_{pv} vanishes, minimizing the force on the pivot. The support rod should be placed at

$$l = \frac{I_p}{ML}. \tag{4}$$

Assuming that the gate is like a long, thin rod, we showed in Example 7.3(d) that $I_p = M(2L)^2/3$, so that $l = 4L/3$.

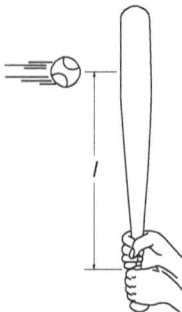

The distance l specified by Eq. (4) in the previous example is called the *center of percussion*. In batting a baseball it is important to hit the ball at the bat's center of percussion to avoid a reaction on the batter's hands and a painful sting.

From Eq. (7.10), the definition of radius of gyration is $k = \sqrt{I_0/M}$. The distance to the center of percussion is therefore $l = k^2/L$, where k is the radius of gyration about the pivot and L is the distance from the pivot to the center of mass.

7.8 Motion Involving Translation and Rotation

Often translation and rotation occur in the same system, as for example a drum rolling down an incline. In such a situation there is no obvious axis as there was in Section 7.6 where we analyzed pure rotation. The problem seems confusing until we recall the theorem in Note 7.1—it is always possible to describe any motion of a rigid body by a translation of its center of mass plus a rotation around its center of mass. By using center of mass coordinates we will find simple expressions for both the translational and rotational motions, as well as the dynamical equation connecting them.

We continue to consider only fixed axis rotation, for which the axis of rotation does not change direction, but now we will let the axis itself translate. Let the z axis be along the axis of rotation. We shall show that L_z, the z component of the angular momentum of the body, can be written as the sum of two terms: L_z is the angular momentum $I_0\omega$ due to rotation of the body around its center of mass, plus the angular momentum $(\mathbf{R} \times M\mathbf{V})_z$ due to motion of the center of mass with respect to the origin of the inertial coordinate system:

$$L_z = I_0\omega + (\mathbf{R} \times M\mathbf{V})_z, \tag{7.11}$$

where $\mathbf{R}$ is a vector from the origin to the center of mass, $\mathbf{V} = \dot{\mathbf{R}}$, and I_0 is the moment of inertia around the center of mass.

To prove Eq. (7.11), we start by considering the body to be an aggregation of N particles with masses m_j ($j = 1, \ldots, N$), located at $\mathbf{r}_j$ with respect to the chosen origin in an inertial system. The angular momentum of the body is

$$\mathbf{L} = \sum_{j=1}^{N} (\mathbf{r}_j \times m_j\dot{\mathbf{r}}_j). \tag{7.12}$$

The center of mass of the body has position vector $\mathbf{R}$:

$$\mathbf{R} = \frac{\Sigma m_j\mathbf{r}_j}{M}, \tag{7.13}$$

where M is the total mass. Let us employ the center of mass coordinates $\mathbf{r}'_j$ that were introduced in Section 4.3:

$$\mathbf{r}_j = \mathbf{R} + \mathbf{r}'_j. \tag{7.14}$$

Combining Eqs. (7.12) and (7.14) gives

$$\mathbf{L} = \sum(\mathbf{R} + \mathbf{r}'_j) \times m_j(\dot{\mathbf{R}} + \dot{\mathbf{r}}'_j)$$
$$= \mathbf{R} \times \sum m_j\dot{\mathbf{R}} + \sum m_j\mathbf{r}'_j \times \dot{\mathbf{R}} + \mathbf{R} \times \sum m_j\dot{\mathbf{r}}'_j + \sum m_j\mathbf{r}'_j \times \dot{\mathbf{r}}'_j. \tag{7.15}$$

This expression looks cumbersome, but we can show that the middle two terms are identically zero and that the first and last terms have simple

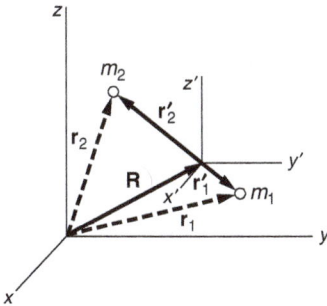

physical interpretations. Starting with the second term, we have

$$\sum m_j \mathbf{r}'_j = \sum m_j(\mathbf{r}_j - \mathbf{R})$$
$$= \sum m_j \mathbf{r}_j - M\mathbf{R}$$
$$= 0$$

by Eq. (7.13). The third term is also zero; because $\sum m_j \mathbf{r}'_j$ is identically zero, its time derivative vanishes.

The first term is

$$\mathbf{R} \times \sum m_j \dot{\mathbf{R}} = \mathbf{R} \times M\dot{\mathbf{R}}$$
$$= \mathbf{R} \times M\mathbf{V},$$

where $\mathbf{V} \equiv \dot{\mathbf{R}}$ is the velocity of the center of mass with respect to the inertial system. Equation (7.15) becomes

$$\mathbf{L} = \mathbf{R} \times M\mathbf{V} + \sum \mathbf{r}'_j \times m_j \dot{\mathbf{r}}'_j. \tag{7.16}$$

The first term of Eq. (7.16) represents the angular momentum due to the center of mass motion. The second term represents angular momentum due to motion around the center of mass. The only way for the particles of a rigid body to move with respect to the center of mass is for the body as a whole to rotate. We shall evaluate the second term for an arbitrary axis of rotation in Chapter 8. In this chapter, however, we are restricting ourselves to fixed axis rotation around the z axis. The z component of Eq. (7.16) is

$$L_z = (\mathbf{R} \times M\mathbf{V})_z + \left(\sum \mathbf{r}'_j \times m_j \dot{\mathbf{r}}'_j\right)_z. \tag{7.17}$$

The second term can be simplified. The body has angular speed ω around its center of mass, and because the origin of $\mathbf{r}'_j$ is the center of mass, the second term is identical in form to the case of pure rotation we treated in Section 7.6:

$$\left(\sum m_j \mathbf{r}'_j \times \dot{\mathbf{r}}'_j\right)_z = \left(\sum m_j \boldsymbol{\rho}'_j \times \dot{\boldsymbol{\rho}}'_j\right)_z$$
$$= \sum m_j \rho'^2_j \omega$$
$$= I_0 \omega,$$

where $\boldsymbol{\rho}'_j$ is the vector to m_j perpendicular from a z axis through the center of mass, and $I_0 = \sum m_j \rho'^2_j$ is the moment of inertia of the body around this axis.

Collecting our results, we have

$$L_z = I_0 \omega + (\mathbf{R} \times M\mathbf{V})_z. \tag{7.18}$$

We have proven the result asserted at the beginning of this section: the angular momentum of a rigid object is the sum of the angular momentum of rotation around its center of mass and the angular momentum of the center of mass around the origin. These two terms are often referred to as the *spin* and *orbital* terms, respectively. The Earth's motion around

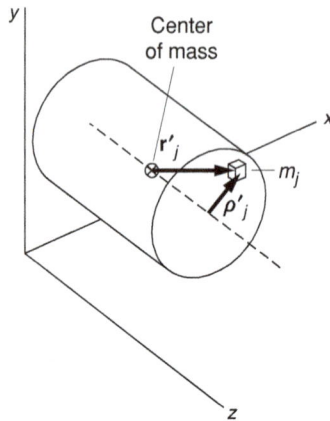

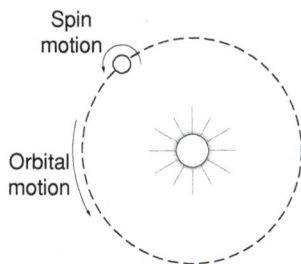

the Sun illustrates the distinction nicely. The daily rotation of the Earth around its polar axis gives rise to the Earth's spin angular momentum, and the Earth's annual revolution around the Sun gives rise to its orbital angular momentum.

An important feature of spin angular momentum is that it is independent of the coordinate system. In this sense it is intrinsic to the body; no change in coordinate system can eliminate spin, whereas orbital angular momentum disappears if the origin is chosen to lie along the line of motion.

Equation (7.18) is valid even if the center of mass is accelerating, because **L** was calculated with respect to an inertial coordinate system.

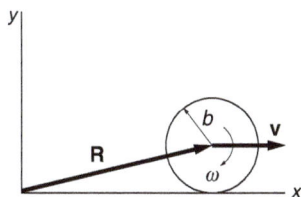

Example 7.13 Angular Momentum of a Rolling Wheel

In this example we apply Eq. (7.18) to the calculation of the angular momentum of a uniform wheel of mass M and radius b that rolls uniformly and without slipping. The moment of inertia of the wheel around its center of mass is $I_0 = \frac{1}{2}Mb^2$. With the motion shown in the sketch, the wheel's angular momentum about the center of mass is

$$L_0 = -I_0\omega$$
$$= -\tfrac{1}{2}Mb^2\omega.$$

L_0 is parallel to the z axis. The minus sign indicates that L_0 is directed into the paper, in the negative z direction.

Because the wheel rolls without slipping, $V = b\omega$, and

$$(\mathbf{R} \times M\mathbf{V})_z = -MbV.$$

The total angular momentum around the origin is then

$$L_z = -\tfrac{1}{2}Mb^2\omega - MbV$$
$$= -\tfrac{1}{2}Mb^2\omega - Mb^2\omega$$
$$= -\tfrac{3}{2}Mb^2\omega.$$

7.8.1 Torque on a Moving Body

Torque also naturally divides itself into two terms. The torque on a body is

$$\tau = \sum \mathbf{r}_j \times \mathbf{f}_j$$
$$= \sum (\mathbf{r}'_j + \mathbf{R}) \times \mathbf{f}_j$$
$$= \sum (\mathbf{r}'_j \times \mathbf{f}_j) + \mathbf{R} \times \mathbf{F}, \qquad (7.19)$$

where $\mathbf{F} = \sum \mathbf{f}_j$ is the total applied force. The first term in Eq. (7.19) is the torque around the center of mass due to the various external forces, and the second term is the torque on the center of mass due to the total external force. For fixed axis rotation Eq. (7.19) can be written

$$\tau_z = \tau_0 + (\mathbf{R} \times \mathbf{F})_z, \tag{7.20}$$

where τ_0 is the z component of the torque around the center of mass. But from Eq. (7.18) for L_z we have

$$\frac{dL_z}{dt} = I_0 \frac{d\omega}{dt} + \frac{d}{dt}(\mathbf{R} \times M\mathbf{V})_z$$

$$= I_0\alpha + (\mathbf{R} \times M\mathbf{a})_z. \tag{7.21}$$

Using $\tau_z = dL_z/dt$, Eqs. (7.20) and (7.21) yield

$$\tau_0 + (\mathbf{R} \times \mathbf{F})_z = I_0\alpha + (\mathbf{R} \times M\mathbf{a})_z$$

$$= I_0\alpha + (\mathbf{R} \times \mathbf{F})_z.$$

Hence

$$\tau_0 = I_0\alpha. \tag{7.22}$$

According to Eq. (7.22), rotational motion around the center of mass depends only on the torque around the center of mass, independent of the translational motion. In other words, Eq. (7.22) is correct even if the axis is accelerating.

Let us recapitulate the line of reasoning in this derivation. Starting with the fundamental equation

$$\boldsymbol{\tau} = \frac{d\mathbf{L}}{dt}$$

and applying it to rotation around an axis fixed in direction, for instance in a rolling body, we showed that

$$L_z = (\mathbf{R} \times M\mathbf{V})_z + I_0\omega$$

$$\frac{dL_z}{dt} = \frac{d}{dt}(\mathbf{R} \times M\mathbf{V})_z + I_0\frac{d\omega}{dt}.$$

Using $d\mathbf{R}/dt = \mathbf{V}$, $Md\mathbf{V}/dt = \mathbf{F}$ and $d\omega/dt = \alpha$

$$\frac{dL_z}{dt} = (\mathbf{R} \times \mathbf{F})_z + I_0\alpha.$$

But $dL_z/dt = \tau_z$ and from Eq. (7.20)

$$\tau_z = \tau_0 + (\mathbf{R} \times \mathbf{F})_z.$$

Comparing our results, it follows that

$$\tau_0 = I_0\alpha.$$

The result $\tau_0 = I_0\alpha$ closely resembles the equation of motion for translation in one dimension $F_z = ma_z$, and demonstrates the close analogies between the equations of motion for translational and rotational motion even though the two modes of motion are totally independent.

The same type of natural separation holds true for the kinetic energy K.

$$
\begin{aligned}
K &= \tfrac{1}{2} \sum m_j v_j{}^2 \\
&= \tfrac{1}{2} \sum m_j (\dot{\boldsymbol{\rho}}_j' + \mathbf{V})^2 \\
&= \tfrac{1}{2} \sum m_j \dot{\rho}_j'^2 + \sum m_j \dot{\boldsymbol{\rho}}_j' \cdot \mathbf{V} + \tfrac{1}{2} \sum m_j V^2 \\
&= \tfrac{1}{2} I_0 \omega^2 + \tfrac{1}{2} M V^2.
\end{aligned}
\tag{7.23}
$$

The first term corresponds to the kinetic energy of spin angular momentum, and the last term arises from the orbital center of mass motion.

Note 7.2 summarizes the dynamical relations governing fixed axis rotation.

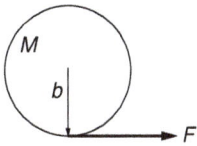

Example 7.14 Disk on Ice
A disk of mass M and radius b is pulled with constant force F by a thin tape wound around its circumference. The disk slides on ice without friction. What is its motion?

We shall solve the problem by two different methods.

Method 1
Analyzing the motion around the center of mass we have

$$
\begin{aligned}
\tau_0 &= bF \\
&= I_0 \alpha
\end{aligned}
$$

or

$$
\alpha = \frac{bF}{I_0}.
$$

The acceleration of the center of mass is

$$
a = \frac{F}{M}.
$$

Method 2
We choose a coordinate system whose origin at point A is on the line of $\mathbf{F}$. The torque around A is

$$
\begin{aligned}
\tau_z &= \tau_0 + (\mathbf{R} \times \mathbf{F})_z \\
&= bF - bF = 0.
\end{aligned}
$$

The torque is zero, as we expect, and angular momentum around the origin is conserved. The angular momentum around A is

$$
\begin{aligned}
L_z &= I_0 \omega + (\mathbf{R} \times M\mathbf{V})_z \\
&= I_0 \omega - bMV.
\end{aligned}
$$

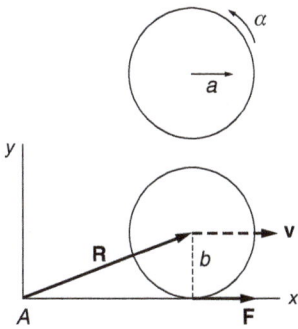

Because $dL_z/dt = 0$, we have

$$0 = I_0\alpha - bMa$$

or

$$\alpha = \frac{bMa}{I_0} = \frac{bF}{I_0},$$

as before.

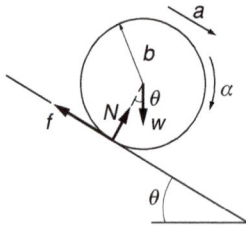

Example 7.15 Drum Rolling Down a Plane

A uniform drum of radius b and mass M rolls without slipping down a plane inclined at angle θ. The moment of inertia of the drum around its axis is $I_0 = Mb^2/2$. Find the drum's acceleration along the plane.

Method 1

The forces on the drum are shown in the diagram. f is the force of friction. The translation of the center of mass along the plane is given by

$$W \sin\theta - f = Ma$$

and the rotation around the center of mass obeys

$$bf = I_0\alpha.$$

For rolling without slipping, we have

$$a = b\alpha.$$

Eliminating f,

$$W \sin\theta - I_0\frac{\alpha}{b} = Ma.$$

Using $I_0 = Mb^2/2$, $\alpha = a/b$, and $W = Mg$, we obtain

$$Mg \sin\theta - \frac{Ma}{2} = Ma$$

or

$$a = \tfrac{2}{3}g \sin\theta.$$

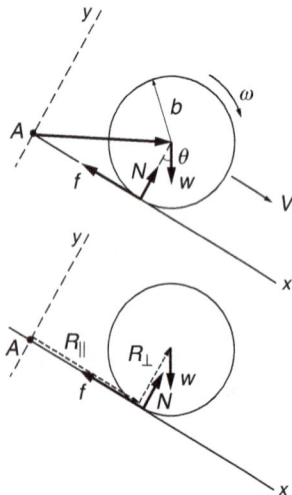

Method 2

Choose a coordinate system whose origin A is on the plane. The torque around A is

$$\begin{aligned}
\tau_y &= \tau_0 + (\mathbf{R} \times \mathbf{F})_z\\
&= -R_\perp f + R_\perp(f - W \sin\theta) + R_\parallel(N - W \cos\theta)\\
&= -bW \sin\theta,
\end{aligned}$$

since $R_\perp = b$ and $W \cos \theta = N$. The angular momentum around A is

$$L_z = -I_0 \omega + (\mathbf{R} \times M\mathbf{V})_z$$
$$= -\tfrac{1}{2} M b^2 \omega - M b^2 \omega$$
$$= -\tfrac{3}{2} M b^2 \omega,$$

where $(\mathbf{R} \times M\mathbf{V})_z = -M b^2 \omega$, as in Example 7.13. Since $\tau_z = dL_z/dt$, we have

$$bW \sin \theta = \frac{3}{2} M b^2 \alpha,$$

or

$$\alpha = \frac{2}{3} \frac{W}{Mb} \sin \theta = \frac{2}{3} \frac{g \sin \theta}{b}.$$

For rolling without slipping, $a = b\alpha$ and

$$a = \tfrac{2}{3} g \sin \theta.$$

Note that the analysis would have been even more direct if we had chosen the origin at the point of contact. In this case we can calculate τ_z directly from

$$\tau_z = \sum (\mathbf{r}_j \times \mathbf{f}_j)_z.$$

Since the unknown forces $\mathbf{f}$ and $\mathbf{N}$ act at the origin, they do not contribute to the torque. The torque is due only to W, and

$$\tau_z = -bW \sin \theta.$$

The moment of inertia about the contact point is $I = M b^2 + I_0 = \tfrac{3}{2} M b^2$ by the parallel axis theorem. Using $a = -\alpha b = \tau_z b/I$, we find $a = \tfrac{2}{3} g \sin \theta$ as before.

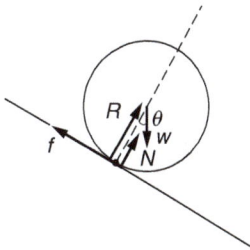

7.9 The Work–Energy Theorem and Rotational Motion

In Chapter 5 we derived the work–energy theorem for a particle

$$K_b - K_a = W_{ba}$$

where

$$W_{ba} = \oint_{\mathbf{r}_a}^{\mathbf{r}_b} \mathbf{F} \cdot d\mathbf{r}.$$

We can generalize this for a rigid body and show that the work–energy theorem divides naturally into two parts, one dealing with translational energy and one dealing with rotational energy.

To derive the translational part, we start with the equation of motion for the center of mass

$$\mathbf{F} = M\frac{d^2\mathbf{R}}{dt^2}$$
$$= M\frac{d\mathbf{V}}{dt}.$$

The work done when the center of mass is displaced by $d\mathbf{R} = \mathbf{V}\,dt$ is

$$\mathbf{F}\cdot d\mathbf{R} = M\frac{d\mathbf{V}}{dt}\cdot\mathbf{V}\,dt$$
$$= d(\tfrac{1}{2}MV^2).$$

Integrating, we obtain

$$\oint_{\mathbf{R}_a}^{\mathbf{R}_b}\mathbf{F}\cdot d\mathbf{R} = \tfrac{1}{2}MV_b^2 - \tfrac{1}{2}MV_a^2. \tag{7.24}$$

Now let us evaluate the work associated with the rotational kinetic energy. The equation of motion for fixed axis rotation around the center of mass is

$$\tau_0 = I_0\alpha$$
$$= I_0\frac{d\omega}{dt}.$$

Rotational kinetic energy has the form $\tfrac{1}{2}I_0\omega^2$, which suggests that we multiply the equation of motion by $d\theta = \omega dt$:

$$\tau_0\,d\theta = I_0\frac{d\omega}{dt}\omega dt$$
$$= d(\tfrac{1}{2}I_0\omega^2).$$

Integrating, we find that

$$\int_{\theta_a}^{\theta_b}\tau_0\,d\theta = \tfrac{1}{2}I_0\omega_b^2 - \tfrac{1}{2}I_0\omega_a^2. \tag{7.25}$$

The integral on the left evidently represents the work done by the applied torque.

The general work–energy theorem for a rigid body is therefore

$$K_b - K_a = W_{ba},$$

where $K = \tfrac{1}{2}MV^2 + \tfrac{1}{2}I_0\omega^2$ and W_{ba} is the total work on the body as it moves from position a to position b. We see from Eqs. (7.24) and (7.25) that the work–energy theorem is composed of two independent theorems, one for translation and one for rotation. In many problems these theorems can be applied separately, as the following example shows.

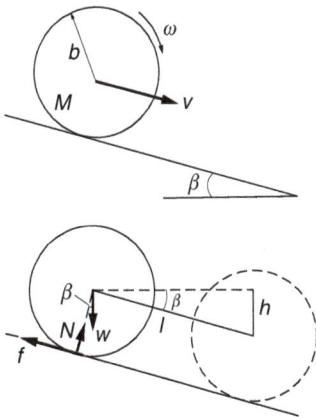

Example 7.16 Drum Rolling Down a Plane: Energy Method

Consider once again a uniform drum of radius b, mass M, weight $W = Mg$, and moment of inertia $I_0 = Mb^2/2$ on a plane of angle β. If the drum starts from rest and rolls without slipping, find the speed V of its center of mass after it has descended a height h.

The forces on the drum are shown in the sketch.

The energy equation for the translational motion is

$$\oint_a^b \mathbf{F} \cdot d\mathbf{r} = \tfrac{1}{2}MV_b^2 - \tfrac{1}{2}MV_a^2$$

or

$$(W \sin\beta - f)l = \tfrac{1}{2}MV^2, \tag{1}$$

where $l = h/\sin\beta$ is the displacement of the center of mass as the drum descends height h.

The energy equation for the rotational motion is

$$\int_{\theta_a}^{\theta_b} \tau \, d\theta = \tfrac{1}{2}I_0\omega_b^2 - \tfrac{1}{2}I_0\omega_a^2,$$

or

$$fb\theta = \tfrac{1}{2}I_0\omega^2,$$

where θ is the rotation angle around the center of mass. For rolling without slipping, $b\theta = l$. Hence

$$fl = \tfrac{1}{2}I_0\omega^2. \tag{2}$$

We also have $\omega = V/b$, so that

$$fl = \frac{1}{2}\frac{I_0V^2}{b^2}.$$

Using this in Eq. (1) to eliminate f gives

$$Wh = \frac{1}{2}\left(\frac{I_0}{b^2} + M\right)V^2$$

$$= \frac{1}{2}\left(\frac{M}{2} + M\right)V^2$$

$$= \tfrac{3}{4}MV^2$$

or

$$V = \sqrt{\frac{4gh}{3}}.$$

An interesting point in this example is that the friction force is not dissipative. From Eq. (1), friction decreases the translational energy by an amount fl. However, from Eq. (2), the torque exerted by friction increases the rotational energy by the same amount. In this motion, friction simply transforms mechanical energy from one mode to another. This is no longer the case if slipping occurs, and then some of the mechanical energy is dissipated as heat.

We conclude this section with an example involving constraints that is easily handled by energy methods.

Example 7.17 The Falling Stick

A stick of length l and mass M, initially upright on a frictionless table, starts falling. The problem is to find the speed of the center of mass as a function of the angle θ from the vertical.

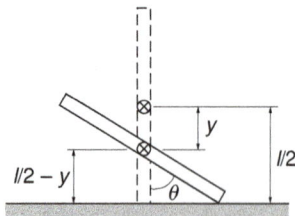

The key lies in realizing that because there are no horizontal forces, the center of mass must fall straight down. Since we must find velocity as a function of position, it is natural to apply energy methods.

The sketch shows the stick after it has rotated through angle θ and the center of mass has fallen distance y.

The initial energy is

$$E = K_0 + U_0$$
$$= \frac{Mgl}{2}.$$

The kinetic energy at a later time is

$$K = \tfrac{1}{2}I_0\dot{\theta}^2 + \tfrac{1}{2}M\dot{y}^2$$

and the corresponding potential energy is

$$U = Mg\left(\frac{l}{2} - y\right).$$

Because there are no dissipative forces, mechanical energy is conserved and $K + U = K_0 + U_0 = Mgl/2$. Hence

$$\tfrac{1}{2}M\dot{y}^2 + \tfrac{1}{2}I_0\dot{\theta}^2 + Mg\left(\frac{l}{2} - y\right) = Mg\frac{l}{2}.$$

We can eliminate $\dot{\theta}$ by using the constraint equation. The sketch shows that

$$y = \frac{l}{2}(1 - \cos\theta).$$

Hence

$$\dot{y} = \frac{l}{2}\sin\theta\,\dot{\theta}$$

and

$$\dot{\theta} = \frac{2}{l\sin\theta}\dot{y}.$$

Using $I_0 = Ml^2/12$, we obtain

$$\tfrac{1}{2}M\dot{y}^2 + \tfrac{1}{2}M\frac{l^2}{12}\left(\frac{2}{l\sin\theta}\right)^2 \dot{y}^2 + Mg\left(\frac{l}{2} - y\right) = Mg\frac{l}{2}$$

or

$$\dot{y}^2 = \frac{2gy}{[1 + 1/(3\sin^2\theta)]},$$

$$\dot{y} = \sqrt{\frac{6gy\sin^2\theta}{3\sin^2\theta + 1}},$$

$$= \sqrt{\frac{3lg(1 - \cos\theta)\sin^2\theta}{3\sin^2\theta + 1}}.$$

7.10 The Bohr Atom

The theory of the hydrogen atom published by the Danish physicist Niels Bohr in 1913 pointed the way to the creation of quantum mechanics in the 1920s. We conclude this chapter with a description of Bohr's theory to illustrate how the concepts of angular momentum and energy developed in this chapter were carried forward to help create a new theory. Our description of the Bohr theory is similar, though not identical, to Bohr's paper that he published in 1913 at the age of 26. This brief account cannot deal adequately with the background to the Bohr theory, but it may give some of the flavor of one of the great chapters in physics. Our discussion is not rigorous: this account is intended to be in the spirit of optional reading rather than an essential step in our development of classical mechanics.

The development of optical spectroscopy in the nineteenth century made available a great deal of experimental data on the structure of atoms. The light from atoms excited by an electric discharge is radiated only at certain discrete wavelengths characteristic of the element involved. In the last half of the nineteenth century tremendous efforts were devoted to measuring the wavelengths and intensities of these spectral lines. The wavelength measurements represented a notable experimental achievement but their interpretation was a notable failure; aside from certain empirical rules that gave no insight into the underlying physical laws, there was no progress in fundamental understanding.

The most celebrated empirical spectral formula was discovered in 1886 by the Swiss high school art teacher Joseph Balmer. He found that the wavelengths of the optical spectrum of atomic hydrogen are given within experimental accuracy by the formula

$$\frac{1}{\lambda} = R_\infty\left(\frac{1}{2^2} - \frac{1}{n^2}\right) \quad n = 3, 4, 5, \ldots,$$

where λ is the wavelength of a particular spectral line, and R_∞ is a constant, named the *Rydberg constant* after the Swedish spectroscopist who modified Balmer's formula to apply to certain other spectra. Numerically, $R_\infty = 109,700$ cm^{-1}. (In this section we shall follow the former tradition of atomic physics by using CGS units.)

Not only did Balmer's formula account for the known lines of hydrogen, $n = 3$ through $n = 6$, it predicted other lines, $n = 7, 8, \ldots$, which were quickly found. Furthermore, Balmer suggested that there might be other lines given by

$$\frac{1}{\lambda} = R_\infty \left(\frac{1}{m^2} - \frac{1}{n^2} \right) \quad m = 3, 4, 5, \ldots \quad n = m+1, m+2, \ldots \quad (7.26)$$

and these, too, were found. (Balmer overlooked the series with $m = 1$, lying in the ultraviolet, which was found in 1916.)

The Balmer formula undoubtedly contained the key to the structure of hydrogen, yet no one was able to create a model for an atom that could radiate such a spectrum.

J. J. Thomson, working in the Cavendish physical laboratory at Cambridge University, surmised the existence of electrons in 1897. This first indication of the divisibility of the atom stimulated further work, and in 1911 New Zealand-born Ernest Rutherford's α scattering experiments at the University of Manchester showed that atoms have a charged core that contains most of the mass. Each atom has an integral number of electrons and an equal number of positive charges in the massive core. However, this planetary model of the atom created a crucial dilemma: according to the laws of electromagnetic theory, the circulating electrons should radiate their energy in a very short time and spiral into the core.

A further development in physics that played an essential role in Bohr's theory was Einstein's theory of the photoelectric effect. In 1905, the same year that he published the special theory of relativity, Einstein proposed that the energy transmitted by light consists of discrete "packages," or quanta. The quantum of light is called a *photon*, and Einstein asserted that the energy of a photon is $E = h\nu$, where ν is the frequency of the light and $h = 6.62 \times 10^{-27}$ erg·s is Planck's constant. (Max Planck had introduced h in 1901 in his theory of radiation from hot bodies.)

Bohr made the following postulates:

1. Atoms cannot possess arbitrary amounts of energy but must exist only in certain *stationary states*. While in a stationary state, an atom does not radiate. By this bold but totally unjustified step, Bohr swept aside the problem of atomic stability.
2. An atom can "jump" from one stationary state a to a lower state b by emitting radiation with energy $E_a - E_b$. The frequency of the emitted "package of radiation" is

$$\nu = \frac{E_a - E_b}{h}.$$

In 1926 the "package" was given the name *photon*.

3. While in a stationary state, the motion of an electron can be described by classical physics, notwithstanding that the idea of energy jumps was totally novel.

4. The angular momentum of the electron is $nh/2\pi$, where n is an integer. In other words, angular momentum is quantized—it can have only certain discrete values.

Since assumption 1 breaks completely with classical physics, assumption 3 hardly seems justified. Bohr recognized this difficulty, and possibly the reason that Bohr continued to apply classical physics to this nonclassical situation was that he felt that at least some of the fundamental concepts of classical physics should carry over into the new physics, and that they should not be discarded until proven to be unworkable.

Bohr did not utilize postulate 4 in his first paper, although he pointed out the possibility of doing so. It has become traditional to treat this postulate as a fundamental assumption.

Let us apply these four postulates to hydrogen. The hydrogen atom consists of a single electron of charge $-e$ and mass m_e, and a nucleus of charge $+e$ and mass M. We assume that the massive nucleus is essentially at rest and that the electron is in a circular orbit of radius r with velocity v. The radial equation of motion is

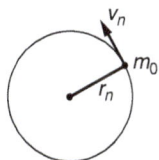

$$-\frac{m_e v^2}{r} = -\frac{e^2}{r^2},$$
(7.27)

where $-e^2/r^2$ is the attractive Coulomb force between the charges. (We are using a system of units in which the force between two charges is $Q_1 Q_2/r^2$.) The energy is

$$E = K + U = \tfrac{1}{2} m_e v^2 - \frac{e^2}{r}.$$
(7.28)

Combining Eqs. (7.27) and (7.28) yields

$$E = -\frac{1}{2} \frac{e^2}{r}.$$
(7.29)

By postulate 4, the angular momentum is $nh/2\pi$, where n is an integer. Labeling the orbit parameters by n, we have

$$\frac{nh}{2\pi} = m_e r_n v_n.$$
(7.30)

Combining Eqs. (7.30) and (7.27) yields

$$r_n = \frac{n^2 h^2}{m_e e^2} \frac{1}{(2\pi)^2},$$
(7.31)

and Eq. (7.29) gives

$$E_n = -\frac{1}{2} \frac{(2\pi)^2 m_e e^4}{n^2 h^2}.$$
(7.32)

If the electron makes a transition from state n to state m, the emitted photon has frequency

$$
\begin{aligned}
v &= \frac{E_n - E_m}{h} \\
&= \frac{(2\pi)^2}{2} \frac{m_e e^4}{h^3} \left(\frac{1}{m^2} - \frac{1}{n^2} \right).
\end{aligned}
$$

The wavelength of the radiation is given by

$$
\begin{aligned}
\frac{1}{\lambda} &= \frac{v}{c} \\
&= \frac{(2\pi)^2}{2c} \frac{m_e e^4}{h^3} \left(\frac{1}{m^2} - \frac{1}{n^2} \right).
\end{aligned}
$$

This is identical in form to the Balmer formula, Eq. (7.26), with the Rydberg constant predicted to depend on fundamental constants according to

$$
R_\infty = \frac{(2\pi)^2}{2c} \frac{m_e e^4}{h^3}.
$$

Because Bohr's prediction of the Rydberg constant agreed with its observed value, his paper was taken seriously, even though it was filled with contradictions.

Note 7.1 Chasles' Theorem

Chasles' theorem asserts that it is always possible to represent an arbitrary displacement of a rigid body by a translation of its center of mass plus a rotation around its center of mass. The proof in this note is rather detailed and an understanding of it is not necessary for following the development of the text. However, the result is interesting and its proof provides a nice exercise in vector methods for those interested.

To avoid algebraic complexities, we consider here a simple rigid body consisting of two masses m_1 and m_2 joined by a massless rigid rod of length l. The position vectors of m_1 and m_2 are $\mathbf{r}_1$ and $\mathbf{r}_2$, respectively, as shown in the sketch. The position vector of the center of mass of the body is $\mathbf{R}$, and $\mathbf{r}'_1$ and $\mathbf{r}'_2$ are the position vectors of m_1 and m_2 with respect to the center of mass. The vectors $\mathbf{r}'_1$ and $\mathbf{r}'_2$ are back to back along the line joining the masses.

In an arbitrary displacement of the body, m_1 is displaced by $d\mathbf{r}_1$ and m_1 is displaced by $d\mathbf{r}_2$. Because the body is rigid, $d\mathbf{r}_1$ and $d\mathbf{r}_2$ are not independent, and we begin our analysis by finding their relation. The distance between m_1 and m_2 is fixed and of length l. Therefore

$$
|\mathbf{r}_1 - \mathbf{r}_2| = l
$$

or

$$
(\mathbf{r}_1 - \mathbf{r}_2) \cdot (\mathbf{r}_1 - \mathbf{r}_2) = l^2. \tag{1}
$$

Taking differentials of Eq. (1), and recalling that $d(\mathbf{A} \cdot \mathbf{A}) = 2\mathbf{A} \cdot d\mathbf{A}$,

$$
(\mathbf{r}_1 - \mathbf{r}_2) \cdot (d\mathbf{r}_1 - d\mathbf{r}_2) = 0. \tag{2}
$$

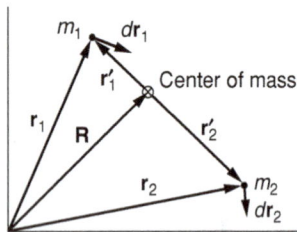

Equation (2) is the "rigid body condition" we seek. There are evidently two ways of satisfying Eq. (2): either $d\mathbf{r}_1 = d\mathbf{r}_2$, or $(d\mathbf{r}_1 - d\mathbf{r}_2)$ is perpendicular to $(\mathbf{r}_1 - \mathbf{r}_2)$.

We now turn to the translational motion of the center of mass. By definition,

$$\mathbf{R} = \frac{m_1\mathbf{r}_1 + m_2\mathbf{r}_2}{m_1 + m_2}.$$

Therefore, the displacement $d\mathbf{R}$ of the center of mass is

$$d\mathbf{R} = \frac{m_1 d\mathbf{r}_1 + m_2 d\mathbf{r}_2}{m_1 + m_2}. \tag{3}$$

If we subtract this translational displacement from $d\mathbf{r}_1$ and $d\mathbf{r}_2$, the residual displacements $d\mathbf{r}_1 - d\mathbf{R}$ and $d\mathbf{r}_2 - d\mathbf{R}$ should give a pure rotation around the center of mass. Before investigating this point, we notice that since

$$\mathbf{r}_1 - \mathbf{R} = \mathbf{r}_1'$$
$$\mathbf{r}_2 - \mathbf{R} = \mathbf{r}_2',$$

the residual displacements are

$$d\mathbf{r}_1 - d\mathbf{R} = d\mathbf{r}_1'$$
$$d\mathbf{r}_2 - d\mathbf{R} = d\mathbf{r}_2'. \tag{4}$$

Using Eq. (3) in Eq. (4) we have

$$d\mathbf{r}_1' = d\mathbf{r}_1 - d\mathbf{R}$$
$$= \left(\frac{m_2}{m_1 + m_2}\right)(d\mathbf{r}_1 - d\mathbf{r}_2) \tag{5}$$

and

$$d\mathbf{r}_2' = d\mathbf{r}_2 - d\mathbf{R}$$
$$= -\left(\frac{m_1}{m_1 + m_2}\right)(d\mathbf{r}_1 - d\mathbf{r}_2). \tag{6}$$

Note that if $d\mathbf{r}_1 = d\mathbf{r}_2$, the residual displacements $d\mathbf{r}_1'$ and $d\mathbf{r}_2'$ are zero and the rigid body translates without rotating.

We must now show that the residual displacements represent a pure rotation around the center of mass to complete the proof. The sketch shows what a pure rotation would look like.

First we show that $d\mathbf{r}_1'$ and $d\mathbf{r}_2'$ are perpendicular to the line $\mathbf{r}_1' - \mathbf{r}_2'$:

$$d\mathbf{r}_1' \cdot (\mathbf{r}_1' - \mathbf{r}_2') = d\mathbf{r}_1' \cdot (\mathbf{r}_1 - \mathbf{r}_2)$$
$$= \left(\frac{m_2}{m_1 + m_2}\right)(d\mathbf{r}_1 - d\mathbf{r}_2) \cdot (\mathbf{r}_1 - \mathbf{r}_2)$$
$$= 0,$$

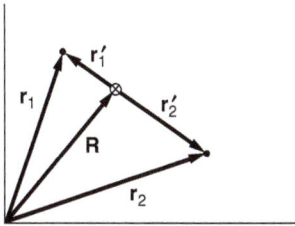

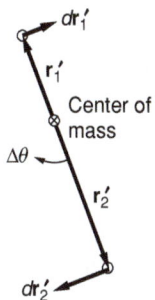

where we have used Eq. (5) and the rigid body condition Eq. (2). Similarly,

$$d\mathbf{r}'_2 \cdot (\mathbf{r}'_1 - \mathbf{r}'_2) = 0.$$

Finally, we require that the residual displacements correspond to rotation through the same angle $\Delta\theta$. With reference to the sketch, this condition in vector form is

$$\frac{d\mathbf{r}'_1}{r'_1} = -\frac{d\mathbf{r}'_2}{r'_2}.$$

Note that

$$\frac{r'_1}{r'_2} = \frac{m_2}{m_1}$$

by definition of the center of mass. Using Eqs. (5) and (6), we have

$$\frac{d\mathbf{r}'_1}{r'_1} = \left(\frac{m_2}{m_1 + m_2}\right)\frac{(d\mathbf{r}_1 - d\mathbf{r}_2)}{r'_1}$$

$$= \left(\frac{m_1}{m_1 + m_2}\right)\frac{(d\mathbf{r}_1 - d\mathbf{r}_2)}{r'_2}$$

$$= -\frac{d\mathbf{r}'_2}{r'_2},$$

completing the proof.

Note 7.2 A Summary of the Dynamics of Fixed Axis Rotation
(a) **Pure rotation around an axis—no translation**

$$L = I\omega$$

$$\tau = I\alpha$$

$$K = \tfrac{1}{2}I\omega^2.$$

(b) **Rotation and translation (subscript 0 refers to the center of mass)**

$$L_z = I_0\omega + (\mathbf{R} \times M\mathbf{V})_z$$

$$\tau_z = \tau_0 + (\mathbf{R} \times \mathbf{F})_z$$

$$\tau_0 = I_0\alpha$$

$$K = \tfrac{1}{2}I_0\omega^2 + \tfrac{1}{2}MV^2.$$

Problems
For problems marked *, refer to page 522 for a hint, clue, or answer.

7.1   *Origins*

(a) Show that if the total linear momentum of a system of particles is zero, the angular momentum of the system is the same around all origins.

(b) Show that if the total force on a system of particles is zero, the torque on the system is the same around all origins.

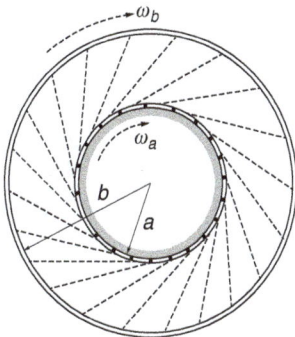

7.2 *Drum and sand**

A drum of mass M_A and radius a rotates freely with initial angular speed $\omega_A(0)$. A second drum with mass M_B and radius $b > a$ is mounted on the same axis and is at rest, although it is free to rotate. A thin layer of sand with mass M_s is distributed on the inner surface of the smaller drum. At $t = 0$, small perforations in the inner drum are opened. The sand starts to fly out at a constant rate $dM/dt = \lambda$ and sticks to the outer drum. Find the subsequent angular velocities of the two drums ω_A and ω_B. Ignore the transit time of the sand.

7.3 *Ring and bug**

A ring of mass M and radius R lies on its side on a frictionless table. It is pivoted to the table at its rim. A bug of mass m walks around the ring with speed v, starting at the pivot. What is the rotational velocity of the ring when the bug is

(a) halfway around?

(b) back at the pivot?

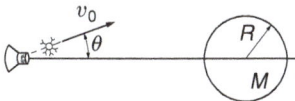

7.4 *Grazing instrument package*

A spaceship is sent to investigate a planet of mass M and radius R. While hanging motionless in space at a distance $5R$ from the center of the planet, the ship fires an instrument package with speed v_0, as shown in the sketch. The package has mass m, which is much smaller than the mass of the spaceship. For what angle θ will the package just graze the surface of the planet?

7.5 *Car on a hill*

A 3000-lb car is parked on a 30° slope, facing uphill. The center of mass of the car is halfway between the front and rear wheels and is 2 ft above the ground. The wheels are 8 ft apart. Find the normal force exerted by the road on the front wheels and on the rear wheels.

7.6 *Man on a railroad car*

A man of mass M stands on a railroad car that is rounding an unbanked turn of radius R at speed v. His center of mass is height L above the car, and his feet are distance d apart. The man is facing the direction of motion. How much weight is on each of his feet?

7.7 *Moment of inertia of a triangle*

Find the moment of inertia of a thin sheet of mass M in the shape of an equilateral triangle around an axis through a vertex, perpendicular to the sheet. The length of each side is L.

7.8 *Moment of inertia of a sphere**

Find the moment of inertia of a uniform sphere of mass M and radius R around an axis through the center.

7.9 *Bar and rollers*

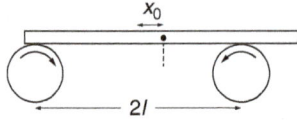

A heavy uniform bar of mass M rests on top of two identical rollers that are continuously turned rapidly in opposite directions, as shown. The centers of the rollers are a distance $2l$ apart. The coefficient of friction between the bar and the roller surfaces is μ, a constant independent of the relative speed of the two surfaces.

Initially the bar is held at rest with its center at distance x_0 from the midpoint of the rollers. At time $t = 0$ it is released. Find the subsequent motion of the bar.

7.10 *Cylinder in groove**

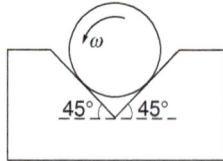

A cylinder of mass M and radius R is rotated in a uniform V groove with constant angular speed ω. The coefficient of friction between the cylinder and each surface is μ. What torque must be applied to the cylinder to keep it rotating?

7.11 *Wheel and shaft**

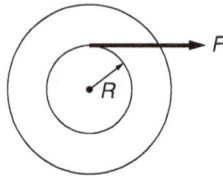

A wheel is attached to a fixed shaft, and the system is free to rotate without friction. To measure the moment of inertia of the wheel–shaft system, a tape of negligible mass wrapped around the shaft is pulled with a known constant force F. When a length L of tape has unwound, the system is rotating with angular speed ω_0. Find the moment of inertia I_0 of the system.

7.12 *Beam and Atwood's machine*

A pivoted beam has a mass m_1 suspended from one end and an Atwood's machine suspended from the other (see left-hand sketch). The frictionless pulley has negligible mass and dimension. Gravity is directed downward, and $m_2 > m_3$.

Find a relation between $m_1, m_2, m_3, l_1,$ and l_2 that will ensure that the beam has no tendency to rotate just after the masses are released.

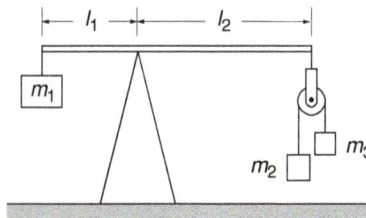

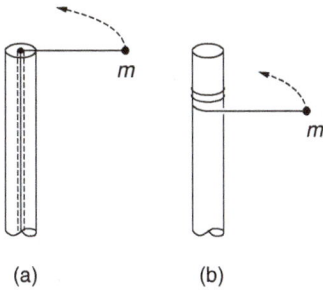

(a)   (b)

7.13 *Mass and post*

A mass m is attached to a post of radius R by a string. Initially it is distance r from the center of the post and is moving tangentially with speed v_0.

Case (a) The string passes through a hole in the center of the post at the top. The string is gradually shortened by drawing it through the hole.

Case (b) The string wraps around the outside of the post.

What quantities are conserved in each case? Find the final speed of the mass when it hits the post for each case.

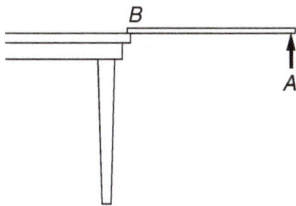

7.14 *Stick on table**

A uniform stick of mass M and length l is suspended horizontally with end B on the edge of a table, and the other end A is held by hand. Point A is suddenly released. At the instant after release:

(*a*) What is the torque around B?

(*b*) What is the angular acceleration around B?

(*c*) What is the vertical acceleration of the center of mass?

(*d*) From (*c*), find by inspection the vertical force at B.

7.15 *Two-disk pendulum*

A pendulum is made of two disks each of mass M and radius R separated by a massless rod. One of the disks is pivoted through its center by a small pin. The disks hang in the same plane and their centers are a distance l apart. Find the period for small oscillations.

7.16 *Disk pendulum*

A physical pendulum is made of a uniform disk of mass M and radius R suspended from a rod of negligible mass. The distance from the pivot to the center of the disk is l. What value of l makes the period a minimum?

7.17 *Rod and springs*

A rod of length l and mass m, pivoted at one end, is held by a spring at its midpoint and a spring at its far end, both pulling in opposite directions. The springs have spring constant k, and at equilibrium their pull is perpendicular to the rod. Find the frequency of small oscillations around the equilibrium position.

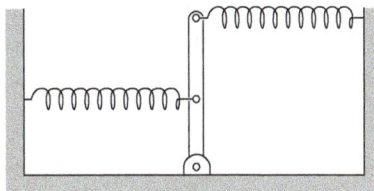

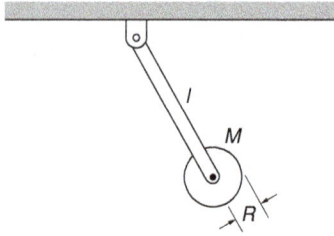

7.18 Rod and disk pendulum

Find the period of a pendulum consisting of a disk of mass M and radius R fixed to the end of a rod of length l and mass m. How does the period change if the disk is mounted to the rod by a frictionless bearing so that it is perfectly free to spin?

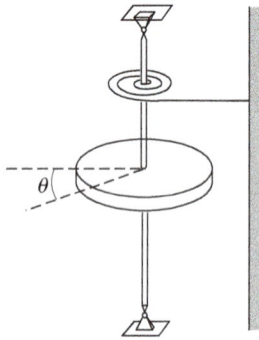

7.19 Disk and coil spring

A solid disk of mass M and radius R is on a vertical shaft. The shaft is attached to a coil spring that exerts a linear restoring torque of magnitude $C\theta$, where θ is the angle measured from the static equilibrium position and C is a constant. Neglect the mass of the shaft and the spring, and assume the bearings to be frictionless.

(*a*) Show that the disk can undergo simple harmonic motion, and find the frequency of the motion.

(*b*) Suppose that the disk is moving according to $\theta = \theta_0 \sin{(\omega t)}$, where ω is the frequency found in part (*a*). At time $t_1 = \pi/\omega$, a ring of sticky putty of mass M and radius R is dropped concentrically on the disk. Find:

(1) The new frequency of the motion.

(2) The new amplitude of the motion.

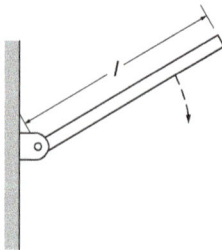

7.20 Falling plank

A thin plank of mass M and length l is pivoted at one end, as shown. The plank is released at 60° from the vertical. What is the magnitude and direction of the force on the pivot when the plank is horizontal?

7.21 Rolling cylinder*

A cylinder of radius R and mass M rolls without slipping down a plane inclined at angle θ. The coefficient of friction is μ.

What is the maximum value of θ for the cylinder to roll without slipping?

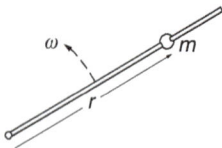

7.22 Bead and rod

A bead of mass m slides without friction on a rod that is made to rotate at a constant angular speed ω. Neglect gravity.

(*a*) Show that $r = r_0 e^{\omega t}$ is a possible motion of the bead, where r_0 is the initial distance of the bead from the pivot.

(*b*) For the motion described in part (*a*), find the force exerted on the bead by the rod.

(*c*) For the motion described above, find the power exerted by the agency that is turning the rod and show by direct calculation that this power equals the rate of change of kinetic energy of the bead.

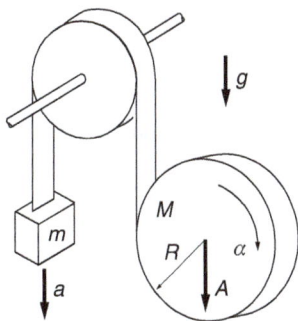

7.23 *Disk, mass, and tape**

A disk of mass M and radius R unwinds from a tape wrapped around it. The tape passes over a frictionless pulley, and a mass m is suspended from the other end. Assume that the disk drops vertically.

(*a*) Relate the accelerations of m and the disk, a and A, respectively, to the angular acceleration α of the disk.

(*b*) Find a, A, and α.

7.24 *Two drums*

Drum A of mass M and radius R is suspended from a drum B also of mass M and radius R, which is free to rotate around its axis. The suspension is in the form of a massless metal tape wound around the outside of each drum, and free to unwind, as shown. Gravity is directed downward. Both drums are initially at rest. Find the initial acceleration of drum A, assuming that it moves straight down.

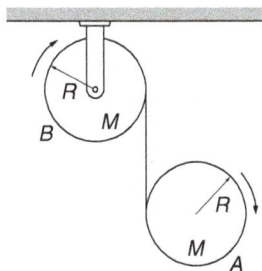

7.25 *Rolling marble**

A marble of mass M and radius R is rolled up a plane of angle θ. If the initial velocity of the marble is v_0, what is the distance l it travels up the plane before it begins to roll back down?

7.26 *Sphere and cylinder*

A uniform sphere of mass M and radius R and a uniform cylinder of mass M and radius R are released simultaneously from rest at the top of an inclined plane. Which body reaches the bottom first if they both roll without slipping?

7.27 *Yo-yo on table*

A yo-yo of mass M has an axle of radius b and a spool of radius R. Its moment of inertia can be taken to be $MR^2/2$. The yo-yo is placed upright on a table and the string is pulled with a horizontal force F as shown. The coefficient of friction between the yo-yo and the table is μ.

What is the maximum value of F for which the yo-yo will roll without slipping?

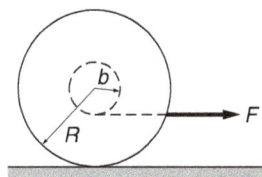

7.28 *Yo-yo pulled at angle*

The yo-yo of the previous problem is pulled so that the string makes an angle θ with the horizontal. For what value of θ does the yo-yo have no tendency to rotate?

7.29 *Yo-yo motion*

A yo-yo of mass M has an axle of radius b and a spool of radius R. Its moment of inertia can be taken to be $MR^2/2$ and the thickness of the string can be neglected. The yo-yo is released from rest.

(*a*) What is the tension in the cord as the yo-yo descends and as it ascends?

(*b*) The center of the yo-yo descends distance h before the string is fully unwound. Assuming that it reverses direction with uniform

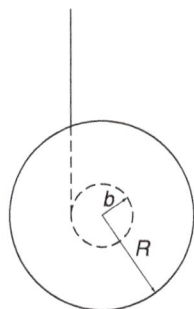

spin velocity, find the average force on the string while the yo-yo turns around.

7.30 *Sliding and rolling bowling ball*
A bowling ball is thrown down the alley with speed v_0. Initially it slides without rolling, but due to friction it begins to roll. Show that its speed when it rolls without sliding is $\frac{5}{7}v_0$.

7.31 *Skidding and rolling cylinder**
A cylinder of radius R spins with angular speed ω_0. When the cylinder is gently laid on a table, it skids for a short time and eventually rolls without slipping. What is the final angular speed ω_f?

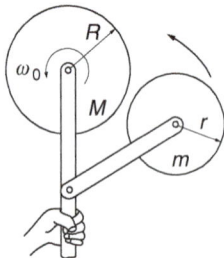

7.32 *Two rubber wheels*
A solid rubber wheel of radius R and mass M rotates with angular speed ω_0 around a frictionless pivot, as shown. A second rubber wheel of radius r and mass m, also mounted on a frictionless pivot, is brought into contact with it. What is the final angular speed of the first wheel?

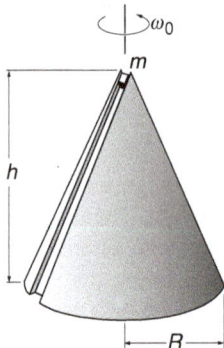

7.33 *Grooved cone and mass*
A cone of height h and base radius R is free to rotate around a fixed vertical axis. It has a thin groove cut in its surface. The cone is set rotating freely with angular speed ω_0, and a small block of mass m is released in the top of the frictionless groove and allowed to slide under gravity. Assume that the block stays in the groove. Take the moment of inertia of the cone around the vertical axis to be I_0.

    (*a*) What is the angular speed of the cone when the block reaches the bottom?

    (*b*) Find the speed of the block in inertial space when it reaches the bottom.

7.34 *Marble in dish**
A marble of radius b rolls back and forth in a shallow dish of radius R, where $R \gg b$. Find the frequency of small oscillations.

7.35 *Cube and drum*
A cubical block of side L rests on a fixed cylindrical drum of radius R. Find the largest value of L for which the block is stable (see left-hand sketch).

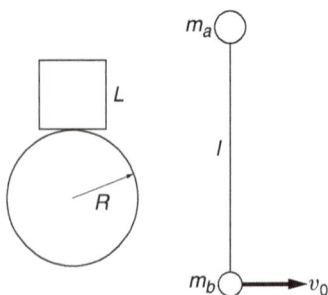

7.36 *Two twirling masses**
Two masses m_a and m_b are connected by a string of length l and lie on a frictionless table. The system is twirled and released with m_a instantaneously at rest and m_b moving with instantaneous velocity v_0 at right angles to the line of centers (see right-hand sketch). Find the subsequent motion of the system and the tension in the string.

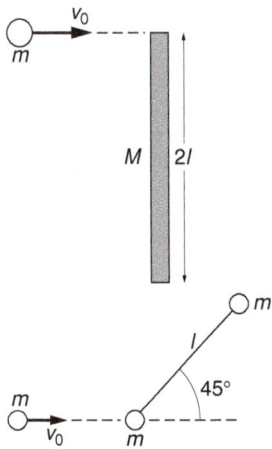

7.37 Plank and ball*

(a) A plank of length $2l$ and mass M lies on a frictionless table. A ball of mass m and speed v_0 strikes its end as shown. Find the final velocity of the ball, v_f, assuming that mechanical energy is conserved and that v_f is along the original line of motion.

(b) Find v_f assuming that the stick is pivoted at the lower end.

7.38 Collision on a table*

A rigid massless rod of length l joins two particles each of mass m. The rod lies on a frictionless table, and is struck by a particle of mass m and velocity v_0, moving as shown. After the collision, the projectile moves straight back.

Find the angular speed of the rod around its center of mass after the collision, assuming that mechanical energy is conserved.

7.39 Child on ice with plank*

A child of mass m runs on ice with velocity v_0 and steps on the end of a plank of length l and mass M that is perpendicular to the child's path, as shown.

(a) Describe quantitatively the motion of the system after the child is on the plank. Neglect friction with the ice.

(b) One point on the plank is at rest immediately after the collision. Where is it?

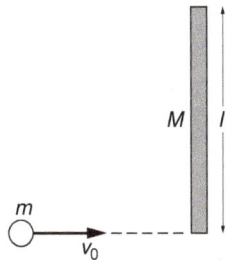

7.40 Toothed wheel and spring

A wheel with fine teeth is attached to the end of a spring with constant k and unstretched length l, as shown. For $x > l$, the wheel slips freely on the surface, but for $x < l$ the teeth mesh with the teeth on the ground so that it cannot slip. The wheel has mass M and radius R. Assume that all the mass of the wheel is in its rim.

(a) The wheel is pulled to $x = l + b$ and released. How close will it come to the wall on its first trip?

(b) How far out will it go as it leaves the wall?

(c) What happens when the wheel next hits the gear track?

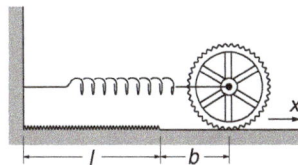

7.41 Leaning plank*

This problem utilizes most of the important laws introduced so far and it is worth a substantial effort. The problem is tricky (although not really complicated), so don't be alarmed if the solution eludes you.

A plank of length $2L$ leans against a wall. It starts to slip downward without friction. Show that the top of the plank loses contact with the wall when it is at two-thirds of its initial height.

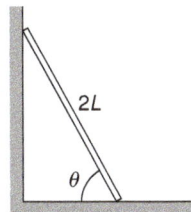

8

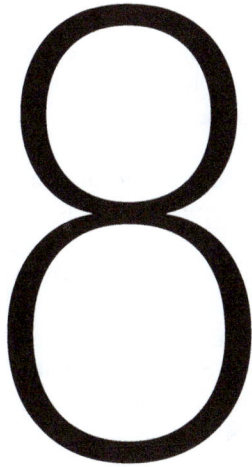

RIGID BODY MOTION

8.1 Introduction

Chapter 7 enabled us to analyze a drum rolling downhill, a yo-yo, and the rotational motion of a body whose axis has a fixed direction. However, to analyze a system even as simple as a bicycle rounding a bend, we need to remove the fixed axis constraint. In this chapter we shall attack the general problem of the motion of rigid bodies that can rotate about any axis. Rather than emphasize the formal mathematical details, we will try to gain insight into the basic principles, mainly by discussing the motion of gyroscopes and other devices that have large spin angular momentum.

Our analysis is based on the elementary concept that angular momentum is a vector. Appreciating the vector nature of angular momentum provides a simple and natural explanation for such a mysterious effect as the precession of a gyroscope, and gives an important insight into the general problem of rigid body motion.

A second topic in this chapter is the conservation of angular momentum. We touched on this conservation law in Chapter 7 but postponed any incisive discussion. Here the challenge is physical subtlety rather than mathematical complexity.

8.2 The Vector Nature of Angular Velocity and Angular Momentum

To describe the most general rotational motion of a body we must introduce suitable coordinates. Recall that in the case of translational motion, our procedure was to employ a Cartesian coordinate system, in which the position vector is $\mathbf{r} = x\hat{\mathbf{i}} + y\hat{\mathbf{j}} + z\hat{\mathbf{k}}$. We then found the velocity and acceleration by successively differentiating $\mathbf{r}$ with respect to time.

It is natural, but incorrect, to attempt to use an analogous procedure for rotational motion using angular coordinates θ_x, θ_y, and θ_z to measure rotation about the x, y, and z axes, respectively, and then define an angular position vector $\boldsymbol{\Theta}$ given by

$$\boldsymbol{\Theta} \stackrel{?}{=} \left(\theta_x\hat{\mathbf{i}} + \theta_y\hat{\mathbf{j}} + \theta_z\hat{\mathbf{k}}\right).$$

Unfortunately, this does not work. It is impossible to employ a vector to describe an angular orientation. The reason is that the addition of true vectors is commutative: $\mathbf{A} + \mathbf{B} = \mathbf{B} + \mathbf{A}$. However, as the following example demonstrates, rotations do not commute: $\theta_x\hat{\mathbf{i}} + \theta_y\hat{\mathbf{j}} \neq \theta_y\hat{\mathbf{j}} + \theta_x\hat{\mathbf{i}}$.

Example 8.1 Rotations through Finite Angles

Consider a can of maple syrup oriented as shown, and let us investigate what happens when we rotate it by an angle of $\pi/2$ around the x axis, and then by $\pi/2$ around the y axis, and compare the result with executing the same rotations but in reverse order.

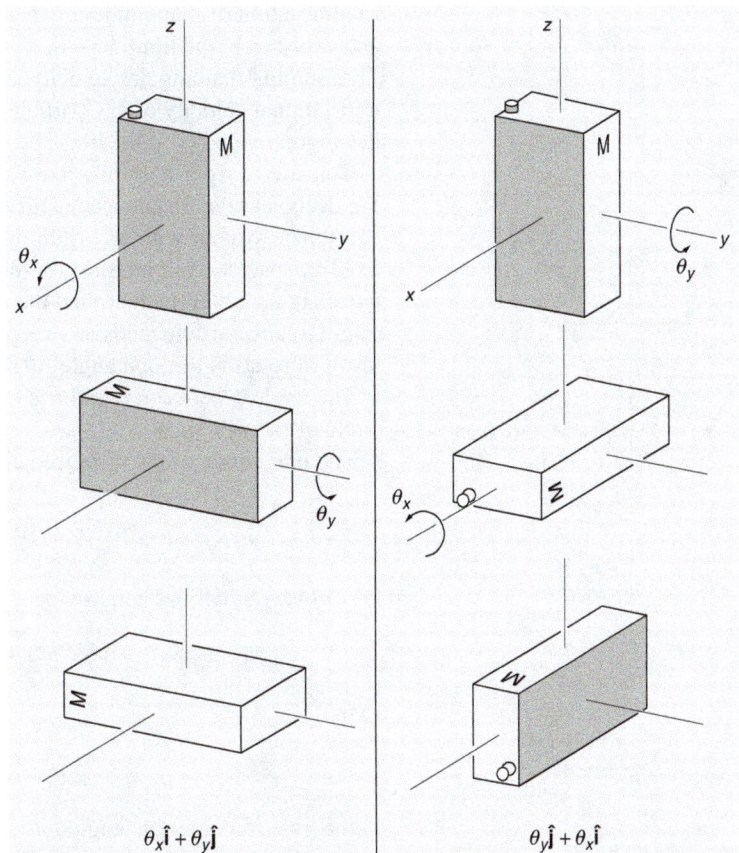

It is evident from the drawings that

$$\theta_x \hat{\mathbf{i}} + \theta_y \hat{\mathbf{j}} \neq \theta_y \hat{\mathbf{j}} + \theta_x \hat{\mathbf{i}}.$$

Fortunately, all is not lost; although angular position cannot be represented by a vector, it turns out that angular velocity, the rate of change of angular position, is a perfectly good vector. We define angular velocity by

$$\boldsymbol{\omega} = \frac{d\theta_x}{dt}\hat{\mathbf{i}} + \frac{d\theta_y}{dt}\hat{\mathbf{j}} + \frac{d\theta_z}{dt}\hat{\mathbf{k}}$$
$$= \omega_x\hat{\mathbf{i}} + \omega_y\hat{\mathbf{j}} + \omega_z\hat{\mathbf{k}}.$$

The essential point is that although rotations through finite angles do not commute, infinitesimal rotations like $\Delta\theta_x$, $\Delta\theta_y$, and $\Delta\theta_z$ do commute. Consequently, $\omega = \lim_{\Delta t \to 0} (\Delta\theta/\Delta t)$ represents a component of a true vector. The reason for this is discussed in Note 8.1 at the end of the chapter. Briefly: the difference in orientations between successive rotations through small angles $\Delta\theta_x$ and $\Delta\theta_y$, and rotations in the reverse order is a

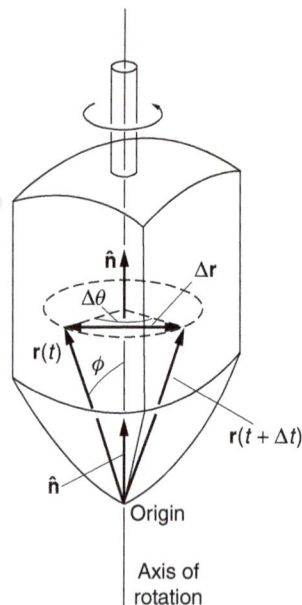

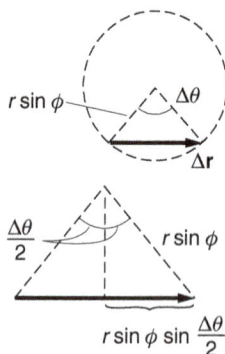

small rotation of magnitude $\Delta\theta_x\,\Delta\theta_y$. This term is second order in angle and vanishes in the limit $\Delta \to 0$.

Assuming that angular velocity is indeed a vector, let us find how the translational velocity of any particle in a rotating rigid body is related to the angular velocity of the body.

Consider a rigid body like the top shown, rotating about some axis. We designate the instantaneous direction of the axis by $\hat{\mathbf{n}}$ and choose a coordinate system with its origin on the axis. The coordinate system is fixed in space and is inertial. As the body rotates, each of its particles moves in a circular path around the axis of rotation. A vector $\mathbf{r}$ from the origin to any particle tends to sweep out a cone. The drawing shows the result of rotation through angle $\Delta\theta$ about the axis along $\hat{\mathbf{n}}$.

The angle ϕ between $\hat{\mathbf{n}}$ and $\mathbf{r}$ is constant, and the tip of $\mathbf{r}$ moves on a circle of radius $r\sin\phi$.

The magnitude of the displacement $|\Delta\mathbf{r}|$ is

$$|\Delta\mathbf{r}| = 2r\sin\phi\sin\left(\frac{\Delta\theta}{2}\right).$$

For $\Delta\theta$ very small, we can use the small-angle approximation

$$\sin(\Delta\theta/2) \approx \Delta\theta/2$$

so that

$$|\Delta\mathbf{r}| \approx r\sin\phi\,\Delta\theta.$$

If the rotation through angle $\Delta\theta$ occurs in time Δt, then $|\Delta\mathbf{r}|/\Delta t \approx r\sin\phi\,(\Delta\theta/\Delta t)$. In the limit $\Delta t \to 0$,

$$\left|\frac{d\mathbf{r}}{dt}\right| = r\sin\phi\,\frac{d\theta}{dt}.$$

In the limit, $d\mathbf{r}/dt$ is tangential to the circle, as shown.

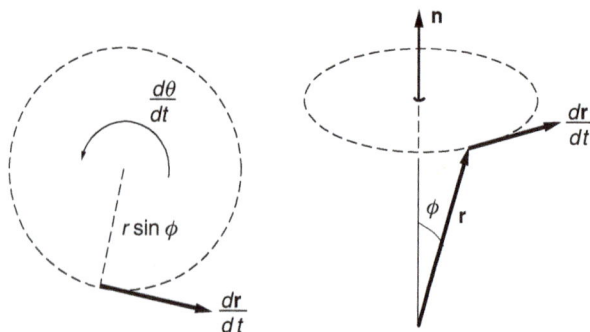

Recalling the definition of the vector cross product (Section 1.4.2), we see that both the magnitude $|d\mathbf{r}/dt| = r\sin\phi\,d\theta/dt$, and the direction of

$d\mathbf{r}/dt$, which is perpendicular to the plane of $\mathbf{r}$ and $\hat{\mathbf{n}}$, are given correctly by $d\mathbf{r}/dt = \hat{\mathbf{n}} \times \mathbf{r}\, d\theta/dt$. Since $d\mathbf{r}/dt = \mathbf{v}$ and $\hat{\mathbf{n}}\, d\theta/dt = \boldsymbol{\omega}$, we have

$$\frac{d\mathbf{r}}{dt} = \mathbf{v} = \boldsymbol{\omega} \times \mathbf{r}. \tag{8.1}$$

Example 8.2 Rotation in the x–y Plane

To connect Eq. (8.1) to a more familiar case—rotation in the x–y plane—we evaluate $\mathbf{v}$ for the rotation of a particle about the z axis. We have $\boldsymbol{\omega} = \omega\hat{\mathbf{k}}$, and $\mathbf{r} = x\hat{\mathbf{i}} + y\hat{\mathbf{j}}$. Hence

$$\mathbf{v} = \boldsymbol{\omega} \times \mathbf{r}$$
$$= \omega\hat{\mathbf{k}} \times \left(x\hat{\mathbf{i}} + y\hat{\mathbf{j}} \right)$$
$$= \omega\left(x\hat{\mathbf{j}} - y\hat{\mathbf{i}} \right).$$

In plane polar coordinates $x = r\cos\theta$, $y = r\sin\theta$, so that

$$\mathbf{v} = \omega r\left(\hat{\mathbf{j}}\cos\theta - \hat{\mathbf{i}}\sin\theta \right).$$

But $\hat{\mathbf{j}}\cos\theta - \hat{\mathbf{i}}\sin\theta$ is a unit vector in the tangential direction $\hat{\boldsymbol{\theta}}$. Therefore,

$$\mathbf{v} = \omega r\hat{\boldsymbol{\theta}}.$$

This is the velocity of a particle moving in a circle of radius r at angular velocity ω.

It can be difficult at first to appreciate the vector nature of angular velocity because we are used to visualizing rotation about a fixed axis, which involves only one component of angular velocity. We are generally much less familiar with simultaneous rotation about several axes.

We have seen that we can treat angular velocity as a vector in the relation $\mathbf{v} = \boldsymbol{\omega} \times \mathbf{r}$. It is important to assure ourselves that this relation remains valid if we resolve $\boldsymbol{\omega}$ into components like any other vector. In other words, if we write $\boldsymbol{\omega} = \boldsymbol{\omega}_1 + \boldsymbol{\omega}_2$, is it true that $\mathbf{v} = (\boldsymbol{\omega}_1 \times \mathbf{r}) + (\boldsymbol{\omega}_2 \times \mathbf{r})$? As the following example shows, the answer is yes.

Example 8.3 The Vector Nature of Angular Velocity

Consider a particle rotating in a vertical plane as shown in the sketch. The angular velocity $\boldsymbol{\omega}$ lies in the x–y plane and makes an angle of $45°$ with the x–y axes. The angular velocity is taken to be constant, so that $\theta = \omega t$.

First we shall calculate $\mathbf{v}$ directly from the relation $\mathbf{v} = d\mathbf{r}/dt$. To find $\mathbf{r}$, note from the sketch that $x = -r\cos\theta/\sqrt{2}$, $y = r\cos\theta/\sqrt{2}$, and $z = r\sin\theta$. Hence

$$\mathbf{r} = r\left(-\frac{\cos\theta}{\sqrt{2}}\,\hat{\mathbf{i}} + \frac{\cos\theta}{\sqrt{2}}\,\hat{\mathbf{j}} + \sin\theta\,\hat{\mathbf{k}} \right).$$

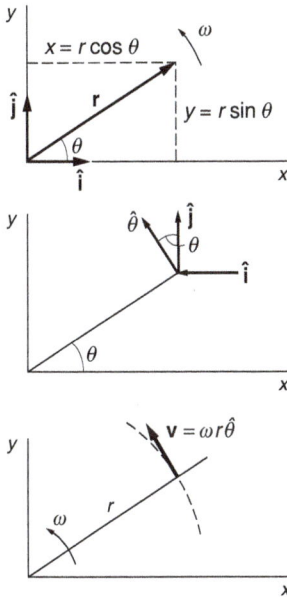

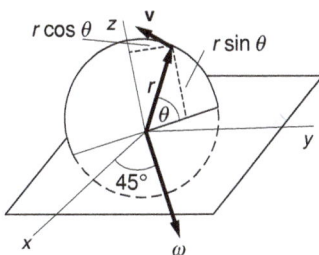

Differentiating with respect to time, and noting that r is constant, we have

$$\frac{d\mathbf{r}}{dt} = \mathbf{v}$$

$$= r\left(\frac{\sin\theta}{\sqrt{2}}\,\hat{\mathbf{i}} - \frac{\sin\theta}{\sqrt{2}}\,\hat{\mathbf{j}} + \cos\theta\,\hat{\mathbf{k}}\right)\frac{d\theta}{dt}$$

$$= \omega r\left(\frac{\sin\theta}{\sqrt{2}}\,\hat{\mathbf{i}} - \frac{\sin\theta}{\sqrt{2}}\,\hat{\mathbf{j}} + \cos\theta\,\hat{\mathbf{k}}\right), \tag{1}$$

where we have used $d\theta/dt = \omega$.

Using the relation $\mathbf{v} = \boldsymbol{\omega} \times \mathbf{r}$ is a simpler way to find the velocity. From the diagram, we see that

$$\boldsymbol{\omega} = \frac{\omega}{\sqrt{2}}\,\hat{\mathbf{i}} + \frac{\omega}{\sqrt{2}}\,\hat{\mathbf{j}},$$

$$\boldsymbol{\omega} \times \mathbf{r} = \begin{vmatrix} \hat{\mathbf{i}} & \hat{\mathbf{j}} & \hat{\mathbf{k}} \\ \frac{\omega}{\sqrt{2}} & \frac{\omega}{\sqrt{2}} & 0 \\ \frac{-r\cos\theta}{\sqrt{2}} & \frac{r\cos\theta}{\sqrt{2}} & r\sin\theta \end{vmatrix}$$

$$= \omega r\left(\frac{\sin\theta}{\sqrt{2}}\,\hat{\mathbf{i}} - \frac{\sin\theta}{\sqrt{2}}\,\hat{\mathbf{j}} + \cos\theta\,\hat{\mathbf{k}}\right),$$

in agreement with Eq. (1).

As we expect, we can treat $\boldsymbol{\omega}$ just like any other vector.

The next example demonstrates how a problem can be greatly simplified by resolving $\boldsymbol{\omega}$ into components along convenient axes. Furthermore, it reveals the fundamental property that angular momentum is not necessarily parallel to angular velocity, in contrast to the case of fixed axis rotation where $\mathbf{L}$ and $\boldsymbol{\omega}$ are parallel and related by $\mathbf{L} = I\boldsymbol{\omega}$.

Example 8.4 Angular Momentum of Masses on a Rotating Skew Rod

Consider a simple rigid body consisting of two particles of mass m separated by a massless rod of length $2l$. The midpoint of the rod is attached to a vertical axis that rotates at angular speed ω around the z axis. The rod is skewed at angle α, as shown in the sketch. The problem is to find the angular momentum of the system.

The most direct method is to calculate the angular momentum from the definition $\mathbf{L} = \Sigma(\mathbf{r}_i \times \mathbf{p}_i)$. Each mass moves in a circle of radius $l\cos\alpha$ with angular speed ω. The linear momentum of each mass is $|\mathbf{p}| = m\omega l\cos\alpha$, and is tangential to the circular path. To calculate the angular momentum of the two masses we shall take the midpoint of

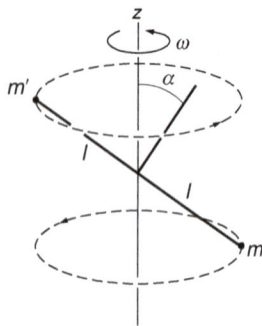

the skew rod as origin, with **r** along the rod and perpendicular to **p**. Hence $|\mathbf{L}| = 2m\omega l^2 \cos\alpha$. **L** is perpendicular to the skew rod and lies in the plane of the rod and the z axis, as shown in the left-hand drawing. **L** turns with the rod, and its tip traces out a circle about the z axis.

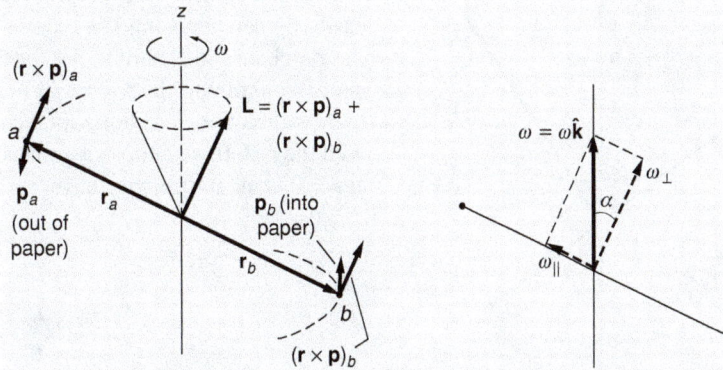

We now turn to a method for calculating **L** that emphasizes the vector nature of $\boldsymbol{\omega}$. First we resolve $\boldsymbol{\omega} = \omega\hat{\mathbf{k}}$ into components $\omega_\perp$ and $\omega_\parallel$, respectively, perpendicular and parallel to the skew rod. From the right-hand drawing, we see that $\omega_\perp = \omega\cos\alpha$, and $\omega_\parallel = \omega\sin\alpha$.

Because the masses are particles with negligible size, $\omega_\parallel$ produces no angular momentum. Consequently, the angular momentum is due entirely to $\omega_\perp$. Because **L** is parallel to $\omega_\perp$, we can use the result from fixed axis rotation $L = I\omega_\perp$, where the moment of inertia about the direction of $\omega_\perp$ is $ml^2 + ml^2 = 2ml^2$. The magnitude of the angular momentum is

$$L = I\omega_\perp$$
$$= 2ml^2\omega_\perp$$
$$= 2ml^2\omega\cos\alpha.$$

L points along the direction of $\omega_\perp$. Hence **L** swings around with the rod; the tip of **L** traces out a circle about the z axis. (We encountered a similar situation in Examples 7.2 and 7.8 with the conical pendulum.) An important feature of this system is that **L** is not parallel to $\boldsymbol{\omega}$, as generally true for non-symmetric bodies.

The dynamics of rigid body motion is governed by $\boldsymbol{\tau} = d\mathbf{L}/dt$, which holds in general for any motion because it is derived from Newton's laws (Chapter 7). We can gain insight into the disarmingly simple rotating skew rod by calculating the torque that causes **L** to change direction.

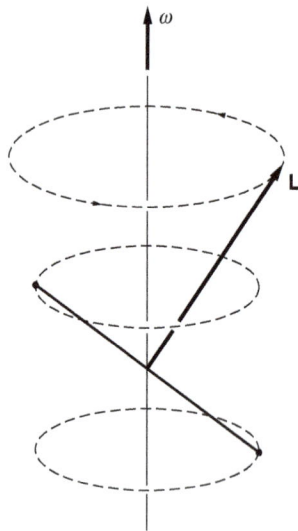

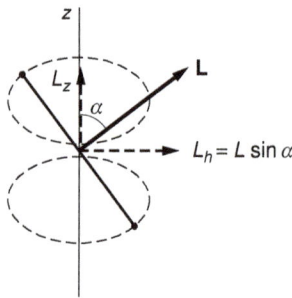

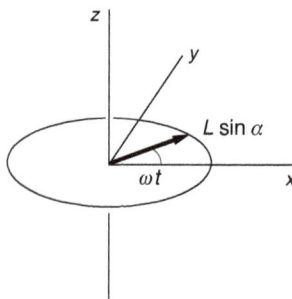

Example 8.5 Torque on the Rotating Skew Rod

In Example 8.4 we showed that the angular momentum of a uniformly rotating skew rod is constant in magnitude but changes in direction. **L** is fixed with respect to the rod and rotates in space with the rod.

The torque on the rod is given by $\boldsymbol{\tau} = d\mathbf{L}/dt$. We can find $d\mathbf{L}/dt$ quite easily by decomposing **L** as shown in the sketch. (We followed a similar procedure in Example 7.9 for the conical pendulum.) The component L_z parallel to the z axis, $L \cos \alpha$, is constant, so there is no torque in the z direction. The horizontal component of **L**, $L_h = L \sin \alpha$, swings with the rod. If we choose x–y axes so that L_h coincides with the x axis at $t = 0$, then at time t we have

$$L_x = L_h \cos \omega t$$
$$= L \sin \alpha \cos \omega t$$
$$L_y = L_h \sin \omega t$$
$$= L \sin \alpha \sin \omega t.$$

Hence

$$\mathbf{L} = L \sin \alpha \left(\hat{\mathbf{i}} \cos \omega t + \hat{\mathbf{j}} \sin \omega t \right) + L \cos \alpha \hat{\mathbf{k}}.$$

The torque is

$$\boldsymbol{\tau} = \frac{d\mathbf{L}}{dt}$$
$$= L\omega \sin \alpha \left(-\hat{\mathbf{i}} \sin \omega t + \hat{\mathbf{j}} \cos \omega t \right).$$

Using $L = 2ml^2 \omega \cos \alpha$ from Example 8.4, we obtain

$$\tau_x = -2ml^2 \omega^2 \sin \alpha \cos \alpha \sin \omega t$$
$$\tau_y = 2ml^2 \omega^2 \sin \alpha \cos \alpha \cos \omega t.$$

Hence

$$\tau = \sqrt{\tau_x^2 + \tau_y^2}$$
$$= 2ml^2 \omega^2 \sin \alpha \cos \alpha$$
$$= \omega L \sin \alpha.$$

Note that $\tau = 0$ for $\alpha = 0$ or $\alpha = \pi/2$. Do you see why? Also, can you see why the torque should be proportional to ω^2?

This analysis may seem roundabout, since the torque can be calculated directly by finding the force on each mass and using $\boldsymbol{\tau} = \Sigma \mathbf{r}_j \times \mathbf{f}_j$. However, the procedure used here is just as quick. Furthermore, it illustrates that angular velocity and angular momentum are *real* vectors that can be resolved into components along any axes we choose.

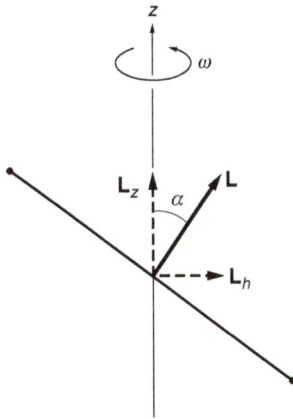

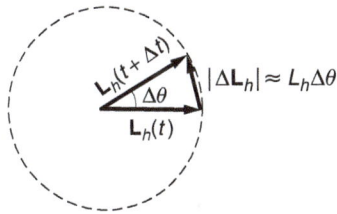

Example 8.6 Torque on the Rotating Skew Rod (Geometric Method)

In Example 8.5 we calculated the torque on the rotating skew rod by resolving $\mathbf{L}$ into components and using $\boldsymbol{\tau} = d\mathbf{L}/dt$. We repeat the calculation here using a geometric argument that emphasizes the connection between torque and the rate of change of $\mathbf{L}$. This method illustrates an approach that will be helpful in analyzing gyroscopic motion.

As in Example 8.5, we begin by resolving $\mathbf{L}$ into a vertical component $L_z = L \cos \alpha$ and a horizontal component $L_h = L \sin \alpha$ as shown in the sketch. Because $\mathbf{L}_z$ is constant, there is no torque about the z axis. $\mathbf{L}_h$ is constant in magnitude but rotates with the rod, and the time rate of change of $\mathbf{L}$ is due solely to this effect.

Once again we are dealing with a rotating vector, as discussed in Section 1.10. Using that idea, we therefore know that $dL_h/dt = \omega L_h$. However, since it is so important to be able to visualize this result, we derive it once more. From the vector diagram we have

$$|\Delta \mathbf{L}_h| \approx |\mathbf{L}_h|\Delta\theta$$

$$\frac{dL_h}{dt} = L_h \frac{d\theta}{dt}$$

$$= L_h \omega.$$

The torque is

$$\tau = \frac{dL_h}{dt}$$

$$= L_h \omega$$

$$= \omega L \sin \alpha,$$

identical to the result we found in Example 8.5. The torque $\boldsymbol{\tau}$ is parallel to $\Delta\mathbf{L}$ in the limit. For the skew rod, $\boldsymbol{\tau}$ is in the tangential direction in the horizontal plane and rotates with the rod.

You may have thought that torque on a rotating system necessarily causes the speed of rotation to change. In this problem, however, the speed of rotation is constant and the torque causes the direction of $\mathbf{L}$ to change. The torque is produced by the forces on the rotating bearing of the skew rod. For a real rod this would have to be an extended structure, something like a sleeve. The torque causes a time-varying load on the sleeve that results in vibration and wear. Because a uniform gravitational field exerts no torque on the skew rod, the rod is said to be *statically balanced*. However, because there is a torque on the skew rod when it rotates, it is not *dynamically balanced*. Rotating machinery must be designed for dynamical balance if it is to run smoothly.

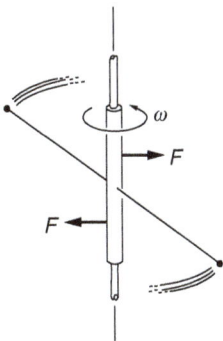

8.3 The Gyroscope

We now turn to some aspects of gyroscope motion that illustrate the basic concepts of angular momentum, torque, and the time derivative of a vector. It might seem frivolous to devote serious effort to understanding a mere toy, but if the toy is a gyroscope, you will find that the experience is worth the effort. We shall discuss each step carefully because this is one area of physics where intuition may not be much help.

We bypass for now the complicated mathematical problem of a general solution for gyroscope motion and initially concentrate on uniform precession. In uniform precession the tip of the gyroscope's axle swings at a constant rate in a horizontal plane. Our aim is to show that uniform precession is consistent with $\tau = d\mathbf{L}/dt$ and Newton's laws. Furthermore, this solution provides an excellent starting point for understanding more general rigid body motion.

The essentials of a gyroscope are a flywheel that spins on an axle and a suspension that allows the axle to assume any orientation. The familiar toy gyroscope shown in the left-hand drawing is adequate for our discussion. The end of the axle rests on a pylon, allowing the axis to take various orientations without constraint.

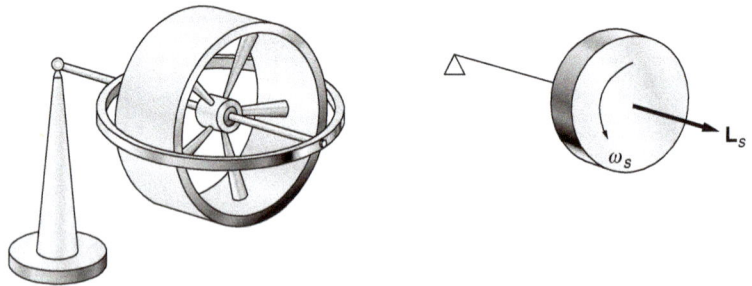

The right-hand drawing is a schematic representation of the gyroscope. The triangle represents the free pivot, and the flywheel spins in the direction shown.

The flywheel is initially brought up to speed, spinning at rate ω_s, most commonly by yanking on a string that is wound around the axle. If the gyroscope is then released horizontally with one end supported by the pivot, its axis wobbles briefly and then the gyroscope settles down to *uniform precession*, in which the axle slowly rotates around the vertical axis with constant angular velocity Ω. One's immediate impulse is to ask why the gyroscope does not fall. A possible answer is suggested by the force diagram. The total vertical force is $N - W$, where N is the vertical force exerted by the pivot and W is the weight. If $N = W$, the center of mass cannot fall. Naturally, if you remove the upward force N, the gyroscope falls like a rock.

This explanation is correct but not really satisfactory because we have asked the wrong question. Instead of wondering why the gyroscope does

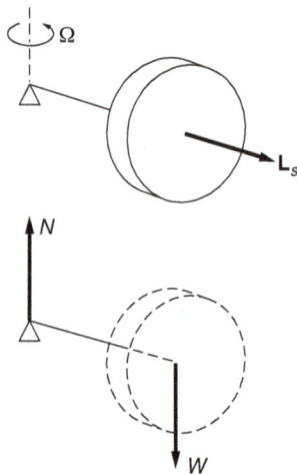

not fall, we should ask why it does not swing from the pivot like a pendulum, under the torque due to its weight.

As a matter of fact, if the gyroscope is released with its flywheel stationary, it starts to behave exactly like a pendulum; instead of processing horizontally, it starts to swing vertically, a behavior we shall look at later. But if the gyroscope is spinning rapidly, it soon begins to precess smoothly.

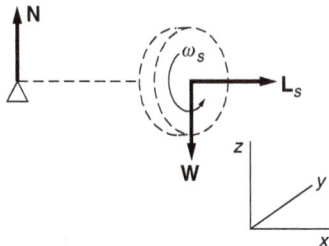

If the axle is held fixed in space, the gyroscope's angular momentum $\mathbf{L}_s$ is due entirely to the spin of its flywheel and is directed along the axle with magnitude $L_s = I_0\omega_s$, where I_0 is the moment of inertia of the flywheel about the axle. When the gyroscope precesses about the z axis, it also has a small orbital angular momentum in the z direction. However, for uniform precession this orbital angular momentum is constant in magnitude and direction and plays no dynamical role. Consequently, we can ignore it here.

$\mathbf{L}_s$ always points along the axle. As the gyroscope precesses, $\mathbf{L}_s$ rotates with it, as shown in sketch (a). We have encountered rotating vectors many times, most recently when discussing the skew rod. If the angular velocity of precession is Ω, the rate of change of $\mathbf{L}_s$ is given by

$$\left|\frac{d\mathbf{L}_s}{dt}\right| = \Omega L_s.$$

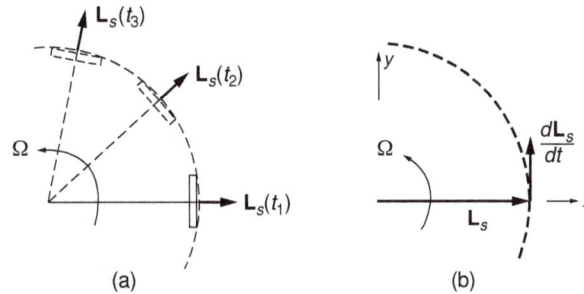

The direction of $d\mathbf{L}_s/dt$ is tangential to the horizontal circle swept out by $\mathbf{L}_s$. At the instant shown in sketch (b), $\mathbf{L}_s$ is in the x direction and $d\mathbf{L}_s/dt$ is in the y direction.

There must be a torque on the gyroscope to account for the change in $\mathbf{L}_s$. The source of the torque is apparent from the force diagram. If we take the pivot as the origin, the torque is due to the weight of the flywheel acting at the center of mass.

The magnitude of the torque is

$$\tau = lW.$$

τ is parallel to $d\mathbf{L}_s/dt$, as we expect.

We can find the rate of precession Ω from the relation

$$\left|\frac{d\mathbf{L}_s}{dt}\right| = \tau.$$

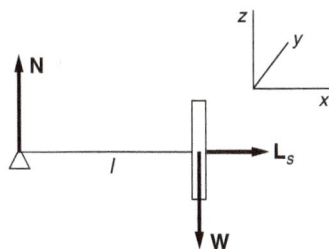

Because $|d\mathbf{L}_s/dt| = \Omega L_s$ and $\tau = lW$, we have

$$\Omega L_s = lW,$$

or

$$\Omega = \frac{lW}{I_0 \omega_s}. \tag{8.2}$$

Alternatively, we could have analyzed the motion about the center of mass (the center of the flywheel). In this case the torque is $\tau_0 = Nl = Wl$ as before, since $N = W$.

Equation (8.2) indicates that the rate of precession Ω increases as the flywheel slows. This effect is easy to see with a toy gyroscope. Obviously Ω cannot increase indefinitely; eventually uniform precession gives way to a violent and erratic motion. This occurs when Ω becomes so large that we cannot neglect small changes in the angular momentum about the vertical axis due to frictional torque. Nevertheless, uniform precession represents an exact solution to the dynamical equations governing the gyroscope, as explained in detail in Note 8.2.

We have assumed that the axle of the gyroscope is horizontal, but the rate of uniform precession is independent of the angle of elevation, as the following example shows.

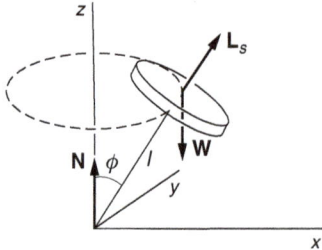

Example 8.7 Gyroscope Precession

Consider a gyroscope in uniform precession with its axle at angle ϕ with the vertical.

The component of $\mathbf{L}_s$ in the x–y plane swings around in space as the gyroscope precesses, while the component parallel to the z axis remains constant.

The horizontal component of $\mathbf{L}_s$ is $L_s \sin\phi$. Hence

$$|d\mathbf{L}_s/dt| = \Omega L_s \sin\phi.$$

The torque due to gravity is horizontal and has magnitude

$$\tau = l \sin\phi \, W.$$

We have

$$\Omega L_s \sin\phi = l \sin\phi \, W$$

$$\Omega = \frac{lW}{I_0 \omega_s}.$$

The precessional velocity is independent of ϕ.

The last example proves that uniform gyroscope precession is consistent with the dynamical equation $\tau = d\mathbf{L}/dt$ but it provides little physical insight as to *why* the gyroscope precesses. Possibly the following example will help provide that insight.

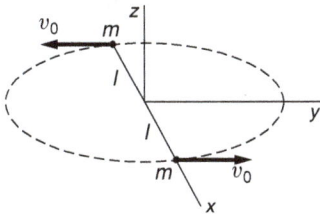

Example 8.8 Why a Gyroscope Precesses

The primary reason that gyroscope precession seems mysterious is that angular momentum is much less familiar than linear momentum. It would be more satisfying if the rotational dynamics of a simple rigid body could be seen to follow directly from Newton's laws. To do this, let us consider a rigid body consisting of two particles of mass m at either end of a rigid massless rod of length $2l$.

Suppose that the rod rotates in free space with angular momentum $\mathbf{L}_s$ along the z direction. The speed of each mass is v_0. We shall show that the effect of applying torque $\boldsymbol{\tau}$ is to cause $\mathbf{L}_s$ to precess with angular velocity $\Omega = \tau/L_s$.

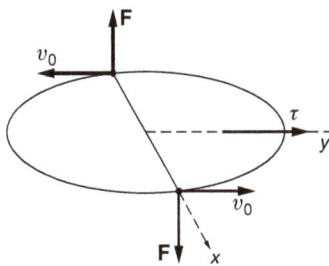

To simplify matters, suppose that the torque is applied only during a short time Δt while the rod is instantaneously oriented along the x axis. We assume that the torque is due to two equal and opposite forces F, as shown. The total force is zero, so the center of mass remains at rest.

The momentum of each mass changes by

$$\Delta \mathbf{p} = m\Delta \mathbf{v} = \mathbf{F}\Delta t.$$

Since $\Delta \mathbf{v}$ is perpendicular to $\mathbf{v}_0$, the velocity of each mass changes direction, as shown, and the rod rotates about a new direction.

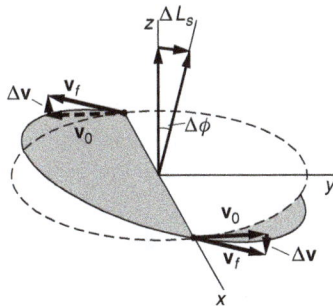

The axis of rotation tilts by the angle

$$\Delta \phi \approx \frac{\Delta v}{v_0}$$
$$= \frac{F\Delta t}{mv_0}.$$

The torque on the system is $\tau = 2Fl$, and the angular momentum is $L_s = 2mv_0 l$. Hence

$$\Delta \phi = \frac{F\Delta t}{mv_0}$$
$$= \frac{2lF\Delta t}{2lmv_0}$$
$$= \frac{\tau \Delta t}{L_s}.$$

The rate of precession during the interval Δt is therefore

$$\Omega = \frac{\Delta \phi}{\Delta t}$$
$$= \frac{\tau}{L_s},$$

which is identical to the result for gyroscope precession. Also, the change in the angular momentum $\Delta\mathbf{L}_s$ is in the y direction, parallel to the torque, as required.

This model gives some insight into why a torque causes a tilt in the axis of rotation of a spinning body. The argument can be elaborated to apply to an extended body like the flywheel of a gyroscope. The result of such an analysis is equivalent to using $\boldsymbol{\tau} = d\mathbf{L}/dt$.

The discussion in this section described uniform precession but this is only a special case of gyroscope motion. At the beginning of our analysis we assumed that the gyroscope was initially precessing smoothly, but other initial conditions lead to other types of motion. For instance, if the free end of the axle is held at rest and suddenly released, the precessional velocity is initially zero. In this case, the gyroscope's center of mass simply starts to fall. It is fascinating to see how this falling motion turns into uniform precession, which we do in Note 8.2 by a straightforward application of $\boldsymbol{\tau} = d\mathbf{L}/dt$. The analysis involves the general relation between $\mathbf{L}$ and $\boldsymbol{\omega}$ that will be developed in Section 8.6.

8.4 Examples of Rigid Body Motion

In this section we analyze some systems that illustrate the behavior of angular momentum in rigid body motion.

Example 8.9 Precession of the Equinoxes

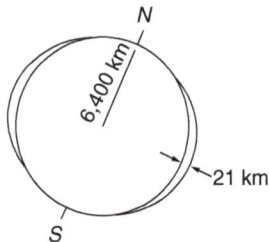

The angular momentum of the Earth has two components: orbital angular momentum arising from its center of mass motion around the Sun, and spin angular momentum due to rotational motion around its polar axis. We are concerned here with the Earth's spin angular momentum, which is tilted by $23\frac{1}{2}^\circ$ from the vertical to the orbital plane (the plane of the *ecliptic*). In the approximation that the Earth is spherical, it experiences no torque from nearby bodies. Consequently, its angular momentum is constant—both its spin and its angular momentum always point in the same direction in space.

If we analyze the Earth–Sun system with more care, taking into account that the Earth is not exactly spherical but slightly oblate, we find that there is a small torque on the Earth. This causes the spin axis to slowly alter its direction, resulting in the phenomenon known as *precession of the equinoxes*.

The torque arises because of the interaction of the Sun and Moon with the non-spherical shape of the Earth. The Earth bulges slightly; its mean radius is approximately 6400 km, but its equatorial radius is

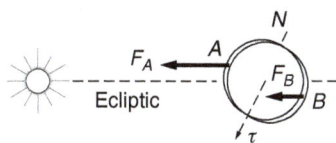

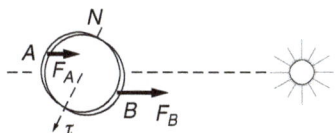

about 21 km larger than its polar radius. Because the Earth's axis of rotation is inclined with respect to the plane of the ecliptic, the gravitational force of the Sun gives rise to a torque.

The sketch represents the winter solstice in December. At that time, the part A of the bulge above the ecliptic is nearer the Sun than the lower part B. The mass at A is therefore attracted more strongly by the Sun than is the mass at B, as indicated. This results in a torque on the Earth perpendicular to the plane of the sketch.

Six months later, at the summer solstice, when the Earth is on the other side of the Sun, B is attracted more strongly than A. However, the torque has the same direction in space as before. Midway between these extremes (the vernal and autumnal equinoxes), the torque is zero. The average torque is perpendicular to the spin angular momentum and lies in the plane of the ecliptic.

The torque causes the spin axis to precess about a normal to the ecliptic. As the spin axis precesses, the torque remains perpendicular to it; the system acts like the gyroscope with tilted axis that we analyzed in Example 8.7.

The period of the precession is 26 000 years. 13 000 years from now, the polar axis will not point toward Polaris, the current north star; it will point $2 \times 23\frac{1}{2}° = 47°$ away. Orion and Sirius, those familiar winter guides, will then shine in the midsummer sky, and the winter solstice will occur in June.

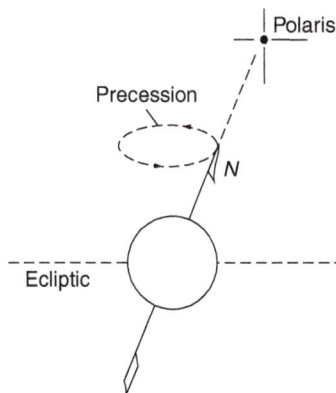

The spring equinox occurs at the instant the Sun is directly over the Equator in its apparent passage from south to north. Due to the precession of the Earth's axis, the position of the Sun at the equinox against the background of fixed stars shifts by 50 seconds of arc each year (1 angular degree = 3600 arc-seconds). This precession of the equinoxes was known to the ancients. It figures in the astrological scheme of cyclic history, which distinguishes twelve ages named by the constellation in which the Sun appears to lie at spring equinox. The present age is Pisces, and in 600 years it will be Aquarius.

Example 8.10 The Gyrocompass

With the advent of the Global Positioning System (GPS), compasses have become less important. But when the GPS satellite signals are inaccessible, as in a submarine or spacecraft, an inertial navigation system can measure the local acceleration and then find velocity and position by integration. To refer the craft's location relative to the Earth, the direction of true north (along the Earth's axis) must be known. The gyrocompass, or "gyro," enables this.

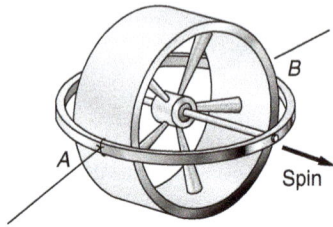

Try the following experiment with a toy gyroscope. Tie strings to the frame of the gyroscope at points A and B on opposite sides midway between the bearings of the spin axis. Hold the strings taut at arm's length with the spin axis horizontal. Now slowly pivot so that the spinning gyroscope moves in a circle. The gyroscope suddenly flips and comes to rest with its spin axis vertical, parallel to your axis of rotation. Rotation in the opposite direction causes the gyro to flip by 180°, making its spin axis again parallel to the rotation axis. (When examined in more detail, the spin axis is found to oscillate about the vertical. Friction in the horizontal axle quickly damps this motion.)

The gyrocompass is based on this effect. A flywheel free to rotate about two perpendicular axes tends to orient its spin axis parallel to the axis of rotation of the system. In the case of a gyro, the "system" is the Earth; the compass comes to rest with its axis parallel to the polar axis.

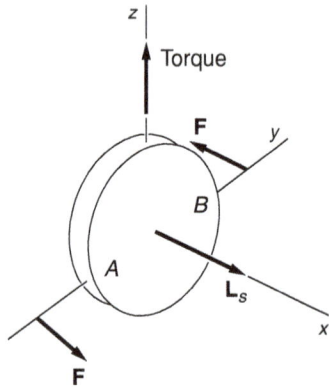

We can understand the motion qualitatively by simple vector arguments. Assume that the axle is horizontal with $\mathbf{L}_s$ pointing along the x axis. Suppose that we attempt to rotate the gyro around the z axis at some rate ω_z by applying a torque τ_z, as shown.

As a result, L_z, the angular momentum along the z axis, starts to increase. If the spin angular momentum $\mathbf{L}_s$ were zero, L_z would be due entirely to rotation of the gyro about the z axis, $L_z = I_z\omega_z$, where I_z is the moment of inertia around the z axis. However, because the flywheel is spinning, another way for L_z to change is for the gyro to rotate around the A–B axis, swinging $\mathbf{L}_s$ toward the z direction. Our experiment shows that if $\mathbf{L}_s$ is large, most of the torque goes into reorienting the spin angular momentum; only a small fraction goes toward rotating the gyro about the z axis.

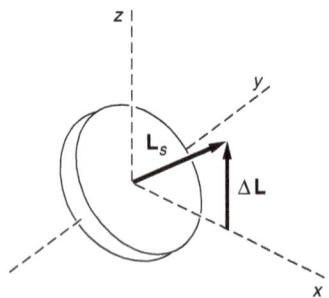

We can see why the effect is so pronounced by considering angular momentum along the y axis. The pivots at A and B allow the system to swing freely about the y axis, so there can be no torque along the y axis. Because L_y is initially zero, it must remain zero. As the gyro starts to rotate about the z axis, L_s starts to acquire a component in the y direction. In order to maintain $L_y = 0$, the gyro and its frame begin to rotate rapidly, or flip, around the y axis. The angular momentum arising from this motion cancels the y component of $\mathbf{L}_s$. When $\mathbf{L}_s$ finally comes to rest parallel to the z axis, the motion of the frame no longer changes the direction of $\mathbf{L}_s$, and its spin axis remains stationary.

The Earth is a rotating system and a gyrocompass on the surface of the Earth will line up with the polar axis indicating true north. A practical gyrocompass is somewhat more complicated, however, since it must continue to indicate true north without responding to the motion of the ship or aircraft that it is guiding. In the next example we solve the dynamical equation for the gyrocompass and show how a gyrocompass fixed to the Earth indicates true north.

Example 8.11 Gyrocompass Motion

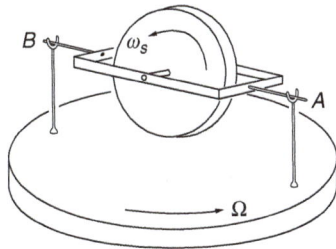

Consider a gyrocompass consisting of a balanced spinning disk held in a light frame supported by a horizontal axle, as shown. The assembly is on a turntable rotating at steady angular velocity Ω. The gyro has spin angular momentum $L_s = I_s \omega_s$ along the spin axis. In addition, it possesses angular momentum due to its bodily rotation about the vertical axis at rate Ω, and also from rotation about the A–B axis.

There can be no torque along the horizontal A–B axis because that axle is pivoted. The angular momentum L_h along the A–B direction is therefore constant, so that $dL_h/dt = 0$.

There are two contributions to dL_h/dt. If θ is the angle between the vertical and the spin axis, and $I_\perp$ is the moment of inertia about the A–B axis, then as θ changes it generates angular momentum $L_h = I_\perp \dot{\theta}$, contributing an amount $I_\perp \ddot{\theta}$ to dL_h/dt.

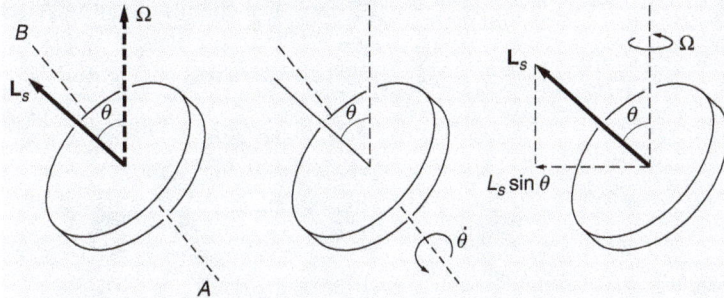

In addition, L_h can change because of a change in the direction of $\mathbf{L}_s$, as we learned from analyzing the precessing gyroscope. The horizontal component of $\mathbf{L}_s$ is $L_s \sin\theta$, and its rate of increase along the A–B axis is $\Omega L_s \sin\theta$.

We have considered the two changes in L_h independently. It is plausible that the total change in L_h is the sum of the two changes; a rigorous justification can be given based on arguments presented in Section 8.7.

Adding the two contributions to dL_h/dt gives

$$\frac{dL_h}{dt} = I_\perp \ddot{\theta} + \Omega L_s \sin\theta.$$

Because $dL_h/dt = 0$, the equation of motion becomes

$$\ddot{\theta} + \left(\frac{L_s\Omega}{I_\perp}\right)\sin\theta = 0.$$

This is identical to the equation for a pendulum, Example 3.10. When the spin axis is near the vertical, $\sin\theta \approx \theta$ and the equation of motion becomes

$$\ddot{\theta} + \left(\frac{L_s\Omega}{I_\perp}\right)\theta = 0.$$

Once again, we have the equation for simple harmonic motion. Consequently, the axis of the gyroscope executes simple harmonic motion given by

$$\theta = \theta_0 \sin\beta t$$

where

$$\beta = \sqrt{\frac{L_s\Omega}{I_\perp}}$$

$$= \sqrt{\frac{\omega_s\Omega I_s}{I_\perp}}.$$

If there is a small amount of friction in the bearings at A and B, the amplitude of oscillation θ_0 will eventually become zero, and the spin axis comes to rest parallel to Ω.

To use the gyro as a compass, fix it to the Earth with the $A-B$ axle vertical and the frame free to turn.

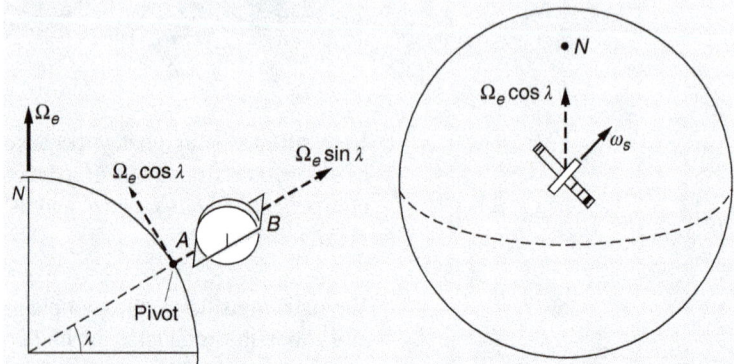

As the drawing shows, if λ is the latitude of the gyro, the component of the Earth's angular velocity Ω_e perpendicular to the $A-B$ axle is the horizontal component $\Omega_e \cos\lambda$. The spin axis oscillates in the horizontal plane about the direction of the north pole, and eventually comes to rest pointing north.

The period of small oscillations is $T = 2\pi/\beta = 2\pi \sqrt{I_\perp/(I_s\omega_s\Omega_e \cos\lambda)}$. For a thin disk $I_\perp/I_s = 1/2$, and $\Omega_e = 2\pi$ rad/day $\approx 7.27 \times 10^{-5}$ rad/s. With a gyro rotating at $\omega_s = 20\,000$ rpm ≈ 2100 rad/s, the period at the Equator is 11 s. Near the north pole the period becomes so long that the gyrocompass is not effective.

Example 8.12 The Stability of Spinning Objects

Angular momentum can make a freely moving object remarkably stable. For instance, spin angular momentum keeps a child's rolling hoop upright even when it hits a bump; instead of falling, the hoop changes direction slightly and continues to roll. The effect of spin on a bullet provides another example. The spiral grooves, or rifling, in a gun's barrel helps to stabilize the bullet by giving the bullet spin angular momentum. If you need proof of the stabilizing effect of rotation, try throwing a Frisbee® without spinning it.

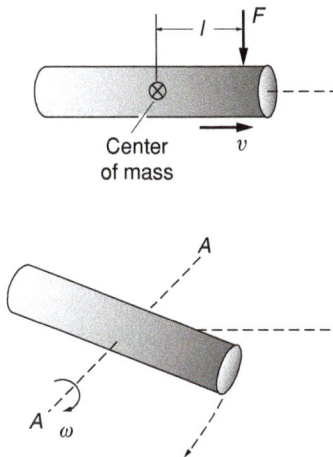

To analyze the stabilizing effect of spin, consider a cylinder of mass M moving parallel to its axis, but not spinning along the axis. Suppose that a small perturbing force F acts on the cylinder for time Δt. Let F be perpendicular to the axis, and the point of application be distance l from the center of mass, as shown.

The torque along the axis A–A through the center of mass is $\tau = Fl$ so that the "angular impulse" is $\tau\Delta t = Fl\Delta t$. Because the cylinder has no spin, the angular momentum acquired around the A–A axis is

$$\Delta L_A = I_A(\omega - \omega_0) = Fl\Delta t.$$

If we assume that the initial angular velocity is 0, then the final angular velocity is

$$\omega = \frac{Fl\Delta t}{I_A}.$$

The effect of the blow is to give the cylinder angular velocity around the transverse axis; it starts to tumble.

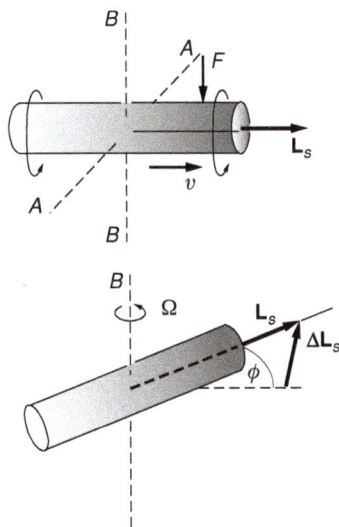

Now consider the same situation, except that the cylinder is rapidly spinning around its long axis with angular momentum $\mathbf{L}_s$.

The situation is similar to that of the gyroscope: torque along the A–A axis causes precession around the B–B axis. The rate of precession while F acts is $dL_s/dt = \Delta L_s$, or

$$\Omega = \frac{Fl}{L_s}.$$

The angle through which the cylinder precesses is

$$\phi = \Omega\Delta t$$

$$= \frac{Fl\Delta t}{L_s}.$$

Instead of starting to tumble, the cylinder slowly changes its orientation while the force is applied, and then stops precessing. The larger the spin, the smaller the angle through which it precesses and the less the effect of the perturbation on its flight.

Note that spin has no effect on the center of mass motion. In both cases, the center of mass acquires velocity $\Delta \mathbf{v} = \mathbf{F}\Delta t/M$.

8.5 Conservation of Angular Momentum

Before tackling the general problem of rigid body motion, let us return to the question of whether or not the angular momentum of an isolated system is conserved. It would be pleasing if this conservation law stemmed directly from Newton's laws, but this is not the case. Nevertheless, the total angular momentum of an isolated system is always conserved. We now turn to see how far this conservation law can be supported.

Consider a system of N particles with masses $m_1, m_2, \ldots, m_j, \ldots, m_N$. We assume that the system is isolated, so that the forces are due entirely to interactions between the particles. The force on particle j is

$$\mathbf{F}_j = \sum_{k=1}^{N} \mathbf{F}_{jk},$$

where $\mathbf{F}_{jk}$ is the force on particle j due to particle k. (In evaluating the sum, we can neglect the term with $k = j$, since $\mathbf{F}_{jj} = 0$, by Newton's third law.)

Let us choose a convenient origin and calculate the torque $\boldsymbol{\tau}_j$ on particle j:

$$\boldsymbol{\tau}_j = \mathbf{r}_j \times \mathbf{F}_j$$
$$= \mathbf{r}_j \times \sum_k \mathbf{F}_{jk}.$$

Let $\boldsymbol{\tau}_{jk}$ be the torque on j due to the particle k:

$$\boldsymbol{\tau}_{jk} = \mathbf{r}_j \times \mathbf{F}_{jk}.$$

Similarly, the torque on k due to j is

$$\boldsymbol{\tau}_{kj} = \mathbf{r}_k \times \mathbf{F}_{kj}.$$

The sum of these two torques is

$$\boldsymbol{\tau}_{jk} + \boldsymbol{\tau}_{kj} = \mathbf{r}_k \times \mathbf{F}_{kj} + \mathbf{r}_j \times \mathbf{F}_{jk}.$$

Because $\mathbf{F}_{jk} = -\mathbf{F}_{kj}$ we have

$$\boldsymbol{\tau}_{jk} + \boldsymbol{\tau}_{kj} = (\mathbf{r}_k \times \mathbf{F}_{kj}) - (\mathbf{r}_j \times \mathbf{F}_{kj})$$
$$= (\mathbf{r}_k - \mathbf{r}_j) \times \mathbf{F}_{kj}$$
$$= \mathbf{r}_{jk} \times \mathbf{F}_{kj},$$

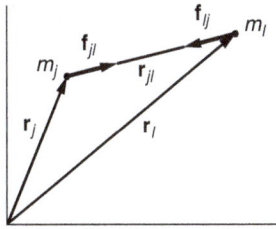

(a)

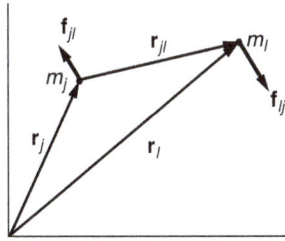

(b)

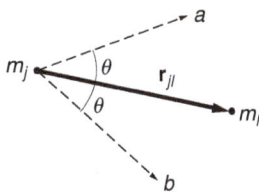

where $\mathbf{r}_{jk}$ is a vector from j to k. If we could prove that $\boldsymbol{\tau}_{jk} + \boldsymbol{\tau}_{kj} = 0$, so that the internal torques cancel in pairs just as the internal forces do, then the total internal torque would vanish, proving that the angular momentum of an isolated system is conserved.

Because neither $\mathbf{r}_{jk}$ nor $\mathbf{F}_{kj}$ is zero, $\mathbf{F}_{kj}$ would need to be parallel to $\mathbf{r}_{jk}$ as shown in figure (a) for the torque $\boldsymbol{\tau}_j = \mathbf{r}_j \times \mathbf{F}_j$ to vanish. For the situation in figure (b), however, the torque is not zero, and angular momentum is not conserved. Nevertheless, the forces are equal and opposite, and linear momentum is conserved.

The situation shown in figure (a) corresponds to the case of *central forces*. We conclude that in the particular case of central force motion the conservation of angular momentum follows from Newton's laws. However, Newton's laws do not explicitly require forces to be central. We must conclude that Newton's laws have no direct bearing on whether or not the angular momentum of an isolated system is conserved because these laws do not exclude the situation shown in figure (b).

It is possible to take exception to the argument above on the following grounds: although Newton's laws do not explicitly require forces to be central, they implicitly make this requirement because in their simplest form Newton's laws deal with particles. Particles are idealized masses that have no size and no structure. In this case, the force between isolated particles must be central, since the only vector defined in a two-particle system is the vector $\mathbf{r}_{jk}$ from one particle to the other.

Suppose that we try to invent a force that lies at angle θ with respect to the interparticle axis, as shown in the diagram. There is no way to distinguish direction a from b; both are at angle θ with respect to $\mathbf{r}_{jk}$. An angle-dependent force cannot be defined using only the single vector $\mathbf{r}_{jk}$; the force between the two particles must be central.

The difficulty in discussing angular momentum in the context of Newtonian ideas is that our understanding of nature now encompasses entities vastly different from simple particles. As an example, perhaps the electron comes closest to the Newtonian idea of a particle. The electron has a well-defined mass and, as far as present knowledge goes, zero radius. In spite of this, the electron has something analogous to internal structure; it possesses spin angular momentum. It seems paradoxical that an object with zero size can nevertheless possess angular momentum, but we must accept this paradox as one of the facts of nature.

Because the spin of an electron defines an additional direction in space, the force between two electrons need not be central. As an example, there might be a force

$$\mathbf{F}_{12} = C\mathbf{r}_{12} \times (\mathbf{S}_1 + \mathbf{S}_2)$$
$$\mathbf{F}_{21} = C\mathbf{r}_{21} \times (\mathbf{S}_1 + \mathbf{S}_2),$$

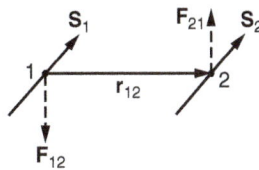

where C is some constant and $\mathbf{S}_i$ is a vector parallel to the angular momentum of the ith electron. The forces are equal and opposite but not central, and they produce a torque.

There are other possibilities for non-central forces. Experimentally, the force between two charged particles moving with respect to each other is not central; the velocities define additional axes on which the force depends. The angular momentum of the two particles is not necessarily conserved. The apparent breakdown of conservation of angular momentum is due to neglect of an essential part of the system: the electromagnetic field. A field refers to a change in the nature of space (or in the case of relativity, of spacetime) by physical interactions such as gravity or electromagnetism. Although fields lie beyond the scope of particle mechanics, it turns out that fields can possess energy, momentum, and angular momentum. When the angular momentum of the field is taken into account, the angular momentum of the entire particle–field system is conserved.

The situation, in brief, is that Newtonian physics is incapable of predicting conservation of angular momentum but no isolated system has yet been encountered experimentally for which the total angular momentum is not conserved. We conclude that conservation of angular momentum is an independent physical law, and until a contradiction is discovered, our physical understanding must be guided by it.

8.6 Rigid Body Rotation and the Tensor of Inertia

The governing equation $\tau = d\mathbf{L}/dt$ for rigid body motion bears a formal resemblance to the translational equation of motion $\mathbf{F} = d\mathbf{P}/dt$. There is, however, an essential difference between them. Linear momentum and center of mass motion are simply related by the vector equation $\mathbf{P} = M\mathbf{V}$ where M is a scalar, a simple number. $\mathbf{P}$ and $\mathbf{V}$ are therefore always parallel. The connection between $\mathbf{L}$ and ω is not so simple. For fixed axis rotation, $\mathbf{L} = I\omega$, and it is tempting to suppose that for any general rotation $\mathbf{L} = I\omega$, where I is a scalar. However, this cannot be correct, since we know from our study of the rotating skew rod, Example 8.4, that $\mathbf{L}$ and ω are not necessarily parallel.

In this section, we shall develop the general relation between angular momentum and angular velocity, and in Section 8.7 we shall attack the problem of solving the equations of motion.

8.6.1 Angular Momentum and the Tensor of Inertia

We begin by showing that to analyze the rotational motion of a rigid body, we really need consider only the angular momentum about the center of mass as origin.

As we discussed in Chapter 7, an arbitrary displacement of a rigid body can be resolved into a displacement of the center of mass plus a rotation about some instantaneous axis through the center of mass. To derive the equations of motion, we start from the general expressions for the angular momentum and torque of a rigid body, Eqs. (7.15) and (7.18):

$$\mathbf{L} = \mathbf{R} \times M\mathbf{V} + \Sigma \mathbf{r}'_j \times m_j \dot{\mathbf{r}}'_j$$
$$\tau = \mathbf{R} \times \mathbf{F} + \Sigma \mathbf{r}'_j \times \mathbf{f}_j,$$

where $\mathbf{r}'_j$ is the position vector of m_j relative to the center of mass and where $\Sigma \mathbf{f}_j$ is the total applied force $\mathbf{F}$. Since $\boldsymbol{\tau} = d\mathbf{L}/dt$, we have

$$\mathbf{R} \times \mathbf{F} + \Sigma \mathbf{r}'_j \times \mathbf{f}_j = \frac{d}{dt}(\mathbf{R} \times M\mathbf{V}) + \frac{d}{dt}(\Sigma \mathbf{r}'_j \times m_j \dot{\mathbf{r}}'_j)$$

$$= \mathbf{R} \times M\mathbf{A} + \frac{d}{dt}(\Sigma \mathbf{r}'_j \times m_j \dot{\mathbf{r}}'_j).$$

Because $\mathbf{F} = M\mathbf{A}$, the terms involving $\mathbf{R}$ cancel, and we are left with

$$\Sigma \mathbf{r}'_j \times \mathbf{f}_j = \frac{d}{dt}(\Sigma \mathbf{r}'_j \times m_j \dot{\mathbf{r}}'_j).$$

In words, the total torque around the center of mass is the rate of change of angular momentum around the center of mass. This relation is independent of the center of mass motion. The angular momentum $\mathbf{L}_{cm}$ about the center of mass is

$$\mathbf{L}_{cm} = \Sigma \mathbf{r}'_j \times m_j \dot{\mathbf{r}}'_j. \tag{8.3}$$

Our task now is to express $\mathbf{L}_{cm}$ for a rigid body in terms of the instantaneous angular velocity $\boldsymbol{\omega}$. Because the length of $\mathbf{r}'_j$ is fixed, the only way for $\mathbf{r}'_j$ to change is by rotation. Consequently, $\mathbf{r}'_j$ is a rotating vector

$$\dot{\mathbf{r}}'_j = \boldsymbol{\omega} \times \mathbf{r}'_j.$$

Therefore

$$\mathbf{L}_{cm} = \Sigma \mathbf{r}'_j \times m_j(\boldsymbol{\omega} \times \mathbf{r}'_j).$$

To simplify the notation, we can adopt center of mass coordinates by writing $\mathbf{L}$ for $\mathbf{L}_{cm}$ and $\mathbf{r}_j$ for $\mathbf{r}'_j$. Our result becomes

$$\mathbf{L} = \Sigma \mathbf{r}_j \times m_j(\boldsymbol{\omega} \times \mathbf{r}_j). \tag{8.4}$$

This result looks complicated. As a matter of fact, it *is* complicated, but we can make it look simple. We will take the pedestrian approach of patiently evaluating the cross products in Eq. (8.4) using Cartesian coordinates. (An elegant way is to use the vector identity $\mathbf{A} \times (\mathbf{B} \times \mathbf{C}) = (\mathbf{A} \cdot \mathbf{C})\mathbf{B} - (\mathbf{A} \cdot \mathbf{B})\mathbf{C}$.)

Because $\boldsymbol{\omega} = \omega_x \hat{\mathbf{i}} + \omega_y \hat{\mathbf{j}} + \omega_z \hat{\mathbf{k}}$, we have

$$\boldsymbol{\omega} \times \mathbf{r} = (z\omega_y - y\omega_z)\hat{\mathbf{i}} + (x\omega_z - z\omega_x)\hat{\mathbf{j}} + (y\omega_x - x\omega_y)\hat{\mathbf{k}}. \tag{8.5}$$

Let us compute one component of $\mathbf{L}$, say L_x. Temporarily dropping the subscript j, we have

$$[\mathbf{r} \times (\boldsymbol{\omega} \times \mathbf{r})]_x = y(\boldsymbol{\omega} \times \mathbf{r})_z - z(\boldsymbol{\omega} \times \mathbf{r})_y. \tag{8.6}$$

If we substitute Eq. (8.5) into Eq. (8.6), the result is

$$[\mathbf{r} \times (\boldsymbol{\omega} \times \mathbf{r})]_x = y(y\omega_x - x\omega_y) - z(x\omega_z - z\omega_x)$$

$$= (y^2 + z^2)\omega_x - xy\omega_y - xz\omega_z.$$

Hence

$$L_x = \Sigma m_j(y_j^2 + z_j^2)\omega_x - \Sigma m_j x_j y_j \omega_y - \Sigma m_j x_j z_j \omega_z. \tag{8.7}$$

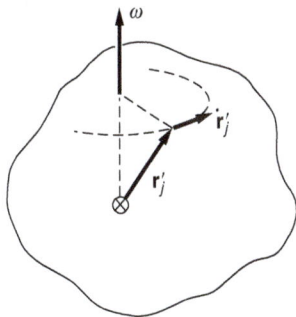

This expression can be written in a more compact form by introducing the following symbols:

$$I_{xx} = \Sigma m_j(y_j{}^2 + z_j{}^2)$$
$$I_{xy} = -\Sigma m_j x_j y_j \qquad\qquad (8.8)$$
$$I_{xz} = -\Sigma m_j x_j z_j.$$

I_{xx} is called a *moment of inertia* and it is the moment of inertia we introduced for fixed axis rotation, $I = \Sigma m_j \rho_j{}^2$, where we take the axis in the x direction so that $\rho_j{}^2 = y_j{}^2 + z_j{}^2$. The quantities I_{xy} and I_{xz} are called *products of inertia*. They are symmetrical in x and y; $I_{xy} = -\Sigma m_j x_j y_j = -\Sigma m_j y_j x_j = I_{yx}$.

To calculate L_y and L_z, we could repeat the above derivation but it is simpler to relabel the coordinates by letting $x \rightarrow y$, $y \rightarrow z$, $z \rightarrow x$. Making these substitutions in Eqs. (8.7) and (8.8) yields

$$L_x = I_{xx}\omega_x + I_{xy}\omega_y + I_{xz}\omega_z \qquad\qquad (8.9a)$$
$$L_y = I_{yx}\omega_x + I_{yy}\omega_y + I_{yz}\omega_z \qquad\qquad (8.9b)$$
$$L_z = I_{zx}\omega_x + I_{zy}\omega_y + I_{zz}\omega_z. \qquad\qquad (8.9c)$$

This array of three equations describes rotation about any axes, and includes fixed axis rotation as a special case. Consider for example rotation about an axis in the z direction, $\boldsymbol{\omega} = \omega\hat{\mathbf{k}}$. Then Eq. (8.9c) reduces to

$$L_z = I_{zz}\omega$$
$$= \Sigma m_j(x_j{}^2 + y_j{}^2)\omega,$$

the result for fixed axis rotation that we derived in Chapter 7.

However, Eqs. (8.9) also show that angular velocity in the z direction can produce angular momentum about any of the three coordinate axes. For example, if $\boldsymbol{\omega} = \omega\hat{\mathbf{k}}$, then $L_x = I_{xz}\omega$ and $L_z = I_{zz}\omega$. In fact, if we look at the set of equations for L_x, L_y, and L_z, we see that in each case the angular momentum along one axis depends on the angular velocities along *all three* axes. Both $\mathbf{L}$ and $\boldsymbol{\omega}$ are ordinary vectors, and $\mathbf{L}$ is proportional to $\boldsymbol{\omega}$ in the sense that doubling the components of $\boldsymbol{\omega}$ doubles the components of $\mathbf{L}$. Nevertheless, as we have already seen from the behavior of the rotating skew rod, $\mathbf{L}$ does not necessarily point in the same direction as $\boldsymbol{\omega}$.

The nine-element array of the moments of inertia and the products of inertia is called the *tensor of inertia*. In addition to mechanics, tensors have an important application in the analysis of curves and surfaces, including the geometry of curved spacetime in general relativity.

Here is an example demonstrating the tensor of inertia.

Example 8.13 Rotating Dumbbell

Consider a dumbbell made of two uniform spheres each of radius b and mass M separated by a thin massless rod. The distance between centers is $2l$. The body is rotating about some axis through its center

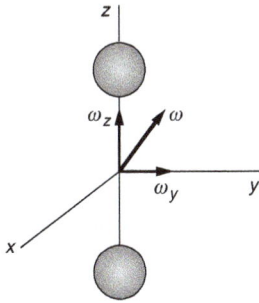

of mass. At a certain instant the rod coincides with the z axis, and ω lies in the y–z plane, $\omega = \omega_y \hat{\mathbf{j}} + \omega_z \hat{\mathbf{k}}$. The problem is to find the instantaneous angular momentum $\mathbf{L}$.

To find $\mathbf{L}$, we need the moments and products of inertia. Fortunately, the products of inertia vanish for a symmetrical body lined up with the coordinate axes. For example, $I_{xy} = -\Sigma m_j x_j y_j = 0$, because for mass m_n located at (x_n, y_n) there is, in a symmetrical body, an equal mass located at $(x_n, -y_n)$; the contributions of these two masses to I_{xy} cancel. In this case Eqs. (8.9) simplify to

$$L_x = I_{xx}\omega_x$$
$$L_y = I_{yy}\omega_y$$
$$L_z = I_{zz}\omega_z.$$

The moment of inertia I_{zz} is just the moment of inertia of two spheres about their diameters:

$$I_{zz} = 2\left(\tfrac{2}{5}Mb^2\right) = \tfrac{4}{5}Mb^2.$$

To calculate I_{yy}, we can use the parallel axis theorem to find the moment of inertia of each sphere about the y axis

$$I_{yy} = 2\left(\tfrac{2}{5}Mb^2 + Ml^2\right)$$
$$= \tfrac{4}{5}Mb^2 + 2Ml^2.$$

Because we are taking ω to lie in the y–z plane, we have $\omega = \omega_y \hat{\mathbf{j}} + \omega_z \hat{\mathbf{k}}$,

$$L_x = 0$$
$$L_y = I_{yy}\omega_y$$
$$L_z = I_{zz}\omega_z.$$

Since $I_{yy} \neq I_{zz}$ it follows that $L_y/L_z \neq \omega_y/\omega_z$ and $\mathbf{L}$ is not parallel to ω, as the drawing shows.

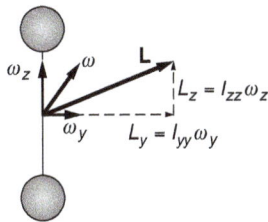

Equations (8.9) are cumbersome and it is convenient to write them using a compact notation:

$$\mathbf{L} = \tilde{\mathbf{I}}\omega. \tag{8.10}$$

This vector equation represents three equations, just as $\mathbf{F} = m\mathbf{a}$ represents three equations. The difference is that m is a simple scalar while $\tilde{\mathbf{I}}$ is the more complicated tensor of inertia.

The nine components of $\tilde{\mathbf{I}}$ can be tabulated in a 3×3 array:

$$\tilde{\mathbf{I}} = \begin{pmatrix} I_{xx} & I_{xy} & I_{xz} \\ I_{yx} & I_{yy} & I_{yz} \\ I_{zx} & I_{zy} & I_{zz} \end{pmatrix}. \tag{8.11}$$

Of the nine components only six can be different, since $I_{yx} = I_{xy}$, $I_{zx} = I_{xz}$, and $I_{yz} = I_{zy}$. The rule for multiplying ω by $\tilde{\mathbf{I}}$ to find $\mathbf{L} = \tilde{\mathbf{I}}\omega$ can be defined using matrix multiplication, writing $\mathbf{L}$ and $\boldsymbol{\omega}$ as column vectors:

$$\begin{pmatrix} L_x \\ L_y \\ L_z \end{pmatrix} = \begin{pmatrix} L_{xx} & L_{xy} & L_{xz} \\ L_{yx} & L_{yy} & L_{yz} \\ L_{zx} & L_{zy} & L_{zz} \end{pmatrix} \begin{pmatrix} \omega_x \\ \omega_y \\ \omega_z \end{pmatrix}. \tag{8.12}$$

The following example is another illustration of the tensor of inertia.

Example 8.14 The Tensor of Inertia for a Rotating Skew Rod

We found the angular momentum of a rotating skew rod from first principles in Example 8.4. Let us now find $\mathbf{L}$ for the skew rod by using $\mathbf{L} = \tilde{\mathbf{I}}\boldsymbol{\omega}$.

A massless rod of length $2l$ separates two equal masses m. The rod is skewed at angle α with the vertical, and rotates around the z axis with angular velocity ω. At $t = 0$ it lies instantaneously in the x–z plane. Using the parameters $\rho = l \cos \alpha$ and $h = l \sin \alpha$, the coordinates of the particles at any other time are:

Particle 1	Particle 2
$x_1 = \rho \cos \omega t$	$x_2 = -\rho \cos \omega t$
$y_1 = \rho \sin \omega t$	$y_2 = -\rho \sin \omega t$
$z_1 = -h$	$z_2 = h.$

The components of $\tilde{\mathbf{I}}$ can now be calculated from their definitions. For example,

$$I_{zz} = m_1(y_1{}^2 + z_1{}^2) + m_2(y_2{}^2 + z_2{}^2)$$
$$= 2m(\rho^2 \sin^2 \omega t + h^2)$$
$$I_{zy} = I_{yz}$$
$$= -m_1 y_1 z_1 - m_2 y_2 z_2$$
$$= 2m\rho h \sin \omega t.$$

The remaining terms are readily evaluated. We find:

$$\tilde{\mathbf{I}} = 2m \begin{pmatrix} \rho^2 \sin^2 \omega t + h^2 & -\rho^2 \sin \omega t \cos \omega t & \rho h \cos \omega t \\ -\rho^2 \sin \omega t \cos \omega t & \rho^2 \cos^2 \omega t + h^2 & \rho h \sin \omega t \\ \rho h \cos \omega t & \rho h \sin \omega t & \rho^2 \end{pmatrix}.$$

The common factor $2m$ multiplies each of the nine terms.

Since $\boldsymbol{\omega} = (0, 0, \omega)$, we have, from Eq. (8.9) or equivalently Eq. (8.12),

$$L_x = 2m\rho h \omega \cos \omega t$$
$$L_y = 2m\rho h \omega \sin \omega t$$
$$L_z = 2m\rho^2 \omega.$$

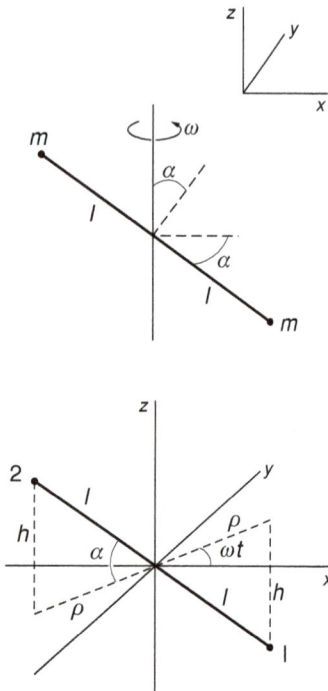

We can find the torque that must be applied to the rod from $\boldsymbol{\tau} = d\mathbf{L}/dt$ by taking the time derivative of $\mathbf{L}$:

$$\tau_x = -2m\rho h\omega^2 \sin \omega t$$
$$\tau_y = 2m\rho h\omega^2 \cos \omega t$$
$$\tau_z = 0.$$

These results are seen to be identical to those in Example 8.5, when we make the substitution $\rho h = l^2 \cos \alpha \sin \alpha$.

8.6.2 Principal Axes

If the symmetry axes of a uniform symmetric body coincide with the coordinate axes, the products of inertia are zero, as we saw for the rotating dumbbell in Example 8.13. In such a case the tensor of inertia takes a simple diagonal form:

$$\tilde{\mathbf{I}} = \begin{pmatrix} I_{xx} & 0 & 0 \\ 0 & I_{yy} & 0 \\ 0 & 0 & I_{zz} \end{pmatrix}. \tag{8.13}$$

Remarkably, for a body of any shape and mass distribution, it is *always* possible to find a set of three perpendicular axes for which the products of inertia vanish. (The proof uses matrix algebra and is given in most texts on advanced dynamics.) Such axes are called *principal axes*. With respect to principal axes, the tensor of inertia has a diagonal form.

For a uniform sphere, any perpendicular axes through the center are principal axes. For a body with cylindrical symmetry, the axis of revolution is a principal axis, and the other two principal axes are mutually perpendicular and lie in a plane through the center of mass perpendicular to the axis of revolution.

Consider a rotating rigid body, and suppose that we introduce a coordinate system 1, 2, 3 that coincides instantaneously with the principal axes of the body.

With respect to this coordinate system, the instantaneous angular velocity has components $\omega_1, \omega_2, \omega_3$, and the components of $\mathbf{L}$ have the simple form

$$L_1 = I_1\omega_1$$
$$L_2 = I_2\omega_2 \tag{8.14}$$
$$L_3 = I_3\omega_3,$$

where I_1, I_2, I_3 are the moments of inertia about the principal axes.

8.6.3 Rotational Kinetic Energy of a Rigid Body

The kinetic energy of a rigid body is

$$K = \tfrac{1}{2}\Sigma m_j v_j{}^2.$$

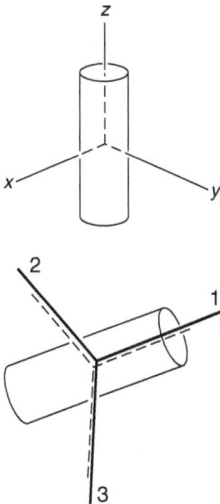

To separate the translational and rotational contributions, we introduce center of mass coordinates:

$$\mathbf{r}_j = \mathbf{R} + \mathbf{r}'_j$$
$$\mathbf{v}_j = \mathbf{V} + \mathbf{v}'_j.$$

We have

$$K = \tfrac{1}{2}\Sigma m_j(\mathbf{V} + \mathbf{v}'_j)^2 = \tfrac{1}{2}\Sigma m_j(\mathbf{V} + \mathbf{v}'_j) \cdot (\mathbf{V} + \mathbf{v}'_j)$$
$$= \tfrac{1}{2}MV^2 + \tfrac{1}{2}\Sigma m_j v'^2_j,$$

where the term $\mathbf{V} \cdot \Sigma m_j\mathbf{v}'_j$ vanishes identically. Using $\mathbf{v}'_j = \omega \times \mathbf{r}'_j$, the kinetic energy of rotation becomes

$$K_{\text{rot}} = \tfrac{1}{2}\Sigma m_j\mathbf{v}'^2_j$$
$$= \tfrac{1}{2}\Sigma m_j(\omega \times \mathbf{r}'_j) \cdot (\omega \times \mathbf{r}'_j).$$

The right-hand side can be simplified with the vector identity $(\mathbf{A} \times \mathbf{B}) \cdot \mathbf{C} = \mathbf{A} \cdot (\mathbf{B} \times \mathbf{C})$. Identifying $\mathbf{A} = \omega, \mathbf{B} = \mathbf{r}'_j$, and $\mathbf{C} = \omega \times \mathbf{r}'_j$, we obtain

$$K_{\text{rot}} = \tfrac{1}{2}\Sigma m_j\omega \cdot [\mathbf{r}'_j \times (\omega \times \mathbf{r}'_j)]$$
$$= \tfrac{1}{2}\omega \cdot [\Sigma m_j\mathbf{r}'_j \times (\omega \times \mathbf{r}'_j)].$$

According to Eq. (8.4), the sum is the angular momentum $\mathbf{L}$. Therefore

$$K_{\text{rot}} = \tfrac{1}{2}\omega \cdot \mathbf{L}. \qquad (8.15)$$

Rotational kinetic energy has a simple form when $\mathbf{L}$ and ω are referred to principal axes. Using Eq. (8.15) we have

$$K_{\text{rot}} = \tfrac{1}{2}\omega \cdot \mathbf{L}$$
$$= \tfrac{1}{2}I_1\omega_1{}^2 + \tfrac{1}{2}I_2\omega_2{}^2 + \tfrac{1}{2}I_3\omega_3{}^2. \qquad (8.16)$$

Alternatively,

$$K_{\text{rot}} = \frac{L_1{}^2}{2I_1} + \frac{L_2{}^2}{2I_2} + \frac{L_3{}^2}{2I_3}. \qquad (8.17)$$

Example 8.15 Why a Flying Saucer is Better Than a Flying Cigar

An early space satellite, cylindrical in shape, was put into orbit spinning around its long axis. To the designer's surprise, even though the spacecraft was torque-free, it began to wobble more and more, until finally it was spinning around a transverse axis.

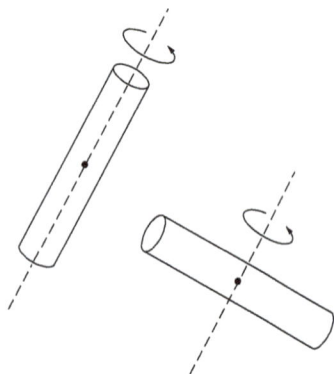

The reason is that although $\mathbf{L}$ is strictly conserved for torque-free motion, kinetic energy of rotation can change if the body is not absolutely rigid. If the satellite rotates slightly off the symmetry axis, each part of the body undergoes a time-varying centripetal acceleration. The spacecraft warps and bends under the fluctuating force, and energy is dissipated by internal friction in the structure. The kinetic energy of rotation decreases to provide the energy converted into heat. From Eq. (8.17), if the body rotates about a principal axis, $K_{rot} = L^2/2I$. K_{rot} is a minimum for the axis with largest moment of inertia, and the motion is stable around that axis. For the cylindrical spacecraft, the initial axis of rotation had the minimum moment of inertia, and the motion was not stable.

A thin disk ("flying saucer") spinning about its cylindrical axis is inherently stable because the other two moments of inertia are only half as large. A cigar-shaped craft is unstable about its long axis and only neutrally stable about the transverse axes; there is no single axis having a maximum moment of inertia.

8.6.4 Rotation about a Fixed Point

We showed at the beginning of this section that in analyzing the motion of a rotating and translating rigid body it is always correct to calculate torque and angular momentum about the center of mass. In some applications, however, one point of a body is fixed in space, like the pivot point of a gyroscope on a pylon. It is often convenient to analyze the motion using the fixed point as origin, because the center of mass motion need not be considered explicitly and the constraint force at the pivot produces no torque.

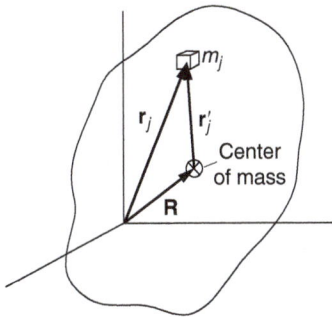

Taking the origin at the fixed point, let $\mathbf{r}_j$ be the position vector of particle m_j and let $\mathbf{R} = X\hat{\mathbf{i}} + Y\hat{\mathbf{j}} + Z\hat{\mathbf{k}}$ be the position vector of the center of mass. The torque about the origin is

$$\boldsymbol{\tau} = \Sigma \mathbf{r}_j \times \mathbf{f}_j,$$

where $\mathbf{f}_j$ is the force on m_j.

If the angular velocity of the body is $\boldsymbol{\omega}$, the angular momentum about the origin is

$$\mathbf{L} = \Sigma \mathbf{r}_j \times m_j \dot{\mathbf{r}}_j$$
$$= \Sigma \mathbf{r}_j \times m_j (\boldsymbol{\omega} \times \mathbf{r}_j).$$

This has the same form as Eq. (8.4). Taking over the results wholesale, we have

$$\mathbf{L} = \tilde{\mathbf{I}} \boldsymbol{\omega}$$

with

$$I_{xx} = \Sigma m_j (y_j'^2 + z_j'^2)$$
$$I_{xy} = -\Sigma m_j x_j' y_j'$$

etc.

where the prime indicates center of mass coordinates. This result is identical in form to Eq. (8.8) but the components of $\tilde{\mathbf{I}}$ are now calculated with respect to the pivot point rather than the center of mass.

If the tensor of inertia $\tilde{\mathbf{I}}_0$ about the center of mass is known, $\tilde{\mathbf{I}}$ about any other origin can be found from a generalization of the parallel axis theorem in Section 7.3.1. Typical results, the proof of which we leave as a problem, are

$$I_{xx} = (I_0)_{xx} + M(Y^2 + Z^2)$$
$$I_{xy} = (I_0)_{xy} - MXY \tag{8.18}$$
etc.

Consider for example a uniform sphere of mass M and radius b centered on the z axis a distance l from the origin. We have $I_{xx} = \frac{2}{5}Mb^2 + Ml^2, I_{yy} = \frac{2}{5}Mb^2 + Ml^2, I_{zz} = \frac{2}{5}Mb^2$.

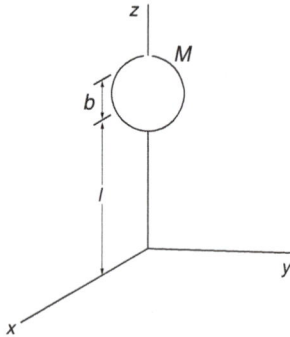

8.7 Advanced Topics in Rigid Body Dynamics

In this section we shall look at a few examples of rigid body motion. However, none of the results will be needed in subsequent chapters and this section can be skipped without loss of continuity.

The fundamental problem of rigid body dynamics is to find how the orientation of a rotating body changes in time, given the torque. The goal is to solve $\boldsymbol{\tau} = d\mathbf{L}/d\mathbf{t}$, which is the rotational analogue to $\mathbf{F} = M\mathbf{a}$. However, the analogy is not really helpful because of the complicated relation $\mathbf{L} = \tilde{\mathbf{I}}\boldsymbol{\omega}$ between angular momentum $\mathbf{L}$ and angular velocity $\boldsymbol{\omega}$. We can make the problem look simpler by using a coordinate system that coincides with the principal axes of the body. The tensor of inertia $\tilde{\mathbf{I}}$ becomes diagonal in form (the off-diagonal products of inertia all vanish), and the components of $\mathbf{L}$ are

$$L_x = I_{xx}\omega_x$$
$$L_y = I_{yy}\omega_y$$
$$L_z = I_{zz}\omega_z.$$

The crux of the problem is that as the body rotates, the principal axes, which are fixed to the body, rotate with it. Because the applied torque $\boldsymbol{\tau}$ is applied in a fixed coordinate system, which we could consider to be our "laboratory system," we require the components of $\mathbf{L}$ along axes having a fixed orientation in space. As the body rotates, its principal axes move out of coincidence with the space-fixed system. The products of inertia are no longer zero in the space-fixed system and, worse yet, the components of the tensor of inertia $\tilde{\mathbf{I}}$ vary with time.

The situation appears hopelessly tangled. Fortunately, if the principal axes do not stray far from the space-fixed system, we can find the motion using simple vector arguments. Leaving the general case for later, we illustrate this approach by finding the torque-free motion of a rigid body.

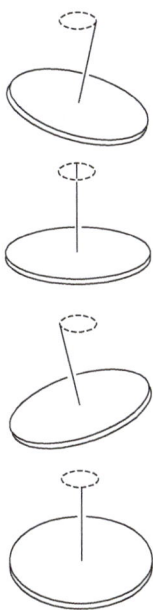

8.7.1 Torque-free Precession: Why the Earth Wobbles

If you drop a spinning coin and give it a slight flip, the coin will fall through the air with a wobbling motion; the symmetry axis tends to rotate in space, as the sketch shows.

Because there is essentially no torque on the freely falling coin, the motion is known as torque-free precession.

Torque-free precession is a characteristic mode of rigid body motion. For example, the spin axis of the Earth moves around the polar axis because of this effect. The physical explanation of the wobbling motion is related to our observation that **L** need not be parallel to ω. If there are no torques on the body, **L** is fixed in space and ω must move, as we shall now show.

To keep the math simple, consider the case of a cylindrically symmetric rigid body like a coin or a spheroid. We shall assume that the precessional motion is small in amplitude so that we can employ small-angle approximations.

Suppose that the body has a large spin angular momentum $L_s = I_s \omega_s$ along the main symmetry axis, where I_s is the corresponding moment of inertia and ω_s is the angular velocity about the symmetry axis. In addition, we shall allow the body to have small angular displacements around the axes transverse to the main symmetry axis.

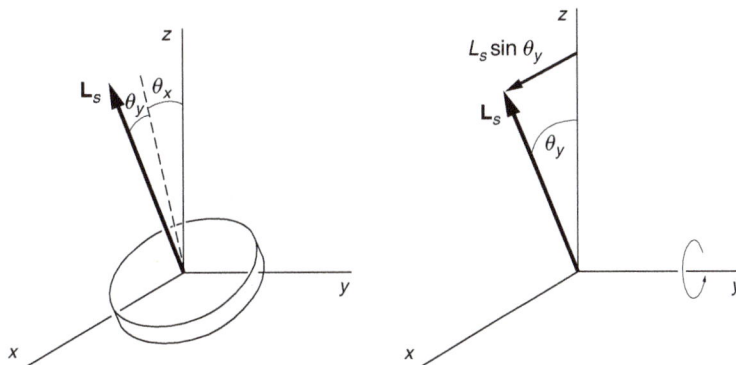

Suppose that $\mathbf{L}_s$ is close to the z axis, directed at angles $\theta_x \ll 1$ and $\theta_y \ll 1$ with respect to the x and y axes. As Note 8.1 explains, rotations about each axis can be considered independently, to first order, for rotations through infinitesimal angles. The contribution to L_x due to rotation about the x axis is $L_x = d(I_{xx}\theta_x)/dt = I_{xx}d\theta_x/dt$. We can treat I_{xx} as a constant because moments of inertia about principal axes are constant to first order for small angular displacements. Similarly, the products of inertia remain zero to first order. (The proofs are left as a problem.) In addition, rotation about y contributes to L_x by giving $\mathbf{L}_s$ a component $L_s \sin \theta_y$ in the x direction. Adding the two contributions, we have

$$L_x = I_{xx}\frac{d\theta_x}{dt} + L_s \sin \theta_y.$$

Similarly,

$$L_y = I_{yy}\frac{d\theta_y}{dt} - L_s\sin\theta_x.$$

By symmetry, $I_{xx} = I_{yy} = I_\perp$. For small angles, $\sin\theta \approx \theta$ and $\cos\theta \approx 1$. Hence

$$L_x = I_\perp\frac{d\theta_x}{dt} + L_s\theta_y \tag{8.19a}$$

$$L_y = I_\perp\frac{d\theta_y}{dt} - L_s\theta_x. \tag{8.19b}$$

To the same order of approximation,

$$L_z = I_s\omega_s. \tag{8.19c}$$

Since the torque is zero, $d\mathbf{L}/dt = 0$. Equation (8.19c) gives $L_s = $ constant, $\omega_s = $ constant, and Eqs. (8.19a) and (8.19b) yield

$$I_\perp\frac{d^2\theta_x}{dt^2} + L_s\frac{d\theta_y}{dt} = 0 \tag{8.20a}$$

$$I_\perp\frac{d^2\theta_y}{dt^2} - L_s\frac{d\theta_x}{dt} = 0. \tag{8.20b}$$

If we let $\omega_x = d\theta_x/dt$, $\omega_y = d\theta_y/dt$, Eqs. (8.20) become

$$I_\perp\frac{d\omega_x}{dt} + L_s\omega_y = 0 \tag{8.21a}$$

$$I_\perp\frac{d\omega_y}{dt} - L_s\omega_x = 0. \tag{8.21b}$$

To solve this pair of coupled equations, we can differentiate Eq. (8.21a) and substitute the value for $d\omega_y/dt$ in Eq. (8.21b), giving

$$\frac{I_\perp^2}{L_s}\frac{d^2\omega_x}{dt^2} + L_s\omega_x = 0$$

or

$$\frac{d^2\omega_x}{dt^2} + \gamma^2\omega_x = 0, \tag{8.22}$$

where

$$\gamma = \frac{L_s}{I_\perp} = \omega_s\frac{I_s}{I_\perp}.$$

Equation (8.22) is the familiar equation for simple harmonic motion. The solution can be written

$$\omega_x = A\sin(\gamma t + \phi), \tag{8.23}$$

where A and ϕ are arbitrary constants. Substituting this in Eq. (8.21a)

$$\omega_y = -\frac{I_\perp}{L_s}\frac{d\omega_x}{dt}$$

$$= \frac{I_\perp}{I_s\omega_s}A\gamma\cos(\gamma t + \phi),$$

or

$$\omega_y = A \cos (\gamma t + \phi). \tag{8.24}$$

By integrating Eqs. (8.23) and (8.24) we obtain

$$\theta_x = \frac{A}{\gamma} \cos (\gamma t + \phi) + \theta_{x0} \tag{8.25a}$$

$$\theta_y = -\frac{A}{\gamma} \sin (\gamma t + \phi) + \theta_{y0}, \tag{8.25b}$$

where θ_{x0} and θ_{y0} are constants of integration. Because we are making the small-angle approximation, we require that $A/\gamma \ll 1$.

Equations (8.25a) and (8.25b) reveal that the spin axis rotates around a fixed direction in space. If we take that direction along the z axis, then $\theta_{x0} = \theta_{y0} = 0$. Assuming that at $t = 0$, $\theta_x = \theta_0$ and $\theta_y = 0$, we have

$$\theta_x = \theta_0 \cos \gamma t \tag{8.26a}$$

$$\theta_y = \theta_0 \sin \gamma t, \tag{8.26b}$$

where we have taken $A/\gamma = \theta_0$ and $\phi = 0$.

Equations (8.26a) and (8.26b) describe torque-free precession in which the spin axis precesses in space at a fixed angle θ_0 with respect to the z axis. The frequency of the precessional motion is $\gamma = \omega_s I_s / I_\perp$. For a body flattened along the axis of symmetry, such as the oblate spheroid in the sketch, $I_s > I_\perp$ and $\gamma > \omega_s$. For a thin coin, $I_s = 2I_\perp$ and $\gamma = 2\omega_s$. Thus, a freely falling coin wobbles twice as fast as it spins.

The Earth is an oblate spheroid and exhibits torque-free precession. The amplitude of the motion is small; the spin axis wanders about the polar axis by about 5 m at the North Pole. Since the Earth itself is spinning at rate ω_s, the apparent rate of precession to an earthbound observer is

$$\gamma' = \gamma - \omega_s$$

$$= \omega_s \left(\frac{I_s - I_\perp}{I_\perp} \right). \tag{8.27}$$

For the Earth, $(I_s - I_\perp)/I_\perp = \frac{1}{300}$, and the precessional motion should have a period of 300 days. In reality, the situation is not quite so ideal. The observed motion is somewhat irregular with an apparent period of about 430 days. The fluctuations arise from the elastic nature of the Earth, which is significant for motions this small.

The discussion of the nutating gyroscope in Note 8.2 is also based on the small-angle approximation and provides another example of the approach used here.

8.7.2 Euler's Equations

We turn now to the task of solving the equation of motion for a rigid body. The equation is simple to write: $\boldsymbol{\tau} = d\mathbf{L}/dt$. To evaluate $d\mathbf{L}/dt$, we shall calculate the change in the components of $\mathbf{L}$ in the time interval from t to $t + \Delta t$ using the small-angle approximation. The results are

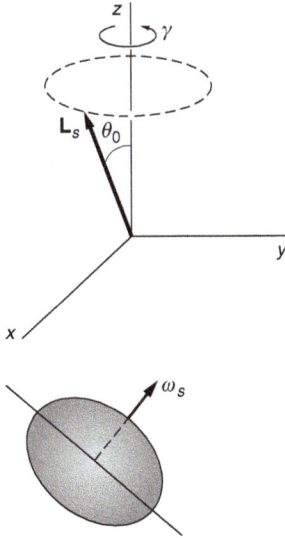

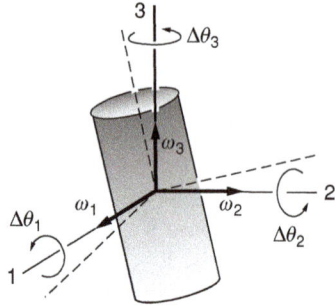

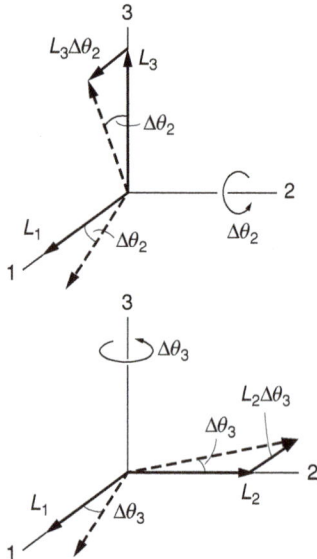

correct only to first order but they will become exact when we take the limit $\Delta t \to 0$.

Let us introduce an inertial coordinate system that coincides with the instantaneous position of the body's principal axes at time t. We label the axes of the inertial system 1, 2, 3. Let the components of the angular velocity ω at time t relative to the 1, 2, 3 system be ω_1, ω_2, ω_3. Because we are using principal axes, the components of **L** are $L_1 = I_1\omega_1, L_2 = I_2\omega_2, L_3 = I_3\omega_3$, where I_1, I_2, I_3 are the respective moments of inertia about the three axes.

In the time interval Δt, the principal axes rotate away from the 1, 2, 3 axes. To first order, the rotation angle about the 1 axis is $\Delta\theta_1 = \omega_1 \Delta t$; similarly, $\Delta\theta_2 = \omega_2 \Delta t$ and $\Delta\theta_3 = \omega_3 \Delta t$. The corresponding change $\Delta L_1 = L_1(t + \Delta t) - L_1(t)$ can be found to first order by treating the three rotations one by one, according to Note 8.1 on infinitesimal rotations. There are two ways L_1 can change. If ω_1 changes, the magnitude of $I_1\omega_1$ also changes. In addition, rotations about the other two axes cause L_2 and L_3 to change direction, and these can contribute to angular momentum along axis 1.

The contribution $\Delta(I_1\omega_1)$ to ΔL_1 is $\Delta(I_1\omega_1) = I_1\Delta\omega_1$ because the components of $\tilde{\mathbf{I}}$ are constant to first order for small angular displacements about the principal axes.

To find the remaining contributions to ΔL_1, first consider rotation about the 2 axis through angle $\Delta\theta_2$. This causes ΔL_1 and ΔL_3 to rotate as shown.

The rotation of L_1 causes no change along the 1 axis, to first order. However, the rotation of L_3 around the 2 axis contributes $L_3\Delta\theta_2 = I_3\omega_3\Delta\theta_2$ along the 1 axis. Similarly, rotation about the 3 axis contributes $-L_2\Delta\theta_3 = -I_2\omega_2\Delta\theta_3$ to ΔL_1.

Adding all the contributions gives

$$\Delta L_1 = I_1 \Delta\omega_1 + I_3\omega_3 \Delta\theta_2 - I_2\omega_2 \Delta\theta_3.$$

Dividing by Δt and taking the limit $\Delta t \to 0$ yields

$$\frac{dL_1}{dt} = I_1\frac{d\omega_1}{dt} + (I_3 - I_2)\omega_3\omega_2.$$

The other components can be treated in a similar fashion or we can simply relabel the subscripts by $1 \to 2, 2 \to 3, 3 \to 1$ to give

$$\frac{dL_2}{dt} = I_2\frac{d\omega_2}{dt} + (I_1 - I_3)\omega_1\omega_3$$

$$\frac{dL_3}{dt} = I_3\frac{d\omega_3}{dt} + (I_2 - I_1)\omega_2\omega_1.$$

Using $\boldsymbol{\tau} = d\mathbf{L}/dt$, we obtain

$$\tau_1 = I_1\frac{d\omega_1}{dt} + (I_3 - I_2)\omega_3\omega_2 \tag{8.28a}$$

$$\tau_2 = I_2\frac{d\omega_2}{dt} + (I_1 - I_3)\omega_1\omega_3 \tag{8.28b}$$

$$\tau_3 = I_3\frac{d\omega_3}{dt} + (I_2 - I_1)\omega_2\omega_1, \tag{8.28c}$$

where τ_1, τ_2, τ_3 are the components of $\boldsymbol{\tau}$ along the axes of the inertial system 1, 2, 3.

Equations (8.28) were first derived by the great mathematician Leonhard Euler in the mid-eighteenth century, and are known as *Euler's equations of rigid body motion*.

Because Euler's equations are tricky to apply, it is important to understand what they mean.

We set up the 1, 2, 3 inertial system to coincide with the instantaneous directions of the body's principal axes at some time t. The components of $\boldsymbol{\tau}$ along the 1, 2, 3 axes at time t are τ_1, τ_2, τ_3. Similarly, $\omega_1, \omega_2, \omega_3$ are the components of $\boldsymbol{\omega}$ along the 1, 2, 3 axes at time t, and $d\omega_1/dt, d\omega_2/dt, d\omega_3/dt$ are the instantaneous rates of change of these components. Euler's equations relate these quantities at time t. To apply Euler's equations at another time t', we have to resolve $\boldsymbol{\tau}$ and $\boldsymbol{\omega}$ along the axes of a new inertial system $1', 2', 3'$ that coincides with the principal axes at t'.

The difficulty is that Euler's equations do not show us how to find the orientation of these coordinate systems in space. Essentially, we have traded one problem for another; we know the disposition of the axes in the familiar x, y, z laboratory coordinate system, but the components of the tensor of inertia vary in an unknown way. In the 1, 2, 3 system, the components of $\bar{\mathbf{I}}$ are constant, but we do not know the orientation of the axes. Consequently, Euler's equations cannot be integrated directly to give angles specifying the orientation of the body relative to the x, y, z laboratory system. Euler overcame this difficulty by expressing $\omega_1, \omega_2, \omega_3$ in terms of a set of angles, known as *Euler's angles*, that relate the principal axes of the rigid body to the axes of the laboratory system.

In terms of these angles, Euler's equations are a set of coupled differential equations. The general equations are fairly complicated and are discussed in advanced texts. Fortunately, in many important applications we can find the motion from Euler's equations by using straightforward geometrical arguments. Here are three examples.

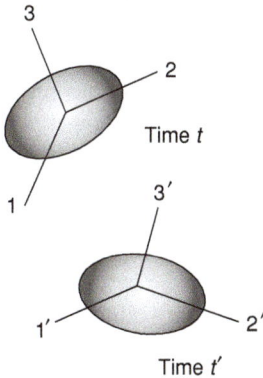

Time t

Time t'

Example 8.16 Dynamical Stability of Rigid Body Motion

In principle, a pencil can be balanced on its point, but in practice, the pencil falls almost immediately. Although a perfectly balanced pencil is in equilibrium, the equilibrium is not stable. If the pencil starts to tip because of some small perturbing force, the gravitational torque causes it to tip even farther; the system continues to move away from equilibrium.

A static system is stable if a displacement from equilibrium gives rise to forces that drive it back toward equilibrium. A dynamical system is stable if it responds to a perturbing force by altering its motion only slightly. In contrast, an unstable dynamical system can have its motion

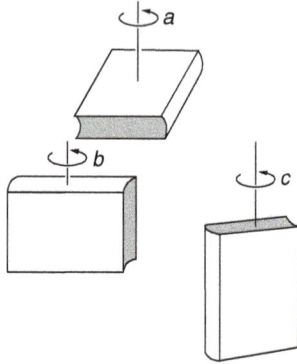

drastically changed by a small perturbing force, possibly leading to catastrophic failure.

A rotating rigid body can exhibit either stable or unstable motion depending on the axis of rotation. As we shall show, the motion is stable for rotation about the axes of maximum or minimum moment of inertia but unstable for rotation about the axis with intermediate moment of inertia. The effect is easy to demonstrate: wrap a book with a rubber band and let it fall spinning about each of its principal axes in turn. Let the moment of inertia be maximum about axis a and minimum about axis c. You will find that the motion is stable if the book is spun about either of these axes. However, if the book is spun about axis b with the intermediate moment of inertia, it tends to flop over as it spins, generally landing on its broad side.

To explain this behavior, we turn to Euler's equations, Eq. (8.28). Suppose that the body initially spins with $\omega_1 = \text{constant} \neq 0$, $\omega_2 = 0$, $\omega_3 = 0$, and that it is suddenly perturbed. Immediately after the perturbation, ω_2 and ω_3 are different from zero but small compared with ω_1. Once the perturbation ends, the motion is torque-free and Euler's equations become:

$$I_1 \frac{d\omega_1}{dt} + (I_3 - I_2)\omega_2\omega_3 = 0 \tag{1}$$

$$I_2 \frac{d\omega_2}{dt} + (I_1 - I_3)\omega_1\omega_3 = 0 \tag{2}$$

$$I_3 \frac{d\omega_3}{dt} + (I_2 - I_1)\omega_1\omega_2 = 0. \tag{3}$$

Because ω_2 and ω_3 are both very small, we can neglect the second term in Eq. (1). Therefore $I_1\, d\omega_1/dt = 0$, so that ω_1 is constant.

If we differentiate Eq. (2) and substitute the value of $d\omega_3/dt$ from Eq. (3), we have

$$I_2 \frac{d^2\omega_2}{dt^2} - \frac{(I_1 - I_3)(I_2 - I_1)}{I_3}\omega_1{}^2\omega_2 = 0$$

or

$$\frac{d^2\omega_2}{dt^2} + A\omega_2 = 0 \tag{4}$$

where

$$A = \frac{(I_1 - I_2)(I_1 - I_3)}{I_2 I_3}\omega_1{}^2.$$

If I_1 is the largest or the smallest moment of inertia, $A > 0$ and Eq. (4) is the equation for simple harmonic motion: ω_2 oscillates at frequency $\sqrt{A}$ with bounded amplitude. It is easy to show that ω_3 also undergoes simple harmonic motion. Since ω_2 and ω_3 are bounded, the motion is stable, corresponding to the torque-free precession we calculated earlier.

If I_1 is the intermediate moment of inertia, $A < 0$. In this case ω_2 and ω_3 tend to increase exponentially with time and the motion is unstable.

Example 8.17 The Rotating Rod

Consider a uniform rod mounted on a horizontal frictionless axle through its center. The axle is carried on a turntable revolving with constant angular velocity $\boldsymbol{\Omega}$, with the center of the rod over the axis of the turntable. Let θ be the angle shown in the sketches. The problem is to find θ as a function of time.

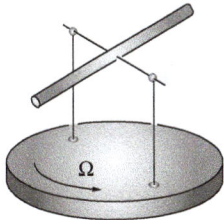

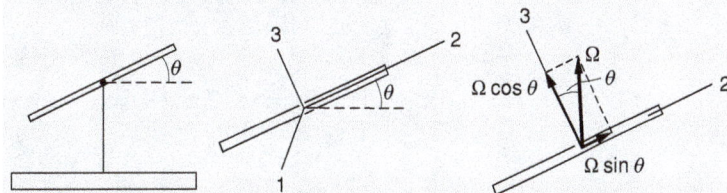

To apply Euler's equations, take principal axis 1 of the rod along the horizontal axle, principal axis 2 along the length of the rod, and principal axis 3 in a vertical plane perpendicular to the rod. $\omega_1 = \dot{\theta}$, and by resolving $\boldsymbol{\Omega}$ along the 2 and 3 directions we find $\omega_2 = \Omega \sin \theta$ and $\omega_3 = \Omega \cos \theta$.

There is no torque about the 1 axis and the first of Euler's equations gives

$$I_1 \ddot{\theta} + (I_3 - I_2)\Omega^2 \sin \theta \cos \theta = 0$$

or

$$\ddot{\theta} + \left(\frac{I_3 - I_2}{2I_1} \right) \Omega^2 \sin 2\theta = 0 \tag{1}$$

using $\sin \theta \cos \theta = \frac{1}{2} \sin 2\theta$.

For oscillations near the horizontal, $\sin 2\theta \approx 2\theta$ and Eq. (1) becomes

$$\ddot{\theta} + \left(\frac{I_3 - I_2}{I_1} \right) \Omega^2 \theta = 0.$$

Recalling that $I_3 > I_2$, we see that θ executes simple harmonic motion with angular frequency $\sqrt{(I_3 - I_2)/I_1}\, \Omega$.

Example 8.18 Euler's Equations and Torque-free Precession

We discussed the torque-free motion of a cylindrically symmetric body in Section 8.7.1 using the small-angle approximation. Here we obtain an exact solution by using Euler's equations.

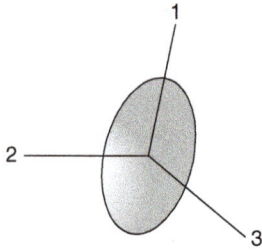

Let the axis of cylindrical symmetry be principal axis 1 with moment of inertia I_1. The other two principal axes are perpendicular to the 1 axis, and $I_2 = I_3 = I_\perp$.

From the first of Euler's equations Eq. (8.28a),

$$\tau_1 = I_1 \frac{d\omega_1}{dt} + (I_3 - I_2)\omega_2\omega_3, \tag{1}$$

we have

$$0 = I_1 \frac{d\omega_1}{dt},$$

which gives

$$\omega_1 = \text{constant} = \omega_s.$$

Principal axes 2 and 3 therefore revolve about the 1 axis at the constant angular velocity ω_s.

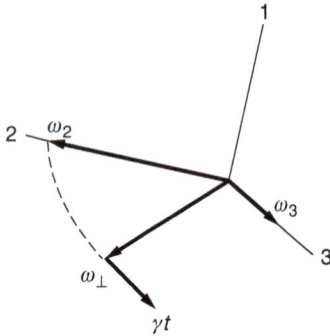

The two remaining Euler's equations are

$$0 = I_\perp \frac{d\omega_2}{dt} + (I_1 - I_\perp)\omega_s\omega_3 \tag{2}$$

$$0 = I_\perp \frac{d\omega_3}{dt} + (I_\perp - I_1)\omega_s\omega_2. \tag{3}$$

Differentiating Eq. (2) and using Eq. (3) to eliminate $d\omega_3/dt$ gives

$$\frac{d^2\omega_2}{dt^2} + \left(\frac{I_1 - I_\perp}{I_\perp}\right)^2 \omega_s{}^2\omega_2 = 0.$$

The angular velocity component ω_2 executes simple harmonic motion with angular frequency

$$\gamma = \left|\frac{I_1 - I_\perp}{I_\perp}\right| \omega_s.$$

Thus ω_2 is given by $\omega_2 = \omega_\perp \cos \gamma t$ where the amplitude $\omega_\perp$ is determined by initial conditions. Then, if $I_1 > I_\perp$, Eq. (2) gives

$$\omega_3 = -\frac{1}{\gamma}\frac{d\omega_2}{dt}$$

$$= \omega_\perp \sin \gamma t.$$

As the drawing shows, ω_2 and ω_3 are the components of a vector $\omega_\perp$ that rotates in the 2–3 plane at rate γ. An observer fixed to the body would see $\omega_\perp$ rotate relative to the body about the 1 axis at angular frequency γ.

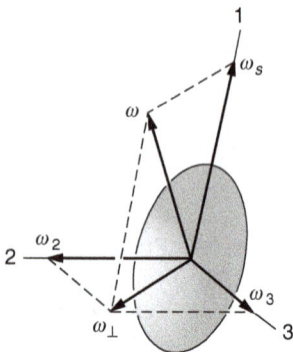

Since the 1, 2, 3 axes are fixed to the body and the body is rotating about the 1 axis at rate ω_s, the rotational speed of $\omega_\perp$ relative to an observer fixed in space is

$$\gamma + \omega_s = \frac{I_1}{I_\perp}\omega_s.$$

Euler's equations have told us how the angular velocity moves relative to the body, but we have yet to find the actual motion of the body in space. Here we must use our ingenuity. We know the motion of ω relative to the body, and we also know that for torque-free motion $\mathbf{L}$ is constant. This is enough to find the actual motion of the body, as we shall now show.

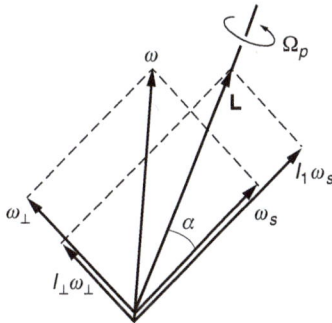

The diagram shows ω and $\mathbf{L}$ at some instant of time. Because $L\cos\alpha = I_1\omega_s$, and ω_s and L are constant, α must be constant as well. Consequently, the relative position of all the vectors in the diagram never changes. The only possible motion is for the diagram to rotate about $\mathbf{L}$ with some "precessional" angular velocity $\mathbf{\Omega}_p$. (Bear in mind that the diagram is moving relative to the body; Ω_p is greater than ω_s.)

The remaining problem is to find $\mathbf{\Omega}_p$. We have shown that ω precesses about ω_s in space at rate $\gamma + \omega_s$. To relate this to $\mathbf{\Omega}_p$, resolve $\mathbf{\Omega}_p$ into a vector $\mathbf{A}$ along ω_s and a vector $\mathbf{B}$ perpendicular to ω_s.

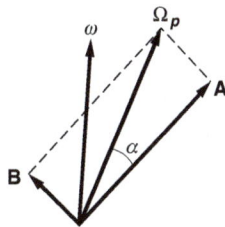

The magnitudes are $A = \Omega_p\cos\alpha$ and $B = \Omega_p\sin\alpha$. The rotation $\mathbf{A}$ turns ω about ω_s, but the rotation $\mathbf{B}$ does not. Hence the rate at which ω precesses about ω_s is $\Omega_p\cos\alpha$. Equating this to $\gamma + \omega_s$,

$$\Omega_p\cos\alpha = \gamma + \omega_s$$

$$= \frac{I_1}{I_\perp}\omega_s$$

or

$$\Omega_p = \frac{I_1\omega_s}{I_\perp\cos\alpha}.$$

The precessional angular velocity $\mathbf{\Omega}_p$ represents the rate at which the symmetry axis rotates about the fixed direction $\mathbf{L}$. It is the frequency of wobble we observe when we flip a spinning coin. Earlier in this section we used the small-angle approximation and found that the rate at which the symmetry axis rotates about a space-fixed direction is $I_1\omega_s/I_\perp$. Our earlier result agrees with the exact result derived here in the limit $\alpha \to 0$.

Note 8.1 Finite and Infinitesimal Rotations

In this Note we shall demonstrate that the order in which rotations are made is important if the rotations are large—finite rotations do not commute—but not if they are small—infinitesimal rotations do commute. By an infinitesimal rotation we mean one for which all powers of the rotation angle beyond the first can be neglected. Commutatitivity is important because it allows us to treat small angular displacements as components of a vector.

In Example 8.1 we demonstrated that if two rotations are performed in opposite order the results are different, so that we cannot write a finite rotation as a vector. To prove this, we shall calculate the effect of successive rotations on a position vector $\mathbf{r}$. Let $\mathbf{r}_\alpha$ be the result of rotating $\mathbf{r}$ through α about $\hat{\mathbf{n}}_\alpha$, and $\mathbf{r}_{\alpha\beta}$ the result of rotating $\mathbf{r}_\alpha$ through β about $\hat{\mathbf{n}}_\beta$. We shall show that

$$\mathbf{r}_{\alpha\beta} \neq \mathbf{r}_{\beta\alpha}.$$

However, for $\alpha \ll 1$ and $\beta \ll 1$, we will find that $\mathbf{r}_{\alpha\beta} = \mathbf{r}_{\beta\alpha}$ to first order in the rotation angles, so there is no problem treating the orientation angle as a vector for infinitesimal rotations. Our calculation is a special case, but it illustrates the essential features of a general proof.

Suppose vector $\mathbf{r}$ is initially along the x axis, $\mathbf{r} = r\,\hat{\mathbf{i}}$. First rotate through angle α about the z axis and then through angle β about the y axis, as follows.

First rotation: through angle α about the z axis.

$$\mathbf{r} = r\,\hat{\mathbf{i}}$$
$$\mathbf{r}_\alpha = r\cos\alpha\,\hat{\mathbf{i}} + u\sin\alpha\,\hat{\mathbf{j}},$$

since $|\mathbf{r}_\alpha| = |\mathbf{r}| = r$.

Second rotation: through angle β about the y axis. The y component $r\sin\alpha\,\hat{\mathbf{j}}$ is unchanged by this rotation.

$$\begin{aligned}\mathbf{r}_{\alpha\beta} &= r\cos\alpha(\cos\beta\,\hat{\mathbf{i}} - \sin\beta\,\hat{\mathbf{k}}) + r\sin\alpha\,\hat{\mathbf{j}}\\ &= r\cos\alpha\cos\beta\,\hat{\mathbf{i}} + r\sin\alpha\,\hat{\mathbf{j}} - r\cos\alpha\sin\beta\,\hat{\mathbf{k}}.\end{aligned} \tag{1}$$

Now go through the same argument but in reverse order. The result is

$$\mathbf{r}_{\beta\alpha} = r\cos\alpha\cos\beta\,\hat{\mathbf{i}} + r\cos\beta\sin\alpha\,\hat{\mathbf{j}} - r\sin\beta\,\hat{\mathbf{k}}. \tag{2}$$

Comparing Eqs. (1) and (2) reveals that $\mathbf{r}_{\alpha\beta}$ and $\mathbf{r}_{\beta\alpha}$ differ in their y and z components. Represent the rotation angles by $\Delta\alpha$ and $\Delta\beta$, as in the sketches, and take $\Delta\alpha \ll 1, \Delta\beta \ll 1$.

If we neglect all terms of second order and higher, then $\sin\Delta\theta \approx \Delta\theta$ and $\cos\Delta\theta \approx 1$, and Eq. (1) becomes

$$\mathbf{r}_{\alpha\beta} = r\,\hat{\mathbf{i}} + r\Delta\alpha\,\hat{\mathbf{j}} - r\Delta\beta\,\hat{\mathbf{k}}, \tag{3}$$

while Eq. (2) becomes

$$\mathbf{r}_{\beta\alpha} = r\,\hat{\mathbf{i}} + r\Delta\alpha\,\hat{\mathbf{j}} - r\Delta\beta\,\hat{\mathbf{k}}. \tag{4}$$

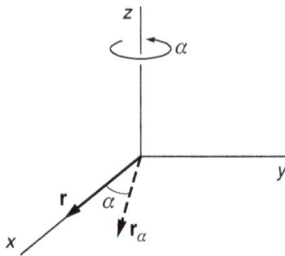

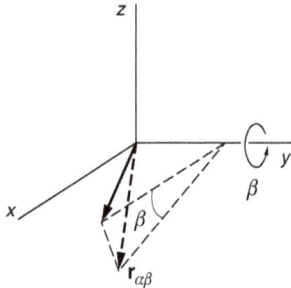

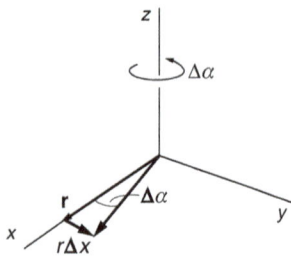

Hence $\mathbf{r}_{\alpha\beta} = \mathbf{r}_{\beta\alpha}$ to first order for small rotations, and the vector

$$\Delta\boldsymbol{\theta} = \Delta\beta\,\hat{\mathbf{j}} + \Delta\alpha\,\hat{\mathbf{k}}$$

is well defined. In particular, the displacement of $\mathbf{r}$ is

$$\begin{aligned}
\Delta\mathbf{r} &= \mathbf{r}_{\text{final}} - \mathbf{r}_{\text{initial}} \\
&= \mathbf{r}_{\alpha\beta} - r\,\hat{\mathbf{i}} \\
&= r\Delta\alpha\,\hat{\mathbf{j}} - r\Delta\beta\,\hat{\mathbf{k}} = \Delta\boldsymbol{\theta} \times \mathbf{r}.
\end{aligned} \tag{5}$$

If the displacement occurs in time Δt, the velocity is

$$\begin{aligned}
\mathbf{v} &= \lim_{\Delta t \to 0} \frac{\Delta\mathbf{r}}{\Delta t} \\
&= \lim_{\Delta t \to 0} \frac{\Delta\boldsymbol{\theta} \times \mathbf{r}}{\Delta t} \\
&= \boldsymbol{\omega} \times \mathbf{r},
\end{aligned}$$

where

$$\boldsymbol{\omega} = \lim_{\Delta t \to 0} \frac{\Delta\boldsymbol{\theta}}{\Delta t}.$$

In our example, $\boldsymbol{\omega} = (d\beta/dt)\,\hat{\mathbf{j}} + (d\alpha/dt)\,\hat{\mathbf{k}}$.

The equality of Eqs. (3) and (4) indicates that the result of two infinitesimal rotations can be found by evaluating the effects of the rotations independently. To first order, the effect of rotating $\mathbf{r} = r\,\hat{\mathbf{i}}$ through $\Delta\alpha$ about z is to generate a y component $r\Delta\alpha\,\hat{\mathbf{j}}$. The effect of rotating $\mathbf{r}$ through $\Delta\beta$ about y is to generate a z component, $-r\Delta\beta\,\hat{\mathbf{k}}$. The total change in $\mathbf{r}$ to first order is the sum of the two effects,

$$\Delta\mathbf{r} = r\Delta\alpha\,\hat{\mathbf{j}} - r\Delta\beta\,\hat{\mathbf{k}},$$

in agreement with Eq. (5).

Note 8.2 More about Gyroscopes

In Section 8.3 we used simple vector arguments to discuss the uniform precession of a gyroscope. However, uniform precession is not the most general form of gyroscope motion. For instance, a gyroscope released with its axle at rest horizontally does not instantaneously start to precess. Instead, the center of mass begins to fall and the falling motion is rapidly converted to an undulatory motion called *nutation*. If the undulations are damped out by friction in the bearings, the gyroscope eventually settles into uniform precession. The purpose of this Note is to show how nutation occurs, using the small-angle approximation. (We used the same approach in Section 8.7.1 to explain torque-free precession.)

Consider a gyroscope consisting of a flywheel of mass M at one end of a shaft of length l whose other end is attached to a universal pivot.

The flywheel is set spinning rapidly and the axle is released from the horizontal. What is the subsequent motion?

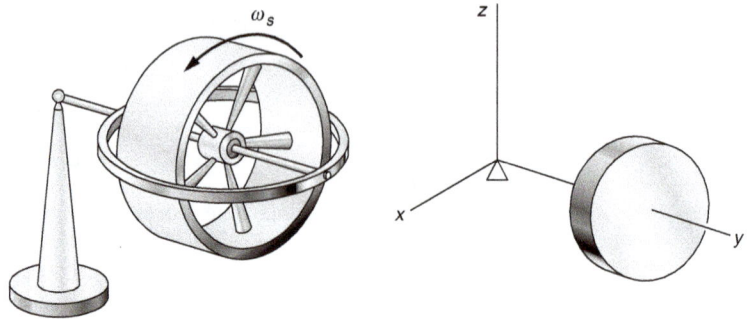

Because it is natural to consider the motion in terms of rotation about the fixed pivot point, we introduce a coordinate system with its origin at the pivot.

Assume for the moment that the gyroscope is not spinning but that the axle is rotating horizontally around the pivot at rate ω_z. To calculate the angular momentum about the origin we need a generalization of the parallel axis theorem. Consider the angular momentum due to rotation of the axle about the z axis at rate ω_z. If the moment of inertia of the disk around a vertical axis through the center of mass is I_{zz}, then the moment of inertia about the z axis through the pivot is $I_{zz} + Ml^2$. (The proof is straightforward, and is left as a problem.) If we let $I_{zz} + Ml^2 = I_p$, then $L_z = I_p\omega_z$. By symmetry, the moment of inertia about the x axis is $I_{xx} + Ml^2 = I_p$, so that $L_x = I_p\omega_x$.

These results are exact when the gyroscope lies along the y axis, as in the drawing, and they are true to first order in angle for small angles of tilt around the y axis.

Now suppose that the flywheel is set spinning at rate ω_s. If the moment of inertia along the axle is I_s, then the spin angular momentum is $L_s = I_s\omega_s$.

There are two separate contributions to the angular momentum associated with small angular displacements from the y axis. The first is from the motion of the system as a whole as it rotates about the pivot at some rate ω_z. This rotation makes a contribution of the form $I\omega_z$, where I is the moment of inertia about the axis of ω_z. The second arises from the change in direction of the spin. As the gyroscope moves away from the y axis, components of $\mathbf{L}_s$ are generated in the x and z directions. For small angular displacements θ, such components will be of the form $L_s\theta$.

For small angular displacements, $\theta_x \ll 1$ about the x axis and $\theta_z \ll 1$ about the z axis, the rotations can be considered independently and their effects added.

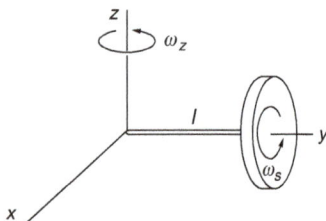

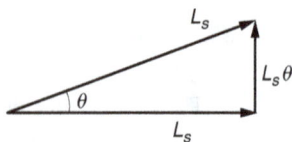

(a)   (b)

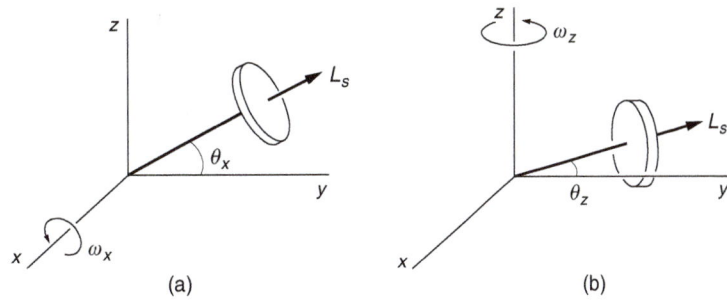

(a) Rotation about the x axis Suppose that the axle has rotated about the x axis through angle $\theta_x \ll 1$, and has instantaneous angular velocity ω_x. Then

$$
\begin{aligned}
L_x &= I_p \omega_x \\
L_y &= L_s \cos \theta_x \approx L_s \\
L_z &= L_s \sin \theta_x \approx L_s \theta_x.
\end{aligned}
\tag{1}
$$

(b) Rotation about the z axis
For rotation by $\theta_z \ll 1$ about the z axis, a similar argument gives

$$
\begin{aligned}
L_x &= -L_s \sin \theta_z \approx -L_s \theta_z \\
L_y &= L_s \cos \theta_z \approx L_s \\
L_z &= I_p \omega_z.
\end{aligned}
\tag{2}
$$

Equations (1) and (2) show that the rotations θ_x and θ_z leave L_y unchanged to first order. However, the rotations give rise to first order contributions to L_x and L_z. From Eqs. (1) and (2) we find

$$
\begin{aligned}
L_x &= I_p \omega_x - L_s \theta_z \\
L_y &= L_s \\
L_z &= I_p \omega_z + L_s \theta_x.
\end{aligned}
\tag{3}
$$

The instantaneous torque about the origin is

$$
\tau_x = -lW,
\tag{4}
$$

where l is the length of the axle and W is the weight of the gyro. Since $\tau = d\mathbf{L}/dt$, Eqs. (3) and (4) give

$$
I_p \dot{\omega}_x - L_s \omega_z = -lW
\tag{5a}
$$

$$
\dot{L}_s = 0
\tag{5b}
$$

$$
I_p \dot{\omega}_z + L_s \omega_x = 0,
\tag{5c}
$$

where we have used $\dot{\theta}_z = \omega_z, \dot{\theta}_x = \omega_x$.

Equation (5b) assures us that the spin is constant, as we expect for a flywheel with good bearings. To solve the remaining equations, we

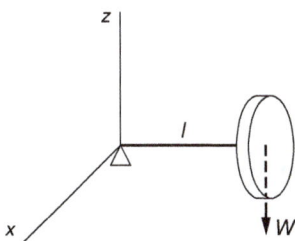

differentiate Eq. (5a) to obtain

$$I_p \ddot{\omega}_x - L_s \dot{\omega}_z = 0.$$

Substituting the result $\dot{\omega}_z = -L_s \omega_x / I_p$ from Eq. (5c) gives

$$\ddot{\omega}_x + \frac{L_s^2}{I_p^2} \omega_x = 0.$$

Letting $\gamma = L_s / I_p = \omega_s I_s / I_p$, this becomes

$$\ddot{\omega}_x + \gamma^2 \omega_x = 0.$$

We again have the familiar equation for simple harmonic motion. The solution is

$$\omega_x = A \cos(\gamma t + \phi), \tag{6}$$

where A and ϕ are arbitrary constants.

We can use Eq. (5a) to find ω_z:

$$\omega_z = \frac{lW}{L_s} + \frac{I_p}{L_s} \dot{\omega}_x.$$

Substituting the result $\dot{\omega}_x = -A\gamma \sin(\gamma t + \phi)$ obtained from Eq. (6) gives

$$\omega_z = \frac{lW}{L_s} - \frac{I_p}{L_s} A\gamma \sin(\gamma t + \phi)$$

$$= \frac{lW}{L_s} - A \, \sin(\gamma t + \phi). \tag{7}$$

We can integrate Eqs. (6) and (7) to obtain

$$\theta_x = B \sin(\gamma t + \phi) + C$$

$$\theta_z = \frac{lW}{L_s} t + B \cos(\gamma t + \phi) + D, \tag{8}$$

where $B = A/\gamma$, and C, D are constants of integration.

The motion of the gyroscope depends on the constants B, ϕ, C, and D in Eq. (8), and these depend on the initial conditions. We consider three separate cases.

Case 1. Uniform Precession If we take $B = 0$ and $C = D = 0$, Eq. (8) gives

$$\theta_x = 0$$

$$\theta_z = lW \frac{t}{L_s}. \tag{9}$$

This corresponds to the case of uniform precession we treated in Section 8.3. The rate of precession is $d\theta_z / dt = lW / L_s$, the same result we found earlier, Eq. (8.2). If the gyroscope is moving in uniform precession at $t = 0$, it will continue to do so. However, if we release the gyroscope from rest at $t = 0$, its initial precession rate is zero. Consequently, we need to examine the solution more closely.

Case 2. Torque-free Precession If we "turn off" gravity so that W is zero, then Eq. (8) gives, with $C = D = 0$,

$$\theta_x = B \sin(\gamma t + \phi) \tag{10a}$$

$$\theta_z = B \cos(\gamma t + \phi). \tag{10b}$$

The tip of the axle moves in a circle about the y axis. The amplitude of the motion depends on the initial conditions. This is identical to the torque-free precession discussed in Section 8.7.1 and Example 8.18.

Case 3. Nutation Suppose that the axle is released from rest along the y axis at $t = 0$. The initial conditions at $t = 0$ on the x motion are $(\theta_x)_0 = 0, (d\theta_x/dt)_0 = 0$. From Eq. (8) we obtain

$$B \sin \phi + C = 0$$

$$B\gamma \cos \phi = 0.$$

Assuming for the moment that B is not zero, we have $\phi = \pi/2, C = -B$. Equation (8) then becomes

$$\theta_z = \frac{lW}{L_s} t - B \sin \gamma t + D.$$

The initial conditions on the z motion at $t = 0$ are $(\theta_z)_0 = 0, (d\theta_z/dt)_0 = 0$, and we obtain

$$D = 0$$

$$-B\gamma + \frac{lW}{L_z} = 0$$

or

$$B = \frac{lW}{\gamma L_s}.$$

Inserting these results in Eq. (8) gives

$$\theta_x = \frac{lw}{\gamma L_s}(\cos \gamma t - 1) \tag{11a}$$

$$\theta_z = \frac{lw}{\gamma L_s}(\gamma t - \sin \gamma t). \tag{11b}$$

The motion described by Eqs. (8.31) is illustrated in the sketch. As time increases, the tip of the axle traces out a cycloidal path. The dipping motion of the axle is called *nutation*. The motion is easy to see with a well-made gyroscope. Note that the initial motion of the axle is vertically down. When the gyroscope is released it starts to fall, but the motion is rapidly reversed and it rises to its initial elevation. Meanwhile, it has precessed slightly. Eventually the nutation dies out due to friction in the pivot, and the damped motion turns into uniform precession, as shown in the sketch.

The axle is left with a slight dip after the nutation is damped; this keeps the total angular momentum about the z axis zero. The rotational energy of precession comes from the fall of the center of mass. Other

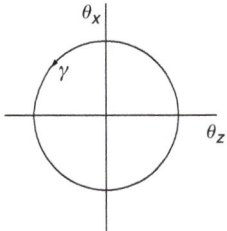

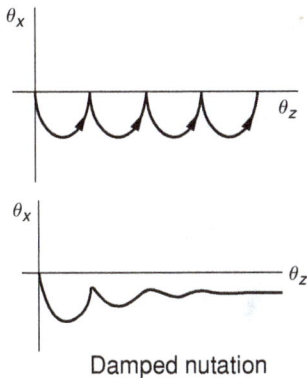

Damped nutation

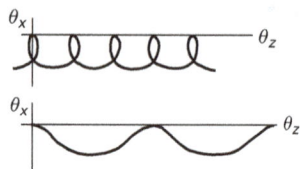

nutational motions are also possible, depending on the initial conditions; the sketches show two possible cases. These can all be described by Eq. (8) by suitable choices of the constants.

We made the approximation that $\theta_x \ll 1, \theta_z \ll 1$, but because of precession, θ_z increases linearly with time and the approximation inevitably breaks down. This is not a problem if we examine the motion for one period of nutation. The nutational motion repeats itself with period $T = 2\pi/\gamma$. If θ_z is small during one period, then we can mentally start the problem over at the end of the period with a new coordinate system having its y axis again along the direction of the axle. The restriction on θ_z is then that $\Omega T \ll 1$, or

$$\frac{2\pi\Omega}{\gamma} \ll 1.$$

Our solution breaks down if the rate of precession becomes comparable to the rate of nutation. More vividly, our approximation is good if the gyroscope nutates many times as it precesses through a full turn.

In a toy gyroscope, friction is so large that it is practically impossible to observe nutation. However, in an air suspension gyroscope, friction is so small that nutation is easy to observe. The rotor of this gyroscope is a massive metal sphere which rests in a close fitting cup. The sphere is suspended on a film of air which flows from an orifice at the bottom of the cup. Torque is applied by the weight of a small mass on a rod protruding radially from the sphere. The pictures are photographs of a stroboscopic light source reflected from a small bead on the end of the rod, showing three modes. By studying the distance between the dots you can discern the variation in speed of the rod through the precession cycle.

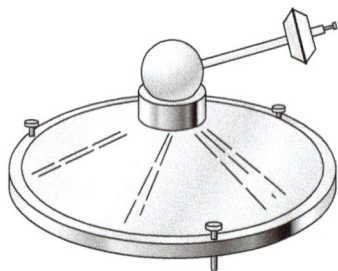

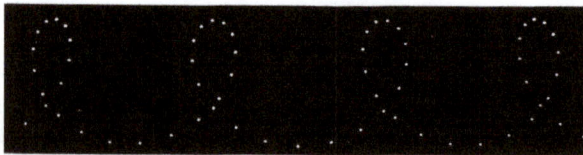

Problems

For problems marked *, refer to page 523 for a hint, clue, or answer.

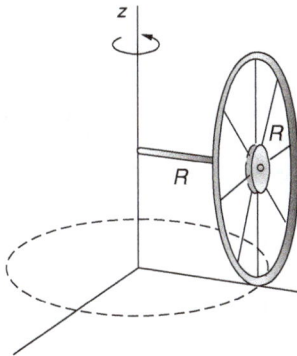

8.1 *Rolling hoop*

A thin hoop of mass M and radius R rolls without slipping about the z axis. It is supported by an axle of length R through its center, as shown. The hoop circles around the z axis with angular speed Ω.

(*a*) What is the instantaneous angular velocity ω of the hoop?

(*b*) What is the angular momentum $\mathbf{L}$ of the hoop? Is $\mathbf{L}$ parallel to ω? (The moment of inertia of a hoop for an axis along its diameter is $\frac{1}{2}MR^2$.)

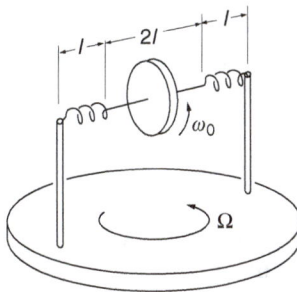

8.2 *Flywheel on rotating table*

A flywheel of moment of inertia I_0 rotates with angular velocity ω_0 at the middle of an axle of length $2l$. Each end of the axle is attached to a support by a spring which is stretched to length l and provides tension T. You may assume that T remains constant for small displacements of the axle. The supports are fixed to a table that rotates at constant angular velocity Ω, where $\Omega \ll \omega_0$. The center of mass of the flywheel is directly over the center of rotation of the table. Neglect gravity and assume that the motion is completely uniform so that nutational effects are absent. The problem is to find the direction of the axle with respect to a straight line between the supports.

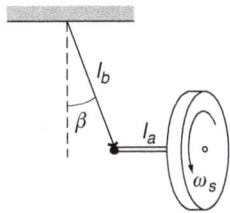

8.3 *Suspended gyroscope*

A gyroscope wheel is at one end of an axle of length l_a. The other end of the axle is suspended from a string of length l_b. The wheel is set into motion so that it executes uniform precession in the horizontal plane. The wheel has mass M and moment of inertia about its center of mass I_0. Its spin angular velocity is ω_s. Neglect the masses of the shaft and string.

Find the angle β that the string makes with the vertical. Assume that β is so small that approximations like $\sin\beta \approx \beta$ are justified.

8.4 *Grain mill**

In an old-fashioned rolling mill, grain is ground by a disk-shaped millstone that rolls in a circle on a flat surface, driven by a vertical shaft. Because of the stone's angular momentum, the contact force with the surface is greater than the weight of the wheel.

Assume that the millstone is a uniform disk of mass M, radius b, and width w, and that it rolls without slipping in a circle of radius R with angular velocity Ω. Find the ratio of the contact force with respect to the surface to the weight of the stone.

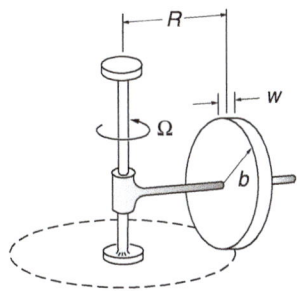

8.5 *Automobile on a curve*

When an automobile rounds a curve at high speed, the loading (weight distribution) on the wheels is markedly changed. For sufficiently high speeds the loading on the inside wheels goes to zero, at which point the car starts to roll over. This tendency can be avoided by mounting a large spinning flywheel on the car.

(a) In what direction should the flywheel be mounted, and what should be the sense of rotation, to help equalize the loading? (Be sure that your method works for the car turning in either direction.)

(b) Show that for a disk-shaped flywheel of mass m and radius R, the requirement for equal loading is that the angular velocity ω of the flywheel is related to the velocity of the car V by

$$\omega = 2V \frac{Mb}{mR^2},$$

where M is the total mass of the car and flywheel, and b is the height of the center of mass of the car (including the flywheel) above the road. Assume that the road is unbanked.

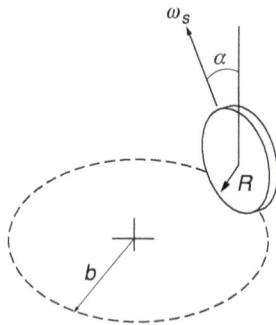

8.6 *Rolling coin**

A coin of radius b and mass M rolls on a horizontal surface at speed V. If the plane of the coin is vertical the coin rolls in a straight line. If the plane is tilted, the path of the coin is a circle of radius R. Find an expression for the tilt angle of the coin α in terms of the given quantities. (Because of the tilt of the coin the circle traced by its center of mass is slightly smaller than R but you can ignore the difference.)

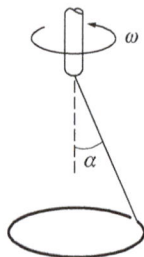

8.7 *Suspended hoop*

A thin hoop of mass M and radius R is suspended from a string through a point on the rim of the hoop. If the support is turned with high angular velocity ω, the hoop will spin as shown, with its plane nearly horizontal and its center nearly on the axis of the support. The string makes angle α with the vertical.

(a) Find, approximately, the small angle β between the plane of the hoop and the horizontal. Assume that the center of mass is at rest.

(b) Find, approximately, the radius of the small circle traced out by the center of mass about the vertical axis.

(c) Find a criterion for the validity of the assumption that motion of the center of mass can be neglected. (With skill you can demonstrate this motion with a rope. It is a favorite cowboy lariat trick.)

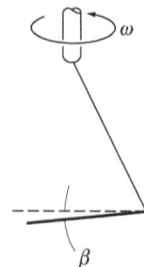

8.8 *Deflected hoop*

A child's hoop of mass M and radius b rolls in a straight line with velocity V. The top of the hoop is given a light tap with a stick at right angles to the direction of motion. The impulse of the blow is I.

(a) Assuming that the spin angular momentum is much larger than any other component of angular momentum, the only effect of the tap is to change the direction that the hoop rolls by some angle Φ. Find Φ.

(b) Find a criterion for the peak applied force F in order for the assumption in part (a) to be valid.

8.9 *Stability of a bicycle*
As a bicycle changes direction, the rider leans inward creating a horizontal torque on the bike. Part of the torque is responsible for the change in direction of the spin angular momentum of the wheels. Consider a bicycle and rider system of total mass M with wheels of mass m and radius b, rounding a curve of radius R at speed V. The center of mass of the system is $1.5b$ from the ground.

(a) Find an expression for the tilt angle α.

(b) Find the value of α, in degrees, if $M = 70$ kg, $m = 2.5$ kg, $V = 30$ km/hour and $R = 30$ m.

(c) What would be the percentage change in α if spin angular momentum were neglected?

8.10 *Measuring latitude with a gyro*
Latitude can be measured with a gyro composed of a spinning disk mounted to pivot freely on an axis through the plane of the disk, with its axle horizontal and lying along the east–west axis.

(a) Show that the gyro can remain stationary when its spin axis is parallel to the polar axis and is at the latitude angle λ with the horizontal.

(b) If the gyro is released with the spin axis at a small angle to the polar axis show that the gyro spin axis will oscillate about the polar axis with a frequency $\omega_{osc} = \sqrt{I_1 \omega_s \Omega_e / I_\perp}$, where I_1 is the moment of inertia of the gyro about its spin axis, $I_\perp$ is its moment of inertia about the fixed horizontal axis, and Ω_e is the Earth's rotational angular velocity.

What value of ω_{osc} is expected for a gyro rotating at 40 000 rpm, assuming that it is a thin disk and that the mounting frame makes no contribution to the moment of inertia?

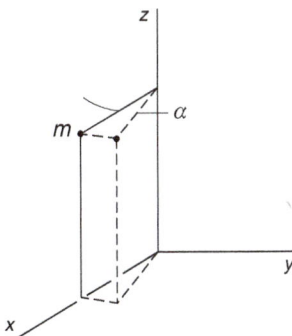

8.11 *Tensor of inertia*
A particle of mass m is located at $x = 2, y = 0, z = 3$.

(a) Find its moments and products of inertia relative to the origin.

(b) The particle undergoes pure rotation about the z axis through a small angle α. Show that its moments and products of inertia are unchanged to first order in α if $\alpha \ll 1$.

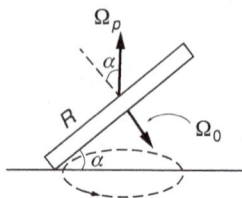

8.12 *Euler's disk**

If you spin a coin or solid uniform disk about a vertical axis on a hard surface, it will eventually lose energy and begin to wobble and perhaps make a buzzing sound as it rolls. With a massive disk—a toy called Euler's disk—the wobbling motion can be astonishing, with the frequency of buzzing increasing as the rotational speed of the disk decreases. The motion seems at first sight to be like gyroscope motion, but it can be argued that Euler's disk is an antigyroscope. The goal of this problem is to understand the underlying physics of the disk's motion.

Consider a thin disk of radius R and mass M that is executing the rapidly buzzing motion. The plane of the disk is at angle α with respect to the table. The axis of the disk precesses around the vertical at rate $\mathbf{\Omega}_p$. The disk also spins about its own axis with spin angular velocity $\mathbf{\Omega}_0$. The result of the two motions is for the contact point on the table to be momentarily at rest.

Find $\mathbf{\Omega}_p$ for a given small angle α.

(a) Find the total angular velocity of the disk and the angular momentum. From this, find the precession rate $\mathbf{\Omega}_p$.

(b) Find the rate at which the disk appears to rotate as viewed from above, when α is very small.

9 NON-INERTIAL SYSTEMS AND FICTITIOUS FORCES

9.1 Introduction

In discussing the principles of dynamics in Chapter 2, we stressed that Newton's second law $\mathbf{F} = m\mathbf{a}$ holds true only in inertial coordinate systems. We have so far avoided non-inertial systems in order not to obscure our goal of understanding the physical nature of forces and accelerations. Because that goal has largely been realized, in this chapter we turn to the use of non-inertial systems with a twofold purpose. By introducing non-inertial systems we can simplify many problems; from this point of view, the use of non-inertial systems represents one more computational tool. However, consideration of non-inertial systems also enables us to explore some of the conceptual difficulties of classical mechanics. Consequently, the second goal of this chapter is to gain deeper insight into Newton's laws, the properties of space, and the meaning of inertia. We start by developing a formal procedure for relating observations in different inertial systems.

9.2 Galilean Transformation

In this section we shall show that any coordinate system moving uniformly with respect to an inertial system is also inertial. This result is so transparent that it hardly warrants formal proof. However, the argument will be helpful in the next section when we analyze non-inertial systems.

Suppose that two physicists, Alice and Bob, set out to observe a series of events such as the position of a body of mass m as a function of time. Each has their own set of measuring instruments and each works in their own laboratory. Alice has confirmed by experiments that Newton's laws hold accurately in her laboratory, and she concludes that her reference frame is therefore inertial. How can she predict whether or not Bob's system is also inertial?

For simplicity, Alice and Bob agree to use Cartesian coordinate systems with identical scale units. In general, their coordinate systems do not coincide. Leaving rotations for later, we suppose for the time being that the systems are in relative motion but that corresponding axes are parallel. Let the position of mass m be given by $\mathbf{r}_\alpha$ in Alice's system, and $\mathbf{r}_\beta$ in Bob's system. If the origins of the two systems are displaced by $\mathbf{S}$, as shown in the sketch, then

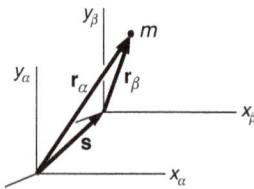

$$\mathbf{r}_\beta = \mathbf{r}_\alpha - \mathbf{S}. \tag{9.1}$$

If Alice sees the mass accelerating at rate $\mathbf{a}_\alpha = \ddot{\mathbf{r}}_\alpha$, she concludes from Newton's second law that there is a force on m given by

$$\mathbf{F}_\alpha = m\mathbf{a}_\alpha.$$

Bob observes m to be accelerating at rate $\mathbf{a}_\beta$, as if it were acted on by a force

$$\mathbf{F}_\beta = m\mathbf{a}_\beta.$$

What is the relation between $\mathbf{F}_\beta$ and the true force $\mathbf{F}_\alpha$ measured in an inertial system?

It is a simple matter to relate the accelerations in the two systems. Successive differentiation of Eq. (9.1) with respect to time yields

$$\mathbf{v}_\beta = \mathbf{v}_\alpha - \mathbf{V} \tag{9.2a}$$

$$\mathbf{a}_\beta = \mathbf{a}_\alpha - \mathbf{A} \tag{9.2b}$$

where $\mathbf{V} = \dot{\mathbf{S}}$ and $\mathbf{A} = \dot{\mathbf{V}} = \ddot{\mathbf{S}}$.

If $\mathbf{V}$ is constant, the relative motion is uniform and $\mathbf{A} = 0$. In this case $\mathbf{a}_\beta = \mathbf{a}_\alpha$ and

$$\mathbf{F}_\beta = m\mathbf{a}_\beta = m\mathbf{a}_\alpha$$
$$= \mathbf{F}_\alpha.$$

The measured force is the same in both systems. The equations of motion in a system moving uniformly with respect to an inertial system are identical to those in the inertial system. It follows that all systems translating uniformly relative to an inertial system are inertial. This simple result leads to an enigma. Although it would be appealing to single out a coordinate system absolutely at rest, there is no dynamical way to distinguish one inertial system from another. Nature provides no clue to absolute rest.

We tacitly made a number of plausible assumptions in the above argument. In the first place, we have assumed that both observers use the same scale for measuring distance. To assure this, Alice and Bob must calibrate their scales with the same standard of length. If Alice determines that the length of a certain rod at rest in her system is L_α, we expect that Bob will measure the same length. This is indeed the case if there is no motion between the two systems. However, it is not true in general. If Bob moves parallel to the rod with uniform velocity v, he will measure a length $L_\beta = L_\alpha \sqrt{1 - v^2/c^2}$, where c is the velocity of light. The contraction of a moving rod, known as the *Lorentz contraction*, follows from the theory of special relativity, as discussed in Section 12.8.2.

A second assumption we made is that time is the same in both systems. If Alice determines that the time interval between two events is T_α, we assumed that Bob will observe the same interval. Here again the assumption breaks down at high velocities. As discussed in Section 12.8.1, Bob finds that the interval he measures is $T_\beta = T_\alpha \sqrt{1 - v^2/c^2}$. Once again nature provides an unexpected result.

The reason these results are so unexpected is that our notions of space and time come chiefly from immediate contact with the world around us, and this never involves velocities even remotely near the velocity of light. If we normally moved with speeds approaching the velocity of light, we would take these results for granted. As it is, even the highest "everyday" velocities are low compared with the velocity of light. For instance, the velocity of an artificial satellite around the Earth is about 8 km/s. In this case $v^2/c^2 \approx 10^{-9}$, and length and time are altered by only 1 part in 10^9.

A third assumption is that the observers agree on the value of the mass. However, mass is defined by experiments that involve both time and distance, and so this assumption must also be examined. According to a further result from special relativity, if an object at rest has mass m_0, the most useful quantity corresponding to mass for an observer moving with velocity v is $m = m_0 / \sqrt{1 - v^2/c^2}$.

Now that we are aware of some of the complexities, let us defer consideration of special relativity until Chapters 12, 13, and 14, and for the time being limit our discussion to situations where $v \ll c$. In this case the Newtonian ideas of space, time, and mass are valid to high accuracy. The following equations then relate measurements made by Alice and Bob, provided that their coordinate systems move with uniform relative velocity $\mathbf{V}$. We choose the origins of the coordinate systems to coincide at $t = 0$ so that $\mathbf{S} = \mathbf{V}t$. Then from Eq. (9.1) we have

$$\mathbf{r}_\beta = \mathbf{r}_\alpha - \mathbf{V}t \tag{9.3a}$$

$$t_\beta = t_\alpha. \tag{9.3b}$$

The time relation Eq. (9.3b) is generally assumed implicitly.

This set of relations, called a *transformation*, gives the prescription for transforming coordinates of an event from one coordinate system to another. Equations (9.3) transform coordinates between inertial systems and are known as the *Galilean transformation*. Because force is unchanged by the Galilean transformation, observers in different inertial systems obtain the same dynamical equations. It follows that the laws of physics look the same in all inertial systems; in other words, they have the same form. Otherwise, different observers would make different predictions; for instance, if one observer predicts the collision of two particles, another observer might not.

The assertion that the forms of the laws of physics are the same in all inertial systems is known as the *principle of relativity*. Although the principle of relativity played only a minor role in the development of Newtonian mechanics, its role in Einstein's theory of relativity is crucial. This is discussed further in Chapter 12, where we show that the Galilean transformation is not universally valid but must be replaced by the more general Lorentz transformation. However, the Galilean transformation is accurate for $v \ll c$, and we shall take it to be exact in this chapter.

9.3 Uniformly Accelerating Systems

Next we turn our attention to how physical laws appear to an observer in a system accelerating at a constant rate $\mathbf{A}$ with respect to an inertial system. To simplify notation we shall drop the subscripts α and β and label quantities in non-inertial systems by primes. Equation (9.2b) then becomes

$$\mathbf{a}' = \mathbf{a} - \mathbf{A},$$

where **A** is the acceleration of the primed system as measured in the inertial system.

In the accelerating system the apparent force is

$$\mathbf{F}' = m\mathbf{a}'$$
$$= m\mathbf{a} - m\mathbf{A}.$$

Because the unprimed system is inertial, $m\mathbf{a}$ is equal to the true force **F** due to physical interactions. Hence

$$\mathbf{F}' = \mathbf{F} - m\mathbf{A}.$$

We can write this as

$$\mathbf{F}' = \mathbf{F} + \mathbf{F}_{\text{fict}},$$

where

$$\mathbf{F}_{\text{fict}} \equiv -m\mathbf{A}.$$

$\mathbf{F}_{\text{fict}}$ is called a *fictitious force*. In words, $\mathbf{F}_{\text{fict}}$ has magnitude mA, and its direction is opposed to the direction of **A**.

The fictitious force experienced in a uniformly accelerating system is uniform and proportional to the mass, like a gravitational force. However, fictitious forces originate in the acceleration of the coordinate system, not in interactions between bodies. The term *fictitious* is intended to emphasize the non-physical nature of $\mathbf{F}_{\text{fict}}$.

Here are three examples illustrating the use of fictitious forces.

Example 9.1 The Apparent Force of Gravity

A small weight of mass m hangs from a string in an automobile that accelerates at rate A. What is the static angle of the string from the vertical, and what is the tension in the string?

We shall analyze the problem both in an inertial frame and in a non-inertial frame accelerating with the car.

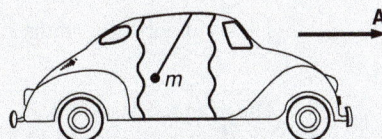

A reminder about signs: a vector is written as positive for the direction shown in the vector diagram, so in this example we write

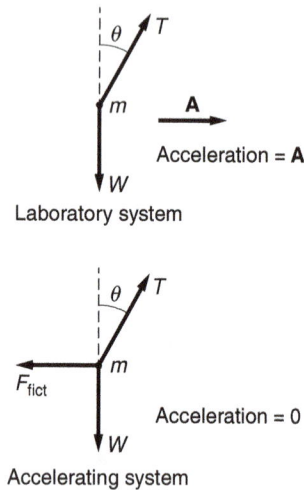

Laboratory system	Accelerating system
$T \cos \theta - W = 0$	$T \cos \theta - W = 0$
$T \sin \theta = mA$	$T \sin \theta - F_{\text{fict}} = 0$
	$F_{\text{fict}} = mA$
$\tan \theta = \dfrac{mA}{W} = \dfrac{A}{g}$	$\tan \theta = \dfrac{A}{g}$
$T = m \sqrt{g^2 + A^2}$	$T = m \sqrt{g^2 + A^2}.$

The physically measurable quantities θ and T are the same regardless of which system we use, as must be the case.

From the point of view of a passenger in the accelerating car, the fictitious force acts like a horizontal gravitational force. The effective gravitational force is the vector sum of the real and fictitious forces. How would a helium-filled balloon held on a string in the accelerating car behave?

The fictitious force in a uniformly accelerating system behaves exactly like a constant gravitational force; the fictitious force is constant and is proportional to the mass. The fictitious force on an extended body therefore acts at the center of mass, as illustrated in the following example.

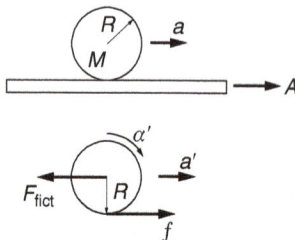

Example 9.2 Cylinder on an Accelerating Plank

A cylinder of mass M and radius R rolls without slipping on a plank that is accelerated at rate A. Find the acceleration of the cylinder.

The force diagram for the horizontal force on the cylinder as viewed in an accelerating system fixed to the plank is shown in the sketch. a' is the acceleration of the cylinder as observed in the accelerating system. f is the friction force, and $F_{\text{fict}} = MA$ with the direction shown.

The equations of motion in the system fixed to the accelerating plank are

$$f - F_{\text{fict}} = Ma'$$
$$Rf = -I_0 \alpha'.$$

The cylinder rolls on the plank without slipping, so

$$\alpha' R = a'.$$

These equations yield

$$Ma' = -I_0 \frac{a'}{R^2} - F_{\text{fict}}$$

$$a' = -\frac{F_{\text{fict}}}{M + I_0/R^2}.$$

Since $I_0 = MR^2/2$, and $F_{\text{fict}} = MA$, we have

$$a' = -\frac{2}{3}A.$$

The acceleration of the cylinder in an inertial system is

$$a = A + a'$$

$$= \frac{1}{3}A.$$

Examples 9.1 and 9.2 can be worked with about the same ease in either an inertial or an accelerating system. Here is a problem that is complicated to solve in an inertial system (try it), but that is almost trivial in an accelerating system.

Example 9.3 Pendulum in an Accelerating Car

Consider again the weight on a string in the accelerating car of Example 9.1, but now assume that the car is at rest with the weight hanging vertically. The car suddenly accelerates at rate A. The problem is to find the maximum angle ϕ through which the weight swings. (ϕ is larger than the equilibrium angle θ found in Example 9.1 due to the sudden acceleration.)

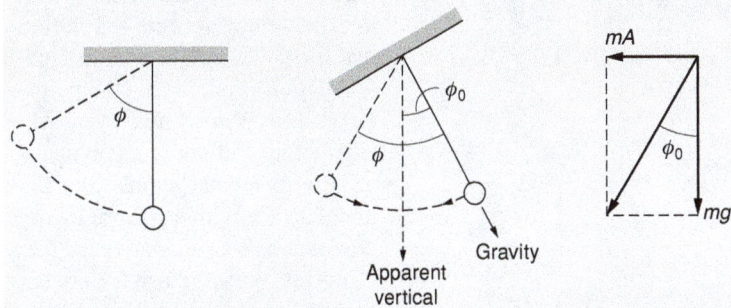

In a system accelerating with the car, the bob behaves like a pendulum in a gravitational field where "down" is at an angle ϕ_0 from the true vertical. From Example 9.1, $\phi_0 = \arctan(A/g)$. The pendulum is initially at rest, so that it swings back and forth with amplitude ϕ_0 about the apparent vertical direction. Hence $\phi = 2\phi_0 = 2\arctan(A/g)$ is the maximum angle of swing from the true vertical.

9.4 The Principle of Equivalence

The laws of physics in a uniformly accelerating system are identical to those in an inertial system provided that we introduce a fictitious force on each particle, $\mathbf{F}_{\text{fict}} = -m\mathbf{A}$. $\mathbf{F}_{\text{fict}}$ is indistinguishable from the force due to a uniform gravitational field $\mathbf{g} = -\mathbf{A}$; both the gravitational force and the fictitious force are constant forces proportional to the mass.

In a local gravitational field $\mathbf{g}$ a free particle of mass m experiences a force $\mathbf{F} = m\mathbf{g}$. Consider the same particle in deep space free of any physical interactions. If it is in a non-inertial system uniformly accelerating at rate $\mathbf{A} = -\mathbf{g}$, it experiences an apparent force $\mathbf{F}_{\text{fict}} = -m\mathbf{A} = m\mathbf{g}$, as before. Is there any way to distinguish physically between these different situations?

The significance of this question was first pointed out by Einstein, who illustrated the problem with the following "gedanken" experiment. (A gedanken, or thought, experiment is meant to be thought about rather than carried out.)

A man is holding an apple in an elevator at rest in a gravitational field g. He lets go of the apple, and it falls with a downward acceleration $a = g$. Now consider the same man in the same elevator, but let the elevator be in free space accelerating upward at rate $a = g$. The man again lets go of the apple, and it again appears to accelerate down at rate g. From the man's point of view the two situations are identical. There is no way to distinguish between acceleration of the elevator and a gravitational field.

The point becomes even more apparent in the case of the elevator freely falling in the gravitational field. The elevator and all its contents accelerate downward at rate g. If the man releases the apple, it will float as if the elevator were motionless in free space. Einstein pointed out that the downward acceleration of the elevator exactly cancels the local gravitational field. From the point of view of an observer in the elevator, there is no way to determine whether the elevator is in free space or whether it is falling in a gravitational field.

This apparently simple idea, known as the *principle of equivalence*, underlies Einstein's general theory of relativity and all other theories of gravitation. We summarize the principle of equivalence as follows: there is no way to distinguish locally between a uniform gravitational acceleration $\mathbf{g}$ and an acceleration of the coordinate system $\mathbf{A} = -\mathbf{g}$.

By saying that there is no way to distinguish locally, we mean that there is no way to distinguish from within a sufficiently confined system. The reason that Einstein put his observer in an elevator was to define such an enclosed system. For instance, if you are in an elevator and observe that free objects accelerate toward the floor at rate a, there are two possible explanations:

1. There is a downward gravitational field g, and the elevator is at rest (or moving uniformly) in the field.
2. There is no gravitational field, but the elevator is accelerating up at rate $a = g$.

To distinguish between these alternatives, you must look out of the elevator. Suppose, for instance, that you see an apple suddenly drop from

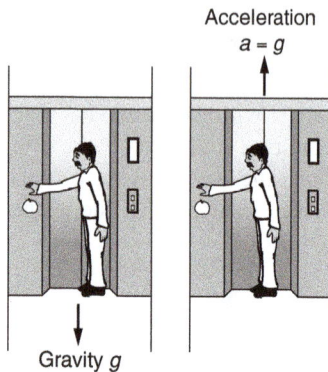

Acceleration
$a = g$

Gravity g

a nearby tree and fall down with acceleration a. The most likely explanation is that you and the tree are at rest in a downward gravitational field of magnitude $g = a$. However, it is conceivable that your elevator and the tree are both at rest on a giant elevator which is accelerating up at rate a.

To choose between these alternatives you must look farther off. If you see that you have an upward acceleration a relative to the fixed stars, that is, if the stars appear to accelerate down at rate a, the only possible explanation is that you are in a non-inertial system; your elevator and the tree are actually accelerating up. The alternative is the impossible conclusion that you are at rest in a gravitational field that extends uniformly *through all of space*. But such fields do not exist; real forces arise from interactions between real bodies, and for sufficiently large separations the forces always decrease. Hence it is most unphysical to invoke a uniform gravitational field extending throughout space.

This, then, is the difference between a gravitational field and an accelerating coordinate system. Real fields are local; at large distances they decrease. An accelerating coordinate system is non-local; the acceleration extends uniformly throughout space. Only for small systems are the two indistinguishable.

Although these ideas may sound somewhat abstract, the next two examples show that they have direct physical consequences.

Example 9.4 The Driving Force of the Tides

The Earth is in free fall toward the Sun, and according to the principle of equivalence it should be impossible to observe the Sun's gravitational force in an earthbound system. However, the equivalence principle applies only to local systems. The Earth is so large that appreciable non-local effects like the tides can be observed. In this example we discuss the origin of the tides to see what is meant by a non-local effect.

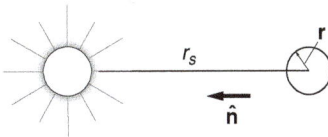

The tides arise because the Sun and the Moon produce an apparent gravitational field that varies from point to point on the Earth's surface. Although the Moon's effect is larger than the Sun's, we shall consider only the Sun for purposes of illustration.

The gravitational field of the Sun at the center of the Earth is

$$\mathbf{G}_0 = \frac{GM_s}{r_s^2}\hat{\mathbf{n}},$$

where M_s is the Sun's mass, r_s is the distance between the center of the Sun and the center of the Earth, and $\hat{\mathbf{n}}$ is the unit vector from the Earth toward the Sun. The Earth accelerates toward the Sun at rate $\mathbf{A} = \mathbf{G}_0$.

If $\mathbf{G}(\mathbf{r})$ is the gravitational field of the Sun at some point $\mathbf{r}$ on the Earth, where the origin of $\mathbf{r}$ is the center of the Earth, then the force on mass

m at $\mathbf{r}$ is

$$\mathbf{F} = m\mathbf{G}(\mathbf{r}).$$

To an earthbound observer, the apparent force is

$$\mathbf{F}' = \mathbf{F} - m\mathbf{A} = m\mathbf{G}(\mathbf{r}) - m\mathbf{G}_0.$$

The apparent field is

$$\mathbf{G}'(\mathbf{r}) = \frac{\mathbf{F}'}{m}$$
$$= \mathbf{G}(\mathbf{r}) - \mathbf{G}_0.$$

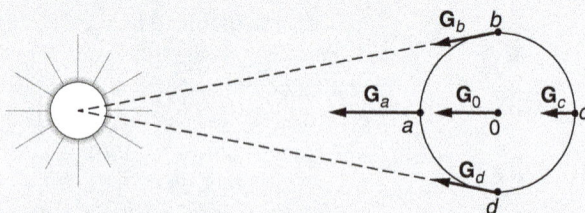

The drawing shows the true field $\mathbf{G}(\mathbf{r})$ at different points on the Earth's surface. (The variations are exaggerated.) $\mathbf{G}_a$ is larger than $\mathbf{G}_0$ because a is closer to the Sun than is the center of the Earth. Similarly, $\mathbf{G}_c$ is less than $\mathbf{G}_0$. The magnitudes of $\mathbf{G}_b$ and $\mathbf{G}_d$ are approximately the same as the magnitude of $\mathbf{G}_0$, but their directions are slightly different.

The sketch shows the apparent field $\mathbf{G}' = \mathbf{G} - \mathbf{G}_0$.

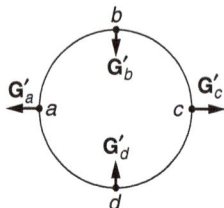

We now evaluate $\mathbf{G}'$ at each of the points indicated.

1. $\mathbf{G}'_a$ and $\mathbf{G}'_c$

The distance from a to the center of the Sun is $r_s - R_e$ where R_e is the Earth's radius. The magnitude of the Sun's field at a is

$$G_a = \frac{GM_s}{(r_s - R_e)^2}.$$

$\mathbf{G}_a$ is parallel to $\mathbf{G}_0$. The magnitude of the apparent field at a is

$$G'_a = G_a - G_0$$
$$= \frac{GM_s}{(r_s - R_e)^2} - \frac{GM_s}{r_s^2}$$
$$= \frac{GM_s}{r_s^2}\left[\frac{1}{[1 - (R_e/r_s)]^2} - 1\right]$$
$$= G_0\left[\frac{1}{[1 - (R_e/r_s)]^2} - 1\right].$$

Since $R_e/r_s = 6.4 \times 10^3 \text{ km}/1.5 \times 10^8 \text{ km} = 4.3 \times 10^{-5} \ll 1$, we have

$$G'_a = G_0 \left[\left(1 - \frac{R_e}{r_s} \right)^{-2} - 1 \right]$$

$$= G_0 \left[1 + 2\frac{R_e}{r_s} + \cdots - 1 \right]$$

$$\approx 2G_0 \frac{R_e}{r_s},$$

where we have neglected terms of order $(R_e/r_s)^2$ and higher.

The analysis at c is similar, except that the distance to the Sun is $r_s + R_e$ instead of $r_s - R_e$. We obtain

$$G'_c = -2G_0 \frac{R_e}{r_s}.$$

Note that $\mathbf{G}'_a$ and $\mathbf{G}'_c$ point radially out from the Earth.

2. $\mathbf{G}'_b$ and $\mathbf{G}'_d$

Points b and d are, to excellent approximation, the same distance from the Sun as is the center of the Earth. However, $\mathbf{G}_b$ is not parallel to $\mathbf{G}_0$; the angle between them is $\alpha \approx R_e/r_s = 4.3 \times 10^{-5} \ll 1$.

To this approximation

$$G'_b \approx G_0 \alpha$$

$$\approx G_0 \frac{R_e}{r_s}.$$

By symmetry, $\mathbf{G}'_d$ is equal and opposite to $\mathbf{G}'_b$. Both $\mathbf{G}'_b$ and $\mathbf{G}'_d$ point toward the center of the Earth.

The sketch shows $\mathbf{G}'(\mathbf{r})$ at various points on the Earth's surface. This diagram is the starting point for analyzing the tides. The forces at a and c tend to lift the oceans, and the forces at b and d tend to depress them. If the Earth were uniformly covered with water, the tangential force components would cause the two tidal bulges to sweep around the globe as the Earth rotates. This picture explains the twice daily ebb and flood of the tides, but the actual motions depend in a complicated way on the response of the oceans as the Earth rotates, and on the local topography.

We can estimate the magnitude of tidal effects quite easily, as the next example shows.

Example 9.5 Equilibrium Height of the Tides
The following argument is based on a model devised by Newton. Pretend that two wells full of water run from the surface of the Earth to the center, where they join. One is along the Earth–Sun axis and the

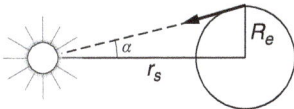

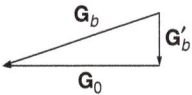

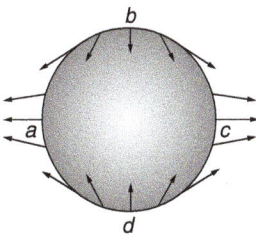

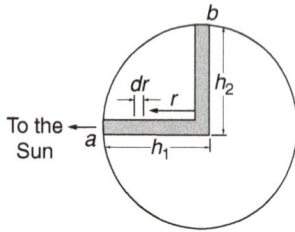

other is perpendicular. For equilibrium, the pressures at the bottom of the wells must be identical.

The pressure due to a short column of water of height dr is $\rho g(r)dr$, where ρ is the density of water and $g(r)$ is the effective gravitational field at r. The condition for equilibrium is

$$\int_0^{h_1} \rho g_1(r)dr = \int_0^{h_2} \rho g_2(r)dr.$$

h_1 and h_2 are the distances from the center of the Earth to the surface of the respective water columns. If we assume that the water is incompressible, then ρ is constant and the equilibrium condition becomes

$$\int_0^{h_1} g_1(r)dr = \int_0^{h_2} g_2(r)dr.$$

The problem is to calculate the difference $h_1 - h_2 = \Delta h_s$, the height of the tide due to the Sun. We assume that the Earth is spherical and neglect effects due to its rotation.

The effective field toward the center of the Earth along column 1 is $g_1(r) = g(r) - G_1'(r)$, where $g(r)$ is the gravitational field of the Earth and $G_1'(r)$ is the effective field of the Sun along column 1. (The negative sign indicates that $G_1'(r)$ is directed radially outward.) In Example 9.4 we evaluated $G_1'(R_e) = G_a' = 2GM_sR_e/r_s^3$. The effective field along column 1 is obtained by substituting r for R_e:

$$G_1'(r) = \frac{2GM_sr}{r_s^3}$$

$$= 2Cr,$$

where $C = GM_s/r_s^3$.

Combining results gives

$$g_1(r) = g(r) - 2Cr.$$

The same reasoning gives

$$g_2(r) = g(r) + G_2'(r)$$

$$= g(r) + Cr.$$

The condition for equilibrium is

$$\int_0^{h_1} [g(r) - 2Cr]dr = \int_0^{h_2} [g(r) + Cr]dr,$$

or, rearranging,

$$\int_0^{h_1} g(r)dr - \int_0^{h_2} g(r)dr = \int_0^{h_1} 2Cr\, dr + \int_0^{h_2} Cr\, dr.$$

Combining the integrals on the left-hand side gives $\int_{h_2}^{h_1} g(r)dr$.

Since h_1 and h_2 are both nearly equal to the Earth's radius, $g(r)$ can be taken as constant in the integral. $g(r) = g(R_e) = g$, the acceleration due to gravity at the Earth's surface. The integrals on the left therefore contribute $g(h_1 - h_2) = g\Delta h_s$. The integrals on the right can be combined by taking $h_1 \approx h_2 \approx R_e$, and they yield $\int_0^{R_e} 3Cr\, dr = \frac{3}{2}CR_e^2$. The final result is

$$g\Delta h_s = \frac{3}{2}CR_e^2.$$

Using $g = GM_e/R_e^2$ and $C = GM_s/r_s^3$, we find

$$\Delta h_s = \frac{3}{2}\frac{M_s}{M_e}\left(\frac{R_e}{r_s}\right)^3 R_e.$$

Using the following data

$$M_s = 1.99 \times 10^{33}\text{ g} \qquad M_e = 5.98 \times 10^{27}\text{ g}$$
$$r_s = 1.49 \times 10^{13}\text{ cm} \qquad R_e = 6.37 \times 10^8\text{ cm,}$$

we obtain

$$\Delta h_s = 24.0\text{ cm.}$$

An identical argument for the Moon gives

$$\Delta h_m = \frac{3}{2}\frac{M_m}{M_e}\left(\frac{R_e}{r_m}\right)^3 R_e.$$

Inserting $M_m = 7.34 \times 10^{25}$ g, $r_m = 3.84 \times 10^{10}$ cm, we obtain $\Delta h_m = 53.5$ cm. We see that the Moon's effect is about twice as large as the Sun's, even though the Sun's gravitational field at the Earth is about 200 times stronger than the Moon's. The reason is that the tidal force depends both on the mass of the Sun or Moon and on the gradient of the gravitational field, which varies as $1/r^3$. Although the Sun is far more massive than the Moon, the Moon's closeness to the Earth means that its gravitational field has much greater variation over the Earth.

The strongest tides, called the spring tides, occur at the new and full moon when the Moon and Sun act along the same line. The weak neap tides occur midway between, at the quarters of the Moon. The ratio of the driving forces in these two cases is

$$\frac{\Delta h_{\text{spring}}}{\Delta h_{\text{neap}}} = \frac{\Delta h_m + \Delta h_s}{\Delta h_m - \Delta h_s} \approx 3.$$

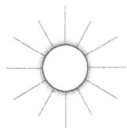

Earth not accelerating

The tides offer convincing evidence that the Earth is in free fall toward the Sun. If the Earth were attracted by the Sun but not in free fall, there would be only a single tide, whereas free fall results in two tides a day, as the sketches illustrate. The fact that we can sense the Sun's gravitational field from a body in free fall does not contradict the principle of equivalence. The height of the tide depends on the ratio of the Earth's radius to

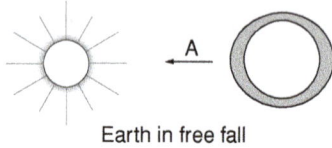

Earth in free fall

the Sun's distance, R_e/r_s. However, for a system to be local with respect to a gravitational field, the variation of the field must be negligible over the dimensions of the system. The Earth would be a local system if R_e were negligible compared to r_s, but then there would be no tides. The tides demonstrate that the Earth is too large to constitute a local system in the Sun's field.

There have been a number of experimental investigations of the principle of equivalence, because in spite of its apparent simplicity, far-reaching conclusions follow from it. For example, the principle of equivalence demands that gravitational force be strictly proportional to inertial mass. An alternative statement is that the ratio of gravitational mass to inertial mass must be the same for all matter, where the gravitational mass is the mass that appears in the expression for gravitational force and the inertial mass is the mass that appears in Newton's second law.

If an object with gravitational mass M_{gr} and inertial mass M_{in} interacts with an object of gravitational mass M_0, we have

$$\mathbf{F} = -\frac{GM_0 M_{gr}\hat{\mathbf{r}}}{r^2}.$$

The acceleration is $\mathbf{F}/M_{in}$, so that

$$\mathbf{a} = -\frac{GM_0}{r^2}\left(\frac{M_{gr}}{M_{in}}\right)\hat{\mathbf{r}}. \tag{9.4}$$

The equivalence principle requires M_{gr}/M_{in} to be the same for all objects because otherwise it would be possible to distinguish locally between a gravitational field and an acceleration. For instance, suppose that for object A, M_{gr}/M_{in} is twice as large as for object B. If we release both objects in an Einstein elevator and they fall with the same acceleration, the only possible conclusion is that the elevator is actually accelerating up. On the other hand, if A falls with twice the acceleration of B, we know that the acceleration must be due to a gravitational field. The upward acceleration of the elevator would be distinguishable from a downward gravitational field, contrary to the principle of equivalence.

The ratio M_{gr}/M_{in} is taken to be 1 in Newton's law of gravitation. Any other choice for the ratio would be reflected in a different value for G, because according to experiment the only requirement is that $G(M_{gr}/M_{in}) = 6.67 \times 10^{-11}$ N $\cdot$ m^2/kg^2.

Newton investigated the equivalence of inertial and gravitational mass by studying the period of a pendulum with interchangeable bobs made of different materials. The equation of motion for the bob in the small-angle approximation is

$$M_{in}l\ddot{\theta} + M_{gr}g\theta = 0.$$

The period of the pendulum is

$$T = \frac{2\pi}{\omega}$$

$$= 2\pi\sqrt{\frac{l}{g}}\sqrt{\frac{M_{in}}{M_{gr}}}.$$

Newton's experiment consisted of looking for a variation in T using different bobs. He found no such change and, from an estimate of the sensitivity of the method, concluded that M_{gr}/M_{in} is constant to better than one part in a thousand for common materials.

The most compelling evidence for the principle of equivalence comes from an experiment devised by the Hungarian physicist Baron Roland von Eötvös at the turn of the twentieth century. (The experiments were completed in 1908 but the results were not published until 1922, three years after von Eötvös' death.) The method and techniques of von Eötvös' experiment were refined by Robert Dicke and his collaborators at Princeton University, and more recently by Eric Adelberger and his colleagues at the University of Washington.

Consider a torsion balance consisting of two masses A and B of different composition at each end of a bar that hangs from a thin elastic fiber capable of producing a restoring torque when twisted. The bar can rotate only about the vertical axis. The masses are attracted by the Earth and also by the Sun. The gravitational force due to the Earth is vertical and causes no rotation of the balance, but as we now show, the Sun's attraction will cause a torque if the principle of equivalence is violated.

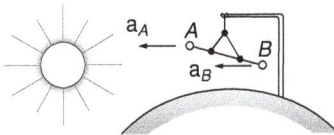

Assume that the Sun is on the horizon, as shown in the sketch, and that the horizontal bar is perpendicular to the Sun–Earth axis. According to Eq. (9.4) the accelerations of the masses due to the Sun are

$$a_A = \frac{GM_s}{r_s^2}\left[\frac{M_{gr}(A)}{M_{in}(A)}\right]$$

$$a_B = \frac{GM_s}{r_s^2}\left[\frac{M_{gr}(B)}{M_{in}(B)}\right],$$

where M_s is the gravitational mass of the Sun, and r_s is the distance between Sun and Earth. The acceleration of the masses in a coordinate system fixed to the Earth are

$$a'_A = a_A - a_0$$

$$a'_B = a_B - a_0,$$

where a_0 is the acceleration of the Earth toward the Sun. (Acceleration due to the rotation of the Earth plays no role and we neglect it.)

If the principle of equivalence is obeyed, $a'_A = a'_B$ and the bar has no tendency to rotate about the fiber. However, if the two masses A and B have different ratios of gravitational to inertial mass, one will accelerate more than the other. The balance will rotate until the restoring torque of the suspension fiber brings it to rest. As the Earth rotates, the apparent direction of the Sun changes; the equilibrium position of the balance would move with a 24 hour period.

Adelberger's apparatus was capable of detecting the deflection caused by a variation of 1 part in 10^{12} in the ratio of gravitational to inertial mass, but no effect was found to this accuracy.

The principle of equivalence is generally regarded as a fundamental law of physics. We have used it to discuss the ratio of gravitational to inertial mass. Surprisingly enough, it can also be used to show that clocks run at different rates in different gravitational fields. A simple argument showing how the principle of equivalence forces us to give up the classical notion of time is presented in Note 9.1.

9.5 Physics in a Rotating Coordinate System

In the previous section we treated physics in a linearly accelerating system. In this section we turn to the more complicated problem of physics in a rotating coordinate system.

But first, a few words on why fictitious forces are useful. We saw in Example 9.1 that physical results are the same whether we use an inertial system or a non-inertial system. In the inertial system, Newton's laws hold without modification. In the non-inertial linearly accelerating system, we added a non-physical fictitious force $-M\mathbf{A}$. Including the fictitious force allowed us to treat the problem just like a problem in an inertial system.

If we tried to treat motion in a rotating coordinate system from the standpoint of an inertial frame, we could easily get bogged down in geometry. We shall see in this section that by adding two fictitious forces, the *centrifugal force* and the *Coriolis force*, we can treat motion in a rotating coordinate system as if we were in an inertial system. The fictitious forces systematically account for the difference between the rotating non-inertial system and an inertial system.

The surface of the Earth provides an excellent example of a rotating coordinate system, and using fictitious forces, we will be able to explain observations on the Earth, for example the precession of the Foucault pendulum and the circular nature of weather systems.

To analyze motion in a rotating coordinate system, we need an equation that relates motion in inertial and rotating systems. Our approach will be to find a general rule for calculating the time derivative of any vector in a coordinate system that is rotating with respect to an inertial system, and then to apply this to relate velocity and acceleration in the two systems.

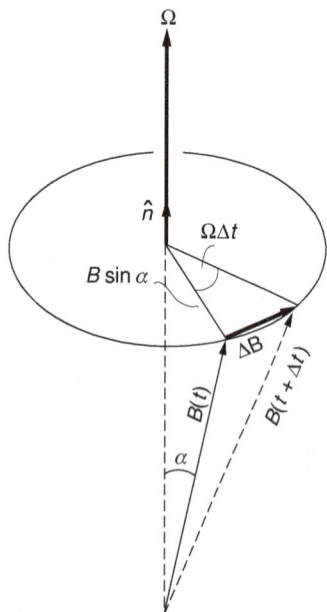

9.5.1 Rate of Change of a Rotating Vector

Consider any vector $\mathbf{B}$ that is rotating at rate Ω about an axis lying in direction $\hat{\mathbf{n}}$. The rotational velocity vector is $\mathbf{\Omega} = \Omega\hat{\mathbf{n}}$. Let α be the angle between $\mathbf{B}$ and $\mathbf{\Omega}$.

The sketch shows $\mathbf{B}(t)$ and $\mathbf{B}(t + \Delta t)$.

The tip of $\mathbf{B}$ sweeps out a circular path with radius $B\sin\alpha$ and the plane of the circle is perpendicular to $\hat{\mathbf{n}}$. In time Δt the tip of $\mathbf{B}$ swings through an angle $\Omega\Delta t$. Consequently, the tip of $\mathbf{B}$ moves by ΔB:

$$\Delta B \approx (B\sin\alpha)\,\Omega\Delta t$$

so that

$$\frac{dB}{dt} = \lim_{\Delta t \to 0} \frac{(B \sin \alpha)\Omega \Delta t}{\Delta t}$$
$$= (B \sin \alpha)\Omega.$$

Note that $(B \sin \alpha) \, \Omega = |\mathbf{\Omega} \times \mathbf{B}|$ and that $\Delta \mathbf{B}$ is perpendicular to $\mathbf{B}$ and $\hat{\mathbf{n}}$. The cross product therefore perfectly describes $d\mathbf{B}/dt$:

$$\frac{d\mathbf{B}}{dt} = \mathbf{\Omega} \times \mathbf{B}. \tag{9.5}$$

This result holds true for *any* vector that undergoes pure rotation around axis $\hat{\mathbf{n}}$ at rate Ω.

9.5.2 Time Derivative of a Vector in a Rotating Coordinate System

Consider any vector $\mathbf{C}$ that is changing at rate $(d\mathbf{C}/dt)_{\text{in}}$ as observed in an inertial system. The problem is to find the time derivative $(d\mathbf{C}/dt)_{\text{rot}}$ as observed in a system rotating at rate $\mathbf{\Omega}$.

Let the base vectors be $(\hat{\mathbf{i}}, \hat{\mathbf{j}}, \hat{\mathbf{k}})$ in the inertial system and $(\hat{\mathbf{i}}', \hat{\mathbf{j}}', \hat{\mathbf{k}}')$ in the rotating system. In the inertial system, $\mathbf{C}$ and its time derivative are

$$\mathbf{C} = C_x \hat{\mathbf{i}} + C_y \hat{\mathbf{j}} + C_x \hat{\mathbf{k}}$$
$$\left(\frac{d\mathbf{C}}{dt}\right)_{\text{in}} = \left(\frac{dC_x}{dt}\hat{\mathbf{i}} + \frac{dC_y}{dt}\hat{\mathbf{j}} + \frac{dC_z}{dt}\hat{\mathbf{k}}\right).$$

Let the components of $\mathbf{C}$ in the rotating system be

$$\mathbf{C} = C_x' \hat{\mathbf{i}}' + C_y' \hat{\mathbf{j}}' + C_z' \hat{\mathbf{k}}'.$$

Keep in mind that a vector is physically measurable, so its magnitude and direction remain the same regardless of what coordinate system we choose to assign its components. In calculating the time derivative of $(d\mathbf{C}/dt)_{\text{rot}}$ we must take into account that the base vectors in the rotating system all rotate with angular velocity $\mathbf{\Omega}$, so from Eq. (9.5)

$$\frac{d\hat{\mathbf{i}}'}{dt} = \mathbf{\Omega} \times \hat{\mathbf{i}}'$$
$$\frac{d\hat{\mathbf{j}}'}{dt} = \mathbf{\Omega} \times \hat{\mathbf{j}}' \tag{9.6}$$
$$\frac{d\hat{\mathbf{k}}'}{dt} = \mathbf{\Omega} \times \hat{\mathbf{k}}'.$$

We have

$$\frac{d\mathbf{C}}{dt} = \left(\frac{dC_x'}{dt}\hat{\mathbf{i}}' + \frac{dC_y'}{dt}\hat{\mathbf{j}}' + \frac{dC_z'}{dt}\hat{\mathbf{k}}'\right) + \left(C_x'\frac{d\hat{\mathbf{i}}'}{dt} + C_y'\frac{d\hat{\mathbf{j}}'}{dt} + C_z'\frac{d\hat{\mathbf{k}}'}{dt}\right). \tag{9.7}$$

The first term on the right is $(d\mathbf{C}/dt)_{\text{rot}}$, the time derivative of $\mathbf{C}$ that would be measured by an observer in the rotating system. Now introduce

Eqs. (9.6) into the second term on the right in Eq. (9.7):

$$\left(C'_x\frac{d\hat{\mathbf{i}}'}{dt} + C'_y\frac{d\hat{\mathbf{j}}'}{dt} + C'_z\frac{d\hat{\mathbf{k}}'}{dt}\right) = \left(C'_x\boldsymbol{\Omega}\times\hat{\mathbf{i}}' + C'_y\boldsymbol{\Omega}\times\hat{\mathbf{j}}' + C'_z\boldsymbol{\Omega}\times\hat{\mathbf{k}}'\right)$$

$$= \boldsymbol{\Omega}\times\left(C'_x\hat{\mathbf{i}}' + C'_y\hat{\mathbf{j}}' + C'_z\hat{\mathbf{k}}'\right)$$

$$= \boldsymbol{\Omega}\times\mathbf{C}.$$

Combining results we have

$$\left(\frac{d\mathbf{C}}{dt}\right)_{\text{in}} = \left(\frac{d\mathbf{C}}{dt}\right)_{\text{rot}} + \boldsymbol{\Omega}\times\mathbf{C}. \qquad (9.8)$$

This equation holds for *any* vector. It is most easily remembered as an operator, analogous to the del operator in Note 5.2. The rule is

$$\left(\frac{d}{dt}\right)_{\text{in}} = \left(\frac{d}{dt}\right)_{\text{rot}} + \boldsymbol{\Omega}\times\underline{}$$

We now use this transformation to find expressions for velocity and acceleration in a rotating system.

9.5.3 Velocity and Acceleration in a Rotating Coordinate System

Applying Eq. (9.8) to the position vector $\mathbf{r}$, we have

$$\left(\frac{d\mathbf{r}}{dt}\right)_{\text{in}} = \left(\frac{d\mathbf{r}}{dt}\right)_{\text{rot}} + \boldsymbol{\Omega}\times\mathbf{r},$$

or

$$\mathbf{v}_{\text{in}} = \mathbf{v}_{\text{rot}} + \boldsymbol{\Omega}\times\mathbf{r}.$$

We can apply Eq. (9.8) once again to find the acceleration $\mathbf{a}_{\text{rot}}$ in the rotating coordinate system. We have

$$\left(\frac{d\mathbf{v}_{\text{in}}}{dt}\right)_{\text{in}} = \left(\frac{d\mathbf{v}_{\text{in}}}{dt}\right)_{\text{rot}} + \boldsymbol{\Omega}\times\mathbf{v}_{\text{in}}$$

$$= \left[\frac{d}{dt}(\mathbf{v}_{\text{rot}} + \boldsymbol{\Omega}\times\mathbf{r})\right]_{\text{rot}} + \boldsymbol{\Omega}\times(\mathbf{v}_{\text{rot}} + \boldsymbol{\Omega}\times\mathbf{r})$$

$$= \left(\frac{d\mathbf{v}_{\text{rot}}}{dt}\right)_{\text{rot}} + \boldsymbol{\Omega}\times\left(\frac{d\mathbf{r}}{dt}\right)_{\text{rot}} + \boldsymbol{\Omega}\times(\mathbf{v}_{\text{rot}} + \boldsymbol{\Omega}\times\mathbf{r})$$

$$= \left(\frac{d\mathbf{v}_{\text{rot}}}{dt}\right)_{\text{rot}} + 2\,\boldsymbol{\Omega}\times\mathbf{v}_{\text{rot}} + \boldsymbol{\Omega}\times(\boldsymbol{\Omega}\times\mathbf{r}).$$

Expressing this in terms of the accelerations $\mathbf{a}_{\text{in}}$ and $\mathbf{a}_{\text{rot}}$, we have

$$\mathbf{a}_{\text{in}} = \mathbf{a}_{\text{rot}} + 2\,\boldsymbol{\Omega}\times\mathbf{v}_{\text{rot}} + \boldsymbol{\Omega}\times(\boldsymbol{\Omega}\times\mathbf{r}). \qquad (9.9)$$

The acceleration viewed in the rotating system is

$$\mathbf{a}_{\text{rot}} = \mathbf{a}_{\text{in}} - 2\,\boldsymbol{\Omega}\times\mathbf{v}_{\text{rot}} - \boldsymbol{\Omega}\times(\boldsymbol{\Omega}\times\mathbf{r}). \qquad (9.10)$$

9.5.4 Fictitious Forces in a Rotating Coordinate System

As viewed in a rotating coordinate system, the equation of motion for mass m is

$$\mathbf{F}_{\text{rot}} = m\,\mathbf{a}_{\text{rot}}.$$

From Eq. (9.10) we have

$$\mathbf{F}_{\text{rot}} = m\,\mathbf{a}_{\text{in}} - 2\,m\,\boldsymbol{\Omega} \times \mathbf{v}_{\text{rot}} - m\,\boldsymbol{\Omega} \times (\boldsymbol{\Omega} \times \mathbf{r}), \qquad (9.11)$$

which we can write as

$$\begin{aligned} \mathbf{F}_{\text{rot}} &= \mathbf{F} + \mathbf{F}_{\text{Coriolis}} + \mathbf{F}_{\text{centrifugal}} \\ &= \mathbf{F} + \mathbf{F}_{\text{fict}} \end{aligned}$$

where $\mathbf{F}$ is the real force and

$$\mathbf{F}_{\text{centrifugal}} = -m\,\boldsymbol{\Omega} \times (\boldsymbol{\Omega} \times \mathbf{r}),$$
$$\mathbf{F}_{\text{Coriolis}} = -2\,m\,\boldsymbol{\Omega} \times \mathbf{v}_{\text{rot}}$$

are fictitious forces. The centrifugal force is easily experienced in circular motion when we twirl a mass on a string. The mass is at rest in the rotating system, so its equation of motion is $F_{\text{centrifugal}} - T = 0$, where T is the tension in the string. The string exerts a pull T on our hand, so we feel that the mass is pulling outward.

The centrifugal force $\mathbf{F}_{\text{centrifugal}} = -m\,\boldsymbol{\Omega} \times (\boldsymbol{\Omega} \times \mathbf{r})$ is perpendicular to the axis of rotation and is directed radially outward. Its magnitude is $m\Omega^2\rho$, where ρ is the perpendicular distance from the axis of rotation to the tip of $\mathbf{r}$. As the sketch shows, $(\boldsymbol{\Omega} \times (\boldsymbol{\Omega} \times \mathbf{r})$ is radially inward; this is the centripetal acceleration, arising because every point at rest in the rotating system is moving in a circular path in the inertial system. The outward fictitious centrifugal force is opposite to the inward centripetal acceleration.

The Coriolis force $\mathbf{F}_{\text{Coriolis}}$ is the fictitious force required to balance the real force that provides the Coriolis acceleration $2\boldsymbol{\Omega} \times \mathbf{v}_{\text{rot}}$. Here is how the Coriolis acceleration arises. The sketch shows $\mathbf{v}_{\text{rot}}$ resolved into vectors $\mathbf{v}_{\text{rot}\parallel}$ and $\mathbf{v}_{\text{rot}\perp}$, parallel and perpendicular to $\boldsymbol{\Omega}$, respectively. Only $\mathbf{v}_{\text{rot}\perp}$ contributes to the cross product. Hence the acceleration is in a plane perpendicular to $\boldsymbol{\Omega}$, so it is convenient to use plane polar coordinates ρ, θ to describe motion in this plane. As shown in the sketch, $\mathbf{v}_{\text{rot}\perp}$ has a radial component $\dot{\rho}$ and a tangential component $\rho\dot{\theta}'$.

The radial component $\dot{\rho}$ contributes $2\,\Omega\dot{\rho}$ in the tangential direction to $\mathbf{a}_{\text{in}}$. This tangential acceleration is simply the Coriolis term we found in Section 1.11.4 for motion in inertial space with angular velocity Ω and radial velocity $\dot{\rho}$.

The tangential component $\rho\dot{\theta}'$ of $\mathbf{v}_{\text{rot}\perp}$ contributes a radial component $2\,\Omega\rho\dot{\theta}'$ to the Coriolis acceleration, directed toward the rotation axis. To see the origin of this term, note that in inertial space the instantaneous

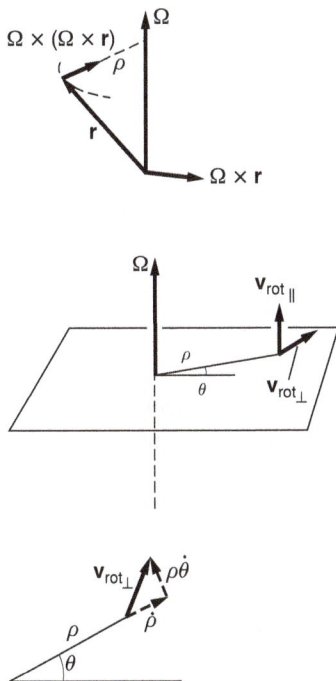

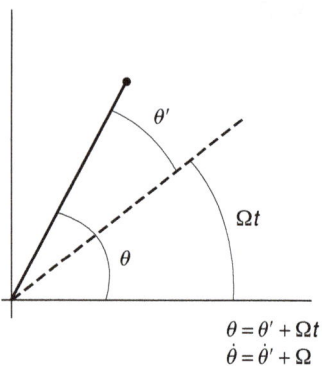

$$\theta = \theta' + \Omega t$$
$$\dot{\theta} = \dot{\theta}' + \Omega$$

angular velocity is $\dot{\theta} = \dot{\theta}' + \Omega$ and the centripetal acceleration term in $\mathbf{a}_{\text{in}}$ is

$$\rho\dot{\theta}^2 = \rho(\dot{\theta}' + \Omega)^2$$
$$= \rho\dot{\theta}'^2 + 2\,\Omega\rho\dot{\theta}' + \rho\Omega^2.$$

The three terms on the right correspond to the three terms on the right of Eq. (9.9). $\rho\dot{\theta}'^2$ is part of $\mathbf{a}_{\text{rot}}$, $2\,\Omega\rho\dot{\theta}'$ follows from $2\,\Omega \times \mathbf{v}_{\text{rot}}$ as we have shown, and $\rho\Omega^2$ comes from $\Omega \times (\Omega \times \mathbf{r})$.

The following examples illustrate the use of rotating coordinates.

Example 9.6 Surface of a Rotating Liquid

A bucket of water spins with angular speed ω. What shape does the water's surface assume? In a coordinate system rotating with the bucket, the problem is purely static. Consider the force on a small volume of water of mass m at the surface of the liquid. For equilibrium, the total force on m must be zero. The forces are the contact force $\mathbf{F}_0$, the weight $\mathbf{W}$, and the radial outward fictitious force $\mathbf{F}_{\text{fict}}$

$$F_0 \cos\phi - W = 0$$
$$-F_0 \sin\phi + F_{\text{fict}} = 0,$$

where $F_{\text{fict}} = m\,\Omega^2 r = m\omega^2 r$, since $\Omega = \omega$ for a coordinate system rotating with the bucket.

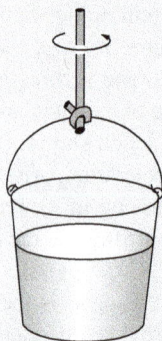
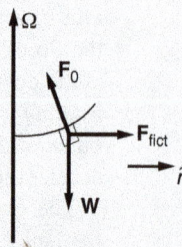
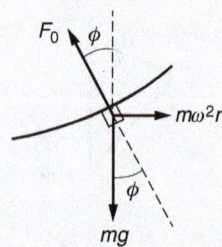

Solving these equations for ϕ yields

$$\phi = \arctan\left(\frac{\omega^2 r}{g}\right).$$

Unlike solids, liquids cannot exert a static force tangential to the surface. Hence $\mathbf{F}_0$, the force on m due to the neighboring liquid, must be perpendicular to the surface. The slope of the surface at any point is therefore

$$\frac{dz}{dr} = \tan\phi$$
$$= \frac{\omega^2 r}{g}.$$

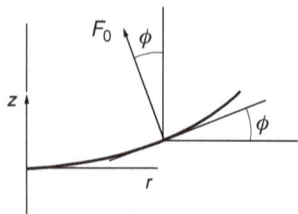

We can integrate to find the equation of the surface $z = f(r)$:

$$\int dz = \frac{\omega^2}{g} \int r \, dr$$

$$z = \frac{1}{2}\frac{\omega^2}{g}r^2,$$

where we have taken $z = 0$ on the axis at the surface of the liquid. The surface is a parabola of revolution.

Example 9.7 A Sliding Bead and the Coriolis Force

A bead slides without friction on a horizontal rigid wire rotating at constant angular speed ω. The problem is to find the force exerted on the bead by the wire. Neglect gravity.

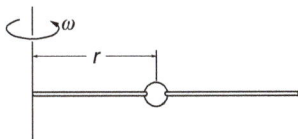

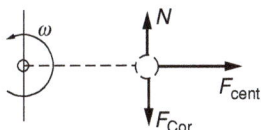

In a coordinate system rotating with the wire the motion is purely radial. The sketch shows the force diagram in the rotating system, as seen from above.

F_{cent} is the centrifugal force and F_{Cor} is the Coriolis force. Since the wire is frictionless, the contact force N is normal to the wire. In the rotating system the equations of motion are

$$F_{cent} = m\ddot{r}$$
$$N - F_{Cor} = 0.$$

Using $F_{cent} = m\omega^2 r$, the first equation gives

$$m\ddot{r} - m\omega^2 r = 0,$$

which has the solution

$$r = Ae^{\omega t} + Be^{-\omega t},$$

where A and B are constants depending on the initial conditions. The tangential equation of motion, which expresses the fact that there is no tangential acceleration in the rotating system, gives

$$N = F_{Cor} = 2m\dot{r}\omega$$
$$= 2m\omega^2(Ae^{\omega t} - Be^{-\omega t}).$$

To complete the problem, we must be given two independent initial conditions to specify A and B, typically $r(0)$ and $\dot{r}(0)$.

Example 9.8 Deflection of a Falling Mass

Because of the Coriolis force, falling objects on the Earth are deflected horizontally. For instance, a mass dropped from a tower lands to the east of a plumb line from the release point. In this example we shall calculate the deflection of a mass m dropped from a tower of height h at the Equator. In the coordinate system r, θ fixed to the Earth (with

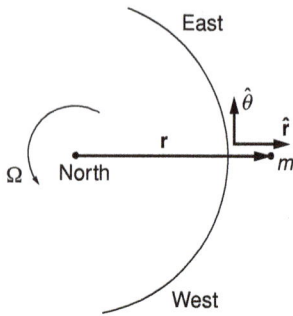

the tangential direction toward the east) the apparent force on m is

$$\mathbf{F} = -mg\,\hat{\mathbf{r}} - 2\,m\,\boldsymbol{\Omega} \times \mathbf{v}_{\mathrm{rot}} - m\,\boldsymbol{\Omega} \times (\boldsymbol{\Omega} \times \mathbf{r}).$$

The gravitational and centrifugal forces are radial, and if m is dropped from rest, the Coriolis force is in the equatorial plane (the plane defined by the great circle of the Equator). Thus the motion of m is confined to the equatorial plane, and we have

$$\mathbf{v}_{\mathrm{rot}} = \dot{r}\,\hat{\mathbf{r}} + r\dot{\theta}\,\hat{\boldsymbol{\theta}}.$$

Using $\boldsymbol{\Omega} \times \mathbf{v}_{\mathrm{rot}} = \Omega\dot{r}\,\hat{\boldsymbol{\theta}} - r\Omega\dot{\theta}\,\hat{\mathbf{r}}$ and $\boldsymbol{\Omega} \times (\boldsymbol{\Omega} \times \mathbf{r}) = \Omega^2 r\,\hat{\mathbf{r}}$, we obtain

$$F_r = -mg + 2\,m\,\Omega\dot{\theta}r + m\,\Omega^2 r,$$
$$F_\theta = -2\,m\,\dot{r}\Omega.$$

The radial equation of motion is

$$m\ddot{r} - mr\,\dot{\theta}^2 = -mg + 2\,m\,\Omega\dot{\theta}r + m\Omega^2 r.$$

To an excellent approximation, m falls vertically and $\dot{\theta} \ll \Omega$. We can therefore omit the terms $m\,r\,\dot{\theta}^2$ and $2\,m\,\Omega\dot{\theta}r$ in comparison with $m\,\Omega^2 r$. Thus

$$\ddot{r} \approx -g + \Omega^2 r. \tag{1}$$

The tangential equation of motion is

$$mr\ddot{\theta} + 2m\dot{r}\dot{\theta} = -2m\dot{r}\Omega.$$

To the same approximation $\dot{\theta} \ll \Omega$ we have

$$r\ddot{\theta} \approx -2\dot{r}\Omega. \tag{2}$$

During the fall, r changes only slightly, from $R_e + h$ to R_e, where R_e is the radius of the Earth, and we can take g to be constant and $r \approx R_e$. Equation (1) becomes

$$\ddot{r} = -g + \Omega^2 R_e$$
$$= -g',$$

where $g' = g - \Omega^2 R_e$ is the acceleration due to the gravitational force minus a centrifugal term. g' is the apparent acceleration due to gravity as observed on the Earth, and since this is customarily denoted by g, we shall drop the prime from here on. The solution of the radial equation of motion $\ddot{r} = -g$ is

$$\dot{r} = -gt$$
$$r = r_0 - \frac{1}{2}gt^2.$$

If we insert $\dot{r} = -gt$ in the tangential equation of motion, Eq. (2), we have

$$r\ddot{\theta} = 2gt\Omega$$

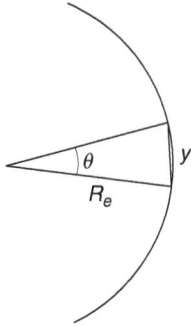

or

$$\ddot{\theta} = \frac{2g\Omega}{R_e}t,$$

where we have used $r \approx R_e$. Hence

$$\dot{\theta} = \frac{g\Omega}{R_e}t^2$$

and

$$\theta = \frac{1}{3}\frac{g\Omega}{R_e}t^3. \tag{3}$$

The horizontal deflection of m is $y \approx R_e\theta$ or

$$y = \frac{1}{3}g\Omega t^3.$$

The time T to fall distance h is given by

$$r - r_0 = -h$$
$$= -\frac{1}{2}gT^2$$

so that

$$T = \sqrt{\frac{2h}{g}} \quad \text{and} \quad y = \frac{1}{3}g\Omega\left(\frac{2h}{g}\right)^{\frac{3}{2}}.$$

$\Omega = 2\pi \, \text{rad/day} \approx 7.3 \times 10^{-5} \, \text{rad/s}$ and $g = 9.8 \, \text{m/s}^2$. For a tower 50 m high, $y \approx 7.7 \times 10^{-3} \, \text{m} = 0.77$ cm. θ is positive, and the deflection is toward the east.

Example 9.9 Motion on the Rotating Earth

A surprising effect of the Coriolis force is that it turns straight line motion on a rotating sphere into circular motion. Consider a velocity **v** tangent to a rotating sphere (like the velocity of wind over the Earth's surface). The horizontal component of the Coriolis force is perpendicular to **v** and its magnitude is independent of the direction of **v**, as we shall prove.

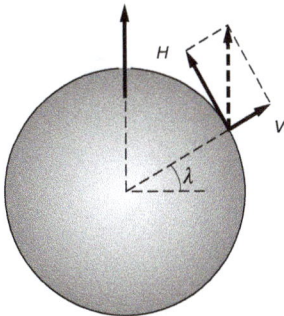

Consider a particle of mass m moving with velocity **v** at latitude λ on the surface of a sphere. The sphere is rotating with angular velocity $\mathbf{\Omega}$. If we decompose $\mathbf{\Omega}$ into a vertical vector $\mathbf{\Omega}_V$ and a horizontal vector $\mathbf{\Omega}_H$ as shown, the Coriolis force is

$$\mathbf{F} = -2m(\mathbf{\Omega} \times \mathbf{v})$$
$$= -2m(\mathbf{\Omega}_V \times \mathbf{v} + \mathbf{\Omega}_H \times \mathbf{v})$$
$$= \mathbf{F}_V + \mathbf{F}_H.$$

$\mathbf{\Omega}_H$ and **v** are horizontal, so that $\mathbf{\Omega}_H \times \mathbf{v}$ is vertical. Thus the horizontal Coriolis force $\mathbf{F}_H$ arises solely from the term $\mathbf{\Omega}_V \times \mathbf{v}$. $\mathbf{\Omega}_V$ is

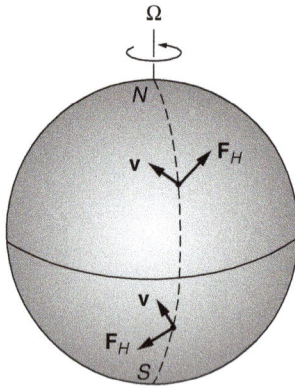

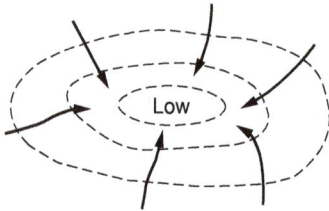

perpendicular to $\mathbf{v}$, and $\boldsymbol{\Omega}_V \times \mathbf{v}$ has magnitude $\Omega_V v$, independent of the direction of $\mathbf{v}$, as we wished to prove.

We can write the result in a more explicit form. If $\hat{\mathbf{r}}$ is a unit vector perpendicular to the surface at latitude λ, $\boldsymbol{\Omega}_V = \Omega \sin \lambda \, \hat{\mathbf{r}}$ and

$$\mathbf{F}_H = -2m\Omega \sin \lambda \, (\hat{\mathbf{r}} \times \mathbf{v}).$$

The magnitude of $\mathbf{F}_H$ is

$$F_H = 2mv\Omega \sin \lambda.$$

$\mathbf{F}_H$ is always perpendicular to $\mathbf{v}$, and in the absence of other horizontal forces it would produce circular motion, clockwise in the northern hemisphere and counterclockwise in the southern. Air flow on the Earth is strongly influenced by the Coriolis force and without it stable circular weather patterns could not form. However, to understand the dynamics of weather systems we must also include other forces, as the next example discusses.

Example 9.10   Weather Systems

Imagine that a region of low pressure occurs in the atmosphere, perhaps because of differential heating of the air. The closed curves in the sketch represent lines of constant pressure, or *isobars*. There is a force on each element of air due to the pressure gradient, and in the absence of other forces winds would blow inward, quickly equalizing the pressure difference.

However, the wind pattern is markedly altered by the Coriolis force. As the air begins to flow inward, it is deflected sideways by the Coriolis force, a fictitious force on the rotating Earth.

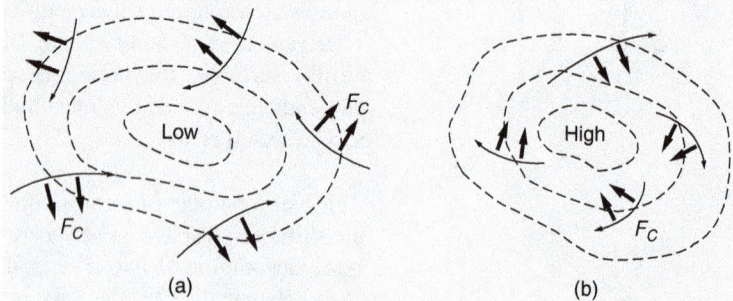

In the northern hemisphere the wind circulates counterclockwise along the isobars about regions of low pressure, (a), and clockwise about highs, (b). The directions of rotation are reversed in the southern hemisphere. The Coriolis force is nearly zero near the Equator (latitude $\approx 0°$), so circular weather systems cannot form there and the weather tends to be uniform.

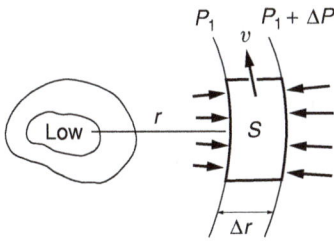

In order to analyze the motion, consider the forces on a parcel of air at latitude λ that is rotating about a low. The pressure force on the face along the isobar P_1 is $P_1 S$, where S is the area of the inner face, as shown in the sketch. The force on the outer face is $(P_1 + \Delta P)S$, and the net pressure force is $(\Delta P)S$ inward.

The Coriolis force is $2mv(\Omega \sin \lambda)$, where m is the mass of the parcel and v its velocity. The air is rotating counterclockwise about the low, so that the Coriolis force is outward. The radial equation of motion for steady circular flow is therefore

$$\frac{mv^2}{r} = (\Delta P)S - 2mv(\Omega \sin \lambda).$$

The volume of the parcel is $\Delta r S$, where Δr is the distance between the isobars, and the mass is $w\Delta r S$, where w is the density of air, assumed constant. Inserting this in the equation of motion and taking the limit $\Delta r \to 0$ yields

$$\frac{v^2}{r} = \frac{1}{w}\frac{dP}{dr} - 2v\Omega \sin \lambda. \tag{1}$$

Air masses do not rotate as rigid bodies. Near the center of the low, where the pressure gradient dP/dr is large, wind velocities are highest. Far from the center, v^2/r is small and can be neglected. Equation (1) predicts that far from the center the wind speed is

$$v = \left(\frac{1}{2\Omega \sin \lambda}\right)\frac{1}{w}\frac{dP}{dr}. \tag{2}$$

The density of air at sea level is 1.3 kg/m^3 and atmospheric pressure is $P_{at} = 10^5$ N/m^2. dP/dr can be estimated by looking at a weather map. Far from a high or low, a typical gradient is 3 millibars over 100 km $\approx 3 \times 10^{-3}$ N/m^3, and at latitude 45° Eq. (2) gives

$$v = 22 \text{ m/s}$$
$$\approx 50 \text{ mi/h.}$$

Near the ground this speed is reduced by friction with the land, but at higher altitudes Eq. (2) can be applied with good accuracy.

A hurricane is an intense compact low in which the pressure gradient can be as high as 30×10^{-3} N/m^3. Hurricane winds are so strong that the v^2/r term in Eq. (1) cannot be neglected. Solving Eq. (1) for v we find

$$v = \sqrt{(r\Omega \sin \lambda)^2 + \frac{r}{w}\frac{dP}{dr}} - r\Omega \sin \lambda. \tag{3}$$

At a distance 100 km from the center of a hurricane (the "eye") at latitude 20°, Eq. (3) predicts a wind speed of 45 m/s $\approx$ 100 mi/h for a pressure gradient of 30×10^{-3} N/m^3, in reasonable agreement with

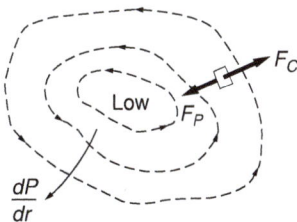

weather observations. At larger radii, the wind speed drops because the pressure gradient decreases.

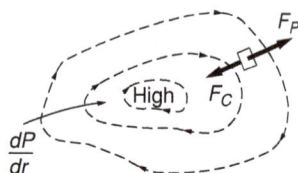

There is an interesting difference between lows and highs. In a low, the pressure force is inward (the pressure gradient dP/dr is outward), and the Coriolis force is outward. In a high, the pressure force is outward and the Coriolis force is inward.

The radial equation of motion for air circulating around a high is

$$\frac{v^2}{r} = 2v\Omega \sin \lambda - \frac{1}{w}\left|\frac{dP}{dr}\right|. \tag{4}$$

Solving Eq. (4) for v yields

$$v = r\Omega \sin \lambda - \sqrt{(r\Omega \sin \lambda)^2 - \frac{r}{w}\left|\frac{dP}{dr}\right|}. \tag{5}$$

We see from Eq. (5) that if $1/w|dP/dr| > r(\Omega \sin \lambda)^2$, the high cannot form; the Coriolis force is too weak to supply the needed centripetal acceleration against the large outward pressure force. For this reason, storms like hurricanes are always low pressure systems; the strong inward pressure force helps hold a low together.

The Foucault pendulum provides one of the most dramatic demonstrations that the Earth is a non-inertial system. The pendulum is simply a heavy bob hanging from a long wire mounted to swing freely in any direction. As the pendulum swings back and forth, the plane of motion precesses slowly about the vertical, taking about a day and a half for a complete rotation in the mid-latitudes. The precession is a result of the Earth's rotation. The plane of motion tends to stay fixed in inertial space while the Earth rotates beneath it.

In the 1850's Foucault hung a pendulum 67 m long from the dome of the Pantheon in Paris. The bob precessed almost a centimeter on each swing, and presented the first direct evidence that the Earth is indeed rotating. The pendulum became the rage of Paris.

The next example uses our analysis of the Coriolis force to calculate the motion of the Foucault pendulum in a simple way.

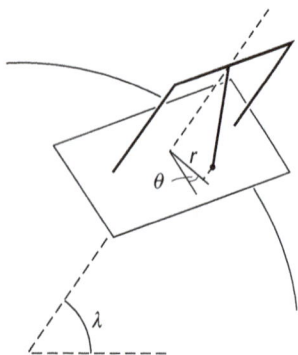

Example 9.11 The Foucault Pendulum

Consider a pendulum of mass m that is swinging with frequency $\gamma = \sqrt{g/l}$, where l is the length of the pendulum. If we describe the position of the pendulum's bob in the horizontal plane by coordinates r, θ, then

$$r = r_0 \sin \gamma t,$$

where r_0 is the amplitude of the motion. In the absence of the Coriolis force, there are no tangential forces and θ is constant. The horizontal

Coriolis force $\mathbf{F}_{CH}$ is

$$\mathbf{F}_{CH} = -2m(\Omega \sin \lambda)\dot{r}\hat{\theta}.$$

The tangential equation of motion $ma_\theta = F_{CH}$ becomes

$$m(r\ddot{\theta} + 2\dot{r}\dot{\theta}) = -2m(\Omega \sin \lambda)\dot{r}$$

or

$$r\ddot{\theta} + 2\dot{r}\dot{\theta} = -2(\Omega \sin \lambda)\dot{r}.$$

The simplest solution to this equation is found by taking $\dot{\theta} = $ constant. The term $r\ddot{\theta}$ then vanishes, and we have

$$\dot{\theta} = -\Omega \sin \lambda.$$

The pendulum precesses uniformly in a clockwise direction. The time for the plane of oscillation to rotate once is

$$T = \frac{2\pi}{\dot{\theta}}$$
$$= \frac{2\pi}{\Omega \sin \lambda}$$
$$= \frac{24\,\text{h}}{\sin \lambda}.$$

At the latitude of Paris, $\approx 49°$, the Foucault pendulum rotates once in 32 h.

At the North Pole (latitude $= 90°$) the period of precession is 24 h; the pendulum rotates clockwise with respect to the Earth at the same rate as the Earth rotates counterclockwise. The plane of motion remains fixed with respect to inertial space. What happens at the Equator?

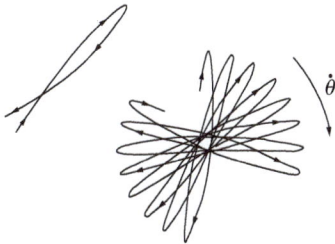

In addition to its dramatic display of the Earth's rotation, the Foucault pendulum embodies a profound mystery. Consider, for instance, a Foucault pendulum at the North Pole. The precession is obviously an artifact; the plane of motion stays fixed while the Earth rotates beneath it. The plane of the pendulum remains fixed relative to the fixed stars. Why should this be? How does the pendulum "know" that it must swing in a plane that is stationary relative to the fixed stars instead of, say, in a plane that rotates at some uniform rate?

Newton described this puzzling question in terms of the following experiment: if a bucket contains water at rest, the surface of the water is flat. If the bucket is set spinning at a steady rate, the water at first lags behind, but gradually, as the water's rotational speed increases, the surface takes on the form of the parabola of revolution discussed in Example 9.6. If the bucket is suddenly stopped, the concavity of the water's surface persists for some time. It is evidently not motion relative to the bucket that is important in determining the shape of the liquid surface. So long as the water rotates, the surface is depressed. Newton concluded that rotational motion is absolute, since by observing the water's surface it is possible to detect rotation without reference to outside objects.

From one point of view there is really no paradox to the absolute nature of rotational motion. The principle of Galilean invariance asserts that there is no way to detect locally the uniform translational motion of a system. However, this does not limit our ability to detect the *acceleration* of a system. A rotating system accelerates in a most non-uniform way. At every point the acceleration is directed toward the axis of rotation; the acceleration shows us the axis. Our ability to detect such an acceleration in no way contradicts Galilean invariance.

Nevertheless, there is an enigma. Both the rotating bucket and the Foucault pendulum maintain their motion *relative to the fixed stars*. How can the fixed stars determine an inertial system? What prevents the plane of the pendulum from rotating with respect to the fixed stars? Why is the surface of the water in the rotating bucket flat only when the bucket is at rest with respect to the fixed stars? Ernst Mach, who in 1883 wrote the first incisive critique of Newtonian physics, put the matter this way. Suppose that we keep a bucket of water fixed and rotate all the stars. Physically there is no way to distinguish this from the original case where the bucket is rotated, and we expect the surface of the water to again assume a parabolic shape. Apparently the motion of the water in the bucket depends on the motion of matter far off in the universe. To put it more dramatically, suppose that we eliminate the stars, one by one, until only our bucket remains. What will happen now if we rotate the bucket? There is no way for us to predict the motion of the water in the bucket—the inertial properties of space might be totally different. We have a most peculiar situation. The local properties of space depend on far-off matter, yet when we rotate the water, the surface *immediately* starts to deflect. There is no time for signals to travel to the distant stars and return. How does the water in the bucket "know" what the rest of the universe is doing?

The principle that the inertial properties of space depend on the existence of far-off matter is known as *Mach's principle*. The principle is accepted by many physicists, but it can lead to strange conclusions. For instance, there is no reason to believe that matter in the universe is uniformly distributed around the Earth; the solar system is located well out in the limb of our galaxy, and matter in our galaxy is concentrated predominantly in a very thin plane. If inertia is due to far-off matter, then we might well expect it to be different in different directions so that the value of mass would depend on the direction of acceleration. No such effects have ever been observed. Inertia remains a mystery.

Note 9.1 The Equivalence Principle and the Gravitational Red Shift

Radiating atoms emit light at only certain characteristic wavelengths. If light from atoms in the strong gravitational field of dense stars is analyzed spectroscopically, the characteristic wavelengths are observed to be slightly increased, shifted toward the red. We can visualize atoms as

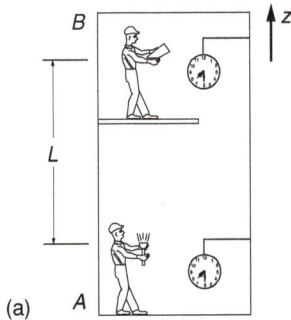

(a)

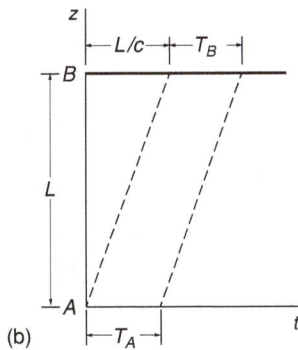

(b)

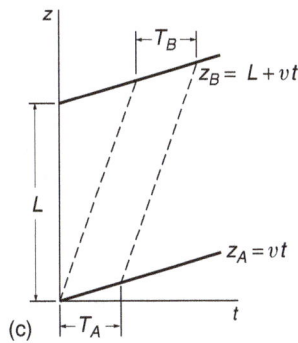

(c)

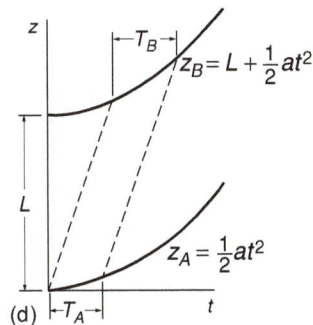

(d)

clocks that "tick" at characteristic frequencies. The shift toward longer wavelengths, known as the gravitational red shift, corresponds to a slowing of the clocks. The gravitational red shift implies that clocks in a gravitational field appear to run slow when viewed from outside the field. As we shall show, the origin of the effect lies in the nature of space, time, and gravity, not in the trivial effect of gravity on mechanical clocks.

The gravitational red shift is different from the Doppler shift due to the relative motion of a source and an observer. There is no relative motion in the gravitational red shift.

It is rather startling to see how the equivalence principle, which is so simple and non-mathematical, leads directly to a connection between space, time, and gravity. To show the connection we must use an elementary result from the theory of relativity; it is impossible to transmit information faster than the velocity of light, $c = 3 \times 10^8$ m/s. However, this is the only relativistic idea needed; our argument is otherwise completely classical.

Consider two scientists Alice and Bob separated by distance L as shown in sketch (a).

Alice sends out a short light pulse at time $t = 0$, and a second pulse shortly thereafter at time T_A. The signals are received by Bob, who uses his own clock to note the interval T_B between pulses. A plot of vertical distance versus time is shown for two light pulses in (b), when the two stations are at rest. The pulses are delayed by the transit time, L/c, but the interval T_B is the same as T_A.

Now consider the situation if both observers move upward uniformly with speed v, as shown in sketch (c). Although both scientists move during the time interval, they move equally, and we still have $T_B = T_A$.

The situation is entirely different if both observers are accelerating upward at uniform rate a as shown in sketch (d). Bob detects the first pulse at time T_1:

$$T_1 = \frac{L + \frac{1}{2}aT_1{}^2}{c}.$$

The numerator is the distance the light pulse travels. Similarly, Bob detects the second pulse at time T_2:

$$T_2 = T_A + \frac{L + \frac{1}{2}aT_2{}^2 - \frac{1}{2}aT_A{}^2}{c}.$$

Note that at time T_A Alice's light source is instantaneously at height $\frac{1}{2}aT_A{}^2$.

Solving these quadratic equations for T_1 and T_2 gives, after some algebra,

$$T_B = T_2 - T_1 = \frac{c}{a}\left[-\sqrt{1 - \frac{2a}{c^2}(L + cT_A - \frac{1}{2}aT_A{}^2)} + \sqrt{1 - \frac{2aL}{c^2}} \right].$$

In the first square root, we can neglect the term $\frac{1}{2}aT_A{}^2$, which is small compared to cT_A.

As is often the case in physics, the exact solution does not give us much insight. To obtain a more useful result, we use the binomial series (Note 1.3) to expand each square root, retaining terms through a^2. Also, we assume that the light pulses are close together, so that $cT_A \ll L$. The first non-vanishing terms are

$$\frac{\Delta T}{T} = \frac{T_B - T_A}{T_A} = \frac{aL}{c^2}.$$

Now, by the principle of equivalence, Alice and Bob cannot distinguish between their upward accelerating system and a system at rest in a downward gravitational field with magnitude $g = a$. If the experiment is repeated in a system at rest in a gravitational field, the equivalence principle therefore requires that $T_B > T_A$, as we just found, and Bob will conclude that Alice's clock is running slow. This is the origin of the gravitational red shift.

Einstein reported this result in 1911, using a derivation involving the Doppler shift. Our derivation in this note uses only kinematics, the principle of equivalence, and the finite speed of light.

On Earth the gravitational red shift is $\Delta T / T = 10^{-16} L$, where L is in meters. The effect, though small, was measured experimentally on the Earth by Pound, Rebka, and Snider at Harvard University. The "clock" was the frequency of a gamma-ray, and by using a technique known as Mössbauer absorption they were able to measure accurately the gravitational red shift due to a vertical displacement of 25 m. The accuracy of the Harvard measurement was 1%, but different experiments since then have confirmed the effect to an accuracy of better than 1 part in 10^8.

Problems

For problems marked *, refer to page 524 for a hint, clue, or answer.

9.1   *Pivoted rod on car*

A uniform thin rod of length L and mass M is pivoted at one end. The pivot is attached to the top of a car accelerating at rate A, as shown.

(*a*) What is the equilibrium value of the angle θ between the rod and the top of the car?

(*b*) Suppose that the rod is displaced a small angle ϕ from equilibrium. What is its motion for small ϕ?

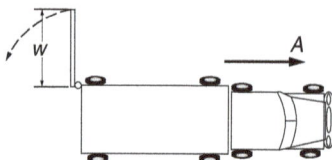

9.2   *Truck door*

A truck at rest has one door fully open, as shown. The truck accelerates forward at constant rate A, and the door begins to swing shut. The door is uniform and solid, has total mass M, height h, and width w. Neglect air resistance.

(*a*) Find the instantaneous angular velocity of the door about its hinges when it has swung through 90°.

(*b*) Find the horizontal force on the door when it has swung through 90°.

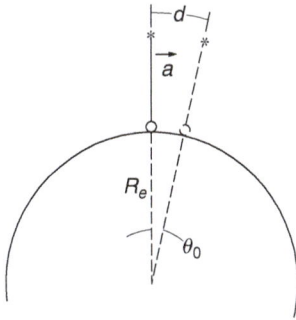

9.3 Pendulum on moving pivot

A pendulum is at rest with its bob pointing toward the center of the Earth. The support of the pendulum is moved horizontally with uniform acceleration a, and the pendulum starts to swing. Neglect rotation of the Earth.

Consider the motion of the pendulum as the pivot moves over a small distance d subtending an angle $\theta_0 \approx d/R_e \ll 1$ at the center of the Earth. Show that if the period of the pendulum is $2\pi \sqrt{R_e/g}$, the pendulum will continue to point toward the center of the Earth, if effects of order θ_0^2 and higher are neglected. (One way to make such a long-period pendulum would be to place the pivot point of a long rod close to its center of mass.)

9.4 Weight on a car's wheels

The center of mass of a 1600 kg car is midway between the wheels and 0.7 m above the ground. The wheels are 2.6 m apart.

(a) What is the minimum acceleration A of the car so that the front wheels just begin to lift off the ground?

(b) If the car decelerates at rate g, what is the normal force on the front wheels and on the rear wheels?

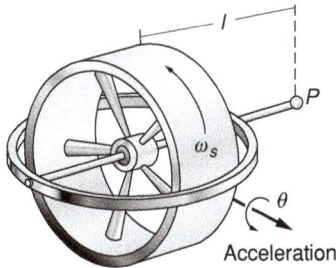

9.5 Gyroscope and acceleration

Gyroscopes can be used to detect acceleration and measure speed. Consider a gyroscope spinning at high speed ω_s. The gyroscope is attached to a vehicle by a universal pivot P. If the vehicle accelerates in the direction perpendicular to the spin axis at rate a, then the gyroscope will precess about the acceleration axis, as shown in the sketch. The total angle of precession, θ, is measured.

Show that if the system starts from rest, the velocity of the vehicle is given by

$$v = \frac{I_s \omega_s}{Ml} \theta,$$

where $I_s \omega_s$ is the gyroscope's spin angular momentum, M is the total mass of the pivoted portion of the gyroscope, and l is the distance from the pivot to the center of mass. (Such a system is called an integrating gyro, since it automatically integrates the acceleration to give the velocity.)

9.6 Spinning top in an elevator

A top of mass M spins with angular speed ω_s about its axis, as shown. The moment of inertia of the top about the spin axis is I_0, and the center of mass of the top is a distance l from the point. The axis is inclined at angle ϕ with respect to the vertical, and the top is undergoing uniform precession. Gravity is directed downward.

The top is in an elevator, with its tip held to the elevator floor by a frictionless pivot. Find the rate of precession, Ω, clearly indicating its direction, in each of the following cases:

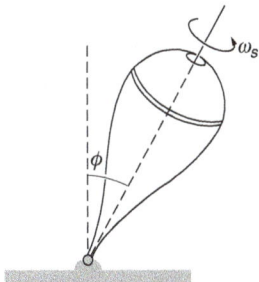

(a) The elevator at rest.

(b) The elevator accelerating down at rate $2g$.

9.7 *Apparent force of gravity*

Find the difference in the apparent force of gravity at the Equator and the poles, assuming that the Earth is spherical.

9.8 *Velocity in plane polar coordinates*

Derive the familiar expression for velocity in plane polar coordinates, $\mathbf{v} = \dot{r}\hat{\mathbf{r}} + r\dot{\theta}\hat{\boldsymbol{\theta}}$, by examining the motion of a particle in a rotating coordinate system in which the velocity is instantaneously radial.

9.9 *Train on tracks**

A 400-ton train runs south at a speed of 60 mi/h at a latitude of $60°$ north.

(a) What is the horizontal force on the tracks?

(b) What is the direction of the force?

9.10 *Apparent gravity versus latitude**

The acceleration due to gravity measured in an earthbound coordinate system is denoted by g. However, because of the Earth's rotation, g differs from the true acceleration due to gravity, g_0. Assuming that the Earth is perfectly spherical, with radius R_e and angular velocity Ω_e, find g as a function of latitude λ. (Assuming the Earth to be spherical is actually not justified—the contribution to the variation of g with latitude due to the polar flattening is comparable to the effect calculated here.)

9.11 *Racing hydrofoil*

A high-speed hydrofoil races across the ocean at the Equator at a speed of 200 mi/h. Let the acceleration of gravity for an observer at rest on the Earth be g. Find the fractional change in gravity, $\Delta g/g$, measured by a passenger on the hydrofoil when the hydrofoil heads in the following directions:

(a) east

(b) west

(c) south

(d) north.

9.12 *Pendulum on rotating platform*

A pendulum is rigidly fixed to an axle held by two supports so that it can swing only in a plane perpendicular to the axle. The pendulum consists of a mass M attached to a massless rod of length l. The supports are mounted on a platform that rotates with constant angular velocity Ω. Find the pendulum's frequency assuming that the amplitude is small.

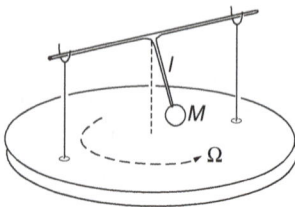

10 CENTRAL FORCE MOTION

10.1 Introduction

Johannes Kepler was the assistant of the sixteenth-century Danish astronomer Tycho Brahe. They had an ideal combination of talents. Brahe had the ingenuity and skill to measure planetary positions to better than $0.01°$, all made by naked eye because the telescope was not invented until a few years after his death. Kepler had the mathematical genius and fortitude to discover that Brahe's measurements could be fitted by three simple empirical laws. The task was formidable. It took Kepler 18 years of laborious calculation to obtain the following three laws of planetary motion, which he stated early in the seventeenth century:

1. Every planet moves in an ellipse with the Sun at one focus.
2. The radius vector from the Sun to a planet sweeps out equal areas in equal times.
3. The period of revolution T of a planet about the Sun is related to the major axis A of the ellipse by $T^2 = kA^3$, where k is the same for all the planets.

Kepler's empirical laws went unexplained until the latter half of the seventeenth century, when Newton's fascination with the problem of planetary motion inspired him to formulate his laws of motion and the law of universal gravitation. Using these mathematical laws, Newton explained Kepler's empirical laws, giving an overwhelming argument in favor of the new mechanics and marking the beginning of modern mathematical physics. Planetary motion and the more general problem of motion under a central force continue to play an important role in many branches of physics and turn up in such topics as particle scattering, atomic structure, and space navigation.

In this chapter we apply Newtonian physics to the general problem of central force motion. We shall start by looking at some of the general features of a system of two particles interacting with a central force $f(r)\hat{\mathbf{r}}$, where $f(r)$ is any function of the distance r between the particles and $\hat{\mathbf{r}}$ is a unit vector along the line of centers. After making a simple change of coordinates, we shall show how to find a complete solution by using the conservation laws of energy and angular momentum. Finally, we shall apply these results to the case of planetary motion, $f(r) \propto 1/r^2$, and show how they predict Kepler's empirical laws.

10.2 Central Force Motion as a One-body Problem

Consider an isolated system consisting of two particles interacting under a central force $f(r)\hat{\mathbf{r}}$. The masses of the particles are m_1 and m_2 and their position vectors are $\mathbf{r}_1$ and $\mathbf{r}_2$. We have

$$\mathbf{r} = \mathbf{r}_1 - \mathbf{r}_2$$
$$r = |\mathbf{r}|$$
$$= |\mathbf{r}_1 - \mathbf{r}_2|.$$

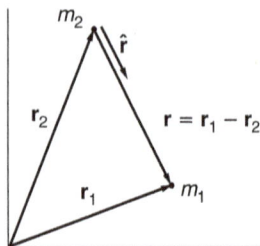

The equations of motion are

$$m_1 \ddot{\mathbf{r}}_1 = f(r)\hat{\mathbf{r}} \tag{10.1a}$$

$$m_2 \ddot{\mathbf{r}}_2 = - f(r)\hat{\mathbf{r}}. \tag{10.1b}$$

From our definition of $\mathbf{r}$, the force is attractive for $f(r) < 0$ and repulsive for $f(r) > 0$. Equations (10.1a) and (10.1b) are coupled together by $\mathbf{r}$; the behavior of $\mathbf{r}_1$ and $\mathbf{r}_2$ depends on $\mathbf{r} = \mathbf{r}_1 - \mathbf{r}_2$. The problem is easier to handle if we replace $\mathbf{r}_1$ and $\mathbf{r}_2$ by $\mathbf{r} = \mathbf{r}_1 - \mathbf{r}_2$ and the center of mass vector $\mathbf{R}$:

$$\mathbf{R} = \frac{m_1 \mathbf{r}_1 + m_2 \mathbf{r}_2}{m_1 + m_2}.$$

The equation of motion for $\mathbf{R}$ is trivial if there are no external forces:

$$\ddot{\mathbf{R}} = 0,$$

which has the simple solution

$$\mathbf{R} = \mathbf{R}_0 + \mathbf{V}t.$$

The constant vectors $\mathbf{R}_0$ and $\mathbf{V}$ depend on the choice of coordinate system and the initial conditions. If we are clever enough to take the origin at the center of mass, $\mathbf{R}_0 = 0$, and if the center of mass is stationary, $\mathbf{V} = 0$.

The equation for $\mathbf{r}$ turns out to be like the equation of motion of a single particle and has a straightforward solution. To find the equation of motion for $\mathbf{r}$ we divide Eq. (10.1a) by m_1 and Eq. (10.1b) by m_2 and subtract to give

$$\ddot{\mathbf{r}}_1 - \ddot{\mathbf{r}}_2 = \left(\frac{1}{m_1} + \frac{1}{m_2} \right) f(r)\hat{\mathbf{r}}$$

or

$$\left(\frac{m_1 m_2}{m_1 + m_2} \right) (\ddot{\mathbf{r}}_1 - \ddot{\mathbf{r}}_2) = f(r)\hat{\mathbf{r}}.$$

Denoting $m_1 m_2/(m_1 + m_2)$ by μ, the *reduced mass*, and using $\ddot{\mathbf{r}}_1 - \ddot{\mathbf{r}}_2 = \ddot{\mathbf{r}}$, we have

$$\mu \ddot{\mathbf{r}} = f(r)\hat{\mathbf{r}}. \tag{10.2}$$

Equation (10.2) is identical to the equation of motion for a particle of mass μ acted on by a force $f(r)\hat{\mathbf{r}}$; no trace of the two-particle problem remains. The two-particle problem has been transformed to a one-particle problem.

Unfortunately, the method cannot be generalized. There is no way to reduce the equations of motion for three or more particles to equivalent one-body equations, and partly for this reason the exact solution of the general three-body problem remains unknown, although there are solutions for a few special cases. We discuss one of them in Example 10.8. If we can solve Eq. (10.2) for $\mathbf{r}$, we can easily work back to find $\mathbf{r}_1$ and $\mathbf{r}_2$ from the relations

$$\mathbf{r} = \mathbf{r}_1 - \mathbf{r}_2$$

$$\mathbf{R} = \frac{m_1\mathbf{r}_1 + m_2\mathbf{r}_2}{m_1 + m_2}.$$

Solving for $\mathbf{r}_1$ and $\mathbf{r}_2$ gives

$$\mathbf{r}_1 = \mathbf{R} + \left(\frac{m_2}{m_1 + m_2}\right)\mathbf{r} \qquad (10.3a)$$

$$\mathbf{r}_2 = \mathbf{R} - \left(\frac{m_1}{m_1 + m_2}\right)\mathbf{r}. \qquad (10.3b)$$

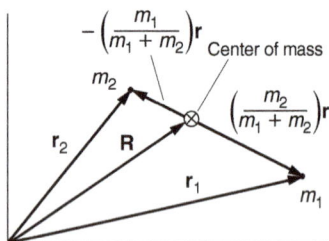

As the sketch shows, $m_2\mathbf{r}/(m_1 + m_2)$ and $-m_1\mathbf{r}/(m_1 + m_2)$ are the position vectors of m_1 and m_2 relative to the center of mass.

10.3 Universal Features of Central Force Motion

Solving the vector equation of motion $\mu\ddot{\mathbf{r}} = f(r)\hat{\mathbf{r}}$ for $\mathbf{r}(t)$ depends on the particular form of $f(r)$, but some properties of central force motion hold true in general regardless of the form of $f(r)$. Constraints imposed by the conservation laws of energy and angular momentum provide a major step toward finding the complete solution. In this section we shall see how to use conservation laws to identify some universal features of the solution and to reduce the vector equation to an equation in a single scalar variable. Conservation of linear momentum adds nothing new, because it is already embodied in the equal and opposite forces on the two masses, and in the uniform motion of the system's center of mass.

Although we shall focus mainly on the gravitational central force $f(r) = -C/r^2$ later in this chapter, the consequences of the conservation laws discussed in this section hold for all central forces, whatever the form of $f(r)$.

10.3.1 Consequences of the Conservation of Angular Momentum

The central force $f(r)\hat{\mathbf{r}}$ is along $\mathbf{r}$ and can exert no torque on the reduced mass μ. Hence the angular momentum $\mathbf{L}$ of μ is constant, both in direction and in magnitude.

A. The motion is confined to a plane

As a proof, $\mathbf{L} = \mathbf{r} \times \mu\dot{\mathbf{r}}$, so it follows that $\mathbf{r}$ is always perpendicular to $\mathbf{L}$ by the properties of the cross product. Because $\mathbf{L}$ is fixed in direction, the plane of the motion is also fixed, and $\mathbf{r}$ can only move in a plane perpendicular to $\mathbf{L}$.

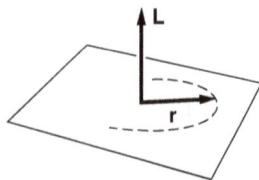

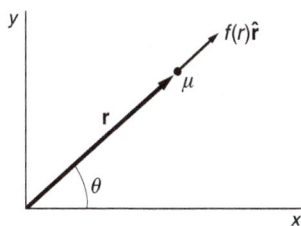

Introducing plane polar coordinates r, θ in the plane of motion, the equation of motion $\mu\ddot{\mathbf{r}} = f(r)\hat{\mathbf{r}}$ becomes

$$\mu(\ddot{r} - r\dot{\theta}^2) = f(r) \tag{10.4a}$$

$$\mu(r\ddot{\theta} + 2\dot{r}\dot{\theta}) = 0. \tag{10.4b}$$

In particular, Eq. (10.4b) is a consequence of the fact that a central force has no tangential component.

B. The law of equal areas

The magnitude L of the angular momentum is constant, and is

$$L = \mu r^2 \dot{\theta}. \tag{10.5}$$

This leads immediately to the law of equal areas, Kepler's second empirical law, that we proved in Example 7.5. Summarizing, the area element in polar coordinates is $dA = r^2 d\theta/2$, so that $dA/dt = r^2\dot{\theta}/2 = L/2\mu =$ constant. The areas swept out by $\mathbf{r}$ are the same for equal time intervals. The law of equal areas holds for any central force and for both closed and open orbits. For the solar system, a planetary orbit is an example of a closed orbit. An open orbit would be like the orbit of a comet entering the solar system, sweeping around the Sun, and heading back out to space, never to return.

10.3.2 Consequences of the Conservation of Energy

The kinetic energy of μ is

$$K = \frac{1}{2}\mu v^2$$

$$= \frac{1}{2}\mu(\dot{r}\hat{\mathbf{r}} + r\dot{\theta}\hat{\theta})^2$$

$$= \frac{1}{2}\mu(\dot{r}^2 + r^2\dot{\theta}^2).$$

We showed in Example 5.9 that all central forces are conservative, so we can associate a potential energy $U(r)$ with $f(r)$:

$$U(r) - U(r_a) = -\int_{r_a}^{r} f(r)dr.$$

The constant $U(r_a)$ is not physically significant, so we can leave r_a unspecified; adding a constant to the energy has no effect on the motion.

From the work–energy theorem,

$$E = K + U(r)$$

$$= \frac{1}{2}\mu v^2 + U(r) \tag{10.6a}$$

$$= \frac{1}{2}\mu(\dot{r}^2 + r^2\dot{\theta}^2) + U(r) \tag{10.6b}$$

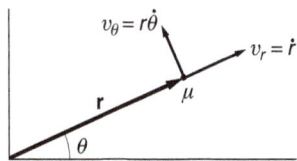

where E, the total mechanical energy, is constant. We can eliminate $\dot{\theta}$ from Eq. (10.6b) by using Eq. (10.5). The result is

$$E = \frac{1}{2}\mu\dot{r}^2 + \frac{1}{2}\frac{L^2}{\mu r^2} + U(r). \tag{10.7}$$

10.3.3 The Effective Potential

Equation (10.7) looks like the energy equation for a particle moving in one dimension; all reference to θ is gone. We can press the parallel further by introducing

$$U_{\text{eff}}(r) \equiv \frac{1}{2}\frac{L^2}{\mu r^2} + U(r), \tag{10.8}$$

so that

$$E = \frac{1}{2}\mu\dot{r}^2 + U_{\text{eff}}(r). \tag{10.9}$$

U_{eff} is called the *effective potential energy*. Often it is referred to simply as the *effective potential*. U_{eff} differs from the true potential $U(r)$ by the term $L^2/2\mu r^2$, called the *centrifugal potential*. Introducing the effective potential is a convenient mathematical trick to make Eq. (10.9) look just like the energy equation for a particle in one dimension. However, the term $L^2/2\mu r^2$ is not a true potential energy related to a force. From Eq. (10.6b), this term is seen to be another way of writing the kinetic energy due to the tangential velocity $r\dot{\theta}$. The term $L^2/2\mu r^2$ is really a kinetic energy, but grouping it with the true potential $U(r)$ helps us write the formal solution of Eq. (10.9) more directly, and it will also help us use simple energy diagrams to describe central force motion qualitatively.

10.3.4 The Formal Solution for Central Force Motion

The formal solution of Eq. (10.9) is

$$\frac{dr}{dt} = \sqrt{\frac{2}{\mu}(E - U_{\text{eff}})} \tag{10.10}$$

or

$$\int_{r_0}^{r} \frac{dr}{\sqrt{(2/\mu)(E - U_{\text{eff}})}} = t - t_0. \tag{10.11}$$

Equation (10.11) formally gives us r as a function of t, although the integral may have to be done numerically in some cases. To find θ as a function of t, rewrite Eq. (10.5) as

$$\frac{d\theta}{dt} = \frac{L}{\mu r^2}. \tag{10.12}$$

Since r is known as a function of t from Eq. (10.11), we can formally integrate to find $\theta(t)$:

$$\theta - \theta_0 = \frac{L}{\mu}\int_{t_0}^{t}\frac{dt}{r^2}. \tag{10.13}$$

Often we are interested in the path of the particle, which means knowing r as a function of θ rather than as a function of time. We often call the trajectory $r(\theta)$ the *orbit* of the particle, even if the trajectory does not close on itself. To eliminate t, combine Eqs. (10.10) and (10.12) using the chain rule:

$$\frac{d\theta}{dr} = \frac{d\theta}{dt}\frac{dt}{dr} = \left(\frac{L}{\mu r^2}\right)\sqrt{\frac{\mu}{2(E - U_{\text{eff}})}}.$$

The formal solution for the orbit is then

$$\theta - \theta_0 = L \int_{r_0}^{r} \frac{dr}{r^2\sqrt{2\mu(E - U_{\text{eff}})}}. \tag{10.14}$$

This completes the formal solution of the central force problem. We can obtain $r(t), \theta(t)$, or $r(\theta)$ as we please; all we need to do is evaluate the appropriate integrals.

10.4 The Energy Equation and Energy Diagrams

In Section 10.3 we found two equivalent ways of writing E, the total energy in the center of mass system. According to Eqs. (10.6a) and (10.9), respectively,

$$E = \frac{1}{2}\mu v^2 + U(r), \tag{10.15a}$$

$$E = \frac{1}{2}\mu \dot{r}^2 + U_{\text{eff}}(r). \tag{10.15b}$$

We generally need to use both these forms in analyzing central force motion. The first form, $\frac{1}{2}\mu v^2 + U(r)$, is handy for evaluating E; all we need to know is the relative speed and position at some instant. However, $v^2 = \dot{r}^2 + (r\dot{\theta})^2$, and the dependence on two coordinates r and θ makes it difficult to visualize the motion. In contrast, the second form, $\frac{1}{2}\mu\dot{r}^2 + U_{\text{eff}}(r)$, depends on the single coordinate r. In fact, it is identical to the equation for the energy of a particle of mass μ constrained to move along a straight line with kinetic energy $\frac{1}{2}\mu\dot{r}^2$ and potential energy $U_{\text{eff}}(r)$. The coordinate θ is completely suppressed—the kinetic energy $\frac{1}{2}\mu(r\dot{\theta})^2$ associated with the tangential motion is accounted for in the effective potential by the relations

$$\frac{1}{2}\mu(r\dot{\theta})^2 = \frac{L^2}{2\mu r^2}$$

$$U_{\text{eff}}(r) = \frac{L^2}{2\mu r^2} + U(r).$$

Equation (10.15b) involves only the radial motion. Consequently, we can use the energy diagram technique developed in Chapter 5 to find the qualitative features of the radial motion. To see how the method works, let's start by looking at a very simple system, two non-interacting particles.

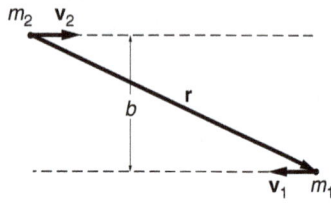

Example 10.1 Central Force Description of Free-particle Motion

Two non-interacting particles m_1 and m_2 move toward each other with velocities $\mathbf{v}_1$ and $\mathbf{v}_2$, respectively. The reduced mass is $\mu = m_1 m_2/(m_1 + m_2)$. Their paths are offset by distance b, as shown in the sketch. We shall develop the equivalent one-body description of this system and its energy diagram.

The relative velocity is

$$\mathbf{v}_0 = \dot{\mathbf{r}}$$
$$= \dot{\mathbf{r}}_1 - \dot{\mathbf{r}}_2$$
$$= \mathbf{v}_1 - \mathbf{v}_2.$$

$\mathbf{v}_0$ is constant because $\mathbf{v}_1$ and $\mathbf{v}_2$ are constant. The energy of the system relative to the center of mass is

$$E = \frac{1}{2}\mu v_0^2 + U(r) = \frac{1}{2}\mu v_0^2,$$

since $U(r) = 0$ for non-interacting particles.

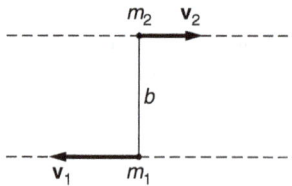

In order to draw the energy diagram we need to find the effective potential

$$U_{\text{eff}} = \frac{L^2}{2\mu r^2} + U(r) = \frac{L^2}{2\mu r^2}.$$

We could evaluate L by direct computation, but it is simpler to use the relation

$$E = \frac{1}{2}\mu \dot{r}^2 + \frac{L^2}{2\mu r^2} = \frac{1}{2}\mu v_0^2.$$

When m_1 and m_2 pass each other, $r = b$ and $\dot{r} = 0$. Hence

$$\frac{L^2}{2\mu b^2} = \frac{1}{2}\mu v_0^2,$$
$$L = \mu b v_0.$$

This result holds for all times because L is constant. Therefore

$$U_{\text{eff}} = \frac{1}{2}\mu v_0^2 \frac{b^2}{r^2}.$$

The energy diagram is shown in the sketch.

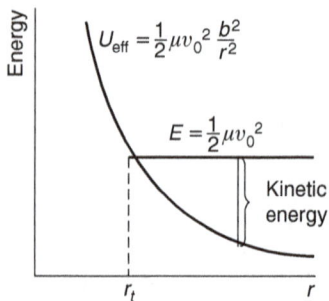

The kinetic energy associated with radial motion is

$$K = \frac{1}{2}\mu \dot{r}^2$$
$$= E - U_{\text{eff}}.$$

K can never be negative, so that the motion is restricted to regions where $E - U_{\text{eff}} \geq 0$. Initially r is very large and $U_{\text{eff}} \approx 0$. As the particles approach, the kinetic energy decreases, vanishing at the *turning point* r_t, where the radial velocity is zero and the motion is purely tangential. At the turning point $E = U_{\text{eff}}(r_t)$, which gives

$$\frac{1}{2}\mu v_0^2 = \frac{1}{2}\mu v_0^2 \frac{b^2}{r_t^2}$$

or

$$r_t = b$$

as we expect, since r_t is the distance of closest approach of the particles; it is the minimum value of r. Once the turning point is passed, r increases and the particles separate. In our one-dimensional picture, the particle μ "bounces off" the barrier of the effective potential.

Despite the colorful language, keep in mind that the "centrifugal potential" $L^2/(2\mu r^2)$ is not related to a real physical force that could accelerate particles according to Newton's second law. The particles in this example do not interact; they move steadily ahead without "bouncing." The purpose of energy diagrams is to show qualitatively the motion as a function of r and in particular to establish the limits of the motion.

Now let us apply energy diagrams to the meatier problem of planetary motion. For the gravitational force, which is always attractive,

$$f(r) = -\frac{Gm_1m_2}{r^2}$$

$$U(\infty) - U(r) = -\int_r^\infty f(r)dr$$

$$= Gm_1m_2 \int_r^\infty \frac{dr}{r^2}$$

$$= \frac{Gm_1m_2}{r}$$

so that

$$U(r) = -\frac{Gm_1m_2}{r}.$$

By the usual convention, we have taken $U(\infty) = 0$. The effective potential energy is

$$U_{\text{eff}} = \frac{L^2}{2\mu r^2} - \frac{Gm_1m_2}{r}.$$

If $L \neq 0$, the repulsive centrifugal potential $L^2/(2\mu r^2)$ dominates at small r, and the attractive gravitational potential $-Gm_1m_2/r$ dominates at large

r. The drawing shows the energy diagram with various values of the total energy E.

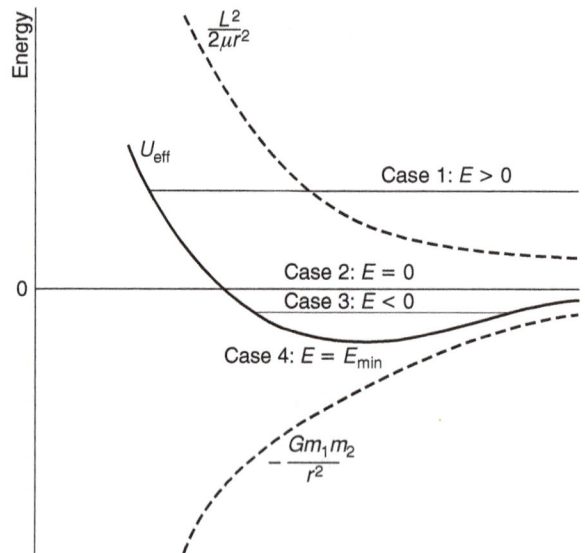

The kinetic energy of radial motion is $K = E - U_{\text{eff}}$, and the motion is restricted to regions where $K \geq 0$. The nature of the motion is determined by the total energy. Here are the various possibilities, as shown in the sketch:

1. $E > 0$: r is unbounded for large values but cannot be less than a certain minimum if $L \neq 0$. The particles are kept apart by the "centrifugal barrier."
2. $E = 0$: this is qualitatively similar to case 1 but on the boundary between unbounded and bounded motion.
3. $E < 0$: the motion is bounded for both large and small r. The two particles form a bound system.
4. $E = E_{\text{min}}$: r is restricted to one value. The particles stay a constant distance from one another.

In Section 10.5 we shall find that case 1 corresponds to motion in a hyperbola; case 2, to a parabola; case 3, to an ellipse; and case 4, to a circle.

There is one other possibility, $L = 0$. In this case the particles accelerate toward each other along a straight line on a collision course, since when $L = 0$ there is no "centrifugal barrier" to keep them apart.

Example 10.2 How the Solar System Captures Comets
Suppose that a comet with $E > 0$ (an extrasolar comet) drifts into the solar system. From our discussion of the energy diagram for motion under a gravitational force, the comet will approach the Sun and then

swing away, never to return. How then can an extrasolar comet with initial energy $E > 0$ become a member of the solar system? For this to happen, its energy would have to be decreased to a negative value. However, the gravitational force is conservative and the total energy of the comet in the Sun's gravity cannot change.

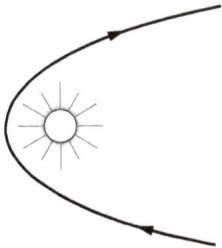

The situation is quite different if a third body is involved. For instance, if the comet is deflected by a massive planet like Jupiter, it can transfer energy to the planet and so become trapped in the solar system.

Suppose that an extrasolar comet is heading outward toward the orbit of Jupiter after swinging around the Sun, as shown in the sketch. Let the velocity of the comet before it starts to interact appreciably with Jupiter be $\mathbf{v}_i$, and let Jupiter's velocity be $\mathbf{V}$. For simplicity we shall assume that the orbits are not appreciably deflected by the Sun during the time of interaction.

In the comet–Jupiter center of mass system Jupiter is essentially at rest because of its much greater mass, and the center of mass velocity of the comet is $\mathbf{v}_{ic} = \mathbf{v}_i - \mathbf{V}$, as shown in (a).

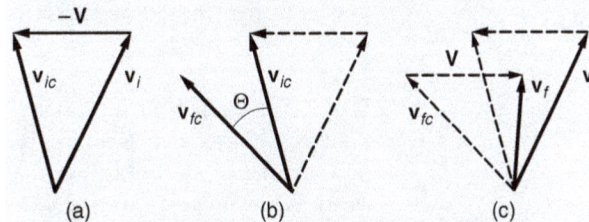

(a)   (b)   (c)

In the center of mass system the path of the comet is deflected by Jupiter, but the final speed is equal to the initial speed v_{ic}. Hence, the interaction merely rotates $\mathbf{v}_{ic}$ through some angle Θ to a new direction $\mathbf{v}_{fc}$, as shown in (b). The final velocity in the space-fixed system is

$$\mathbf{v}_f = \mathbf{v}_{fc} + \mathbf{V}.$$

Figure (c) shows $\mathbf{v}_f$ and, for comparison, $\mathbf{v}_i$. For the deflection shown, $v_f < v_i$, and the comet's energy has decreased. Conversely, if the deflection is in the opposite direction, interaction with Jupiter would increase the energy, possibly freeing a bound comet from the solar system. A large proportion of known comets have energies near zero (bounded elliptic orbit if $E \lesssim 0$ or unbounded hyperbolic orbit if $E \gtrsim 0$). The interaction of a comet with Jupiter is therefore often sufficient to change the orbit from unbound to bound, or vice versa.

This mechanism for transferring energy to or from a planet can be used to accelerate an interplanetary spacecraft. By picking the orbit cleverly, the spacecraft can "hop" from planet to planet with a great saving in fuel.

The process we have described may seem to contradict the idea that the gravitational force is strictly conservative. Only gravity acts on the comet and yet its total energy can change. The reason is that the comet experiences a time-dependent gravitational force, and time-dependent forces are intrinsically non-conservative. Nevertheless, the total energy of the entire system is conserved, as we expect; in the comet–Jupiter system, the excess energy is taken up by a slight change in the motion of Jupiter.

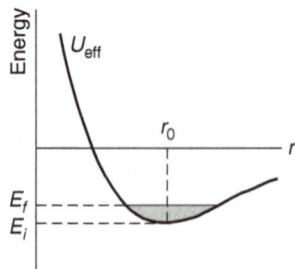

Example 10.3 Perturbed Circular Orbit

A satellite of mass m orbits the Earth in a circle of radius r_0. One of its engines is fired briefly toward the center of the Earth, changing the energy of the satellite but not its angular momentum. The problem is to find the new orbit.

The energy diagram shows the initial energy E_i and the final energy E_f. Note that firing the engine radially does not change the effective potential because L is not altered. Since the Earth's mass M_e is much greater than m, the reduced mass is nearly m and the Earth is effectively fixed.

If E_f is not much greater than E_i, the energy diagram shows that r never differs much from r_0. Rather than solve the planetary motion problem exactly, as we shall do in the next section, we instead approximate $U_{\text{eff}}(r)$ in the neighborhood of r_0 by a parabolic potential.

As we know from our analysis of small oscillations of a particle about equilibrium, Section 6.2, the resulting radial motion of the satellite will be simple harmonic motion about r_0 to good accuracy. The effective potential is, with $C \equiv GmM_e$,

$$U_{\text{eff}}(r) = -\frac{C}{r} + \frac{L^2}{2mr^2}.$$

The minimum of U_{eff} is at $r = r_0$. Since the slope is zero there, we have

$$\left.\frac{dU_{\text{eff}}}{dr}\right|_{r_0} = \frac{C}{r_0^2} - \frac{L^2}{mr_0^3} = 0,$$

which gives

$$L = \sqrt{mCr_0}. \tag{1}$$

This result can also be found by applying Newton's second law to circular motion. From Section 6.2, the frequency of oscillation of the system, which we shall denote by β, is

$$\beta = \sqrt{\frac{k}{m}},$$

where

$$k = \left.\frac{d^2 U_{\text{eff}}}{dr^2}\right|_{r_0}. \tag{2}$$

This is readily evaluated to yield

$$\beta = \sqrt{\frac{C}{mr_0{}^3}} = \frac{L}{mr_0{}^2}. \tag{3}$$

The radial position is then given by

$$\begin{aligned} r &= r_0 + A \sin \beta t + B \cos \beta t \\ &= r_0 + A \sin \beta t. \end{aligned} \tag{4}$$

We have taken $B = 0$ in the complete solution to satisfy the initial condition $r(0) = r_0$. Although we could calculate the amplitude A in terms of E_f, we shall not bother with the algebra here except to note that $A \ll r_0$ for E_f nearly equal to E_i.

To find the new orbit, we must eliminate t and express r as a function of θ. For the circular orbit,

$$\dot{\theta} = \frac{L}{mr_0{}^2}, \tag{5}$$

or

$$\theta = \left(\frac{L}{mr_0{}^2}\right) t \equiv \beta t. \tag{6}$$

Comparing Eqs. (3) and (5) shows, surprisingly, that the frequency of rotation $\dot{\theta}$ is equal to the frequency of radial oscillation β.

Equation (6) is accurate enough for our purposes, even though the radius oscillates slightly after the engine is fired; t occurs only in a small correction term to r in Eq. (4), and we are neglecting terms of order A^2 and higher.

From Eqs. (1) and (5) the frequency of rotation of the satellite around the Earth can be written

$$\dot{\theta} = \frac{L}{mr_0{}^2} = \frac{\sqrt{mCr_0}}{mr_0{}^2} = \sqrt{\frac{C}{mr_0{}^3}}.$$

If we substitute Eq. (6) in Eq. (4), we obtain

$$r = r_0 + A \sin \theta. \tag{7}$$

The new orbit is shown as the solid line in the sketch.

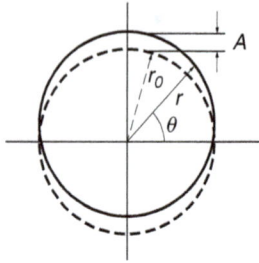

The orbit looks almost circular, but it is no longer centered on the Earth. As we shall show in Section 10.5, the exact orbit for $E = E_f$ is an ellipse described by the equation

$$r = \frac{r_0}{1 - (A/r_0) \sin \theta}.$$

If $A/r_0 \ll 1$,

$$r \approx r_0 \left(1 + \frac{A}{r_0} \sin \theta \right)$$

$$= r_0 + A \sin \theta.$$

To first order in A, Eq. (7) is the equation of an ellipse. However, the exact calculation is harder to derive (and to digest) than the approximate result we found in this example with the help of the energy diagram.

10.5 Planetary Motion

In this section we solve the chapter's main problem: finding the orbit for a planet of mass m moving about a star of mass M under the gravitational interaction

$$U(r) = -G\frac{Mm}{r} \equiv -\frac{C}{r}. \tag{10.16}$$

Our results would also apply to a satellite of mass m orbiting a planet of mass M, or even to a binary star system with two stars of masses m and M. Further, we shall use our results to show how Newtonian mechanics can account for Kepler's empirical laws of planetary motion.

Because we are often interested in a satellite of mass m orbiting the Earth (mass M_e), it is handy in this case to express C in more familiar terms. At the Earth's surface ($r = R_e$) the acceleration due to gravity is $g = GM_e/R_e^2$, so C can be written

$$C = GmM_e = mgR_e^2 \tag{10.17}$$

for a satellite orbiting the Earth.

Inserting the potential $U(r)$ from Eq. (10.16) into the orbit equation, Eq. (10.14), we obtain

$$\theta - \theta_0 = L \int \frac{dr}{r \sqrt{(2\mu Er^2 + 2\mu Cr - L^2)}},$$

where θ_0 is a constant of integration. Note 10.1 shows how the integral over r can be evaluated by converting it to a standard form, with the result

$$r = \frac{(L^2/\mu C)}{1 - \sqrt{1 + (2EL^2/\mu C^2)} \sin(\theta - \theta_0)}. \tag{10.18}$$

The usual conventions are to take $\theta_0 = -\pi/2$ and to introduce the parameters

$$r_0 \equiv \frac{L^2}{\mu C} \tag{10.19}$$

$$\epsilon \equiv \sqrt{1 + \frac{2EL^2}{\mu C^2}}. \tag{10.20}$$

Physically, r_0 is the radius of the circular orbit corresponding to the given values of L, μ, and C. The dimensionless parameter ϵ, called the *eccentricity*, characterizes the shape of the orbit, as we shall see. With these replacements, Eq. (10.18) becomes

$$r = \frac{r_0}{1 - \epsilon \cos \theta}. \tag{10.21}$$

Equation (10.21) looks more familiar in Cartesian coordinates $r = \sqrt{x^2 + y^2}$, $r \cos \theta = x$. Rewriting Eq. (10.21) in the form $r - \epsilon r \cos \theta = r_0$, we have

$$\sqrt{x^2 + y^2} - \epsilon x = r_0$$

or

$$(1 - \epsilon^2)x^2 - 2r_0\epsilon x + y^2 = r_0^2. \tag{10.22}$$

This quadratic form describes the *conic sections*—hyperbola, parabola, ellipse, circle—traced out by a plane cutting a cone at various angles.

The shape of the orbit depends on ϵ, hence also on E through Eq. (10.20). Here are the possibilities:

1. $\epsilon > 1$ hence $E > 0$; the system is unbounded: The coefficients of x^2 and y^2 are unequal and opposite in sign; the equation has the form $y^2 - Ax^2 - Bx = \text{constant}$, which is the equation of a *hyperbola*.
2. $\epsilon = 1$ hence $E = 0$; the system is on the border between bounded and unbounded: Eq. (10.22) becomes

$$x = \frac{y^2}{2r_0} - \frac{r_0}{2},$$

which is the equation of a *parabola*.
3. $0 \le \epsilon < 1$ hence $-\mu C^2/2L^2 \le E < 0$; the system is bounded: The coefficients of x^2 and y^2 are unequal but of the same sign; the equation has the form $y^2 + Ax^2 - Bx = \text{constant}$, which is the equation of an *ellipse*. The term linear in x means that the geometric center of the ellipse is not at the origin of coordinates. As proved in Note 10.2, one focus of the ellipse is at the center of mass, which we have taken to be the origin.

When $\epsilon = 0$, E has its lowest possible value $-\mu C^2/2L^2$. The equation of the orbit becomes $x^2 + y^2 = r_0^2$; the ellipse degenerates to a *circle*, $r = \text{constant}$.

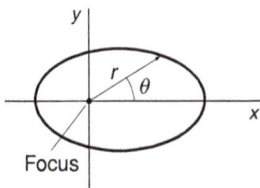

Focus

10.5.1 Hyperbolic Orbits

To calculate a real orbit we need to derive the initial conditions from experimentally accessible parameters. For example, if the orbit is unbounded, we might know the energy and the initial trajectory, obtained from observation of the position at two or more times.

In this section we shall show how to use experimental parameters to describe a hyperbolic orbit, which could apply to the motion of an extra-solar comet about the Sun. Another application could be the trajectory of a charged particle scattering off an atomic nucleus, because the electric force and the gravitational force are both $\propto 1/r^2$.

Let v_0 be the speed of μ when it is far from the origin, and let the projection of its initial path pass the origin at distance b, as shown; b is commonly called the *impact parameter*.

The angular momentum L and energy E are

$$L = \mu v_0 b$$

$$E = \frac{1}{2}\mu v_0^2.$$

For an inverse-square force $U(r) = -C/r$, the equation of the orbit is

$$r = \frac{r_0}{1 - \epsilon \cos \theta}, \tag{10.23}$$

where

$$r_0 = \frac{L^2}{\mu C} = \frac{\mu v_0^2 b^2}{C}$$

$$= \frac{2Eb^2}{C} \tag{10.24}$$

and

$$\epsilon = \sqrt{1 + \frac{2EL^2}{\mu C^2}}$$

$$= \sqrt{1 + \left(2Eb/C^2\right)^2}. \tag{10.25}$$

We have used $L^2 = 2\mu E b^2$ from Eq. (10.24).

When $\theta = \pi$, $r = r_{\min}$, and from the orbit equation (10.23)

$$r_{\min} = \frac{r_0}{1 + \epsilon}$$

$$= \frac{2Eb^2/C}{1 + \sqrt{1 + (2Eb/C)^2}}.$$

For $E \to \infty$, $r_{\min} \to b$. Hence $0 < r_{\min} < b$.

The half-angle θ_a between the asymptotes can be found from Eq. (10.23) for the orbit by letting $r \to \infty$.

We find

$$\cos \theta_a = \frac{1}{\epsilon}. \tag{10.26}$$

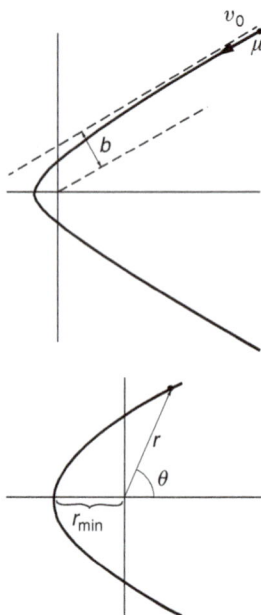

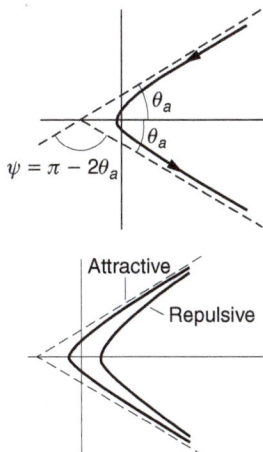

In the interaction, μ is deflected through the angle $\psi = \pi - 2\theta_a$; in atomic and nuclear physics, ψ is called the *scattering angle*. According to the sketch, ψ is related to θ_a, and hence to the eccentricity. The scattering angle ψ is

$$\psi = \pi - 2\theta_a$$

so that

$$\cos\theta_a = \cos(\pi/2 - \psi/2)$$
$$= \sin(\psi/2)$$

and using Eqs. (10.25) and (10.26) it follows that

$$\sin(\psi/2) = 1/\epsilon$$
$$= \frac{1}{\sqrt{1 + (2Eb/C)^2}}. \tag{10.27}$$

The scattering angle ψ approaches $180°$ if $(2Eb/C)^2 \ll 1$.

Example 10.4 Rutherford (Coulomb) Scattering

In Rutherford's classic experiment (1909) fast α-rays (doubly charged helium nuclei) emitted by radioactive decay products of radium bombarded thin gold foils. After being deflected ("scattered") by the foil, the α-rays were allowed to strike a zinc sulfide phosphor screen viewed through a microscope, giving off momentary flashes of light to mark where the α-rays hit. The time-consuming experiment, carried out with the help of Hans Geiger and Ernest Marsden (an undergraduate), consisted of measuring the relative number of α-rays scattered through various angles ranging from a few degrees to $150°$. The foil was so thin that it was unlikely for an α-ray to be scattered by more than one gold nucleus, simplifying the analysis.

Rutherford calculated the scattering of the α-rays according to the Coulomb potential $U(r) = C'/r$, and found good agreement with experiment. The α-rays followed hyperbolic orbits even when r_{min} was 3×10^{-15} m (at $150°$), far less than the radius of the atom $\lesssim 150 \times 10^{-12}$ m, proving that most of the mass of an atom must be concentrated in a small volume, the nucleus. These results disproved the earlier "plum pudding" model of the atom that had negative electrons distributed through a sphere of positive charge and therefore did not possess a massive scattering center that could cause large scattering angles.

To analyze the scattering using hyperbolic orbits, we start from Eq. (10.27), which relates the scattering angle ψ to the eccentricity

$$\sin(\psi/2) = \frac{1}{\sqrt{1 + (2Eb/C')^2}}. \tag{1}$$

Rutherford was unable to determine whether the gold nuclei attracted ($C' < 0$) or repelled ($C' > 0$) the α-rays. According to Eq. (1) the scattering angle depends on $(2Eb/C')^2$, making it impossible to measure the algebraic sign of the strength parameter C'.

Suppose that a narrow beam of α-rays traveling parallel to the x axis is incident on the foil, and let the uniform flux in the beam be $\mathcal{N}$ m^{-2}s^{-1}. Imagine a geometric annulus of radius b and width Δb centered on, and perpendicular to, the x axis, and located far from the scattering center, as shown in the sketch. The rate n_s at which incident α-rays pass through the annulus per second is

$$n_s = 2\pi b \Delta b \mathcal{N}. \tag{2}$$

An α-ray passing through the annulus will have an impact parameter between b and $b + \Delta b$, and will be scattered through an angle between ψ and $\psi + \Delta\psi$. Taking differentials of Eq. (1),

$$\cos(\psi/2)\Delta\psi = -2\frac{(2E/C')^2 b\Delta b}{[1 + (2Eb/C')^2]^{3/2}}$$

which can be rewritten using Eq. (1) to give

$$\cos(\psi/2)\Delta\psi = -2(2E/C')^2 \sin^3(\psi/2) b\Delta b. \tag{3}$$

We can disregard the negative sign, which simply tells us that ψ increases as b decreases.

Simplifying, let the phosphor scintillation screen be a hollow sphere of radius R centered on the origin. (The macroscopic R is far greater than the submicroscopic scattering geometry.) The α-rays scattered through angles between ψ and $\psi + \Delta\psi$ strike the sphere in a ring of radius $R\sin\psi$ and width $R\Delta\psi$, hence with area $\Delta A = R^2 \sin\psi\Delta\psi$. The flux of scattered α-rays striking ΔA is then $n_s/\Delta A$ m^{-2}s^{-1}. Relative to the incident flux $\mathcal{N}$ we have

$$\begin{aligned}
\frac{n_s/\Delta A}{\mathcal{N}} &= \frac{\cos(\psi/2)\Delta\psi}{2(2E/C')^2 R^2 \sin\psi\Delta\psi \sin^3(\psi/2)} \\
&= \frac{1}{4(2E/C')^2 R^2 \sin^4(\psi/2)}
\end{aligned} \tag{4}$$

where we have used the identity $\sin\psi = 2\sin(\psi/2)\cos(\psi/2)$.

Equation (4) shows the strong dependence of Coulomb scattering on the scattering angle ψ. A large proportion of the α-rays are scattered through fairly small angles, but there is a small but non-zero probability of scattering through very large angles, $\psi \to 180°$, as observed by Geiger and Marsden.

When high energy α-rays scattered off "light" elements (elements whose nuclei have fewer than 10–15 protons), Rutherford observed deviations from his scattering calculation. He realized that for these

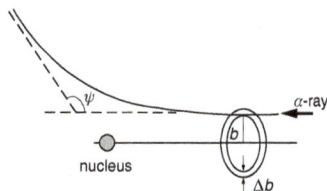

nuclei the α-rays could move close to the nucleus itself, against a repulsive Coulomb force weaker than for a gold nucleus. The observations allowed estimates of the radius R of the nucleus, generally expressed today as $R = R_0 A^{1/3}$, where $R_0 \approx 1.4 \times 10^{-15}$ m and where the mass number A is the total number of protons and neutrons in the nucleus. The expression is consistent with the density of nuclear matter being constant.

In the 1920's, physicists realized that the newly developed quantum mechanics was the correct tool for accurately describing phenomena on the atomic or subatomic scale. Much to Rutherford's delight, the quantum mechanical result for Coulomb scattering turned out to be nearly the same as the classical account, especially for scattering from heavy elements such as gold.

10.5.2 Elliptic Orbits and Planetary Motion

Elliptic orbits are so important in astronomy and astrophysics that it is worth looking at their properties in more detail. For elliptic orbits, $E < 0$, and the eccentricity is $0 \le \epsilon < 1$. In Cartesian coordinates, Eq. (10.22) for the orbit is

$$(1 - \epsilon^2)x^2 - 2r_0\epsilon x + y^2 = r_0{}^2$$

and we see that the ellipse is symmetric about the x axis in our chosen coordinate system. Further, the term linear in x shows that the ellipse is displaced from the origin along the x axis. Note 10.2 proves that a focus of the ellipse is at the origin, so that for the solar system, the Sun is at a focus of the ellipse, in accord with Kepler's laws.

The length of the major axis is

$$A = r_{\min} + r_{\max}$$

$$= r_0 \left(\frac{1}{1 + \epsilon} + \frac{1}{1 - \epsilon} \right)$$

$$= \frac{2r_0}{1 - \epsilon^2}. \qquad (10.28)$$

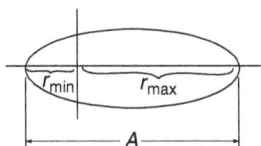

Expressing r_0 and ϵ in terms of E, L, μ, C by Eqs. (10.19) and (10.20) gives

$$A = \frac{2r_0}{1 - \epsilon^2}$$

$$= \frac{2L^2/(\mu C)}{1 - [1 + 2EL^2/(\mu C^2)]}$$

$$= \frac{C}{(-E)}. \qquad (10.29)$$

The length of the major axis is independent of L; orbits with the same major axis have the same energy. For instance, all the orbits in the sketch correspond to the same value of E although they have different values of L.

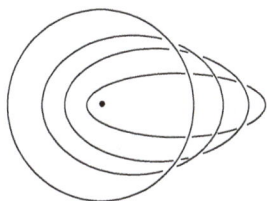

Substituting $E = -C/A$ in the energy equation $E = (1/2)\mu v_0{}^2 - C/r$, we obtain the useful relation

$$v^2 = \frac{2C}{\mu}\left(\frac{1}{r} - \frac{1}{A}\right),\tag{10.30}$$

which gives the orbital speed v at any radial position r on the orbit.

The Period of an Elliptic Orbit

To find the period T of an elliptic orbit, a direct method is to integrate Eq. (10.11) for $r(t)$. Taking $U_{\text{eff}} = (1/2)L^2/(\mu r^2) - C/r$,

$$t_b - t_a = \mu \int_{r_a}^{r_b} \frac{r\,dr}{\sqrt{(2\mu E r^2 + 2\mu C r - L^2)}}.$$

Integrating by parts, with $E < 0$,

$$t_b - t_a = \left.\frac{\sqrt{(2\mu E r^2 + 2\mu C r - L^2)}}{2E}\right|_{r_a}^{r_b}$$

$$-\left(\frac{\mu C}{2E}\right)\frac{1}{\sqrt{-2\mu E}}\arcsin\left.\left(\frac{-2\mu E r - \mu C}{\sqrt{\mu^2 C^2 + 2\mu E L^2}}\right)\right|_{r_a}^{r_b}.$$

For a complete period, $t_b - t_a = T$. The first term vanishes because $r_b = r_a$, and in the second term, the arcsin changes by 2π. The result is

$$T = \left(\frac{\pi\mu C}{-E}\right)\frac{1}{\sqrt{-2\mu E}}$$

or

$$T^2 = \frac{\pi^2 \mu C^2}{-2E^3}.$$

Finally, using Eq. (10.29),

$$T^2 = \frac{\pi^2 \mu}{2C}A^3.\tag{10.31}$$

Incidentally, we have just proved Kepler's third law, $T^2 = kA^3$, where k is essentially the same for all planets about the Sun. Table 10.1 lists A^3/T^2 for several planets. Despite variations of ≈ 100 in the major axis and ≈ 1000 in the period, the value of A^3/T^2 is constant to within 0.05%.

Table 10.1*

Planet	ϵ	A, km	T, s	A^3/T^2
Mercury	0.206	1.16×10^8	7.62×10^6	2.69×10^{10}
Earth	0.017	2.99×10^8	3.16×10^7	2.68×10^{10}
Mars	0.093	4.56×10^8	5.93×10^7	2.70×10^{10}
Jupiter	0.048	1.557×10^9	3.743×10^8	2.69×10^{10}
Neptune	0.007	9.05×10^9	5.25×10^9	2.69×10^{10}

*Source: G. Woan, *The Cambridge Handbook of Physics Formulas*, Cambridge University Press (2003).

A simpler way to calculate the period is to start from Eq. (10.12) for angular momentum

$$L = \mu r^2 \frac{d\theta}{dt},$$

which can be written

$$\frac{L}{2\mu} dt = \frac{1}{2} r^2 d\theta.$$

But $(1/2)r^2 d\theta$ is the area element in polar coordinates, so integrating over a complete period T sweeps out the area of the ellipse.

$$\frac{L}{2\mu} T = \text{area of ellipse} = \pi ab.$$

Here a is the semi-major axis, $a = A/2$, so that from Eq. (10.29),

$$a = \frac{C}{-2E}.$$

From Note 10.2, the semi-minor axis b is

$$b = \frac{L}{\sqrt{-2\mu E}}.$$

Using these values,

$$T^2 = \frac{\pi^2 \mu}{2C} A^3$$

as before.

Orbit Eccentricities

The ratio r_{max}/r_{min} is

$$\frac{r_{max}}{r_{min}} = \frac{r_0/(1-\epsilon)}{r_0/(1+\epsilon)}$$

$$= \frac{1+\epsilon}{1-\epsilon}.$$

When ϵ is near zero, $r_{max}/r_{min} \approx 1$ and the ellipse is nearly circular. When ϵ is near 1, the ellipse is very elongated. The shape of the ellipse is determined entirely by ϵ; r_0 only supplies the scale.

Table 10.2 gives the eccentricities of the orbits of the planets, the body Pluto, and Halley's comet. The table reveals why the Ptolemaic theory of circles moving on circles was reasonably successful in dealing with early observations. All the planetary orbits, except those of Mercury and the body Pluto, have eccentricities near zero and are nearly circular. Mercury is never far from the Sun and is hard to observe, and Pluto (no longer classed as a planet) was not discovered until 1930, so that neither of these was an impediment to the Ptolemaists. Mars has the most eccentric orbit of the easily observable planets, and its motion was a stumbling block to the Ptolemaic theory. Kepler discovered his laws of planetary

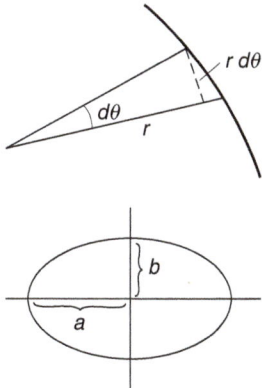

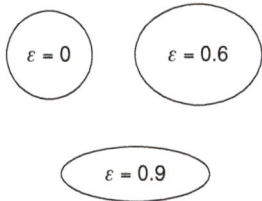

Table 10.2[*]

Planet	ϵ	Planet/body	ϵ
Mercury	0.206	Saturn	0.055
Venus	0.007	Uranus	0.051
Earth	0.017	Neptune	0.007
Mars	0.093	Pluto	0.252
Jupiter	0.048	Halley's comet	0.967

[*]Note 10.2 derives further geometric properties of elliptic orbits.

motion by trying to fit his calculations to Brahe's accurate observations of Mars' orbit.

Orbits

Example 10.5 Geostationary Orbit

For communications purposes, satellites are typically placed in a circular *geosynchronous* orbit. If the orbit is in the equatorial plane of the Earth, it is called *geostationary*. A satellite's orbital speed in a geostationary orbit is set to match the angular velocity Ω_e of the rotating Earth, so that as seen from the Earth the satellite is stationary above a fixed point on the Equator. For a satellite of mass m in a geostationary circular orbit, Eq. (10.30) gives, with $A = 2r, \mu \approx m, C = mgR_e{}^2$, and $v = r\Omega_e$,

$$v^2 = (r\Omega_e)^2 = \frac{2C}{m}\left(\frac{1}{r} - \frac{1}{2r}\right)$$

$$r^3 = \frac{gR_e{}^2}{\Omega_e{}^2}.$$

With $\Omega_e \approx 2\pi/86\,400$ rad/s,

$$r \approx 42\,250 \text{ km}$$

so that the satellite's altitude h above the Earth is

$$h = (42\,250 - 6400) = 35\,850 \text{ km} \approx 22\,280 \text{ mi}.$$

Its orbital speed in the geostationary orbit is $v = r\Omega_e = 3070$ m/s $\approx$ 6870 mi/hr.

Orbit transfer maneuvers are frequently needed in astronautics. For example, in the Apollo flights to the Moon the vehicle was first put into near Earth orbit and then transferred to a trajectory toward the Moon. In order to transfer a spacecraft from one orbit to another, its velocity must be altered at a point where the old and new orbits intersect. The next two examples look at the physical principles of satellite launch and orbit transfer.

Example 10.6 Satellite Orbit Transfer 1

The most energy-efficient way to put a satellite into circular orbit is to launch it into an elliptical transfer orbit whose apogee is at the desired final radius. When the satellite is at apogee, it is accelerated tangentially into the circular orbit. In this example we shall look at the energy required for the first step of the process: putting a satellite into an elliptical transfer orbit. Example 10.7 considers the second step, transferring the satellite into a circular geostationary orbit.

The problem is to find the energy E_{launch} to launch the satellite. E_{launch} is the difference between the satellite's energy E_{orb} in its transfer orbit, and its initial energy E_{ground} on the ground just before launch. To calculate these we will need to find the eccentricity and angular momentum of the orbit, and the satellite's speed at apogee and perigee.

Suppose that the satellite has mass $m = 2000$ kg. Because $m \ll M_e$, we treat the Earth as a fixed center of force and take the reduced mass $\mu \approx m$. The radius of the Earth is taken to be $R_e = 6400$ km. Assume that at perigee the satellite has an altitude of 1100 km above the Earth. The elliptic orbit is chosen so that at apogee the satellite's altitude is 35 850 km, the altitude of the desired geostationary orbit. (The figure is not to scale.)

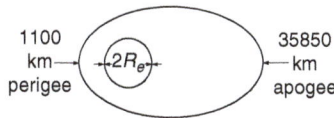

How much energy E_{launch} is required to launch the satellite, and what is its energy in elliptic orbit E_{orb}, its eccentricity ϵ, its angular momentum L, and its orbital speed at perigee and at apogee?

The energy of the satellite on the ground just before launch is

$$E_{\text{ground}} = U(R_e) + K_0$$

where $U(R_e) = -C/R_e = -mgR_e^2/R_e = -mgR_e$ and K_0 is the initial kinetic energy due to the Earth's rotation. If the launch is from the Equator, $K_0 = (1/2)mv_0^2 = (1/2)m(R_e\Omega_e)^2$, where $\Omega_e = 2\pi/86\,400$ rad/s is the angular speed of the Earth. Combining the potential and kinetic energies,

$$\begin{aligned}
E_{\text{ground}} &= -mgR_e + (1/2)m(R_e\Omega_e)^2 \\
&= mR_e(-g + (1/2)R_e\Omega_e^2) \\
&= (2000)(6.4 \times 10^6)[-9.8 + 0.5(6.4 \times 10^6)(2\pi/86\,400)^2] \\
&= (2000)(6.4 \times 10^6)[-9.8 + 0.017] \\
&= -1.25 \times 10^{11}\,\text{J}.
\end{aligned}$$

According to this result, launching from the Equator decreases E_{ground} by less than 0.2%, which might not seem significant. But for a given maneuver, the amount of fuel consumed is roughly proportional to the required change in velocity, so the Earth's rotation helps save fuel. The efficiency of an orbital maneuver is judged by the $\Delta \mathbf{v}$ required, the smaller the better.

The major axis A of the orbit is $(1100 + 6400 + 35\,850 + 6400)$ km $= 5.0 \times 10^7$ m, so we can find E_{orb} from Eq. (10.29).

$$
\begin{aligned}
E_{\text{orb}} &= -\frac{C}{A} \\
&= -\frac{mgR_e^2}{A} \\
&= -\frac{(2 \times 10^3)(9.8)(6.4 \times 10^6)^2}{5.0 \times 10^7} \\
&= -1.61 \times 10^{10} \text{ J.}
\end{aligned}
$$

The energy needed to launch the satellite, neglecting losses due to inefficiencies of the rocket engines, is therefore

$$
E_{\text{launch}} = E_{\text{orb}} - E_{\text{ground}} = 1.09 \times 10^{11} \text{ J.}
$$

Turning now to the problem of finding the angular momentum, we can use Eq. (10.20) to find L from the eccentricity, and we can find the eccentricity from the dimensions of the orbit. Using $r_{\text{min}} = r_0/(1 + \epsilon)$ and $r_{\text{max}} = r_0/(1 - \epsilon)$ we can solve for the eccentricity. We have

$$
r_0 = (1 + \epsilon)r_{\text{min}} = (1 - \epsilon)r_{\text{max}}
$$

from which we find

$$
\begin{aligned}
\epsilon &= \frac{r_{\text{max}} - r_{\text{min}}}{r_{\text{max}} + r_{\text{min}}} \\
&= \frac{r_{\text{max}} - r_{\text{min}}}{A} \\
&= \frac{3.5 \times 10^7 \text{ m}}{5.0 \times 10^7 \text{ m}} \\
&= 0.70.
\end{aligned}
$$

From the definition of ϵ, Eq. (10.20),

$$
\begin{aligned}
\epsilon^2 &= 1 + \frac{2E_{\text{orb}}L^2}{mC^2} \\
&= 1 + \frac{2E_{\text{orb}}L^2}{m(mgR_e^2)^2}
\end{aligned}
$$

which yields

$$
L = 1.43 \times 10^{14} \text{ kg} \cdot \text{m}^2/\text{s.}
$$

At perigee $r_p = 1100 + 6400 = 7.500 \times 10^6$ m and at apogee $r_a = 35\,850 + 6400 = 4.225 \times 10^7$ m. We know the angular momentum, and because the velocity is perpendicular to the radius vector at the extremes of the orbit, we can immediately find the speed at these points. At perigee,

$$L = mr_p v_p$$

$$v_p = \frac{L}{mr_p} = \frac{1.43 \times 10^{14}}{(2000)(7.500 \times 10^6)}$$

$$= 9530 \, \text{m/s} \approx 21\,300 \, \text{mi/hr}.$$

Similarly, at apogee,

$$v_a = \frac{L}{mr_a} = \frac{1.43 \times 10^{14}}{(2000)(4.225 \times 10^7)}$$

$$= 1690 \, \text{m/s} \approx 3800 \, \text{mi/hr}.$$

Alternatively, we could use Eq. (10.30), which gives the same results.

Example 10.7 Satellite Orbit Transfer 2

We now want to transfer the 2000 kg satellite of Example 10.6 into a circular geostationary orbit. As we showed in Example 10.5, its orbital speed in the geostationary orbit will be 3070 m/s but from Example 10.6 its speed at apogee in the elliptic orbit is only 1690 m/s. The rocket engine must therefore give a burst to increase the speed.

If the satellite is moving with velocity $\mathbf{v}$ and the engine boosts the velocity by $\Delta\mathbf{v}$, the increase in energy is then

$$\Delta E = (1/2)m(\mathbf{v} + \Delta\mathbf{v})^2 - (1/2)m(\mathbf{v})^2$$

$$= (1/2)m(\mathbf{v} \cdot \Delta\mathbf{v} + \Delta\mathbf{v}^2).$$

The increase in energy is maximum when the velocity boost is parallel to the orbit. The fuel needed to inject a satellite into a circular orbit is a minimum if the satellite is first launched into an elliptic orbit with apogee at the desired final radius, and then transferred to the desired circular orbit by boosting its speed at apogee, as shown in the sketch. Plans for efficient orbital transfer were first put forth by Walter Hohmann in 1925, a forward-looking German scientist interested in the possibilities of space flight.

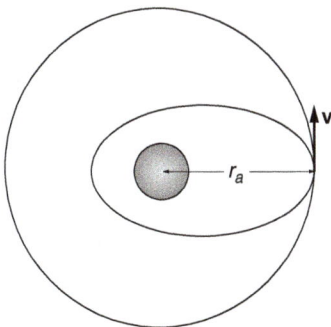

Using Eq. (10.29), $E = -C/A$, the change in energy going from an orbit with major axis A_i to an orbit with major axis A_f is

$$\Delta E = -C\left(\frac{1}{A_f} - \frac{1}{A_i}\right) = -mgR_e^2\left(\frac{1}{A_f} - \frac{1}{A_i}\right). \tag{1}$$

From Example 10.6, $A_i = 5.0 \times 10^7$ m. For the geostationary orbit, $A_f = 2 \times 42\,250$ km $= 8.45 \times 10^7$ m. Substituting in Eq. (1) gives

$$\Delta E = -(2000)(9.8)(6.4 \times 10^6)^2 \left(\frac{1}{8.45 \times 10^7} - \frac{1}{5.0 \times 10^7} \right)$$

$$= 6.6 \times 10^9 \text{ J}.$$

This increases the orbital energy from -16.1×10^9 J for the elliptical orbit to -9.5×10^9 J for the geostationary orbit.

Similar considerations apply when a spacecraft on a space mission returns to Earth. First it is slowed enough to be captured in a circular orbit, and then at the proper time it is transferred to an elliptic orbit that intersects the Earth.

Example 10.8 Trojan Asteroids and Lagrange Points

The Trojan asteroids are a remarkable feature of the solar system. There is a clump of several hundred asteroids traveling along Jupiter's orbit preceding the planet, and another clump on the orbit trailing Jupiter an equal distance behind. Furthermore, the Sun, Jupiter, and each clump are located at the vertices of an equilateral triangle. The asteroids are called Trojan asteroids, named after characters in Homer's *Iliad*, the story of some events in the Greek siege of Troy. Trojan asteroids have been observed associated with other planets, including at least one on the Earth's orbit.

The problem of three gravitating masses has never been solved in general, but this example discusses a special restricted case with known solution. Consider a planet (mass M_p) orbiting the Sun (mass M_s) and an asteroid (mass m), all located at the vertices of an equilateral triangle. Assume that the asteroid has small mass, so that its gravitational force has a negligible effect on the motion of the Sun and the planet. Assume also that the Sun and the planet are in circular orbit about their center of mass. This is nearly the case for Jupiter, because of its small eccentricity ($\epsilon = 0.048$).

The sketch shows the geometry, with R_0 the length of each side of the triangle. We take the origin of the coordinate system to be at the center of mass of the Sun and planet (m is assumed to be small in our treatment, $m \ll M_p$). Then, by definition of the center of mass,

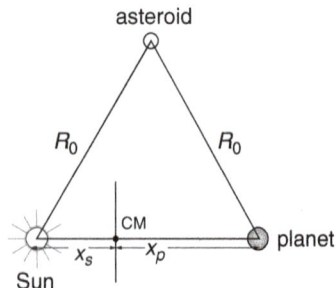

$$0 = \frac{M_p x_p - M_s x_s}{M_p + M_s}. \tag{1}$$

Using $x_p + x_s = R_0$, we can solve for x_p and x_s:

$$x_p = \frac{M_s R_0}{M_s + M_p} \tag{2a}$$

$$x_s = \frac{M_p R_0}{M_s + M_p}. \tag{2b}$$

The equation of motion for the planet in circular orbit with angular velocity Ω about the center of mass is

$$M_p\, x_p\, \Omega^2 = \frac{GM_pM_s}{R_0{}^2}$$

$$\Omega^2 = \frac{G(M_s + M_p)}{R_0{}^3} \tag{3}$$

where we have used Eq. (2a) for x_p.

Let $\mathbf{r}_1$ be the vector from the asteroid to the Sun, $\mathbf{r}$ the vector from the asteroid to the center of mass, and $\mathbf{r}_2$ the vector from the asteroid to the planet, as shown. Because the triangle is equilateral, $|\mathbf{r}_1| = |\mathbf{r}_2| = R_0$. The gravitational force $\mathbf{F}$ on the asteroid due to the Sun and the planet is

$$\mathbf{F} = \frac{GmM_s}{r_1{}^3}\, \mathbf{r_1} + \frac{GmM_p}{r_2{}^3}\, \mathbf{r_2}$$

$$= \left(\frac{Gm}{R_0{}^3}\right)\left(M_s\, \mathbf{r_1} + M_p\, \mathbf{r_2}\right).$$

Using

$$\mathbf{r}_1 = \mathbf{r} - x_s\, \hat{\mathbf{i}}$$

$$\mathbf{r}_2 = \mathbf{r} + x_p\, \hat{\mathbf{i}}$$

we have

$$\mathbf{F} = \left(\frac{Gm}{R_0{}^3}\right)\left[(M_s + M_p)\mathbf{r} + (M_p x_p - M_s x_s)\hat{\mathbf{i}}\right]$$

$$= \left(\frac{Gm(M_s + M_p)}{R_0{}^3}\right)\mathbf{r}$$

$$= m\Omega^2\mathbf{r}$$

where $M_p x_p - M_s x_s = 0$ according to Eq. (1), and where we have used the result for Ω^2 from Eq. (3).

Our result shows that $\mathbf{F}$ is radially inward, directed toward the center of mass and that the asteroid's angular velocity about the center of mass is the same as the angular velocity of the Sun–planet system.

In a coordinate system rotating with angular velocity Ω about the center of mass, the inward gravitational force on the asteroid is balanced by

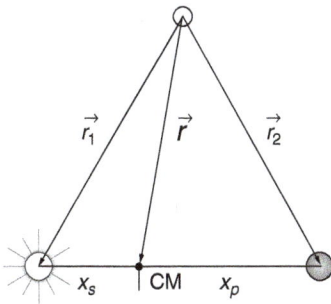

the outward fictitious centrifugal force, so the vertex of the triangle is an equilibrium point where the net force on the asteroid in the rotating system is zero. The net force on the Sun and on the planet is similarly zero. The Sun, planet, and asteroid therefore retain their triangular configuration as the system circles about the center of mass.

The equilibrium points of a three-body system in circular motion were discovered by the Italian mathematician Joseph Louis Lagrange toward the end of the eighteenth century but the first observation of a Trojan asteroid was not made until early in the twentieth century, in the Sun–Jupiter system. Lagrange calculated that with three gravitating masses in circular motion, the third mass can be at one of five special locations, now called *Lagrange points*, where it is in equilibrium and can in principle remain in a fixed configuration relative to the Sun and planet. The Lagrange point at the vertex of the triangle in this example is called L_4, and the symmetric one trailing the planet is L_5. The other three Lagrange points L_1, L_2, L_3 are collinear with the Sun and the planet.

An asteroid near L_4 or L_5 is stable, moving harmonically (like $A \sin \omega t + B \cos \omega t$) for small displacements from equilibrium. The motion of an asteroid near L_1, L_2, or L_3 can vary both harmonically and exponentially for small displacements, so the asteroid is only conditionally stable at the three collinear Lagrange points. For space research, an artificial satellite placed at L_1, L_2, or L_3 could stay in a fixed configuration with only modest use of fuel, if the initial conditions are chosen to minimize the exponential terms.

Lagrange made other substantial contributions to mechanics. He reformulated Newtonian mechanics in a more powerful form that eliminates reference to forces in favor of the more fundamental concepts of kinetic and potential energy.

Example 10.9 Cosmic Keplerian Orbits and the Mass of a Black Hole

The crowning achievement of Newtonian mechanics was the explanation of Kepler's laws of planetary motion from the laws of motion and the law of universal gravitation. These laws also appear to hold on a galactic or cosmic scale far larger than our solar system, and their application can give us information about a remarkable object, a *black hole*.

A black hole is so compact and so massive that not even light can escape its gravitational field. It is non-Newtonian, and must be described using concepts of space and time from general relativity. Nevertheless, the gravitational field outside the black hole follows Newton's inverse-square law. There is now overwhelming evidence that a massive black hole is at the center of our galaxy, based on observations that stars follow elliptic orbits about a mass invisible to optical telescopes. The black hole, in the constellation Sagittarius (the "Archer"), is named Sagittarius A-star (abbreviated Sgr A*). No radiation can escape, but radiation is emitted by material falling in. A strong image of Sgr A* has been detected in the radio-frequency region of the electromagnetic spectrum. Our Sun is rotating about the galactic center at 1000 km/s, but measurements show that Sgr A* is nearly stationary, evidence that it is at or near the galactic center.

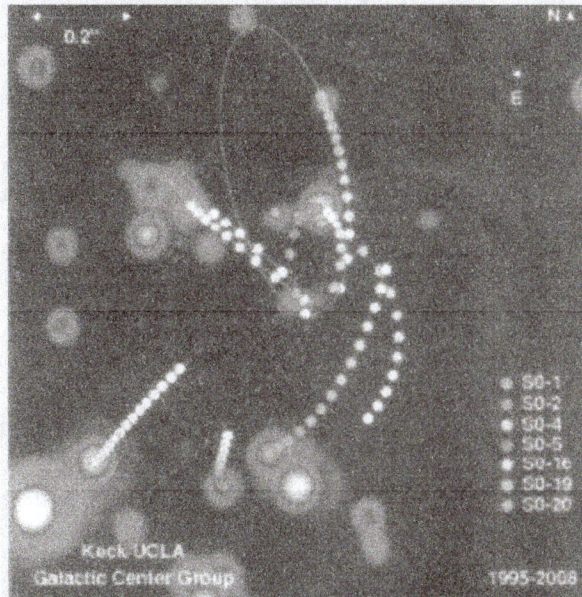

Partial orbits, marked by dots, of seven stars near Sgr-A* at the center of our galaxy. The data were taken at uniform time intervals between 1995 and 2008 at the Very Large Telescope of the European Southern Observatory in northern Chile. (Figure courtesy of A. G. Hey.)

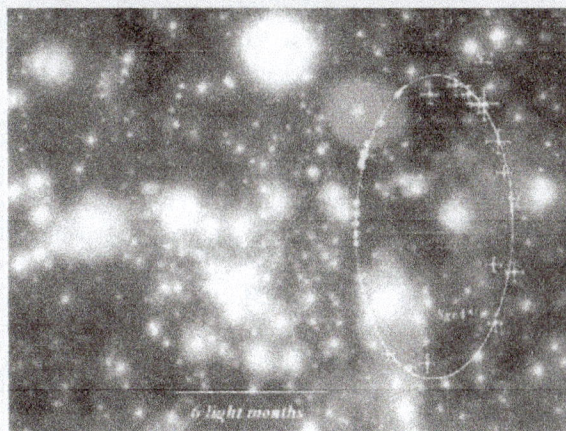

The cosmic elliptic orbit of star S2, superimposed on the background star field at the galactic center but magnified by 100. The observations extend over the entire orbital period of 15.8 years. The plane of the orbit does not lie in the plane of the paper. The eccentricity of S2's orbit is approximately 0.87. (Figure courtesy of R. Genzel.)

S2's orbit is huge: the major axis A is ≈ 11 light days $= 2.9 \times 10^{11}$ km. The eccentricity of the orbit is $\epsilon = 0.88$. To put the orbit's size in perspective, it takes light only 86 minutes to travel across the full diameter of Jupiter's orbit.

We can now estimate the mass of the black hole relative to the mass of the Sun, using the measured values and Kepler's third law $M \propto A^3/T^2$ for a massive attractor

$$\frac{M_{\text{black hole}}}{M_{\text{Sun}}} = \frac{(2.9 \times 10^{11})^3/(5.0 \times 10^8)^2 \, \text{km}^3 \cdot \text{s}^{-2}}{2.7 \times 10^{10} \, \text{km}^3 \cdot \text{min}^{-2}}$$
$$\approx 4 \times 10^6,$$

where we have used $A^3/T^2 = 2.7 \times 10^{10}$ for the Sun from Table 10.1. Sgr A* has a mass approximately 4×10^6 times the mass of our Sun. Estimates of the size of Sgr A*, based on radio observations, suggest that it is incredibly dense, hundreds of times denser than water. Sgr-A* is actually a modest black hole, in comparison to some known to be billions of times more massive than our Sun.

The discussion in this example draws on Mark J. Reid, Is there a super-massive black hole at the center of the Milky Way?, *J. Modern Physics D* 18, 889 (2009).

10.6 Some Concluding Comments on Planetary Motion

The description of two particles interacting under an inverse-square gravitational force according to Newtonian mechanics predicts planetary motion surprisingly well considering that our solar system consists of eight planets and numerous smaller bodies. We might therefore expect to observe large departures from Kepler's laws. Fortunately, the mass of the Sun is so large compared to the masses of the planets that the influence of the planets on one another is minor, and the effects can be treated as small corrections. Techniques for such calculations, called *perturbation theory*, were highly developed in the nineteenth century, and in 1930 observed perturbations in the orbit of Neptune helped Clyde Tombaugh in Arizona discover Pluto.

In the nineteenth century, fundamental problems arose with the laws of mechanics themselves. Astronomical observations showed that the perihelion of Mercury's orbit is slowly precessing, although for an ideal two-body system the perihelion should remain stationary. "Slowly" here means only 574 seconds of arc per century, small but accurately measurable. Taking perturbations of other planets into account predicted a precession rate of 531 seconds of arc per century. Although this was very close to the observed rate, Einstein was not satisfied. The failure of classical physics to predict Mercury's motion precisely was an essential piece of evidence in the development of his general theory of relativity. In 1915, Einstein used general relativity to predict an additional precession of 43 seconds of arc per century, resolving the earlier discrepancy and marking a great success for general relativity and giving Einstein great confidence in the truth of his revolutionary new theory.

Ancient peoples viewed celestial events like comets and eclipses with both wonder and foreboding, because they thought that any changes in the supposedly immutable heavens foretold upcoming great events, good or bad. Newtonian mechanics swept away these ideas and became widely accepted by the public soon after it was introduced. Salon philosophers saw that Newtonian mechanics described the universe as a mechanical system, and they began to picture the universe as a great deterministic mechanical clock, which, once wound and set to initial conditions, would run on its predetermined way to the end of time.

A fundamental problem with the deterministic clock model came to light early in the twentieth century when the French physicist and mathematician Henri Poincaré discovered the phenomenon of chaos in mechanics. Newtonian mechanics is remarkably accurate, able to predict eclipses from thousands of years in the past to thousands of years in the future. In a chaotic system, however, a small change in the initial conditions can lead to exponentially divergent behavior at later times. The change grows like $e^{\Lambda t}$, where Λ is known as the *Lyapounov characteristic exponent*. Because a chaotic system is not calculable, we cannot make accurate predictions of its future behavior. Weather systems on the Earth appear to be chaotic, with small changes in one place possibly causing large effects in another (the "butterfly effect"). Even planetary motion, the cornerstone of Newtonian mechanics, can exhibit chaotic behavior. The characteristic exponent Λ for the Earth's orbit is in the neighborhood of 4 to 5 million years. Luckily for us, the effect is evidently not large; life has existed on the Earth for 600 million years, but during that time the Earth's orbit has not changed enough to cause the Earth to become either too hot or too cold to sustain life.

Four centuries have passed since Kepler discovered his empirical laws. Soon after, Galileo used an early telescope to observe that the moons of Jupiter constituted a solar system in miniature, and then Newton formulated the laws of dynamics and gravitation. Today, the dynamics of chaos and the dynamics of intense gravitational fields near black holes are among the frontier topics of science. Physics, it would seem, is inexhaustible.

Note 10.1 Integrating the Orbit Integral

In this Note we integrate the orbit integral

$$\theta - \theta_0 = L \int \frac{dr}{r\sqrt{(2\mu Er^2 + 2\mu Cr - L^2)}}. \tag{1}$$

Make the substitution

$$r = \frac{1}{s - \alpha}$$

$$dr = -\frac{ds}{(s - \alpha)^2}$$

$$\frac{dr}{r} = -\frac{ds}{(s - \alpha)}$$

where α is a constant to be determined.

With this substitution, the integral in Eq. (1) becomes

$$\theta - \theta_0 = L \int \frac{ds}{(s - \alpha) \sqrt{\frac{2\mu E}{(s-\alpha)^2} + \frac{2\mu C}{(s-\alpha)} - L^2}}$$

$$= -L \int \frac{ds}{\sqrt{2\mu E + 2\mu C(s - \alpha) - L^2(s - \alpha)^2}}$$

$$= -L \int \frac{ds}{\sqrt{2\mu E + 2\mu C s - 2\mu C \alpha - L^2 s^2 + 2L^2 \alpha s - L^2 \alpha^2}}.$$

Now we choose $\alpha = -\mu C/L^2$, which makes the term linear in s vanish, leaving

$$\theta - \theta_0 = -L \int \frac{ds}{\sqrt{2\mu E - 2\mu C \alpha - L^2 s^2 - L^2 \alpha^2}}$$

$$= -L \int \frac{ds}{\sqrt{2\mu E + \frac{2(\mu C)^2}{L^2} - L^2 s^2 - \frac{(\mu C)^2}{L^2}}}$$

$$= -L \frac{ds}{\sqrt{2\mu E + \frac{(\mu C)^2}{L^2} - L^2 s^2}}$$

$$= -L^2 \int \frac{ds}{\sqrt{2\mu E L^2 + (\mu C)^2 - L^4 s^2}}.$$

This integral can be put into the standard form for the arcsin, $\sin \alpha = \int_0^\alpha dx/\sqrt{1 + x^2}$, so we have

$$\theta - \theta_0 = -\arcsin\left(s \sqrt{\frac{L^4}{2\mu E L^2 + (\mu C)^2}} \right)$$

$$\sin(\theta - \theta_0) = -\frac{sL^2}{\sqrt{2\mu E L^2 + (\mu C)^2}}$$

and using $s = 1/r + \alpha = 1/r - \mu C/L^2$,

$$\sin(\theta - \theta_0) = \frac{\mu C - L^2/r}{\sqrt{2\mu E L^2 + (\mu C)^2}}. \tag{2}$$

Finally, solving Eq. (2) for r gives

$$\frac{L^2}{r} = \mu C - \sqrt{2\mu EL^2 + (\mu C)^2} \sin(\theta - \theta_0)$$

$$r = \frac{L^2}{\mu C - \sqrt{2\mu EL^2 + (\mu C)^2} \sin(\theta - \theta_0)}$$

$$= \frac{(L^2/\mu C)}{1 - \sqrt{1 + (2EL^2/\mu C^2)} \sin(\theta - \theta_0)}$$

in agreement with Eq. (10.18).

Note 10.2 Properties of the Ellipse

The equation in polar coordinates of an ellipse is

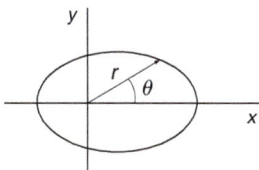

$$r = \frac{r_0}{1 - \epsilon \cos \theta}. \tag{1}$$

Converting to Cartesian coordinates $r = \sqrt{x^2 + y^2}$, $x = r \cos \theta$, Eq. (1) becomes

$$(1 - \epsilon^2)x^2 - 2r_0 x + y^2 = r_0^2. \tag{2}$$

The ellipse corresponds to the case $0 \leq \epsilon < 1$. The ellipse described by Eqs. (1) and (2) is symmetrical about the x axis, but its center does not lie at the origin.

We can use Eq. (1) to determine the important dimensions of the ellipse. The maximum value of r, which occurs at $\theta = 0$, is

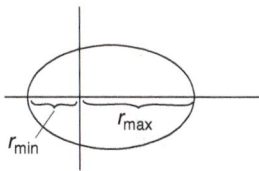

$$r_{max} = \frac{r_0}{1 - \epsilon}.$$

The minimum value of r, which occurs at $\theta = \pi$, is

$$r_{min} = \frac{r_0}{1 + \epsilon}.$$

The major axis A is

$$A = r_{max} + r_{min}$$

$$= r_0 \left(\frac{1}{1 - \epsilon} + \frac{1}{1 + \epsilon} \right)$$

$$= \frac{2r_0}{1 - \epsilon^2}. \tag{3}$$

The semi-major axis a is

$$a = \frac{A}{2}$$

$$= \frac{r_0}{1 - \epsilon^2}. \tag{4}$$

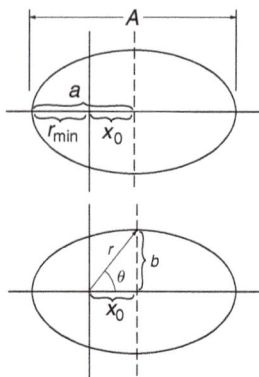

The distance from the origin to the center of the ellipse is

$$x_0 = a - r_{min}$$

$$= r_0\left(\frac{1}{1 - \epsilon^2} - \frac{1}{1 + \epsilon}\right)$$

$$= \frac{r_0\epsilon}{1 - \epsilon^2}. \tag{5}$$

Comparing Eqs. (4) and (5), we see that $\epsilon = x_0/a$.

To find the length of the semi-minor axis $b = \sqrt{r^2 - x_0^2}$, note that the tip of the semi-minor axis has angular coordinates given by $\cos\theta = x_0/r$. We have

$$r = \frac{r_0}{1 - \epsilon\cos\theta}$$

$$= \frac{r_0}{1 - \epsilon x_0/r}$$

or

$$r = r_0 + \epsilon x_0 = r_0\left(1 + \frac{\epsilon^2}{1 - \epsilon^2}\right)$$

$$= \frac{r_0}{1 - \epsilon^2}.$$

Hence

$$b = \sqrt{r^2 - x_0^2} = \left(\frac{r_0}{1 - \epsilon^2}\right)\sqrt{1 - \epsilon^2}$$

$$= \frac{r_0}{\sqrt{1 - \epsilon^2}}.$$

Finally, we shall prove that the origin lies at a focus of the ellipse. According to the definition of an ellipse, the sum of the distances from the two foci to a point on the ellipse is a constant. We start by assuming that one focus is at the origin, and by symmetry the other focus is therefore at $2x_0$, because the distance from the first focus to the center of the ellipse is x_0.

Let r and r' be the distances from the foci to a point on the ellipse, as shown in the sketch. We shall now show that $r + r' = $ constant, justifying our initial assumption.

By the law of cosines,

$$r'^2 = r^2 + 4x_0^2 - 4rx_0\cos\theta. \tag{6}$$

From Eq. (1) we find that

$$r\cos\theta = \frac{r - r_0}{\epsilon}.$$

Equation (6) becomes

$$r'^2 = r^2 + 4x_0^2 - \frac{4rx_0}{\epsilon} + \frac{4r_0x_0}{\epsilon}.$$

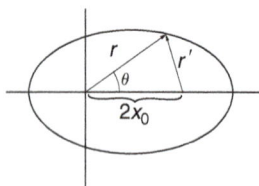

Using the relation $x_0 = r_0\epsilon/(1 - \epsilon^2)$ from Eq. (5) gives

$$r'^2 = r^2 - \left(\frac{4r_0}{1 - \epsilon^2}\right)r + \frac{4r_0^2\epsilon^2}{(1 - \epsilon^2)^2} + \frac{4r_0^2}{(1 - \epsilon^2)}$$

$$= r^2 - \left(\frac{4r_0}{1 - \epsilon^2}\right)r + \frac{4r_0^2}{(1 - \epsilon^2)^2}.$$

The right-hand side is a perfect square, so

$$r' = \pm\left(r - \frac{2r_0}{1 - \epsilon^2}\right)$$

$$= \pm(r - A).$$

Since $A > r$, we must choose the negative sign to keep $r' > 0$. Therefore

$$r' + r = A$$

$$= \text{constant},$$

which supports our assumption that a focus is at the origin.

To conclude, we list a few of our results in terms of E, l, μ, C for the inverse-square force problem $U(r) = -C/r$. When using these formulas, E must be taken to be a negative number. From Eqs. (10.19) and (10.20),

$$r_0 = \frac{l^2}{\mu C}$$

and

$$\epsilon = \sqrt{1 + 2El^2/(\mu C^2)}.$$

Hence

$$\text{semi-major axis } a = \frac{r_0}{1 - \epsilon^2} = \frac{C}{-2E}$$

$$\text{semi-minor axis } b = \frac{r_0}{\sqrt{1 - \epsilon^2}} = \frac{1}{\sqrt{-2\mu E}}$$

$$\frac{\text{semi-minor axis}}{\text{semi-major axis}} = \frac{b}{a} = \sqrt{1 - \epsilon^2} = \sqrt{\frac{-2E}{\mu C^2}}$$

and the distance of the focus from the origin is

$$x_0 = \frac{r_0\epsilon}{1 - \epsilon^2} = \left(\frac{C}{-2E}\right)\sqrt{1 + \frac{2El^2}{\mu C^2}}.$$

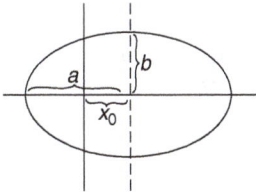

Problems

For problems marked *, refer to page 524 for a hint, clue, or answer.

10.1   *Equations of motion*
       Obtain Eqs. (10.4a) and (10.4b) by differentiating Eqs. (10.5) and (10.6b) with respect to time.

10.2 r^3 *central force**

A particle of mass 50 g moves under an attractive central force of magnitude $4r^3$ dynes. The angular momentum is equal to 1000 g·cm²/s.

(*a*) Find the effective potential energy.

(*b*) Indicate on a sketch of the effective potential the total energy for circular motion.

(*c*) The radius of the particle's orbit varies between r_0 and $2r_0$. Find r_0.

10.3   *Motion with* $1/r^3$ *central force*

A particle moves in a circle under the influence of an inverse-cube law force. Show that the particle can also move with uniform radial velocity, either in or out. (This is an example of neutral stability. Any slight perturbation to the circular orbit will start the particle moving radially, and it will continue to move uniformly.) Find θ as a function of r for motion with uniform radial velocity v.

10.4   *Possible stable circular orbits*

For what values of n are circular orbits stable with the potential energy $U(r) = -A/r^n$, where $A > 0$?

10.5   *Central spring force*

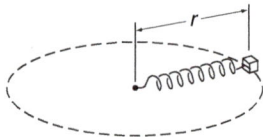

A 2-kg mass on a frictionless table is attached to one end of a massless spring. The other end of the spring is held by a frictionless pivot. The spring produces a force of magnitude $3r$ newtons on the mass, where r is the distance in meters from the pivot to the mass. The mass moves in a circle and has a total energy of 12 J.

(*a*) Find the radius of the orbit and the velocity of the mass.

(*b*) The mass is struck by a sudden sharp blow, giving it instantaneous velocity of 1 m/s radially outward. Show the state of the system before and after the blow on a sketch of the energy diagram.

(*c*) For the new orbit, find the maximum and minimum values of r.

10.6 r^4 *central force*

A particle of mass m moves under an attractive central force Kr^4 with angular momentum l. For what energy will the motion be circular, and what is the radius of the circle? Find the frequency of radial oscillations if the particle is given a small radial impulse.

10.7   *Transfer to escape*

A rocket is in elliptic orbit around the Earth. To put it into an escape orbit, its engine is fired briefly, changing the rocket's velocity by ΔV. Where in the orbit, and in what direction, should the firing occur to attain escape with a minimum value of ΔV?

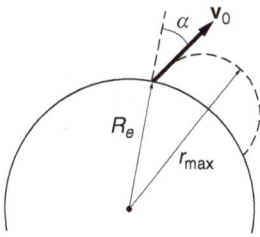

10.8 *Projectile rise**

A projectile of mass m is fired from the surface of the Earth at an angle α from the vertical. The initial speed v_0 is equal to $\sqrt{GM_e/R_e}$. How high does the projectile rise? Neglect air resistance and the Earth's rotation.

10.9 *Halley's comet*

Halley's comet is in an elliptic orbit about the Sun. The eccentricity of the orbit is 0.967 and the period is 76 years. The mass of the Sun is 2×10^{30} kg, and $G = 6.67 \times 10^{-11}$ N·m²/kg².

(*a*) Using these data, determine the distance of Halley's comet from the Sun at perihelion and at aphelion.

(*b*) What is the speed of Halley's comet when it is closest to the Sun?

10.10 *Satellite with air friction**

(*a*) A satellite of mass m is in circular orbit about the Earth. The radius of the orbit is r_0 and the mass of the Earth is M_e. Find the total mechanical energy of the satellite.

(*b*) Now suppose that the satellite moves in the extreme upper atmosphere of the Earth where it is retarded by a constant feeble friction force f. The satellite will slowly spiral toward the Earth. Since the friction force is weak, the change in radius will be very slow. We can therefore assume that at any instant the satellite is effectively in a circular orbit of average radius r. Find the approximate change in radius per revolution of the satellite, Δr.

(*c*) Find the approximate change in kinetic energy ΔK of the satellite per revolution.

10.11 *Mass of the Moon*

Before landing astronauts on the Moon, the Apollo 11 space vehicle was put into orbit about the Moon. The mass of the vehicle was 9979 kg and the period of the orbit was 120 min. The maximum and minimum distances from the center of the Moon were 1861 km and 1838 km. Assuming the Moon to be a uniform spherical body, what is the mass of the Moon according to these data? $G = 6.67 \times 10^{-11}$ N·m²/kg².

10.12 *Hohmann transfer orbit*

A space vehicle is in circular orbit about the Earth. The mass of the vehicle is 3000 kg and the radius of the orbit is $2R_e = 12\,800$ km. It is desired to transfer the vehicle to a circular orbit of radius $4R_e$.

(*a*) What is the minimum energy expenditure required for the transfer?

(*b*) An efficient way to accomplish the transfer is to use a semi-elliptical orbit (known as a Hohmann transfer orbit), as shown. What velocity changes are required at the points of intersection, A and B?

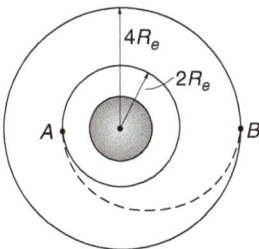

10.13 *Lagrange point L_1*

    The Lagrange point L_1 is on a line between the Sun and Jupiter, at approximately 5.31×10^{10} m from Jupiter. The Sun–Jupiter distance is 7.78×10^{11} m, the mass of the Sun is 1.99×10^{30} kg, and the mass of Jupiter is 1.90×10^{27} kg. The period of Jupiter is 4330 days.

    An asteroid of small mass is located at L_1.

    (*a*) Write the equation of motion for the asteroid in equilibrium in the rotating system.

    (*b*) Using the numerical data, show that the equation of motion is satisfied to good accuracy.

    (*c*) There are three Lagrange points on the Sun–Jupiter axis. Show, on physical grounds, where the other two can be found. (Part *c* is qualitative: the exact solution requires finding the three real roots of a fifth-order polynomial.)

10.14 *Speed of S2 around Sgr A**

    Using the data in Example 10.9, what is the maximum speed of the star S2 as it orbits the black hole Sgr A*? For comparison, the speed of the Earth about the Sun is 30 km/s.

10.15 *Sun–Earth mass ratio*

    Kepler's laws also apply to the motion of satellites around a planet. The table shows A^3/T^2 for a number of Earth satellites. The ratio A^3/T^2 is constant to a fraction of a percent, although the periods vary by ≈ 35.

    Using these data and taking the major diameter of Earth's solar orbit to be $A_e = 2.99 \times 10^8$ km, calculate the mass of the Sun relative to the mass of the Earth, M_s/M_e.

SATELLITE	ϵ	A, km	T, s	A^3/T^2
Amsat-Oscar 7 (1974)	1.28×10^{-3}	1.566×10^4	6.894×10^3	8.08×10^4
Geotail (1992)	0.83	1.21×10^5	1.485×10^5	8.03×10^4
Apostar 1A (1996)	1.30×10^{-4}	8.433×10^4	8.616×10^4	8.08×10^4
Integral (2002)	0.897	1.679×10^5	2.420×10^5	8.08×10^4
Cosmos 2431 (2007)	1.94×10^{-3}	5.102×10^4	4.06×10^4	8.07×10^4

Based on UCS Satellite Database, Union of Concerned Scientists, with the diameter of the Earth taken to be 12,757 km.

11 THE HARMONIC OSCILLATOR

11.1 Introduction

The harmonic oscillator plays a loftier role in physics than one might guess from its humble origin: a mass bouncing at the end of a spring. The harmonic oscillator underlies the creation of sound by musical instruments, the propagation of waves in media, the analysis and control of vibrations in machinery and airplanes, and the time-keeping crystals in digital watches. Furthermore, the harmonic oscillator arises in numerous atomic and optical quantum scenarios, in quantum systems such as lasers, and it is a recurrent motif in advanced quantum field theories. In short, if there were a competition for a logo for the universality of physics, the harmonic oscillator would make a pretty strong contender.

We encountered simple harmonic motion—the periodic motion of a mass attached to a spring—in Chapter 3. The treatment there was highly idealized because it neglected friction and the possibility of a time-dependent driving force. It turns out that friction is essential for the analysis to be physically meaningful and that the most interesting applications of the harmonic oscillator generally involve its response to a driving force. In this chapter we will look at the harmonic oscillator including friction, a system known as the *damped harmonic oscillator*, and then examine how the system behaves when driven by a periodic applied force, a system called the *driven harmonic oscillator.*

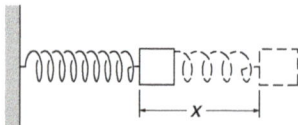

11.2 Simple Harmonic Motion: Review

To establish the notation we briefly review simple harmonic motion, the motion of the ideal harmonic oscillator introduced in Section 3.7: a mass m moves under the influence of the spring force $F_{spring} = -kx$, where x is the displacement from equilibrium. The equation of motion is $m\ddot{x} = -kx$, which is written in the standard form

$$\ddot{x} + \omega_0^2 x = 0 \tag{11.1}$$

where

$$\omega_0 = \sqrt{\frac{k}{m}}.$$

The solution is

$$x = X_0 \cos(\omega_0 t + \phi), \tag{11.2}$$

where X_0 and ϕ are arbitrary constants that can be chosen to make the general solution meet any two given independent initial conditions. Typically these are the position and velocity at a time taken to be $t = 0$. Note that we are now denoting the natural frequency by the symbol ω_0 rather than ω, as in Section 3.7. The solution can be cast in a different form by using the trigonometric identity $\cos(\alpha + \beta) = \cos\alpha\cos\beta - \sin\alpha\sin\beta$. Applying this to Eq. (11.2) casts the solution into the form

$$x = B \cos\omega_0 t + C \sin\omega_0 t \tag{11.3}$$

where

$$X_0 = \sqrt{B^2 + C^2},$$

$$\phi = \arctan\left(-\frac{B}{C}\right). \qquad (11.4)$$

We shall generally use Eq. (11.2) as the standard form for the motion of the ideal (frictionless) harmonic oscillator.

11.2.1 Nomenclature
In the expression

$$x = X_0 \cos(\omega_0 t + \phi)$$

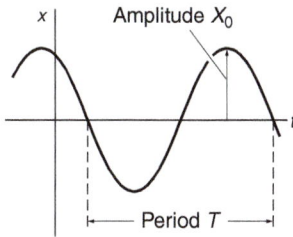
Amplitude X_0
Period T

X_0 is the *amplitude* of the motion (the distance from zero displacement to a maximum) while ω_0 is the *frequency* (more precisely, the *angular frequency*) of the oscillator. Angular frequency is measured in units of radians per second: $\omega_0 = \sqrt{k/m}$ rad/s. Because the radian is dimensionless, angular frequency is generally written in units of s^{-1}. The *circular frequency* ν is the frequency expressed as revolutions per second or cycles per second. $\nu = \omega_0/2\pi$ hertz, where one hertz (1 Hz) = 1 cycle per second. The quantity $\omega_0 t + \phi$ is the *phase* of the oscillation at time t, and ϕ is known as the *phase constant*. The *period* of the motion—the time for the system to execute one complete cycle—is $T = 2\pi/\omega_0$.

Example 11.1 Incorporating Initial Conditions
At time $t = 0$ the position of the mass of a harmonic oscillator is observed to be $x(0)$ and its velocity is $v(0)$. From Eq. 11.2, the displacement and velocity are given by

$$x = X_0 \cos(\omega_0 t + \phi)$$
$$v = -\omega_0 X_0 \sin(\omega_0 t + \phi).$$

Evaluating x and v at $t = 0$ gives

$$x(0) = X_0 \cos\phi$$
$$v(0) = -\omega_0 X_0 \sin\phi.$$

The complete solution is given by Eq. (11.2) with

$$X_0 = \sqrt{x(0)^2 + (v(0)/\omega_0)^2}$$

$$\phi = \arctan\left(\frac{-v(0)}{\omega_0 x(0)}\right).$$

Many other initial conditions are possible. Any two independent pieces of information are sufficient to provide a complete solution. These could be the position at two times or the position at one time, the velocity at another. Values of the position and the acceleration at a given time would not be sufficient, because they are related by $m\ddot{x} = -kx$, so their values are not independent.

11.2.2 Energy of the Harmonic Oscillator

A harmonic oscillator possesses kinetic energy from its translational motion, and potential energy from its spring. The kinetic energy is

$$K(t) = \tfrac{1}{2}mv^2 = \tfrac{1}{2}m\omega_0^2 X_0^2 \sin^2(\omega_0 t + \phi)$$
$$= \tfrac{1}{2}kX_0^2 \sin^2(\omega_0 t + \phi), \tag{11.5}$$

where we have used $v = \dot{x} = -\omega_0 X_0 \sin(\omega_0 t + \phi)$, and $m = k/\omega_0^2$.

The potential energy, which is taken as zero for the unstretched spring, is

$$U(t) = \tfrac{1}{2}kx^2$$
$$= \tfrac{1}{2}kX_0^2 \cos^2(\omega_0 t + \phi). \tag{11.6}$$

The total energy of the harmonic oscillator is therefore

$$E = K(t) + U(t) = \tfrac{1}{2}kX_0^2[\cos^2(\omega_0 t + \phi) + \sin^2(\omega_0 t + \phi)]$$
$$= \tfrac{1}{2}kX_0^2. \tag{11.7}$$

The total energy is constant, a familiar feature of motion in systems where the forces are conservative.

11.3 The Damped Harmonic Oscillator

The ideal harmonic oscillator is frictionless. It turns out that friction is often essential and omitting it can lead to absurd predictions. Let us therefore examine the effect of a viscous friction force $f_{\text{fric}} = -bv$, as discussed in Section 3.6. This type of friction is most often encountered, so our analysis will therefore be widely applicable. For example, in the case of oscillations in electromagnetic circuits, the electrical resistance of the circuit is precisely analogous to a viscous retarding force.

The total force acting on the mass m is

$$F = F_{\text{spring}} + f_{\text{fric}}$$
$$= -kx - bv.$$

The equation of motion is

$$m\ddot{x} = -kx - b\dot{x},$$

which can be written in the standard form

$$\ddot{x} + \gamma\dot{x} + \omega_0^2 x = 0 \tag{11.8}$$

where

$$\gamma = b/m, \tag{11.9}$$

and $\omega_0^2 = k/m$, as before.

This is the first time we have encountered a differential equation in the form of Eq. (11.8). A method for finding its solution is described in Note 11.2. However, we shall attempt to guess the solution from the physics of the situation because this can yield insights that the formal solution may hide.

If friction were negligible the motion would be given by

$$x = X_0 \cos(\omega_0 t + \phi).$$

On the other hand, if the spring force were negligible the mass would move according to $v = v_0 e^{-(b/m)t}$, as shown in Section 3.6. We might therefore guess that the solution to Eq. (11.8) is of the form

$$x = X_0 e^{-\alpha t} \cos(\omega_1 t + \phi) \tag{11.10}$$

where, if our guess is correct, the constants α and ω_1 can be chosen to make this trial solution satisfy Eq. (11.8). X_0 and ϕ are arbitrary constants for satisfying the initial conditions. Substituting the trial solution in the equation of motion Eq. (11.8), we find that the equation is satisfied provided that

$$\alpha = \gamma/2, \tag{11.11}$$

$$\omega_1 = \sqrt{\omega_0^2 - (\gamma/2)^2}, \tag{11.12}$$

where $\gamma = b/m$ and $\omega_0 = \sqrt{k/m}$, as before.

This solution is valid when $\omega_0^2 - \gamma^2/4 > 0$. Other cases are discussed in Note 11.2.

The motion described by Eq. (11.10) is known as *damped harmonic motion*. Several examples are shown in the sketches for increasing values of $\gamma/(2\omega_1)$. The motion is reminiscent of the undamped harmonic motion described in the last section. To emphasize this, we can rewrite Eq. (11.10) as

$$x = X_0 e^{-(\gamma/2)t} \cos(\omega_1 t + \phi), \tag{11.13}$$

or

$$x = X(t) \cos(\omega_1 t + \phi), \tag{11.14}$$

where

$$X(t) = X_0 e^{-(\gamma/2)t}. \tag{11.15}$$

The motion is similar to the undamped case except that the amplitude decreases exponentially in time and the frequency of oscillation ω_1 is less than the undamped frequency ω_0. The motion is periodic because the zero crossings of $X_0 e^{-(\gamma/2)t} \cos(\omega_1 t + \phi)$ are separated by equal time intervals $T = 2\pi/\omega_1$, but the peaks do not lie exactly halfway between them.

The essential features of the motion depend on the ratio ω_1/γ. If $\omega_1/\gamma \gg 1$, the amplitude decreases only slightly during the time the cosine makes many zero crossings; in this regime, the motion is called *lightly damped*. If ω_1/γ is comparatively small, $X(t)$ tends rapidly to zero while the cosine makes only a few oscillations. This motion is called *heavily damped*. For light damping, $\omega_1 \approx \omega_0$, but for heavy damping

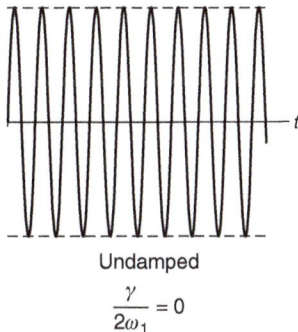

Undamped

$$\frac{\gamma}{2\omega_1} = 0$$

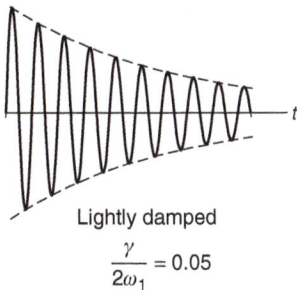

Lightly damped

$$\frac{\gamma}{2\omega_1} = 0.05$$

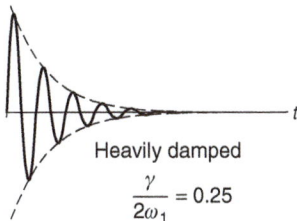

Heavily damped

$$\frac{\gamma}{2\omega_1} = 0.25$$

ω_1 can be significantly smaller than ω_0. If $\omega_1 < \gamma/2$, the trial solution (11.10) fails, and the motion is not oscillatory. The system is described as *overdamped*.

11.3.1 Energy Dissipation in the Damped Oscillator

Friction dissipates mechanical energy, so the energy of a damped oscillator must decay in time. To evaluate the kinetic energy we first find the velocity by differentiating Eq. (11.13). The result is

$$v = -X_0 e^{-(\gamma/2)t} \left[\omega_1 \sin(\omega_1 t + \phi) + \frac{\gamma}{2} \cos(\omega_1 t + \phi) \right]. \qquad (11.16)$$

We will be most interested in systems with light damping, where $\omega_1 \gg \gamma/2$, so that $\omega_1 \approx \omega_0$. This allows us to make an approximation that simplifies the arithmetic and reveals some universal features

$$\omega_1^2 = \omega_0^2 - (\gamma/2)^2 \approx \omega_0^2. \qquad (11.17)$$

With our approximation that $\omega_1 \gg \gamma/2$, the second term in the bracket of Eq. (11.16) can be neglected, giving

$$v = V_0 e^{-(\gamma/2)t} \sin(\omega_0 t + \phi), \qquad (11.18)$$

where

$$V_0 = \omega_0 X_0.$$

In this case, the potential energy is

$$U(t) = \frac{1}{2} k X_0^2 e^{-\gamma t} \cos^2(\omega_0 t + \phi), \qquad (11.19)$$

and the kinetic energy is

$$K(t) = \frac{1}{2} m V_0^2 = \frac{1}{2} m \omega_0^2 X_0^2 e^{-\gamma t} \sin^2(\omega_0 t + \phi)$$

$$= \frac{1}{2} k X_0^2 e^{-\gamma t} \sin^2(\omega_0 t + \phi). \qquad (11.20)$$

The total energy is

$$E(t) = \frac{1}{2} k X_0^2 e^{-\gamma t}. \qquad (11.21)$$

The decay of the total energy is described by a simple differential equation:

$$\frac{dE}{dt} = -\gamma E,$$

which has the solution

$$E = E_0 e^{-\gamma t}, \qquad (11.22)$$

where E_0 is the energy at time $t = 0$.

The energy's decay is characterized by the time $\tau = 1/\gamma$ in which the energy decreases from its initial value by a factor of $e^{-1} \approx 0.368$. τ is often called the *damping time* of the system. In the limit of zero damping,

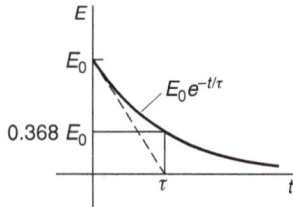

$\gamma \to 0$, $\tau \to \infty$ and E is constant. The system behaves like an undamped oscillator.

Note that we could have found the same result directly from the work–energy theorem. The rate at which work is done on the system by friction is $v f_{\text{fric}} = -bv^2$. Using the expression for velocity in Eq. (11.17), making the approximation $\omega_1 = \omega_0$, and replacing $\sin^2(\omega_0 t + \phi)$ by its average value of $1/2$, we have

$$\frac{dE}{dt} = -bv^2 = -\frac{b}{2}V_0^2 e^{-\gamma t} = -\frac{b}{2}\frac{k}{m}X_0^2 e^{-\gamma t}$$
$$= -\gamma E,$$

as we expect.

Example 11.2 Physical Limitations to Damped Motion

According to Eq. (11.13), once an oscillator is set into motion it will oscillate forever, even though its amplitude steadily decreases. To understand the limits to such an unlikely prediction, we need to examine the physics of the harmonic oscillator in more detail.

The equation of motion (11.12) describes an isolated oscillator acted on only by the viscous damping force. The idea of a truly isolated system is fundamentally non-physical, because it would be out of contact with all its surroundings, including any measurement apparatus. In reality, systems are always in contact with their surroundings, and if they are in equilibrium they can be characterized by a temperature T. We introduced the relation between random thermal motion of atoms in a gas and the temperature of the gas in Section 5.9. The connection between random motion and temperature is universal and eventually thermal effects become important.

The equipartition theorem, introduced in Section 5.11, predicts the average thermal energy of a system that is in thermal equilibrium at temperature T. The mean kinetic energy of a particle of mass m in equilibrium at temperature T is $\frac{1}{2}m\overline{v^2} = \frac{3}{2}kT$, where k, known as Boltzmann's constant, has the value $k \approx 1.38 \times 10^{-23}$ m^2 kg/s^{-2}.

More generally, the equipartition theorem asserts that if the energy of a system can be written in the form of a sum of terms that are quadratic in form, for instance a free particle for which the kinetic energy is $m(v_x^2 + v_y^2 + v_z^2)/2$, then in thermal equilibrium the system possesses, on the average, energy of $\frac{1}{2}kT$ for each of the quadratic terms. Thus, the mean kinetic energy of an atom in a gas is $\frac{3}{2}kT$. A harmonic oscillator with energy $\frac{1}{2}kx^2 + \frac{1}{2}mv^2$ possesses mean thermal energy kT. When the energy of an oscillator decays to the point where it becomes comparable with kT, the energy stops decreasing and simply fluctuates around this average value.

To reduce thermal fluctuations, ultra-precise measurements are often made at very low temperature, but this strategy eventually fails when quantum effects become important. According to the laws of quantum physics, the energy of a harmonic oscillator E_{ho} cannot be arbitrarily small because it must obey the quantum rule

$$E_{ho} = (n + \tfrac{1}{2})\hbar\omega_0,$$

where $\hbar \approx 1.055 \times 10^{-34} \text{m}^2\text{kg/s}$. $\hbar$ (pronounced "h bar") is Planck's constant h divided by 2π and n is a non-negative integer: $n = 0, 1, 2, \dots$ There is a minimum or "ground state" energy of $\frac{1}{2}\hbar\omega_0$ that is present even when $n = 0$. This energy arises from the intrinsic quantum fluctuations present in all systems. For most purposes, in mechanical systems these fluctuations are so small that they can be overlooked. Nevertheless, research in quantum physics has reached such sensitivity that mechanical quantum fluctuations can be observed and pose a fundamental limit to precision measurements.

11.3.2 The Q of an Oscillator

The degree of damping of an oscillator is often characterized by a dimensionless parameter Q, known as the *quality factor*, defined by

$$Q = \frac{\text{average energy stored in the oscillator}}{\text{average energy dissipated during 1 radian of motion}}. \quad (11.23)$$

"Average" is used here to mean the time average over one cycle of motion, for which $\langle \sin^2 \theta \rangle = \langle \cos^2 \theta \rangle = 1/2$.

The energy dissipated per radian is the energy lost during the time it takes the system to oscillate through one radian. During the period $T = 2\pi/\omega_0$, the system oscillates through 2π radians. Thus the time to oscillate through one radian is $T/2\pi = 1/\omega_0$.

From Eq. (11.22), the energy decays at the rate $\dot{E} = -\gamma E$. Consequently, the energy lost during time $\Delta t = 1/\omega_0$ is

$$\Delta E \approx \frac{dE}{dt}\Delta t = \gamma E \frac{1}{\omega_0},$$

so that the quality factor is given by

$$Q = \frac{E}{\Delta E} = \frac{E}{\gamma E/\omega_0} = \frac{\omega_0}{\gamma}. \quad (11.24)$$

A lightly damped oscillator has $Q \gg 1$ while a heavily damped system that loses energy rapidly has a low Q. A tuning fork has a Q of a thousand or so, whereas a superconducting microwave cavity can have a Q in excess of 10^7 and some systems have a $Q > 10^9$. In the limit of zero damping, $Q \to \infty$.

Example 11.3 The Q of Two Simple Oscillators

A musician's tuning fork rings at A above middle C, 440 Hz. A sound level meter indicates that the sound intensity decreases by a factor of 5 in 4 seconds. What is the Q of the tuning fork?

The sound intensity from the tuning fork is proportional to the energy of oscillation. Since the energy of a damped oscillator decreases as $e^{-\gamma t}$, we can find γ by taking the ratio of the energies at $t = 0$ and at $t = 4$ s:

$$5 = \frac{E(0)e^{(0)}}{E(0)e^{-4\gamma}} = e^{4\gamma}.$$

Hence

$$4\gamma = \ln 5 = 1.6$$

$$\gamma = 0.4 \text{ s}^{-1},$$

and

$$Q = \frac{\omega_0}{\gamma} = \frac{2\pi(440 \text{ s}^{-1})}{0.4 \text{ s}^{-1}}$$

$$\approx 7000.$$

The energy loss is due primarily to the heating of the metal as it bends. Air friction and energy loss to the mounting point also contribute. The symmetrical design of a tuning fork minimizes loss to the mount. Incidentally, if you try this experiment, bear in mind that the ear is a poor sound level meter because it does not respond linearly to sound intensity; its response is more nearly logarithmic.

A rubber band exhibits a much lower Q than a tuning fork, primarily because of the internal friction generated by the coiling of the long-chain molecules. In one experiment, a paperweight suspended from a hefty rubber band had a period of 1.2 s and the amplitude of oscillation decreased by a factor of 2 after three periods. What is the estimated Q of this system?

From Eq. (11.15) the amplitude is $X(t) = X_0 e^{(-\gamma/2)t}$. The period of the oscillator is $T = 1.2$ s and so the frequency is $\omega_0 = 2\pi/T = 5.24 \text{ s}^{-1}$. The ratio of the amplitudes at $t = 0$ and at $t = 3(1.2 \text{ s}) = 3.6$ s is

$$2 = \frac{X(0)}{X(3.6 \text{ s})} = \frac{X_0 e^{(0)}}{X_0 e^{-3.6(\gamma/2)}}.$$

Solving, we have

$$1.8\gamma = \ln 2 = 0.69$$

or

$$\gamma = 0.39 \text{ s}^{-1}.$$

Therefore

$$Q = \frac{\omega_0}{\gamma} = \frac{5.24 \text{ s}^{-1}}{0.39 \text{ s}^{-1}}$$

$$= 13.$$

It may seem questionable to use the light damping result, $Q = \omega_0/\gamma$, when Q is so low. The approximations involved introduce errors of order $(\gamma/\omega_0)^2 = (1/Q)^2$. For $Q > 10$ the error is less than 1 percent.

Note that the damping constants for the tuning fork and for the rubber band are very nearly the same. The tuning fork has a much higher Q, however, because it goes through many more cycles of oscillation in one damping time and loses correspondingly less of its energy per cycle.

Example 11.4 Graphical Analysis of a Damped Oscillator

The illustration is taken from a photograph of an oscilloscope trace of the displacement of an oscillating system versus time. We immediately recognize that the system is a damped harmonic oscillator. The frequency ω_1 and quality factor Q can be found from the photograph.

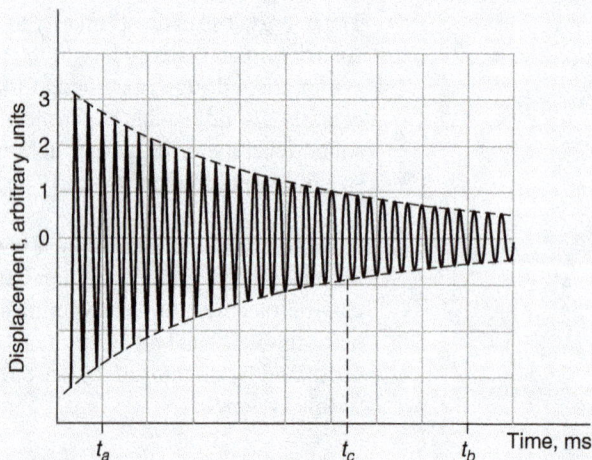

The time interval from t_a to t_b is 8 ms. There are 28.5 cycles (complete periods) in this interval. (Check this for yourself from the illustration.) The period of oscillation is $T = 8 \times 10^{-3} \text{ s}/28.5 = 2.81 \times 10^{-4} \text{ s}$. The angular frequency is $\omega_0 = 2\pi/T = 22\,400 \text{ s}^{-1}$. The corresponding circular frequency is $\nu = \omega_0/2\pi = 3560 \text{ Hz}$.

In order to calculate the quality factor $Q = \omega_1/\gamma$, the damping constant must be known. From Eq. (11.15) the amplitude is $X_0 e^{-(\gamma/2)t}$. This function describes the *envelope* of the displacement curve, which has been drawn with a dashed curve on the photograph. At time t_a the envelope has magnitude $X_a = 2.75$ units. When the envelope decays by a factor $e^{-1} = 0.368$, its magnitude is 1.01 units. From the photograph this occurs at $t_c = 5.35$ ms, measured from t_a. Hence $e^{-(\gamma/2)t_c} = e^{-1}$, or $\gamma = 2/t_c = 374 \text{ s}^{-1}$. The quality factor is $Q = \omega_1/\gamma = 60$.

Now for a word about the system. This is not a mechanical oscillator, nor even an electrical oscillator. The signal is produced by radiating electrons in a small volume of hydrogen gas. The signal was greatly amplified for oscilloscope display. Furthermore, the atoms were actually radiating at 9.2×10^9 Hz. Since this is much too high for the oscilloscope to follow, the frequency was translated to a lower value by electronic means. This did not affect the shape of the envelope, and our measured value of γ is correct. If we use the true value of the frequency of the atomic system, we find that the actual Q is

$$Q = \frac{2\pi\nu}{\gamma} = \frac{2\pi(9.2 \times 10^9)}{374} = 1.6 \times 10^8.$$

Such a high Q is not unusual in atomic systems.

11.4 The Driven Harmonic Oscillator

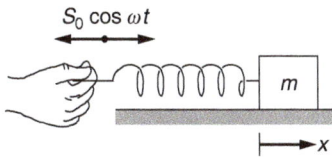

$S_0 \cos \omega t$

The most interesting applications of harmonic oscillators generally involve their behavior when they are subject to a time-varying force $F(t)$, particularly when the force is periodic. Such a system is called a *driven harmonic oscillator*. For a mass on a spring, a force could be applied by moving the end of the spring. To be concrete, let the end of the spring move according to $S = S_0 \cos \omega t$, as shown in the sketch. The force on the mass is $-k(x - S_0 \cos \omega t)$, where x is the position of the mass measured from equilibrium. The spring force is therefore

$$F_{\text{spring}} = -k(x - S_0 \cos \omega t) = -kx + F_0 \cos \omega t,$$

where $F_0 = kS_0$. We assume that there is also a damping force $-bv$, so that the equation of motion is

$$m\ddot{x} = -b\dot{x} - kx + F_0 \cos \omega t,$$

which is conveniently written in the standard form

$$\ddot{x} + \gamma\dot{x} + \omega_0^2 x = \frac{F_0}{m} \cos \omega t, \tag{11.25}$$

where $\gamma = b/m$ and $\omega_0 = \sqrt{k/m}$, as before. A formal method for solving Eq. (11.25) is presented in Note 11.3, but once again, it is worth trying to guess the solution. The right-hand side of Eq. (11.25) varies as $\cos \omega t$, so that it is tempting to try $x = X_0 \cos \omega t$. However, the first derivative term on the left, $\gamma\dot{x}$, introduces a $\sin \omega t$ time dependence that is absent on the right. To deal with this, let's try

$$x = X_0 \cos(\omega t + \phi). \tag{11.26}$$

This indeed satisfies Eq. (11.24), provided that X_0 and ϕ have the values

$$X_0 = \frac{F_0}{m} \frac{1}{[(\omega_0{}^2 - \omega^2)^2 + (\omega\gamma)^2]^{1/2}}, \tag{11.27}$$

$$\phi = \arctan\left(\frac{\gamma\omega}{\omega_0{}^2 - \omega^2}\right). \tag{11.28}$$

When ω is close to ω_0, the amplitude X_0 is large and the phase varies rapidly. It is therefore reasonable to make the approximation

$$\omega_0^2 - \omega^2 = (\omega_0 + \omega)(\omega_0 - \omega) \approx 2\omega_0(\omega_0 - \omega),$$

and take $\omega \approx \omega_0$ elsewhere. With these approximations,

$$X_0 = \frac{F_0}{2m\omega_0} \frac{1}{[(\omega_0 - \omega)^2 + (\gamma/2)^2]^{1/2}}, \tag{11.29}$$

$$\phi = \arctan\left(\frac{\gamma/2}{\omega_0 - \omega}\right). \tag{11.30}$$

We will also need an expression for the velocity, which is

$$v = -V_0 \sin(\omega t + \phi), \tag{11.31}$$

where

$$V_0 = \omega X_0. \tag{11.32}$$

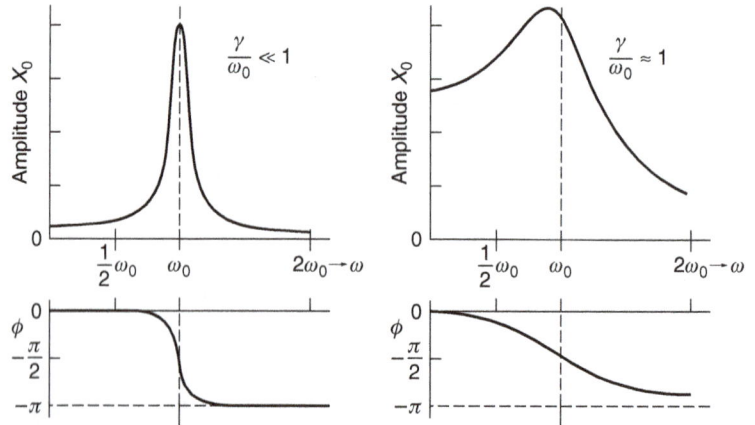

Plots of the amplitude X_0 (upper) and phase ϕ (lower) versus the driving frequency ω are shown in the figures, for small damping (left) and large damping (right). Note that the phase changes by π as ω ranges from $\omega \ll \omega_0$ to $\omega \gg \omega_0$. The frequency range over which the amplitude and phase change significantly depends on the ratio ω_0/γ.

Example 11.5 Driven Harmonic Oscillator Demonstration

Break a long rubber band and suspend something like a heavy pocket knife from one end, holding the other end in your hand. The resonant frequency ω_0 is easily determined by observing the free motion. Now slowly jiggle your hand at a frequency $\omega < \omega_0$: the weight will move in phase with your hand. If you jiggle the system with $\omega > \omega_0$, you will find that the weight moves in the opposite direction to your hand. For a given amplitude of motion of your hand, the weight moves with decreasing amplitude as ω is increased above ω_0. If you try to jiggle the system at resonance $\omega = \omega_0$, the amplitude increases so much that the weight either flies up in the air or hits your hand. In either case the system no longer behaves like a simple oscillator.

11.4.1 Energy Stored in a Driven Harmonic Oscillator

Energy considerations simplified the discussion of the isolated oscillator in Section 11.2.2, and they are even more useful for the driven oscillator. From Eqs. (11.26) and (11.31) we have

$$
\begin{aligned}
E(t) &= K(t) + U(t) \\
&= \tfrac{1}{2}X_0^2[m\omega_0^2\sin^2(\omega t + \phi) + k\cos^2(\omega t + \phi)] \\
&= \tfrac{1}{2}kX_0^2,
\end{aligned}
$$

where we have again made the approximation $\omega \approx \omega_0$.

Energy flows into the driven oscillator from work by the driving force and is dissipated by the damping force. From the work–energy theorem,

$$
W^{df} = \Delta E + W^{nc}
$$

where in a given time interval W^{df} is the work by the driving force, ΔE is the change in the mechanical energy of the oscillator, and W^{nc} is the non-conservative work by the damping force. After steady conditions have been reached, the mechanical energy of the oscillator is constant $\Delta E = 0$, leaving

$$
W^{df} = W^{nc}.
$$

In steady conditions, the rate of work by the driving force is equal to the rate that work is dissipated by the damping.

The mechanical energy of the oscillator is constant but mechanical energy is not conserved because the oscillator is not isolated. Work is done on the oscillator by the driving force and work is done by the oscillator against the viscous damping force. If we enlarged our system to include the driving force the total energy of the enlarged system would be conserved. The total energy of the enlarged system includes whatever source of energy provides the driving force, the mechanical energy of the oscillator, and the heat energy generated by the damping.

The oscillator is at rest with zero mechanical energy when the driving force is first applied. The oscillator's stored energy builds to its final steady value in the initial period during which some of the work by the driving force goes into the stored mechanical energy. The period of change is called the *transient*, which we shall discuss in Section 11.5.

Using the values of X_0 and V_0 given by Eqs. (11.27) and (11.32), the stored energy is

$$E(\omega) = \frac{1}{8} \frac{F_0^2}{m} \frac{1}{(\omega - \omega_0)^2 + (\gamma/2)^2}. \tag{11.33}$$

We can rewrite $E(\omega)$ as the product of three factors:

$$E_0 = \frac{1}{2} \frac{F_0^2}{m\omega_0^2},$$

$$\left(\frac{\omega_0}{\gamma}\right)^2 = Q^2,$$

$$g(\omega) = \frac{(\gamma/2)^2}{(\omega_0 - \omega)^2 + (\gamma/2)^2},$$

which allow $E(\omega)$ to be written as

$$E(\omega) = E_0 Q^2 g(\omega). \tag{11.34}$$

E_0 is twice the kinetic energy of a free mass m that is driven by the force $F_0 \cos \omega t$. "Twice" is because a free particle lacks the potential energy that the oscillating mass possesses.

The quality factor $Q = \omega_0/\gamma$ was introduced in Section 11.3.2, where we used it to describe the decay time for the energy of the oscillator, as counted by the number of oscillations.

11.4.2 Resonance

The function $g(\omega)$ is called the *lineshape function* (also, the *Lorentzian*) because it first arose in the analysis of the shape of the spectral lines radiated by atoms. The lineshape function describes the frequency dependence of the oscillator's energy when excited by a periodic driving force. The peak in the vicinity of ω_0 is called a *resonance*, ω_0 is called the *resonance frequency*, and the curve itself is often referred to as a *resonance curve*.

At resonance, $g(\omega) = 1$. The curve decreases to one-half its peak value when $\omega_\pm - \omega_0 = \pm\gamma/2$. The frequency width of the curve at half its maximum value is called the *resonance width* $\Delta\omega$, often abbreviated FWHM (full width at half maximum). Since $\omega_+ - \omega_- = 2(\gamma/2) = \gamma$, we have

$$\Delta\omega = \gamma. \tag{11.35}$$

As γ decreases the curve becomes narrower, the range of frequency over which the system responds significantly becomes smaller, and the oscillator becomes increasingly selective in frequency.

The maximum value of the stored energy is

$$E_{\max} = Q^2 E_0. \tag{11.36}$$

This result gives some insight into the usefulness of the harmonic oscillator. Q^2 can be enormous and so the oscillator can amplify the effect of a very small periodic force by storing the energy it delivers each cycle.

In Section 11.3.2, the quality factor

$$Q = \omega_0/\gamma \tag{11.37}$$

was introduced to characterize the time for the free oscillator to dissipate its energy. The time for the energy to decay by a factor of e^{-1} is $\tau = 1/\gamma$, so that we have

$$\tau = Q/\omega.$$

In this section, the quality factor has taken on quite a different meaning. Because the width of the lineshape function is $\Delta\omega = \gamma$, we could rewrite Eq. (11.23) as

$$Q = \frac{\text{resonance frequency}}{\text{frequency width of resonance curve}} = \frac{\omega_0}{\Delta\omega}. \tag{11.38}$$

The drawings show lineshape curves with different Q. It is apparent that the system with $Q = 20$ is considerably more selective than the system with $Q = 4$. As pointed out in Example 11.4, certain atomic systems can have a Q greater than 10^8. The sharpness of the resonance curve means that the system will not respond substantially unless driven very near its resonance frequency. This frequency selectivity underlies the use of harmonic oscillators to serve as frequency standards or clocks, for instance the oscillating quartz crystals in digital watches.

It is evident that the response of an oscillator in time and its response in frequency are intimately related. However, before discussing the implications, we need to complete the solution for the driven oscillator, because the solution so far does not tell the whole story.

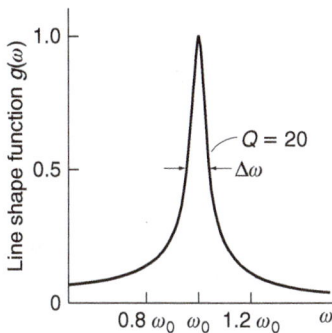

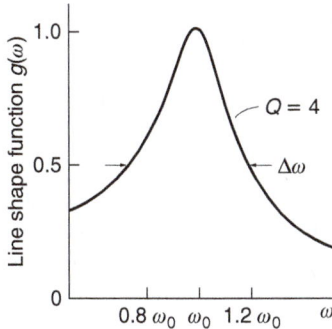

11.5 Transient Behavior

The solution for the motion of the driven harmonic oscillator $x = X_0 \cos(\omega t + \phi)$ satisfies the equation of motion Eq. (11.25) but it is incomplete because it is unable to accommodate the arbitrary initial conditions of a real problem. For instance, if the mass is released from rest at $t = 0$, for which the initial conditions are $x(0) = 0$ and $v(0) = 0$, there is no way that our solution could describe it, because X_0 and ϕ already have definite values given by Eqs. (11.29) and (11.30).

Fortunately, the fix is simple. Note that the left-hand sides for the motion of the undriven, or free, damped oscillator, Eq. (11.8), and the equation for the driven oscillator, Eq. (11.25), are identical. The difference is that the right-hand side of Eq. (11.8) is zero, while the right-hand side of Eq. (11.25) is the driving term $F_0 \cos \omega t / m$. Consequently, if x_{free} is a solution to Eq. (11.8) and x_{driven} is a solution to Eq. (11.25), then

$$x(t) = x_{\text{free}}(t) + x_{\text{driven}}(t) \tag{11.39}$$

is also a solution to Eq. (11.25).

Inserting the solution for x_{free} from Eq. (11.10) and the solution for x_{driven} from Eq. (11.26), we have

$$x(t) = X_f e^{-(\gamma/2)t} \cos(\omega_0 t + \phi_f) + X_0 \cos(\omega t + \phi), \tag{11.40}$$

where X_f and ϕ_f are arbitrary constants, and X_0 and ϕ are given by Eqs. (11.29) and (11.30). (In the first term on the right we made the approximation $\omega_1 \approx \omega_0$.)

The first term in Eq. (11.40), the *transient*, decreases exponentially with time, and eventually dies away, leaving the completely determined *steady-state* behavior $X_0 \cos(\omega t + \phi)$.

Example 11.6 Harmonic Analyzer

A device that analyzes the spectrum of a time-varying signal composed of many frequencies is called a harmonic analyzer. A harmonic analyzer measures the response to a driving signal at a resonant frequency that can be selected. A simple example is an old-fashioned dial-tuned radio that selects one of a multitude of broadcast frequencies by changing the resonant frequency of an electrical circuit.

We take the signal to be $(F_0/m) \cos \omega_0 t$. The response of the harmonic oscillator is given by Eqs. (11.26), (11.27), and (11.28). The phase constant of the oscillator is given by Eq. (11.28), which at resonance yields $\phi = \pm \arctan \infty$. The ambiguity in sign is because the phase changes by π as the frequency passes through resonance. We shall take $\phi = -\pi/2$. The steady-state oscillator response to the driving field is thus $X_0 \sin \omega_0 t$ so that Eq. (11.40) takes the form

$$x(t) = X_f e^{-(\gamma/2)t} \cos(\omega_0 t + \phi_f) + X_0 \sin \omega_0 t,$$

where, from Eq. (11.27) at resonance,

$$X_0 = \frac{F_0}{m \omega_0 \gamma}.$$

For the mass to start from rest at $t = 0$, we require $x(0) = 0$ and $\dot{x}(0) = 0$. The result is (assuming $\omega_0 \gg \gamma/2$)

$$x(t) = X_0(1 - e^{-(\gamma/2)t}) \sin \omega_0 t. \tag{11.41}$$

The sketches show plots of $x(t)$ for low and high values of Q. The mass, initially at rest, builds up to its final amplitude over a time that depends on the Q of the oscillator. The characteristic time for the build-up is the damping time $\tau = 1/\gamma$. It is desirable for the analyzer to respond quickly, particularly if the signal amplitude is time-varying, and this

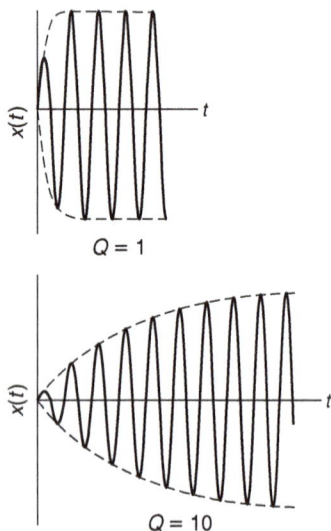

$Q = 1$

$Q = 10$

requires a short damping time. Because $Q = \omega_0 \tau$, low Q is desirable for this application. As the sketch shows, the system with $Q = 1$ reaches steady state in less than 2 cycles but the system with $Q = 10$ takes more than 10 cycles to reach steady state.

On the other hand, if the oscillator is intended to resolve small differences in frequency, the resonance linewidth $\Delta\omega$ must be small. Because $Q = \omega_0/\Delta\omega$, high Q is desirable, but the response will be slow. There is a trade-off between speed of response and spectral resolution.

11.6 Response in Time and Response in Frequency

The smaller the damping of a free oscillator, the more slowly its energy is dissipated. The same oscillator, when driven, becomes increasingly more selective in frequency as the damping is decreased. As we shall now show, the time dependence of the damped free oscillator and the frequency dependence of the driven oscillator are intimately related.

Recall from Eq. (11.22) that the energy of a free oscillator is

$$E(t) = E_0 e^{-\gamma t}.$$

The damping time is $\tau = 1/\gamma$.

Next, consider the response in frequency of the same oscillator when it is driven by a force $F_0 \cos \omega t$. From Eq. (11.35) the width of the resonance curve is

$$\Delta\omega = \gamma.$$

The damping time $\tau = 1/\gamma$ and the resonance curve width $\Delta\omega$ obey

$$\tau \, \Delta\omega = 1. \tag{11.42}$$

According to this result it is impossible to design an oscillator for which the damping time and the resonance width are both arbitrary; if we choose one, the other is automatically fixed by Eq. (11.42).

Equation (11.42) has many implications for the design of mechanical and electrical systems. Any element that is highly frequency selective will oscillate for a long time if it is accidentally perturbed. Furthermore, such an element will take a long time to reach the steady state when a driving force is applied because the effects of the initial conditions die out only slowly. More generally, Eq. (11.42) plays a fundamental role in quantum mechanics; it is closely related to one form of the Heisenberg uncertainty principle.

Example 11.7 Vibration Attenuator

The phenomenon of resonance has both positive and negative aspects in practice. By operating at the resonance frequency of a system we can obtain a response of large amplitude for a very small driving force. Organ pipes utilize this principle effectively, and resonant electric circuits enable us to tune our radios to the desired frequency. On the negative side, we do not want motions of large amplitude in the springs of an automobile or in the crankshaft of its engine. To reduce undue response at resonance a dissipative friction force is needed.

The problem of isolating a body from its surroundings arises in the design of sensitive experimental apparatus and in numerous everyday situations, for instance isolating an automobile body from the effects of a bumpy road. The basic strategy is to use a spring to cushion the disturbance. This strategy can be made to work very well, but, as we shall see, it also holds the possibility of making matters worse.

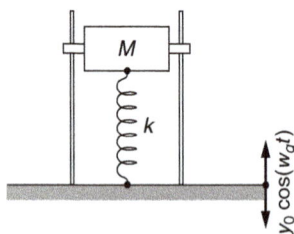

To illustrate the principles, we idealize the system as a simple mass M that rests on top of a spring with constant k, whose lower end is attached to the floor, which may be vibrating. We assume that the system is constrained so that the only important motion is vertical. The floor vibrates at frequency ω_g with amplitude y_0, so that its position is given by

$$y = y_0 \cos(\omega_g t).$$

We denote the displacement of M from its equilibrium position by x. The equation of motion for M is

$$M\ddot{x} = -k(x - y) = -k(x - y_0 \cos(\omega_g t))$$

which can be written in the standard form

$$\ddot{x} + \omega_0^2 x = \omega_0^2 y_0 \cos(\omega_g t),$$

where $\omega_0 = \sqrt{k/M}$, as usual.

The steady-state motion is given by $x = x_0 \cos(\omega_g t)$, where

$$x_0 = y_0 \frac{\omega_0^2}{\omega_0^2 - \omega_g^2}. \tag{1}$$

The effectiveness of the vibration attenuator depends on the ratio $|x_0|/|y_0|$, which we denote by $\mathcal{F}$:

$$\mathcal{F} = \frac{|x_0|}{|y_0|}.$$

The object of the spring suspension is to make $\mathcal{F}$ as small as possible. In the absence of damping, $\mathcal{F}$ is, from Eq. (1),

$$\mathcal{F} = \left| \frac{\omega_0^2}{(\omega_0^2 - \omega_g^2)} \right|.$$

For $\omega_g \ll \omega_0$, $\mathcal{F} \approx 1$, and the vibration is essentially transmitted without reduction. However, for $\omega_g \gg \omega_0$, $\mathcal{F} < 1$, and the vibration is attenuated. Thus, for the vibration attenuator to be effective, its resonance frequency must be low compared to the driving frequency.

Our system suffers from a fatal flaw; if vibration occurs near the resonant frequency $\omega_g \approx \omega_0$, the vibration attenuator becomes a vibration amplifier. To avoid this, some damping mechanism must be provided. Often this is accomplished with a device called a *dashpot* (in an automobile, a *shock absorber*, which consists of a piston in a cylinder of oil). The dashpot provides a viscous retarding force $-bv$, where $v = \dot{x} - \dot{y}$ is the relative velocity of its ends.

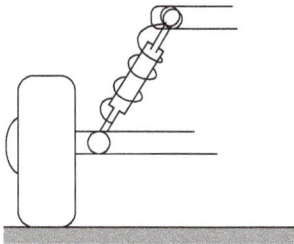

The equation of motion of M is

$$M\ddot{x} = -k(x - y) - b(\dot{x} - \dot{y}).$$

Taking $y = y_0 \cos \omega_g t$ and $v = -\omega_g y_o \sin \omega_g t$, we obtain

$$\ddot{x} + \gamma \dot{x} + \omega_o{}^2 x = \omega_o{}^2 y_o \cos \omega_g t - \gamma \omega_g y_0 \sin \omega_g t$$

where $\gamma = b/M$ and $\omega_0{}^2 = k/M$. This is like the equation of a driven damped oscillator, Eq. (11.25), except that the motion of the floor on the dashpot has introduced an additional driving term $\gamma \omega y_0 \sin \omega t$. We can guess that the solution has the form $x = x_0 \cos(\omega_g t + \phi)$ or use the method described in Note 11.3 to formally derive the solution. Either method yields

$$\mathcal{F} = \left[\frac{\omega_0{}^4 + (\omega_g \gamma)^2}{(\omega_0{}^2 - \omega_g^2)^2 + (\omega_g \gamma)^2} \right]^{\frac{1}{2}}.$$

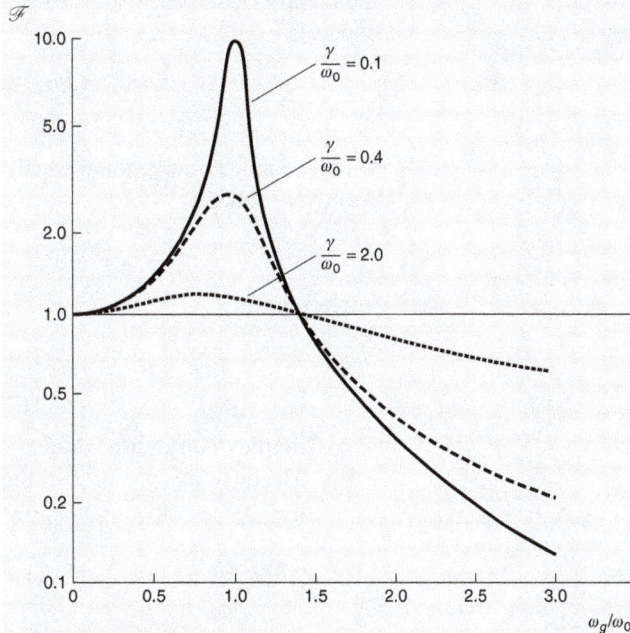

The graph shows $\mathcal{F}$ versus ω_g/ω_0 for various values of γ/ω_0. For ω_g/ω_0 less than about 1.5, $|x_0|/|y_0| > 1$. The vibration is actually enhanced, showing that even with damping it is essential to reduce the resonance frequency below the driving frequency. When ω_g/ω_0 is greater than 1.5, $\mathcal{F} < 1$. For these higher frequencies, the vibration isolation is more effective the smaller the damping. However, small damping increases the danger from vibrations near resonance.

If a smooth turnpike ride is the chief consideration in an auto, one wants a massive car with weak damping and soft springs. Such a car is difficult to control on a bumpy road where resonance could be excited. The best suspensions are heavily damped and feel rather stiff. The danger in driving with defective shock absorbers is that the car may be thrown out of control if it is excited at resonance by bumps.

Note 11.1 Complex Numbers

All the equations of motion in this chapter can be solved simply by using complex variables. This Note summarizes the definitions and algebra of complex numbers and the two following Notes lay out how to solve the equations of motion using complex numbers.

1. Basic properties

Every complex number z can be written in the Cartesian form $x + iy$, where $i^2 = -1$. x is the *real* part of z, and y is the *imaginary* part.

The sum of two complex numbers $z_1 = x_1 + iy_1$ and $z_2 = x_2 + iy_2$ is the complex number $z_1 + z_2 = (x_1 + x_2) + i(y_1 + y_2)$. The product of z_1 and z_2 is

$$z_1 z_2 = (x_1 + iy_1)(x_2 + iy_2) = x_1 x_2 + ix_1 y_2 + iy_1 x_2 + i^2 y_1 y_2$$
$$= (x_1 x_2 - y_1 y_2) + i(x_1 y_2 + y_1 x_2).$$

If two complex numbers are equal, the real and imaginary parts are respectively equal:

$$x_1 + iy_1 = x_2 + iy_2$$

implies that

$$x_1 = x_2$$
$$y_1 = y_2.$$

2. Complex conjugate

$z^* \equiv x - iy$ is the *complex conjugate* of $z = x + iy$. The quantity $|z| = \sqrt{zz^*}$ is the magnitude of z:

$$|z| = \sqrt{zz^*}$$
$$= \sqrt{(x + iy)(x - iy)}$$
$$= \sqrt{x^2 + y^2}.$$

3. de Moivre's theorem

de Moivre's theorem states that $e^{i\theta} = \cos\theta + i\sin\theta$. This is proved by using the power series representation

$e^x = 1 + x + (1/2)x^2 + (1/3!)x^3 + \cdots$. Using $i^2 = -1, i i^3 = -i$, etc., we have

$$e^{i\theta} = 1 + i\theta - (1/2)\theta^2 + (1/3!)(i\theta)^3 + \cdots$$
$$= 1 - (1/2)\theta^2 + \cdots + i(\theta - (1/3!)\theta^3 + \cdots).$$

Comparing these expressions with the power series expansions for $\cos\theta$ and $\sin\theta$ in Chapter 1, Note 1.3, completes the proof.

4. Standard form Any complex number can be written in the standard form $x + iy$, where x and y are real. Because $i^2 = -1$, i can never appear in an expression to a power higher than the first. Here is an example:

$$\frac{(a + ib)}{(c + id)} = \frac{(a + ib)}{(c + id)} \frac{(c - id)}{(c - id)}$$
$$= \frac{(a + ib)(c - id)}{(c^2 + d^2)}$$
$$= \frac{(ac + bd) + i(bc - ad)}{(c^2 + d^2)}$$

where we have multiplied the numerator and the denominator by the complex conjugate of the denominator.

5. Polar representation Every complex number z can be written in the polar form $re^{i\theta}$. r is a real number, the *modulus*, and θ is the *argument*. To transform from Cartesian to polar form we use de Moivre's theorem:

$$re^{i\theta} = r\cos\theta + ir\sin\theta$$
$$= x + iy,$$

from which it follows that

$$x = r\cos\theta$$
$$y = r\sin\theta$$

and

$$r = \sqrt{x^2 + y^2}$$
$$\theta = \arctan\frac{y}{x}.$$

We see that $r = |z| = \sqrt{x^2 + y^2}$.

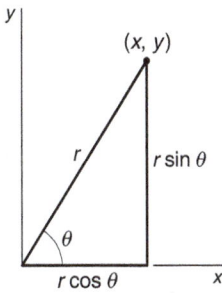

Note 11.2 Solving the Equation of Motion for the Damped Oscillator

The equation of motion is

$$\ddot{x} + \gamma\dot{x} + \omega_0{}^2 x = 0. \tag{1}$$

To cast this into complex form we introduce the companion equation

$$\ddot{y} + \gamma\dot{y} + \omega_0{}^2 y = 0. \tag{2}$$

Multiplying Eq. (2) by i and adding it to Eq. (1) yields

$$\ddot{z} + \gamma\dot{z} + \omega_0{}^2 z = 0. \tag{3}$$

Note that either the real or imaginary part of z is an acceptable solution for the equation of motion.

All the coefficients of the derivatives of z are constants and so a natural choice for the solution of Eq. (3) is

$$z = z_0 e^{\alpha t}, \tag{4}$$

where z_0 and α are constants. With this trial solution Eq. (3) yields

$$\alpha^2 z_0 e^{\alpha t} + \alpha\gamma z_0 e^{\alpha t} + \omega_o{}^2 z_0 e^{\alpha t} = 0.$$

Cancelling the common factor $z_0 e^{\alpha t}$, we have

$$\alpha^2 + \alpha\gamma + \omega_0{}^2 = 0, \tag{5}$$

which has the solution

$$\alpha = -\frac{\gamma}{2} \pm \sqrt{\left(\frac{\gamma}{2}\right)^2 - \omega_0{}^2}. \tag{6}$$

Let the two roots be α_1 and α_2. The solution can be written as

$$z = z_A e^{\alpha_1 t} + z_B e^{\alpha_2 t},$$

where z_A and z_B are constants.

There are three possible forms of the solution, depending on whether α is real or complex. We consider these solutions in turn.

Case 1: light damping: $(\gamma/2)^2 \ll \omega_0{}^2$.

In this case $\sqrt{(\gamma/2)^2 - \omega_0^2}$ is imaginary and we can write

$$\alpha = -\frac{\gamma}{2} \pm i\sqrt{\omega_0^2 - \left(\frac{\gamma}{2}\right)^2} \tag{7}$$
$$= -\frac{\gamma}{2} \pm i\omega_1$$

where

$$\omega_1 = \sqrt{\omega_0{}^2 - (\gamma/2)^2}.$$

The solution is

$$z = e^{-(\gamma/2)t}(z_1 e^{i\omega_1 t} + z_2 e^{-i\omega_1 t}), \tag{8}$$

where z_1 and z_2 are complex constants. In order to find the real part of z we write the complex numbers in Cartesian form

$$x + iy = e^{-(\gamma/2)t}[(x_1 + iy_1)(\cos\omega_1 t + i\sin\omega_1 t)$$
$$+ (x_2 + iy_2)(\cos\omega_1 t - i\sin\omega_1 t)].$$

The real part can be rearranged

$$x = Ae^{-(\gamma/2)t} \cos(\omega_1 t + \phi)$$

where A and ϕ are new arbitrary constants. This is the result quoted in Eq. (11.10). The imaginary part of z, which is also an acceptable solution, has the same form.

Case 2: heavy damping: $(\gamma/2)^2 > \omega_0^2$.

In this case, $\sqrt{(\gamma/2)^2 - \omega_0^2}$ is real and Eq. (5) has the solution

$$\alpha = -\frac{\gamma}{2} \pm \frac{\gamma}{2} \sqrt{1 - \frac{\omega_0^2}{(\gamma/2)^2}}.$$

Both roots are negative, and we can write

$$z = z_1 e^{-|\alpha_1|t} + z_2 e^{-|\alpha_2|t}. \tag{9}$$

The exponentials are real. The real part of z is

$$x = Ae^{-|\alpha_1|t} + Be^{-|\alpha_2|t}. \tag{10}$$

This solution has no oscillatory behavior and is known as *overdamped*.

Case 3: critical damping: $\gamma^2/4 = \omega_0^2$.

If $\gamma^2/4 = \omega_o^2$ we have only the single root

$$\alpha = -\frac{\gamma}{2}.$$

The corresponding solution is

$$x = Ae^{-(\gamma/2)t}. \tag{11}$$

However, this solution is incomplete. Mathematically, the solution of a second-order linear differential equation must always involve two arbitrary constants. Physically, the solution must have two constants to allow us to specify the initial position and initial velocity of the oscillator. As described in texts on differential equations, the second solution can be found by using a "variation of parameters" trial solution.

$$x = u(t)e^{(-\gamma/2)t}.$$

Substituting in Eq. (1) and recalling that $\gamma = 2\omega_o$ for this case, we find that $u(t)$ must satisfy the equation

$$\ddot{u} = 0.$$

Hence

$$u = a + bt$$

and the general solution for critical damping is therefore

$$x = (A + Bt)e^{-(\gamma/2)t}. \tag{11.43}$$

Note 11.3 Solving the Equation of Motion for the Driven Harmonic Oscillator

The equation of motion is

$$\ddot{x} + \gamma \dot{x} + \omega_0{}^2 x = \frac{F_0}{m} \cos \omega t. \tag{1}$$

The companion equation is

$$\ddot{y} + \gamma \dot{y} + \omega_0{}^2 y = \frac{F_0}{m} \sin \omega t. \tag{2}$$

Multiplying Eq. (2) by i and adding to Eq. (1) yields

$$\ddot{z} + \gamma \dot{z} + \omega_0{}^2 z = \frac{F_0}{m} e^{i\omega t}. \tag{3}$$

z must vary as $e^{i\omega t}$, so we try

$$z = z_0 e^{i\omega t}.$$

Inserting this in Eq. (3) gives

$$(-\omega^2 + i\omega\gamma + \omega_0{}^2) z_0 e^{i\omega t} = \frac{F_0}{m} e^{i\omega t}$$

or

$$z_0 = \frac{F_0}{m} \left(\frac{1}{\omega_0^2 - \omega^2 + i\omega\gamma} \right).$$

We can put z_0 into Cartesian form by multiplying numerator and denominator by the complex conjugate of the denominator. This gives

$$z_0 = \frac{F_0}{m} \left(\frac{(\omega_0^2 - \omega^2) - i\omega\gamma}{(\omega_0^2 - \omega^2)^2 + (\omega\gamma)^2} \right).$$

In polar form, $z_0 = R e^{i\phi}$, where

$$R = \sqrt{z_0 z_0^*}$$

$$= \frac{F_0}{m} \sqrt{\frac{1}{(\omega_0{}^2 - \omega^2)^2 + (\omega\gamma)^2}} \tag{4}$$

and

$$\phi = \arctan \left(\frac{\omega\gamma}{\omega^2 - \omega_0{}^2} \right). \tag{5}$$

The complete solution is

$$z = R e^{i\phi} e^{i\omega t},$$

which has the real part

$$x = R \cos(\omega t + \phi).$$

Problems

*For problems marked *, refer to page 524 for a hint, clue, or answer.*

11.1 *Time average of* $\sin^2$
Show by direct calculation that $\overline{\langle \sin^2(\omega t)\rangle} = \frac{1}{2}$, where the time average is taken over any complete period $t_1 \le t \le t_1 + 2\pi/\omega$.

11.2 *Time average of* $\sin \times \cos$
Show by direct calculation that $\overline{\langle \sin(\omega t) \cos(\omega t)\rangle} = 0$ when the average is over a complete period.

11.3 *Damped mass and spring*
A 0.3-kg mass is attached to a spring and oscillates at 2 Hz with a Q of 60. Find the spring constant and damping constant.

11.4 *Phase shift in a damped oscillator*
In an undamped free harmonic oscillator the motion is given by $x = A \sin \omega_0 t$. The displacement is maximum exactly midway between the zero crossings.

In a damped oscillator the motion is no longer sinusoidal, and the maximum is advanced before the midpoint of the zero crossings. Show that the maximum is advanced by a phase angle ϕ given approximately by

$$\phi = \frac{1}{2Q},$$

where we assume that Q is large.

11.5 *Logarithmic decrement*
The *logarithmic decrement* δ is defined to be the natural logarithm of the ratio of successive maximum displacements (in the same direction) of a free damped oscillator. Show that $\delta = \pi/Q$.

11.6 *Parameters of a damped oscillator*
Find the spring constant k and damping constant b of a damped oscillator having a mass of 5 kg, frequency of oscillation 0.5 Hz, and logarithmic decrement 0.02.

11.7 *Critically damped oscillator*
If the damping constant of a free oscillator is given by $\gamma = 2\omega_0$, the system is said to be critically damped.

(*a*) Show by direct substitution that in this case the motion is given by

$$x = (A + Bt)e^{-(\gamma/2)t},$$

where A and B are constants.

(*b*) A critically damped oscillator is at rest at equilibrium. At $t = 0$ it is given a blow of total impulse I. Sketch the motion, and find the time at which the velocity starts to decrease.

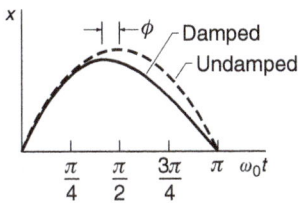

11.8 *Scale spring constant**
A mass of 10 kg falls 50 cm onto the platform of a spring scale, and sticks. The platform eventually comes to rest 10 cm below its initial position. The mass of the platform is 2 kg.

(*a*) Find the spring constant.

(*b*) It is desired to put in a damping system so that the scale comes to rest in minimum time without overshoot. This means that the scale must be critically damped (see Note 11.2). Find the necessary damping constant and the equation for the motion of the platform after the mass hits.

11.9 *Velocity and driving force in phase**
Find the driving frequency for which the velocity of a driven damped oscillator is exactly in phase with the driving force.

11.10 *Grandfather clock*
The pendulum of a grandfather clock activates an escapement mechanism every time it passes through the vertical. The escapement is under tension (provided by a hanging weight) and gives the pendulum a small impulse a distance *l* from the pivot. The energy transferred by this impulse compensates for the energy dissipated by friction, so that the pendulum swings with a constant amplitude.

(*a*) What is the impulse needed to sustain the motion of a pendulum of length L and mass m, with an amplitude of swing θ_0 and quality factor Q?

(*b*) Why is it desirable for the pendulum to engage the escapement as it passes vertical rather than at some other point of the cycle?

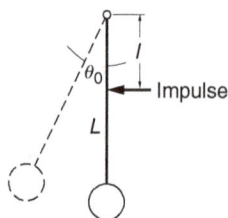

11.11 *Average stored energy*
Show that for a lightly damped driven oscillator

$$\frac{\text{average energy stored in the oscillator}}{\text{average energy dissipated per radian}} \approx \frac{\omega_o}{\gamma} = Q.$$

11.12 *Cuckoo clock**
A small cuckoo clock has a pendulum 25 cm long with a mass of 10 g and a period of 1 s. The clock is powered by a 200-gram weight which falls 2 m between the daily windings. The amplitude of the swing is 0.2 rad. What is the Q of the clock? How long would the clock run if it were powered by a battery with 1 J capacity?

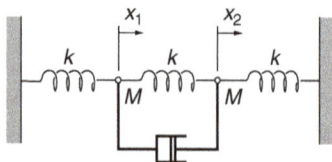

11.13 *Two masses and three springs**
Two identical masses M are hung between three identical springs. Each spring is massless and has spring constant k. The masses are connected as shown to a dashpot of negligible mass. Neglect gravity.

The dashpot exerts a force bv, where v is the relative velocity of its two ends. The force opposes the motion. Let x_1 and x_2 be the displacements of the two masses from equilibrium.

(a) Find the equation of motion for each mass.

(b) Show that the equations of motion can be solved in terms of the new dependent variables $y_1 = x_1 + x_2$ and $y_2 = x_1 - x_2$.

(c) Show that if the masses are initially at rest and mass 1 is given an initial velocity v_0, the motion of the masses after a sufficiently long time is

$$x_1 = x_2$$
$$= \frac{v_0}{2\omega} \sin \omega t.$$

Evaluate ω.

11.14 *Motion of a driven damped oscillator*

The motion of a damped oscillator driven by an applied force $F_0 \cos \omega t$ is given by $x_a(t) = X_0 \cos(\omega t + \phi)$, where X_0 and ϕ are given by Eqs. (11.29) and (11.30). Consider an oscillator that is released from rest at $t = 0$. Its motion must satisfy $x(0) = 0, v(0) = 0$, but after a very long time, we expect that $x(t) = x_a(t)$. To satisfy these conditions we can take as the solution

$$x(t) = x_a(t) + x_b(t),$$

where $x_b(t)$ is the solution to the equation of motion of the free damped oscillator, Eq. (11.10).

(a) Show that if $x_a(t)$ satisfies the equation of motion for the driven damped oscillator, then so does $x(t) = x_a(t) + x_b(t)$, where $x_b(t)$ satisfies the equation of motion of the free damped oscillator, Eq. (11.10).

(b) Choose the arbitrary constants in $x_b(t)$ so that $x(t)$ satisfies the initial conditions. (Note that A and ϕ here are arbitrary.)

(c) Sketch the resulting motion for the case where the oscillator is driven at resonance.

12 THE SPECIAL THEORY OF RELATIVITY

12.1 Introduction

In the centuries following publication of the *Principia*, Newtonian dynamics was accepted whole-heartedly not only because of its enormous success in explaining planetary motion but also in accounting for all motions commonly encountered on the Earth. Physicists and mathematicians (often the same people) created elegant reformulations of Newtonian physics and introduced more powerful analytical and calculational techniques, but the foundations of Newtonian physics were assumed to be unassailable. Then, on June 30 1905, Albert Einstein presented his special theory of relativity in his publication *The Electrodynamics of Moving Bodies*. The English translation, available on the web, is reprinted from *Relativity: The Special and General Theory*, Albert Einstein, Methuen, London (1920). The original publication is *Zur Elektrodynamik bewegter Körper*, Annalen der Physik 17 (1905). Einstein's paper transformed our fundamental view of space, time, and measurement.

The reason that Newtonian dynamics went unchallenged for over two centuries is that although we now realize that it is only an approximation to the laws of motion, the approximation is superb for motion with speed much less than the speed of light, $c \approx 3 \times 10^8$ m/s. Relativistic modifications to observations of a body moving with speed v typically involve a factor of v^2/c^2. Most familiar phenomena involve speeds $v \ll c$. Even for the high speed of an Earth-orbiting satellite, $v^2/c^2 \approx 10^{-10}$. There is one obvious exception to this generalization about speed: light itself. Thus, it is hardly surprising that the problems that triggered Einstein's thinking concerned not mechanics but light, problems that grew out of Einstein's early fascination with Maxwell's electromagnetic theory—the theory of light.

12.2 The Possibility of Flaws in Newtonian Physics

The German physicist and philosopher Ernst Mach first pointed out the possibility of flaws in Newtonian thought. Although Mach proposed no changes to Newtonian dynamics, his analysis impressed the young Einstein and was crucial in the revolution shortly to come. Mach's 1883 text *The Science of Mechanics* incorporated the first incisive critique of Newton's ideas about dynamics. Mach carefully analyzed Newton's explanation of the dynamical laws, taking care to distinguish between definitions, derived results, and statements of physical law. Mach's approach is now widely accepted; our discussion of Newton's laws in Chapter 2 is very much in Mach's spirit.

The Science of Mechanics raised the question of the distinction between absolute and relative motion. According to Mach, the fundamental weakness in Newtonian dynamics was Newton's conception of space and time. Newton avowed that he would forgo abstract speculation ("I do not make hypotheses") and deal only with observable facts, but he was not totally faithful to this resolve. In particular, consider the following description of time that appears in the *Principia*. (The excerpt is condensed.) *Absolute, true and mathematical time, of itself and by its own true nature, flows uniformly on, without regard to anything external. Relative, apparent and common time is some sensible and external measure of absolute time estimated by the motions of bodies, whether accurate or inequable, and is commonly employed in place of true time; as an hour, a day, a month, a year.*

Mach commented "it would appear as though Newton in the remarks cited here still stood under the influence of medieval philosophy, as though he had grown unfaithful to his resolve to investigate only actual facts." Mach went on to point out that since time is necessarily measured by the repetitive motion of some physical system, for instance the pendulum of a clock or the revolution of the Earth about the Sun, then the properties of time must be connected with the laws that describe the motions of physical systems. Simply put, Newton's idea of time without clocks is metaphysical; to understand the properties of time we must observe the properties of clocks. As a prescient question, we might inquire whether a time interval observed on a moving clock has the same value as the interval observed on a clock at rest. A simple question? Yes indeed, except that the idea of absolute time is so natural that the eventual consequences of Mach's critique, the relativistic description of time, still comes as something of a shock to students of science.

There are similar weaknesses in the Newtonian view of space. Mach argued that since position in space is determined using measuring rods, the properties of space can be understood only by investigating the properties of meter sticks. For example, does the length of a meter stick observed while it is moving agree with the length of the same meter stick at rest? To understand space we must look to nature, not to Platonic ideals.

Mach's special contribution was to examine the most elemental aspects of Newtonian thought, to look critically at matters that might seem too simple to discuss, and to insist that correctly understanding nature means turning to experience rather than invoking mental abstractions. From this point of view, Newton's assumptions about space and time must be regarded merely as postulates. Newtonian mechanics follows from these postulates, but other assumptions are possible and from them different laws of dynamics could follow.

Mach's critique had no immediate effect but its influence was eventually profound. The young Einstein, while a student at the Polytechnic Institute in Zurich in the period 1897–1900, was much attracted by Mach's work and by Mach's insistence that physical concepts be defined in terms of observables. However, the most urgent reason for superseding Newtonian physics was not Mach's critique but Einstein's recognition that there were inconsistencies in interpreting the results of Maxwell's electromagnetic theory, notwithstanding that Maxwell's theory was considered the crowning achievement of classical physics.

The crucial event that triggered the theory of special relativity and decisively altered physics is generally taken to be the Michelson–Morley experiment, though it is not clear precisely what role this experiment actually played in Einstein's thinking. Nevertheless, most treatments of special relativity take it as the point of departure and we shall follow this tradition.

12.3 The Michelson–Morley Experiment

The problem that Michelson attacked was to detect the effect of the Earth's motion on the speed of light. Briefly, Maxwell's electromagnetic theory (1861) predicted that electromagnetic disturbances in empty space would propagate at 3×10^8 m/s—the speed of light. The evidence was overwhelming that light consisted of electromagnetic waves, but there was a serious conceptual difficulty.

The only waves then known to physics propagated in matter—solid, liquid, or gas. A sound wave in air, for example, consists of alternate regions of higher and lower pressure propagating with a speed of 330 m/s, somewhat less than the speed of molecular motion. The speed of mechanical waves in a metal bar is higher, typically 5000 m/s. The speed of sound increases with the rigidity of the material or the strength of the "spring forces" between neighboring atoms.

Electromagnetic wave propagation seemed to be fundamentally different. By analogy with mechanical waves in matter, electromagnetic waves were assumed to propagate through space as vibrations in a medium called the *ether* that supported electromagnetic wave propagation. Unfortunately, the ether had to possess contradictory properties; immensely rigid to allow light to propagate at 3×10^8 m/s while so insubstantial that it did not interfere with the motion of the planets.

One consequence of the ether hypothesis is that the speed of light should depend on the observer's motion relative to the ether. Maxwell suggested an astronomical experiment to detect this effect. The motion of the planet Jupiter through space relative to the Earth should affect the speed with which its light reaches us. The periodic eclipses of the moons of Jupiter create a clock. The clock should appear to periodically advance or fall behind, as the speed of light increases or decreases as Jupiter approaches to and recedes from the Earth. The effect turned out to be too small to be measured accurately. Nevertheless, Maxwell's proposal was historically important: it stimulated Albert A. Michelson, a young U.S. Navy officer at Annapolis, to invent a laboratory experiment for measuring the Earth's motion through the ether.

The following explanation of the Michelson–Morley experiment assumes some familiarity with optical interference. If you do not yet know about interference, you can skip the description and take the conclusion on faith: the speed of light is always the same, regardless of the relative motion of the source and the observer.

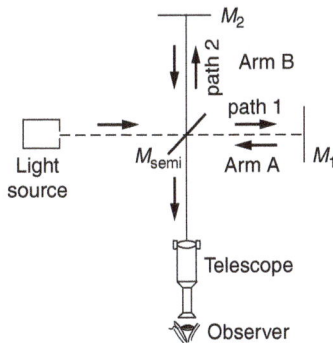

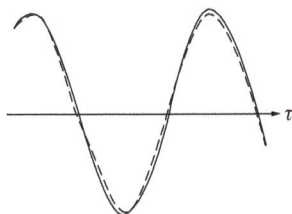

(a) in phase

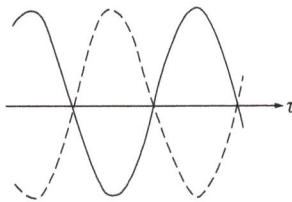

(b) 180° out of phase

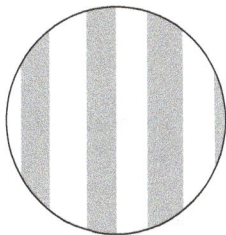

Michelson's apparatus was an optical interferometer. As shown in the drawing, light from a source is split into two beams by a semi-silvered mirror M_{semi} that reflects half the light and transmits half. Half of the beam from the light source travels straight ahead on path 1, passing through M_{semi} until it is reflected by mirror M_1. It then returns to mirror M_{semi}, and half is reflected to the observer. The remainder of the beam from the light source is the half that is reflected by M_2, along path 2. It is reflected by mirror M_2, which directs it to the observer after passing again through M_{semi}. Thus beams 1 and 2 each have 1/4 the intensity of the initial beam.

If beams 1 and 2 travel the same distance, they arrive at the observer in phase so that their electric fields add. The observer sees light. However, if the path lengths differ by *half* a wavelength, the fields arrive out of phase and cancel so that no light reaches the observer. In practice, the two beams are slightly misaligned and the observer sees a pattern of bright and dark interference fringes.

If the length of one of the arms is slowly changed, the fringe pattern moves. Changing the difference in path lengths by one wavelength shifts the pattern by one fringe.

The motion of the Earth through the ether should cause a difference between the times for light to transit the two arms of the interferometer, just as if there were a small change in the distance. The difference in transit time depends on the orientation of the arms with respect to the velocity of the Earth through the ether.

We suppose that the laboratory moves through the ether with speed v and that arm A lies in the direction of motion while arm B is perpendicular. According to the ether hypothesis, an observer moving toward the source of a light signal with speed v will observe the signal to travel with speed $c + v$, while for motion away from the source the speed is $c - v$.

If the length of the arms from the partially silvered mirror M_{semi} to their ends is l, then the time interval for the light to go from M_{semi} to M_1 and return along arm A is τ_A, where

$$\tau_A = \frac{l}{c+v} + \frac{l}{c-v} = \frac{2l}{c}\left(\frac{1}{1 - v^2/c^2}\right).$$

Because $v^2/c^2 \ll 1$, we can simplify the result using the Taylor series expansion in Note 1.3: $1/(1-x) = 1 + x + x^2 + \cdots$. Letting $x = v^2/c^2$ we have

$$\tau_A \approx \frac{2l}{c}\left(1 + \frac{v^2}{c^2}\right).$$

Arm B is perpendicular to the motion so the speed of light is not affected by the motion. However, there is nevertheless a time delay due to the motion of M_{semi} as the light traverses the arm. Denoting the round trip time by τ_B, then during that interval M_{semi} moves a distance $v\tau_B$.

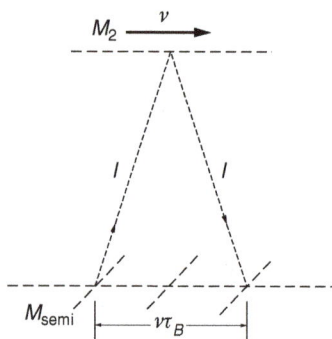

Consequently, the light travels along the hypotenuse of the right triangles shown in the sketch, and the distance traveled is $2\sqrt{l^2 + (v\tau_B/2)^2}$.

Consequently,

$$\tau_B = \frac{2}{c}\sqrt{l^2 + (v\tau_B/2)^2},$$

which gives

$$\tau_B = \frac{2l}{c}\frac{1}{\sqrt{1 - v^2/c^2}}.$$

Using the approximation $1/\sqrt{1-x} = 1 + (1/2)x + (1/8)x^2 + \cdots$, and keeping the first term, we have

$$\tau_B = \frac{2l}{c}\left(1 + \frac{1}{2}v^2/c^2\right).$$

The difference in time for the two paths is

$$\Delta\tau = \tau_A - \tau_B \approx \frac{l}{c}\left(\frac{v^2}{c^2}\right).$$

The frequency of light ν is related to its wavelength λ and the speed of light by $\nu = c/\lambda$. The interference pattern shifts by one fringe for each cycle of delay. Consequently, the number of fringe shifts caused by the time difference is

$$N = \nu\Delta T = \frac{l}{\lambda}\left(\frac{v^2}{c^2}\right).$$

The orbital speed of the Earth around the Sun gives $v/c \approx 10^{-4}$. Taking the path length $l = 1.2$ m, and using sodium light for which $\lambda = 590 \times 10^{-9}$ m, Michelson predicted a fringe shift of $N = 0.02$. In his initial attempt in 1881, Michelson searched for a fringe shift as the rotation of the Earth changed the direction of motion through the ether, but could detect none to within experimental accuracy.

In 1887 Michelson repeated the experiment in collaboration with the chemist Edward Morley using an apparatus mounted on a granite slab 35 cm thick that floated on mercury and could be continuously rotated. The path length was extended by a factor of 10 using repeated reflections between the mirrors. However, again no fringe shift. The Michelson–Morley experiment has been refined and repeated over the years but no effect of motion through the ether has ever been detected. We are forced to recognize that the speed of light is unaffected by motion of the observer through the ether. Ironically, Michelson, who conceived and executed the experiment for which he is famous, viewed it as a failure. He set out to see the effect of motion through the ether but could not detect any.

Various attempts to explain the null result of the Michelson–Morley experiment introduced such complexity as to threaten the foundations of electromagnetic theory. One attempt was the hypothesis proposed by the

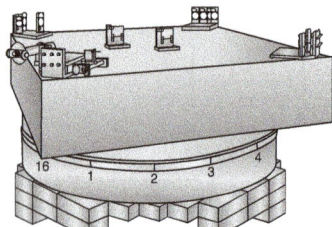

Irish physicist G.F. FitzGerald and the Dutch physicist H.A. Lorentz that motion of the Earth through the ether caused a shortening of one arm of the Michelson interferometer (the "Lorentz–FitzGerald contraction") by exactly the amount required to eliminate the fringe shift. Other theories that involved such artifacts as drag of the ether by the Earth were even less productive. The elusive nature of the ether remained a troubling enigma.

12.4 The Special Theory of Relativity

It is an indication of Einstein's genius that the troublesome problem of the ether pointed the way not to complexity and elaboration but to a simplification that unified the fundamental concepts of physics. Einstein regarded the difficulty with the ether not as a fault in electromagnetic theory but as an error in basic dynamical principles. He presented his ideas in the form of two postulates, prefacing them with a note on simultaneity and how to synchronize clocks.

12.4.1 Synchronizing Clocks

Before presenting his theory of space and time, Einstein considered the elementary process of comparing measurements of time by different observers having identical clocks. For the measurements to agree, the clocks must be *synchronized*—they must be adjusted to agree on the time of a single event. In Newtonian physics, if a flash of light occurs, the flash arrives simultaneously at all synchronized clocks, wherever their locations.

The Newtonian procedure would work if the speed of light were infinite or so large that it could be regarded as infinite. However, if one accepts that signals can propagate no faster than the speed of light, the procedure is wrong in principle. For instance, a signal from the Moon to the Earth takes about one second. One might attempt to synchronize a clock on the Moon with a clock on the Earth by advancing the Moon clock by one second. With this adjustment, the Moon clock would always appear to agree with the Earth clock. However, for the observer on the Moon, the Earth clock would always lag the Moon clock by two seconds. Thus the clocks would be synchronized for one observer but not the other.

Einstein proposed a simple procedure for synchronizing clocks so that all observers agree on the time of an event. Observer A sends observer B a signal at time T_A. Observer B notes that the signal arrives at time T_B on the local clock. B immediately sends a signal back to A who detects it at time $T'_A = T_A + \Delta T$. The clocks are synchronized if B's clock reads $T_B = T_A + \Delta T/2$. Interpreting the times reported by different observers requires knowing their positions, but everyone would agree on the time of an event.

Einstein thought about time measurements in terms of railway clocks at different stations for which light propagation times are of no practical importance. Today, Einstein's procedure for synchronizing clocks is crucially important: it is essential for comparing atomic clocks in international time standards laboratories, as well as for keeping the Internet synchronized and for maintaining the voltage–current phase across the national power grid.

12.4.2 The Principle of Relativity

The special theory of relativity rests on two postulates. The first, known as the principle of relativity, is that the laws of physics have the same form with respect to all inertial systems. In Einstein's words: "The laws by which the states of physical systems undergo change are not affected, whether these changes of state be referred to the one or the other of two systems of co-ordinates in uniform translatory motion." The principle of relativity was hardly novel; Galileo is credited with first pointing out that there is no dynamical way to determine whether one is moving uniformly or is at rest, and Newton gave it rigorous expression in his dynamical laws in which acceleration, not velocity, is paramount. If the principle of relativity were not true, energy and momentum might be conserved in one inertial system but not in another. The principle of relativity played only a minor role in the development of classical mechanics: Einstein elevated it to a keystone of dynamics. He extended the principle to include not only the laws of mechanics but also the laws of electromagnetic interaction and *all* the laws of physics. Furthermore, in his hands the principle of relativity became a powerful tool for discovering the correct form of physical laws.

We can only guess at the sources of Einstein's inspiration, but they must have included the following consideration. If the speed of light were not a universal constant, that is, if the ether could be detected, then the principle of relativity would fail; a special inertial frame would be singled out, the one at rest in the ether. However, the form of Maxwell's equations, as well as the failure of any experiment to detect motion through the ether, cause us to conclude that the speed of light is independent of the motion of the source. Our inability to detect absolute motion, either with light or with Newtonian dynamics, forces us to accept that absolute motion has no role in physics.

The Universal Speed

The second postulate of relativity is that the speed of light is a universal constant, the same for all observers. "Any ray of light moves in the stationary system of co-ordinates with the determined speed c, whether the ray be emitted by a stationary or by a moving body." Einstein argued that because the speed of light c predicted by electromagnetic theory involves no reference to a medium, then no matter how we measure the speed of light the result will *always* be c, independent of our motion. This is in contrast to the behavior of sound waves, for example, where the observed speed of the wave depends on the motion of the observer through the medium. The idea of a universal speed was indeed a bold hypothesis, contrary to all previous experience and, for many of Einstein's contemporaries, defying common sense. But common sense can be a poor guide. Einstein once quipped that common sense consists of the prejudices one learns before the age of eighteen.

Rather than regarding the absence of the ether as a paradox, Einstein saw that the concept of a universal speed preserved the simplicity of the principle of relativity. His view was essentially conservative; he insisted on maintaining the principle of relativity that the ether would destroy. The urge toward simplicity appeared to be fundamental to Einstein's personality. The special theory of relativity was the simplest way to preserve the unity of classical physics.

To summarize, the postulates of special relativity are: *The laws of physics have the same form in all inertial systems. The speed of light in empty space is a universal constant, the same for all observers regardless of their motion.*

These postulates require us to revise our ideas about space and time, and this has immediate consequences for physics. The mathematical expression of kinematics and dynamics in the special theory of relativity is embodied in the Lorentz transformation—a simple prescription for relating events in different inertial systems.

12.5 Transformations

In the world of relativity, a transformation is a set of equations that relate observations in one coordinate system to observations in another. As you will see, the logic of special relativity is reasonably straightforward and the mathematics is not arcane. Nevertheless, the reasoning is likely to seem perplexing because of the underlying question "Isn't this a peculiar way to do physics?" The answer is "Yes! This is a most peculiar way to do physics!" Rather than examining forces, conservation laws, dynamical equations, and other staples of Newtonian physics, Einstein merely discussed how things look to different observers.

Einstein was the first person to use transformation theory to discover new physical behavior, in particular, to create the theory of special relativity. From two simple assumptions, he derived a new way to look at space and time and discovered a new system of dynamics.

Special relativity can be written with all the elegance of a beautiful mathematical theory but its most attractive attribute is that it not only looks beautiful, it works beautifully. The theory of special relativity is among the most carefully studied theories in physics and its predictions have always been correct within experimental error.

The heart of special relativity is the Lorentz transformation, but to introduce Einstein's approach let us first look at the corresponding procedure for Newtonian physics where the transformation is known as the Galilean transformation.

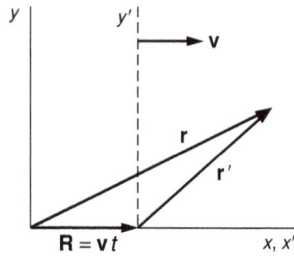

12.5.1 The Galilean Transformation

We will frequently refer to observations in two standard inertial systems: $S = (x, y, z, t)$ and $S' = (x', y', z', t')$. S' moves with respect to S at speed v in the x direction. Alternatively, S moves with respect to S' at speed v in the negative x direction. For convenience, we take the origins to coincide at $t = 0$, and take the x and x' axes to be parallel.

If a particular point in space has coordinates $\mathbf{r} = (x, y, z)$ in S, the coordinates in S' are $\mathbf{r}' = (x', y', z')$. These are related by

$$\mathbf{r}' = \mathbf{r} - \mathbf{R},$$

where

$$\mathbf{R} = \mathbf{v}t.$$

Since $\mathbf{v}$ is in the x direction, we have

$$\begin{aligned} x' &= x - vt, \\ y' &= y \\ z' &= z \\ t' &= t. \end{aligned} \tag{12.1}$$

The fourth equation $t' = t$, listed merely for completeness, is taken for granted in Newtonian dynamics, and follows immediately from the Newtonian concept of "ideal" time.

Equations (12.1) are known as the *Galilean transformation*. Because the laws of Newtonian mechanics hold in all inertial systems, the form of the laws is unaffected by this transformation. More concretely, there is no way to distinguish between systems on the basis of the motion they predict. The following example illustrates what this means.

Example 12.1 Applying the Galilean Transformation

Consider how we might discover the law of force between two isolated bodies from observations of their motion. For example, the problem might be to discover the law of gravitation from data on the elliptical orbit of one of Jupiter's moons. If m_1 and m_2 are the masses of the moon and of Jupiter, respectively, and if $\mathbf{r_1}$ and $\mathbf{r_2}$ are their positions relative to an observer on the Earth, we have

$$\begin{aligned} m_1 \ddot{\mathbf{r}}_1 &= \mathbf{F}(r) \\ m_2 \ddot{\mathbf{r}}_2 &= -\mathbf{F}(r), \end{aligned}$$

where we assume that the force $\mathbf{F}$ between the bodies is a central force that depends only on the separation $r = |\mathbf{r_2} - \mathbf{r_1}|$. From our observations of $\mathbf{r_1}(t)$ we can evaluate $\ddot{\mathbf{r}}_1$, from which we obtain the value of $\mathbf{F}$. Suppose the data reveal that $\mathbf{F}(r) = -Gm_1 m_2 \hat{\mathbf{r}}/r^2$.

Now let us look at the problem from the point of view of an observer in a spacecraft that is moving with constant speed far from the Earth.

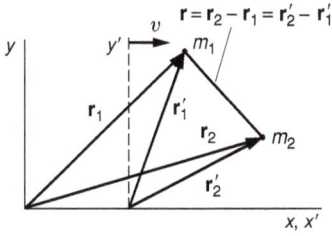

According to the principle of relativity this observer must obtain the same force law as the earthbound observer. The situation is represented in the drawing. x, y is the earthbound system, x', y' is the spacecraft system, and $\mathbf{v}$ is the relative velocity of the two systems. Note that the vector $\mathbf{r}$ from m_1 to m_2 is the same in both coordinate systems.

In the x', y' system the observer sees that the moon is accelerating at rate $\ddot{\mathbf{r}}_1'$ and concludes that the force is

$$\mathbf{F}'(r) = m_1 \ddot{\mathbf{r}}_1'.$$

A fundamental property of the Galilean transformation is that acceleration is unaltered. Here is the formal proof: because $\dot{\mathbf{v}} = 0$, we have

$$\mathbf{r}_1 = \mathbf{r}_1' + \mathbf{v}t$$

$$\dot{\mathbf{r}}_1 = \dot{\mathbf{r}}_1' + \mathbf{v}$$

$$\ddot{\mathbf{r}}_1 = \ddot{\mathbf{r}}_1'.$$

Consequently,

$$\mathbf{F}'(r') = m_1 \ddot{\mathbf{r}}_1'$$
$$= m_1 \ddot{\mathbf{r}}_1$$
$$= \mathbf{F}(r)$$
$$= -\frac{G m_1 m_2}{r^2} \hat{\mathbf{r}}.$$

The law of force is identical in the two systems. This is what we mean when we say that two inertial systems are equivalent. If the form of the law, or the value of G, were not identical we could make a judgment about the speed of a coordinate system in empty space by investigating the law of gravitation in that system. The inertial systems would not be equivalent.

Example 12.1 is almost trivial because the force depends on the separation of the two particles, a quantity that is unchanged (invariant) under the Galilean transformation. All forces in Newtonian physics are due to interactions between particles, interactions that depend on the *relative* coordinates of the particles. Consequently, they are invariant under the Galilean transformation.

What happens to the equation for a light signal under the Galilean transformation? The following example shows the difficulty.

Example 12.2 Describing a Light Pulse by the Galilean Transformation

At $t = 0$ a pulse of light is emitted from the origin of the S system, and travels along the x axis at speed c. The equation for the location of the pulse along the x axis is $x = ct$.

In the S' system, the equation for the wavefront along the x' axis is

$$x' = x - vt$$
$$= (c - v)t,$$

where v is the relative velocity of the two systems. The speed of the pulse in the S' system is

$$\frac{dx'}{dt} = c - v.$$

But this result is contrary to the postulate that the speed of light is always c, the same for all observers.

Because the Galilean transformation is incompatible with the principle that the speed of light is always c, our task is to find a transformation that is compatible. Before undertaking this, it is useful to think carefully about the nature of measurement.

The Galilean transformation relates the spatial coordinates of an event measured by observers in two inertial systems moving with relative speed v. By an "event" we mean the unique values of a set of coordinates in space and time. Physically meaningful measurements invariably involve more than a single event. For instance, measuring the length of a rod involves placing the rod along a calibrated scale such as a meter stick, and recording the position at each end. Consequently, length involves two measurements. If the rod is at rest along the axis in the S system, the coordinates of its end points might be x_a and x_b, where $x_b = x_a + L$. According to an observer in the S' system, the x' coordinates are given by Eq. (12.1): $x'_a = x_a - vt$ and $x'_b = x_b - vt$. Since $x_b = x_a + L$, we have $L = x_b - x_a$ and $L' = x'_b - x'_a = L$. The two observers agree on the length.

In this simple exercise in measurement we have made a natural assumption: the measurements are made *simultaneously*. This is not important in the S system because the rod is at rest. However, in the S' system the rod is moving. If the end points were recorded at different times, the value for L' would have been incorrect. We have used the Galilean assumption $t' = t$, which implies that if measurements are simultaneous in one coordinate system they are simultaneous in all coordinate systems. This would be the case if the speed of light were infinite, but the finite speed of light profoundly affects our idea of simultaneity. We therefore digress briefly to examine the nature of simultaneity.

12.6 Simultaneity and the Order of Events

We have an intuitive idea of simultaneity: two events are simultaneous if their time coordinates have the same value. However, as the following example shows, events that are simultaneous in one coordinate system are not necessarily simultaneous when observed in a different coordinate system.

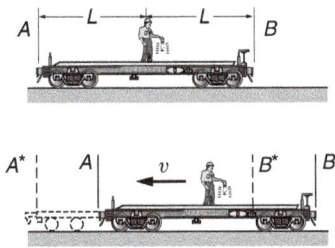

Example 12.3 Simultaneity

A railwayman stands at the middle of a flatcar of length $2L$. He flicks on his lantern and a light pulse travels out in all directions with the velocity c.

Light arrives at the ends of the car after a time interval L/c. In this system, the flatcar's rest system, the light arrives simultaneously at the end points A and B.

Now let us observe the same situation from a frame moving to the right with velocity v. In this frame the flatcar moves to the left with velocity v. As observed in this frame the light still has velocity c, according to the second postulate of special relativity. However, during the transit time, A moves to A^* and B moves to B^*. It is apparent that the pulse arrives at B^* before A^*; the events are not simultaneous in this frame.

Just as events that are simultaneous in one inertial system may not be simultaneous in another, it can be shown that events that are spatially coincident— having the same coordinates in space—in one system may not appear to be coincident in another. We shall show later that two events can be classified as either *spacelike* or *timelike*. For spacelike events it is impossible to find a coordinate system in which the events coincide in space, though there is a system in which they are simultaneous in time. For timelike events it is impossible to find a coordinate system in which the events are simultaneous in time, though there is a system in which they coincide in space.

At this point we need a systematic way to solve the problem of relating observations made in different inertial systems in a fashion that obeys the principle of relativity. This task constitutes the core of special relativity.

12.7 The Lorentz Transformation

The failure of the Galilean transformation to satisfy the postulate that the speed of light is a universal constant constituted a profound dilemma. Einstein solved the dilemma by introducing a new transformation law for relating the coordinates of events as observed in different inertial systems. He introduced a system designed to ensure that a signal moving at the speed of light in one system would be observed to move at the same speed in the other, irrespective of the relative motion. Such a "fix" took some courage because to alter a transformation law is to alter the fundamental relation between space and time.

Let us refer once more to our standard systems, the S system (x, y, z, t), and the S' system (x', y', z', t'). The system S' moves with velocity v along the positive x axis, and the origins coincide at $t = t' = 0$. We take the most general transformation relating the coordinates of a given event

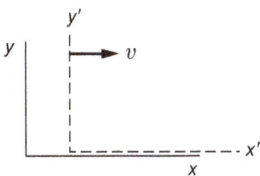

in the two systems to be of the form

$$x' = Ax + Bt \qquad (12.2a)$$
$$y' = y \qquad (12.2b)$$
$$z' = z \qquad (12.2c)$$
$$t' = Cx + Dt. \qquad (12.2d)$$

Some comments on Eqs. (12.2): the transformation equations are linear because a nonlinear transformation could yield an acceleration in one system even if the velocity were constant in the other. Further, we leave the y' and z' axes unchanged, by symmetry.

Here is one model to justify the assumption that $y' = y$ and $z' = z$: consider two trains on parallel tracks. Each train has an observer holding a paint brush at the same height in their system, say at 1 m above the floor of the train. Each train is close to a wall. The trains approach at relative speed v, and each observer holds the brush to the wall, leaving a stripe. Observer 1 paints a blue stripe and observer 2 paints a yellow stripe.

Suppose that observer 1 sees that the height of observer 2 has changed, so that the blue stripe is below the yellow stripe. Observer 2 would have to see the same phenomenon except that it is now the yellow stripe that is below the blue stripe. Because their conclusions are contradictory they cannot both be right. Since there is no way to distinguish between the systems, the only conclusion is that both stripes are at the same height. We conclude that distance perpendicular to the direction of motion is unchanged by the motion of the observer.

We can evaluate the four constants A, B, C, D in Eqs. (12.2) by comparing coordinates for four events. These could be:

(1) The origin of S is observed in S':
$S : (x = 0, t)$; $S' : (x' = -vt', t')$.
From Eqs. (12.2a) and (12.2d), $-vt' = 0 + Bt$, and $t' = 0 + Dt$.
Consequently, $B = -vD$.

(2) The origin of S' is observed in S:
$S' : (x' = 0, t')$; $S : (x = +vt, t)$.
From Eq. (12.2a), $0 = Avt + Bt$.
Consequently, $B = -vA$, and using result (1), if follows that $D = A$.

(3) A light pulse is emitted from the origin at $t = 0, t' = 0$ and is observed later along the x and x' axes.
$S : (x = ct, t)$; $S' : (x' = ct', t')$.
From Eqs. (12.2a) and (12.2d), $ct' = ctA + Bt$, and $t' = ctC + Dt$
Using $D = A$ and $B = -vA$, it follows that
$C = -(v/c^2)A$.

(4) A light pulse is emitted from the origin at $t = 0, t' = 0$ and is observed later along the y axis in the S system. In S, $x = 0$, $y = ct$, but in S' the pulse has both x' and y' coordinates.

$$S : (x = 0, y = ct, t); \qquad S' : (x' = -vt', y' = \sqrt{(ct)^2 - (-vt')^2}, t').$$

From Eqs. (12.2b) and (12.2d), $ct = \sqrt{(ct')^2 - (-vt')^2}$ and $t' = Dt$ which give

$$D = 1/\sqrt{1 - v^2/c^2}.$$

(We selected the positive sign for the square root because otherwise t and t' would have opposite signs.) The factor $1/\sqrt{1 - v^2/c^2}$ occurs so frequently that it is given a special symbol:

$$\gamma \equiv \frac{1}{\sqrt{1 - v^2/c^2}}.$$

Note that $\gamma \geq 1$ and that as $v \to c$, $\gamma \to \infty$.

Substituting our results in Eqs. 12.2 yields

$$x' = \gamma (x - vt) \tag{12.3a}$$

$$y' = y \tag{12.3b}$$

$$z' = z \tag{12.3c}$$

$$t' = \gamma \left(t - vx/c^2\right). \tag{12.3d}$$

The transformation from S' to S can be found by letting $v \to -v$:

$$x = \gamma (x' + vt') \tag{12.4a}$$

$$y = y' \tag{12.4b}$$

$$z = z' \tag{12.4c}$$

$$t = \gamma \left(t' + vx'/c^2\right). \tag{12.4d}$$

Equations (12.3) and (12.4) are the prescription for relating the coordinates of an event in different inertial systems so as to satisfy the postulates of special relativity. They are called the *Lorentz transformation* after the physicist Hendrik Lorentz who first wrote them, though in a very different context.

The Lorentz transformation equations have a straightforward physical interpretation. The factor γ is a scaling factor that ensures that the speed of light is the same in both systems. The factor vt in Eq. (12.3a) reveals that system S' is moving in the positive x direction, with speed v. The factor vx/c^2 in Eq. (12.3d) is a little more subtle. The clock synchronization algorithm requires that the time registered on a clock be corrected for the transit time $\tau_{transit}$ from the event point. If the point is moving with speed v, then the transit time correction must be adjusted correspondingly. The additional distance traveled is $d = v\tau_{transit}$, where $\tau_{transit} = x/c$. Hence, the time in Eq. (12.3d) needs to be corrected by the quantity $d/c = vx/c^2$.

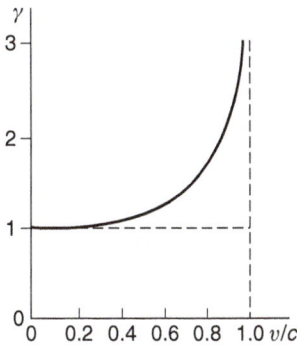

In the limit $v/c \to 0$ (or alternatively $c \to \infty$), where $\gamma \to 1$, the Lorentz transformation becomes identical to the Galilean transformation. However, in the general case, the Lorentz transformation requires a rethinking of the concepts of space and time.

Before looking into the consequences of this rethinking, let us examine how the Lorentz transformation demonstrates why Michelson's experiment had to give a null result.

12.7.1 Michelson–Morley Revisited

With the Lorentz transformation in hand we can understand why the Michelson–Morley experiment failed to display any fringe shift as the apparatus was rotated. We introduce again the two reference systems $S : (x, y, z, t)$ and the system $S' : (x', y'z', t')$ moving with relative speed v along x. Their origins coincide at $t = t' = 0$. A pulse of light is emitted at $t = 0$ in the S system and spreads spherically. The locus of the pulse is given by $x^2 + y^2 + z^2 = (ct)^2$. We leave it as an exercise to show that the Lorentz transformation, Eqs. (12.3), predicts that in the "moving" S' system the locus of the pulse is given by $x'^2 + y'^2 + z'^2 = (ct')^2$. Observers in the two frames see the same phenomenon: a pulse spreading in space with the speed of light c. There is no trace of a reference to their relative speed v.

The Michelson–Morley experiment was designed to show the difference in the speed of light between directions parallel and perpendicular to the Earth's motion but according to the second postulate of the special theory of relativity—the speed of light is the same for all observers—there should be none. The Lorentz transformation shows explicitly that there is none.

12.8 Relativistic Kinematics

Because the principles of special relativity require us to rethink basic ideas of measurement and observation, they have important consequences for dynamics. The goal for the rest of this chapter is to learn how the principles of special relativity are employed to relate measurements in different inertial systems. The motivation for this is to some extent practical: relativistic kinematics is essential in areas of physics ranging from elementary particle physics to cosmology and also to technologies such as the global positioning system. More fundamentally, the study of relativistic transformations leads to new physics, most famously the relation $E = mc^2$, and to an elegant unified approach to dynamics and electromagnetic theory.

Predictions of the Lorentz transformation often defy intuition because we lack experience moving at speeds comparable to the speed of light. Two surprising predictions are that a moving clock runs slow and a moving meter stick contracts. These follow from the Lorentz transformation of time and space intervals. We will also derive these results by geometric arguments that may help provide intuition about this unexpected behavior.

Caution: in the discussion to follow, *either S or S′* may be the rest system for an observer, and in addition there is the possibility of introducing other systems. We need to be clear not only about the physical phenomena taking place but the system from which it is being observed.

12.8.1 Time Dilation

A clock is at rest in S at some location x. The clock's rate is determined by the interval τ_0 between its ticks. The problem is to find the corresponding interval observed in the $S′$ system, in which the clock is moving with speed $-v$.

Successive ticks of the clock in the rest system S are

$$\text{tick 1 (event 1)}: \quad t$$
$$\text{tick 2 (event 2)}: \quad t + \tau_0.$$

The corresponding times observed in the moving system $S′$ are, from Eq. (12.3d),

$$t' = \gamma(t - vx/c^2)$$
$$t' + \tau_0' = \gamma(t + \tau_0 - vx/c^2).$$

Subtracting, we obtain

$$\tau_0' = \gamma\tau_0. \tag{12.5}$$

Because $\gamma \geq 1$, the time interval observed in the moving system is longer than in the clock's rest system. Thus, the moving clock runs slow. As $v \to c$, time stands still.

This result, known as *time dilation*, is hardly intuitive and so it may be instructive to derive it by a different approach.

Let us consider an idealized clock in which the timing element consists of two parallel mirrors with a light pulse bouncing between them. (Our discussion follows *Introduction to Electrodynamics*, David J. Griffiths, Prentice Hall, Upper Saddle Ridge New Jersey, 1999.)

Each round trip of the light constitutes a clock tick. The clock is mounted vertically on a railway car that moves with speed v, as shown. An observer on the railway car monitors the rate of the ticks. If the distance between the mirrors is h, then the time interval between ticks is

$$\tau_0 = 2h/c.$$

In this calculation, the railway car is the rest system S for the clock.

An observer on the ground system $S′$ also monitors the rate of ticks of the clock. $S′$ is moving at speed $-v$ with respect to the rest system on the railway car. For this observer, the time interval for the light, up or down,

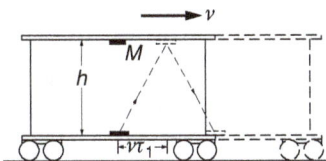

is $\tau_1 = \sqrt{h^2 + (v\tau_1)^2}/c$. Solving for τ_1, the roundtrip time τ'_0 is

$$\tau'_0 = 2\tau_1 = (2h/c)\frac{1}{\sqrt{1 - v^2/c^2}}.$$

Recalling that $\gamma = 1/\sqrt{1 - v^2/c^2}$, we have

$$\tau'_0 = \gamma\tau_0$$

in agreement with Eq. (12.5).

Example 12.4 The Role of Time Dilation in an Atomic Clock

Possibly you have looked through a spectroscope at the light from an atomic discharge lamp. Each line of the spectrum is the light emitted when an atom makes a transition between two of its internal energy states. The lines have different colors because the frequency ν of the light is proportional to the energy change ΔE in the transition. If ΔE is of the order of electron-volts, the emitted light is in the optical region ($\nu \approx 10^{15}$ Hz). There are some transitions, however, for which the energy change is so small that the emitted radiation is in the microwave region ($\nu \approx 10^{10}$ Hz). These microwave signals can be detected and amplified with available electronic instruments. Since the oscillation frequency depends almost entirely on the internal structure of the atom, the signals can serve as a frequency reference to govern the rate of an atomic clock. Atomic clocks are highly stable and relatively immune to external influences.

Each atom radiating at its natural frequency serves as a miniature clock. The atoms are frequently in a gas and move randomly with thermal velocities. Because of their thermal motion, the clocks are not at rest with respect to the laboratory and the observed frequency is shifted by time dilation.

Consider an atom that is radiating its characteristic frequency ν_0 in the rest frame. We can think of the atom's internal harmonic motion as akin to the swinging motion of the pendulum of a grandfather clock: each cycle corresponds to a complete swing of the pendulum. If the period of the swing is τ_0 seconds in the rest frame, the period in the laboratory is $\tau = \gamma\tau_0$. The observed frequency in the laboratory system is

$$\nu = \frac{1}{\tau} = \frac{1}{\gamma\tau_0} = \frac{\nu_0}{\gamma}$$

$$= \nu_0\sqrt{1 - \frac{v^2}{c^2}}.$$

The shift in the frequency is $\Delta\nu = \nu - \nu_0$. If $v^2/c^2 \ll 1$, $\gamma \approx 1 - \frac{1}{2}v^2/c^2$, and the fractional change in frequency is

$$\frac{\Delta\nu}{\nu_0} = \frac{\nu - \nu_0}{\nu_0} = -\frac{1}{2}\frac{v^2}{c^2}. \tag{1}$$

A handy way to evaluate the term on the right is to multiply numerator and denominator by the mass of the atom M:

$$\frac{\Delta \nu}{\nu_0} = -\frac{\frac{1}{2}Mv^2}{Mc^2}.$$

$\frac{1}{2}Mv^2$ is the kinetic energy due to thermal motion of the atom. This energy increases with the temperature of the gas, and according to our treatment of the ideal gas in Section 5.9

$$\frac{1}{2}M\bar{v^2} = \frac{3}{2}kT,$$

where $\bar{v^2}$ is the average squared velocity, $k = 1.38 \times 10^{-23}$ J/deg is Boltzmann's constant, and T is the absolute temperature.

In the atomic clock known as the hydrogen maser, the reference frequency arises from a transition in atomic hydrogen. M is close to the mass of a proton, 1.67×10^{-27} kg, and using $c = 3 \times 10^8$ m/s, we find

$$\frac{\Delta \nu}{\nu} = \frac{\frac{3}{2}kT}{Mc^2} = -\frac{\frac{3}{2}(1.38 \times 10^{-23})T}{(1.67 \times 10^{-27})(9 \times 10^{16})}$$
$$= 1.4 \times 10^{-13}T.$$

At room temperature, $T = 300$ K (300 degrees on the absolute temperature scale $\approx 27\,^\circ$C), we have

$$\frac{\Delta \nu}{\nu} = -4.2 \times 10^{-11}.$$

This is a sizable effect in modern atomic clocks. In order to correct for time dilation to an accuracy of 1 part in 10^{13}, it is necessary to know the temperature of the hydrogen atoms to an accuracy of 1 K. However, if one wishes to compare frequencies to parts in 10^{15}, the absolute temperature must be known to within a millikelvin, a much harder task.

The creation of techniques to cool atoms to the microkelvin regime has opened the way to a new generation of atomic clocks. These clocks, operating at optical rather than microwave frequencies, have achieved a stability greater than 1 part in 10^{17}—equivalent to a difference of about 1 second over the age of the Earth.

12.8.2 Length Contraction

A rod at rest in S has length L_0. What is the length observed in the system S' that is moving with speed $-v$ along the direction of the rod?

The rod lies along the x axis and its ends are at x_a and x_b, where $x_b = x_a + L_0$. The measurement involves two events but because the rod is at rest in S the times are unimportant, so we can take the observations in S to be simultaneous at time t. The length in S is found from the coordinates of two events:

$$\text{event } 1 : (x_a, t)$$
$$\text{event } 2 : (x_b, t).$$

The length observed in the rest frame S is $x_b - x_a = L_0$. The problem is to find the length observed in system S' where the rod is moving with speed $-v$.

A natural, but *wrong!*, approach to finding the coordinates in S' would be to use Eq. (12.3a) to find values for x'_b and x'_a and subtract. This would give $L'_0 = x'_b - x'_a = \gamma L_0$. The result is wrong because the times for the two events in S' are not identical, as can be seen from Eq. (12.3d).

Meaningful measurements of the dimensions of a moving object must be made simultaneously. We must therefore find the correspondence between values of x' and x at the same time t' in the S' system. This is readily accomplished by applying the Lorentz transformation to relate events in S to those in S'. Equation (12.4a) gives $x = \gamma(x' - vt')$. Consequently,

$$x_b = \gamma(x'_b - vt')$$
$$x_a = \gamma(x'_a - vt').$$

Subtracting, we obtain $x_b - x_a = \gamma(x'_b - x'_a)$, so that $L_0 = \gamma L'$ and

$$L' = L_0/\gamma = \sqrt{1 - v^2/c^2}\, L_0. \tag{12.6}$$

The rod appears to be contracted. As $v \to c$, $L' \to 0$. The contraction occurs along the direction of motion only: if the rod lay along the y axis, we would use the transformation $y' = y$ and conclude that $L' = L_0$.

As in the case of time dilation, we have a non-intuitive result. This, too, can be understood using a geometrical argument.

An observer on a train could measure the length of the train car L_0 by bouncing light between mirrors at each end and measuring the roundtrip time τ_0:

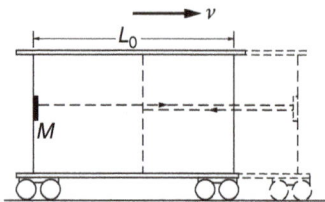

$$\tau_0 = 2\frac{L_0}{c}.$$

The observer on board concludes that the length of the car is

$$L_0 = \frac{c}{2}\tau_0. \tag{12.7}$$

An observer on the ground also measures the length of the car L' as the train goes by at speed $+v$ by measuring the time for a pulse to make a roundtrip between the ends. As seen by the ground observer, the time τ_+ for the pulse to travel from the rear mirror to the front is longer than L'/c, because the front mirror moves slightly ahead during the transit time. The distance traveled is $L' + v\tau_+$. Consequently, $\tau_+ = (L' + v\tau_+)/c$ so that $\tau_+ = L'/(c - v)$. Similarly, the time for the return trip is $\tau_- = L'/(c + v)$.

The roundtrip for the light pulse is

$$\tau_0' = \tau_+ + \tau_- = L'\left(\frac{1}{c-v} + \frac{1}{c+v}\right) = \frac{2L'}{c}\left(\frac{1}{1-v^2/c^2}\right).$$

Consequently,

$$L' = \frac{c}{2}\tau_0'(1 - v^2/c^2).$$

Comparing this with Eq. (12.7), we have

$$L' = L_0\frac{\tau_0}{\tau_0'}(1 - v^2/c^2).$$

Taking the value of τ_0/τ_0' from Eq. (12.5), we have

$$L' = L_0\sqrt{1 - v^2/c^2}.$$

Because $L' < L_0$, the ground observer sees the length of the car contracted by the factor $\sqrt{1 - v^2/c^2}$.

12.8.3 Proper Time and Proper Length

We introduced the symbols τ_0 and L_0 to denote time and length intervals observed in the rest frame of the events. These quantities are referred to as *proper*: τ_0 is the *proper time* and L_0 is the *proper length*.

Proper time τ is the time measured by a clock in its own rest system, which might for example be a clock carried aboard a spacecraft. According to Eq. (12.5), a time interval $\Delta t'$ measured in a moving frame is always *greater* than the proper time interval $\Delta\tau$:

$$\Delta t' = \gamma\Delta\tau = \frac{\Delta\tau}{\sqrt{1 - v^2/c^2}} \geq \Delta\tau.$$

Similarly, proper length is the length of an object measured in its own rest frame, for example a meter stick carried aboard a spacecraft. According to Eq. (12.6), the length L' measured in a moving frame is always *less* than the proper length L_0:

$$L' = \frac{L_0}{\gamma} = \sqrt{1 - v^2/c^2}L_0 \leq L_0.$$

12.8.4 Are Relativistic Effects Real?

Time and distance are such intuitive concepts that it may be difficult, at least at first, to accept that the predictions of special relativity are "real" in the familiar sense of physical reality. We shall look at some examples where time dilation and length contraction unquestionably occur. Paradoxes immediately come to mind, for instance the pole-vaulter paradox: a farmer has a barn with a door at each end. A pole-vaulter runs through the barn gripping a horizontal pole longer than the barn. The farmer wants to slam the doors with the pole inside. The farmer instructs the runner to go so fast that the length contraction permits the pole to fit.

The moment the runner is inside, the farmer slams the doors. The paradox is that from the runner's point of view, the pole is unchanged but the length of the barn has contracted. Rather than making the task of fitting in the barn easier, running makes it harder!

The paradox hinges on the difference between Newtonian and relativistic concepts of simultaneity. The runner will not agree that the doors were both shut at the same time, and it will be left as a problem to show that from the runner's point of view the pole was never totally in the barn.

The first dramatic experimental demonstration of time dilation occurred in an early study of cosmic rays. The experiment also demonstrated that although time dilation and length contraction appear to be fundamentally different phenomena, they are essentially two sides of the same coin.

Example 12.5 Time Dilation, Length Contraction, and Muon Decay

The negatively charged muon (symbol μ^-) is an elementary particle related to the electron: it carries one unit of negative charge, same as the electron, and it has a positively charged antiparticle μ^+ analogous to the positron, the electron's antiparticle. The muon differs from the electron most conspicuously in its mass, which is about 205 times the electron's mass, and in being unstable. Electrons are totally stable, but the muon decays into an electron and two neutrinos.

The decay of the muon is typical of radioactive decay processes: if there are $N(0)$ muons at $t = 0$, the number at time t is

$$N(t) = N(0)e^{-t/\tau}$$

where τ is a time constant characteristic of the decay. It is easy to show that the average time before a given muon decays is τ, and so τ is known as the "lifetime" of the particle. For muons, $\tau = 2.2$ μs. (Caution: the symbol μ stands for "micro", 10^{-6}, as well as for muon. One needs to keep one's wits about symbols in physics.) If the muons travel with speed v, the average distance they travel before decaying is $\langle L \rangle = v\tau$.

Muons were discovered in research on cosmic rays. They are created at high altitudes by high energy protons streaming toward the Earth. The protons are quickly lost in the atmosphere by collisions, but the muons continue to sea level with very little loss. The early experiments ran into a paradox. If one assumes that the muons travel at high speed, close to the speed of light, then the maximum average distance that they travel before decay should be no bigger than $\langle L \rangle = c\tau$. Consequently, after traveling distance L, the flux of muons should be decreased by a factor of at least $\exp -L/\langle L \rangle$. For a 2.2 μs lifetime, $\langle L \rangle = 660$ m. In the initial experiment (B. Rossi and D. B. Hall, Physical Review, 59, 223 (1941)), the flux was monitored on a mountain top in Colorado and at a site 2000 m below. The flux was expected to decrease by a factor of exp -2000 m/660 m = 0.048. However, the observed loss ratio was much smaller. The paradox was that the muons behaved as if on their journey to Earth they lived for about three times their known lifetime.

The resolution of the paradox was the realization that the muons actually lived that long. The quantity γ was determined from measurements of the muon energy. When time dilation was taken into account, the lifetimes were calculated to increase by a factor close to the observed value.

The concept of proper time is another way to look at this. The muons carry their own "clock" that determines their decay rate. Their clock, in the muon rest frame, measures proper time, so the decay rate measured by a ground observer is longer. Using modern particle accelerators, muons can be created with much higher energies than obtained with cosmic rays, leading to correspondingly larger values of γ. In one experiment (R. M. Casey, *et al.*, Phys. Rev. Letters, 82, 1632 (1997)) the lifetime was extended so much that useful signals could be observed for up to 440 µs, 200 times the muon lifetime. Do moving clocks "really" run slow? The answer depends on how you wish to interpret the experiment. In a coordinate system moving with the muons (in the muon rest system), the particles decay with their natural decay rate. However, in this system the muons "see" that the thickness of the atmosphere is smaller than seen by a ground-based observer. Lorentz contraction reduced the path length from 2000 m to $2000/\gamma$ m. The fraction of muons that penetrated through is the same as if the problem were viewed from a ground-based coordinate system.

We see that once we accept the postulates of relativity we are forced to abandon the intuitive idea of simultaneity. Nevertheless, the Lorentz transformation, which embodies the postulates of relativity, allows us to calculate the times of events in two different systems.

Example 12.6 An Application of the Lorentz Transformation

A light pulse is emitted at the center of a railway car $x = 0$ at time $t = 0$. How do we find the time of arrival of the light pulse at each end of the railway car, which has length $2L$? The problem is trivial in the rest frame. The two events are

$$\text{Event 1: The pulse arrives at end A} \begin{cases} x_1 = -L \\ t_1 = \dfrac{L}{c} = T \end{cases}$$

$$\text{Event 2: The pulse arrives at end B} \begin{cases} x_2 = L \\ t_2 = \dfrac{L}{c} = T. \end{cases}$$

To find the time of the events in system S' moving with respect to the railway car, we apply the Lorentz transformation for the time coordinates.

Event 1:

$$t_1' = \gamma\left(t_1 - \frac{vx_1}{c^2}\right)$$

$$= \gamma\left(T + \frac{vL}{c^2}\right)$$

$$= \frac{1}{\sqrt{1 - v^2/c^2}}\left(T + \frac{v}{c}T\right)$$

$$= T\sqrt{\frac{1 + v/c}{1 - v/c}}.$$

Event 2:

$$t_2' = \gamma\left(t_2 - \frac{vx_2}{c^2}\right)$$

$$= T\sqrt{\frac{1 - v/c}{1 + v/c}}.$$

In the moving system, the pulse arrives at B (event 2) earlier than it arrives at A, as we anticipated.

Simultaneity is not a fundamental property of events; it depends on the coordinate system. Is it possible to find a coordinate system in which any two events are simultaneous? The following example proves what was asserted in Section 12.6: there are two classes of events. For two given events, we can find either a coordinate system in which the events are simultaneous in time or one in which the events occur at the same point in space—but not both.

Example 12.7 The Order of Events: Timelike and Spacelike Intervals

Two events A and B on the x axis have the following coordinates in S:

Event $A : (x_A, t_A)$; Event $B : (x_B, t_B)$.

The distance between the events is $L = x_B - x_A$ and the time T separating the events is $T = t_B - t_A$.

The distance between the events in the x', y' system, as described by the Lorentz transformation, is

$$L' = \gamma(L - vT).$$

$$T' = \gamma\left(T - \frac{vL}{c^2}\right).$$

Because v is always less than c, it follows that if $L > cT$ then L' is always positive, while T' can be positive, negative, or zero. Such an interval is called *spacelike*, since it is impossible to choose a system in which the events occur at the same place, though it is possible for them to be simultaneous, namely, in a system moving with $v = c^2T/L$. On the other hand, if $L < cT$, then T' is always positive and the events can never appear to be simultaneous, but L' can be positive, negative, or zero. The interval is then known as *timelike*, because it is impossible to find a coordinate system in which the events occur at the same time.

12.9 The Relativistic Addition of Velocities

The closest star is more than four light years away and our galaxy is roughly 10 000 light years across. Consequently, any method for traveling faster than light could be priceless for galactic exploration. Toward this goal, suppose we build a spaceship, the Starship Sophie, that can achieve a speed of $0.900c$. The crew of the Sophie then launches a second ship, Starship Surprise, that can reach $0.800c$. According to Newtonian rules, the Surprise should fly away at $1.700c$. Let's see what happens relativistically.

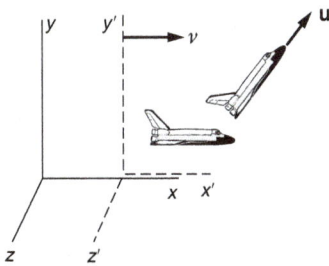

We designate our rest system (x, y, z, t) by S and spaceship Sophie's system (x', y', z', t') by S'. S' moves with velocity v along the x axis. The velocity of the Surprise, as observed from the Sophie, is $\mathbf{u}' = (u'_x, u'_y)$. Our task is to find the velocity $\mathbf{u}$ of the Surprise that we observe in our rest system S.

From the definition of velocity, in S' we have

$$u'_x = \lim_{\Delta t' \to 0} \frac{\Delta x'}{\Delta t'}$$

$$u'_y = \lim_{\Delta t' \to 0} \frac{\Delta y'}{\Delta t'}$$

$$u'_z = \lim_{\Delta t' \to 0} \frac{\Delta z'}{\Delta t'}.$$

The corresponding components in S are

$$u_x = \lim_{\Delta t \to 0} \frac{\Delta x}{\Delta t}$$

$$u_y = \lim_{\Delta t \to 0} \frac{\Delta y}{\Delta t}$$

$$u_z = \lim_{\Delta t \to 0} \frac{\Delta z}{\Delta t}.$$

The problem is to relate displacements and time intervals in S to those in S'. From the Lorentz transformation Eqs. (12.3) we have

$$\Delta x = \gamma(\Delta x' + v\Delta t')$$
$$\Delta y = \Delta y'$$
$$\Delta z = \Delta z'$$
$$\Delta t = \gamma\left(\Delta t' + (v/c^2)\Delta x'\right).$$

Hence

$$\frac{\Delta x}{\Delta t} = \frac{\gamma(\Delta x' + v\Delta t')}{\gamma[\Delta t' + (v/c^2)\Delta x']}$$
$$= \frac{\Delta x'/\Delta t' + v}{1 + (v/c^2)(\Delta x'/\Delta t')}.$$

Next we take the limit $\Delta t' \to 0$. Using $u'_x = \lim_{\Delta t' \to 0} \Delta x'/\Delta t'$, we obtain

$$u_x = \frac{u'_x + v}{1 + vu'_x/c^2}. \tag{12.8a}$$

Similarly,

$$u_y = \frac{u'_y}{\gamma[1 + vu'_x/c^2]} \tag{12.8b}$$

and

$$u_z = \frac{u'_z}{\gamma[1 + vu'_x/c^2]}. \tag{12.8c}$$

Equations (12.8) are the relativistic rules for adding velocities. For $v \ll c$, we obtain the Galilean result $\mathbf{u} = \mathbf{v} + \mathbf{u}'$.

The transformation from S to S' is

$$u'_x = \frac{u_x - v}{1 - vu_x/c^2}. \tag{12.9a}$$
$$u'_y = \frac{u_y}{\gamma[1 - vu_x/c^2]}. \tag{12.9b}$$
$$u'_z = \frac{u_z}{\gamma[1 - vu_x/c^2]}. \tag{12.9c}$$

Returning to the problem of the two starships, let $u'_x = 0.800c$ be the speed of the Surprise relative to the Sophie and $v = 0.900c$ be the speed of the Sophie relative to us. The velocity of the Surprise relative to us is, from Eq. (12.8a),

$$u_x = \frac{0.900c + 0.800c}{1 + (0.900)(0.800)}$$
$$= \frac{1.700c}{1.720} = 0.988c.$$

The speed of the Surprise is less than c. Equation (12.8a) reveals that we cannot exceed the speed of light by changing reference frames.

Taking the limiting case $u'_x = c$, the final velocity in the rest system is then

$$u_x = \frac{c + v}{1 + vc/c^2}$$
$$= c,$$

independent of v. This agrees with the postulate that we built into the Lorentz transformation: the speed of light is the same for all observers. Furthermore, it suggests that the speed of light is the ultimate speed allowed by the theory of relativity.

Example 12.8 The Speed of Light in a Moving Medium

As an exercise in the relativistic addition of velocities, let us find how a moving medium, such as flowing water, influences the speed of light.

The speed of light in matter is less than c. The index of refraction, n, is used to specify the speed in a medium:

$$n = \frac{c}{\text{velocity of light in the medium}}.$$

$n = 1$ corresponds to empty space; in ordinary matter $n > 1$. The slowing can be appreciable: for water $n = 1.3$.

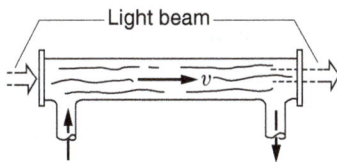
Light beam

The problem is to find the speed of light through a moving liquid. For instance, consider a tube filled with water. If the water is at rest, the speed of light in the water with respect to the laboratory is $u = c/n$. What is the speed of light when the water is flowing with speed v?

Consider the speed of light in water as observed in a coordinate system $S' = (x', y')$ moving with the water. The speed in S' is

$$u' = \frac{c}{n}.$$

The speed in the laboratory is, by Eq. (12.8a),

$$u = \frac{u' + v}{1 + u'v/c^2}$$

$$= \frac{c/n + v}{1 + v/nc}$$

$$= \frac{c}{n}\left(\frac{1 + nv/c}{1 + v/nc}\right).$$

If we expand the factor on the right-hand side and neglect terms of order $(v/c)^2$ and smaller, we obtain

$$u = \frac{c}{n}\left(1 + \frac{nv}{c} - \frac{v}{nc}\right)$$

$$= \frac{c}{n} + v\left(1 - \frac{1}{n^2}\right).$$

The light appears to be "dragged" by the fluid, but not completely. Only the fraction $f = 1 - 1/n^2$ of the fluid velocity is added to the speed of light c/n. This effect was observed experimentally in 1851 by Fizeau, although it was not explained satisfactorily until the advent of relativity.

12.10 The Doppler Effect

The Doppler effect is the change in the frequency of a wave due to motion between the source and observer. It causes the familiar drop in pitch of the horn of a truck or the whistle of a train as they pass by. For astronomers and astrophysicists the Doppler effect provides an invaluable tool for measuring the speed of far off objects by the shift in the spectral wavelengths they emit. All our knowledge about how fast the universe is expanding comes from observations of the Doppler effect in spectral lines. More prosaically, the Doppler effect is at the heart of reliable and cheap radar speed monitors.

The relativistic Doppler effect differs from the classical effect in a pleasing manner: it is simpler. Furthermore, it displays a phenomenon absent in classical behavior, the transverse Doppler effect that causes a frequency shift in light from a moving source, as seen by an observer transverse to the path.

To start, we review the classical Doppler effect in sound.

12.10.1 The Doppler Effect in Sound

Sound travels through a medium, such as air, with a speed w determined by the properties of the medium, independent of the motion of the source.

Consider sound waves from a source moving with velocity w through the medium toward an observer at rest. For now, we shall restrict ourselves to the case where the observer is along the line of motion. We will picture sound as regular series of pulses separated by time $\tau_0 = 1/\nu_0$, where ν_0 is the number of pulses per second generated by the source. The distance between pulses is $w\tau_0 = w/\nu_0$, which we designate by λ. We could equally well picture the disturbance as a sine wave, in which case ν_0 corresponds to the frequency of sound and the distance between successive crests is the wavelength $\lambda = w/\nu_0$.

If the source moves toward the observer at speed v, then the distance between successive pulses is $\lambda_D = \lambda - v\tau_0 = \lambda - v/\nu_0$. Hence

$$\frac{w}{\nu_0'} = \frac{w}{\nu_0} - \frac{v}{\nu_0},$$

$$\nu_0' = \nu_0 \left(\frac{1}{1 - v/w} \right) \qquad \text{(moving source)}. \qquad (12.10)$$

The shift in frequency $(\Delta \nu)\lambda_D = \nu_0' - \nu_0$ is known as the *Doppler shift*.

The situation is somewhat different if the observer is moving toward the source at speed w. Previously, the observer was at rest in the medium: now the observer is moving through the medium. The relative velocity between source and observer is the same, w. The distance between wave fronts is unchanged, but the relative speed of arrival is now $w + v$. Consequently, the frequency is $\nu_0' = (w + v)/\lambda$, which can be written

$$\nu_0' = \nu_0(1 + v/w) \qquad \text{(moving observer)}. \qquad (12.11)$$

Equations (12.11) and (12.12) are identical to first order in the ratio v/w, but they differ in the second order. The second-order difference could

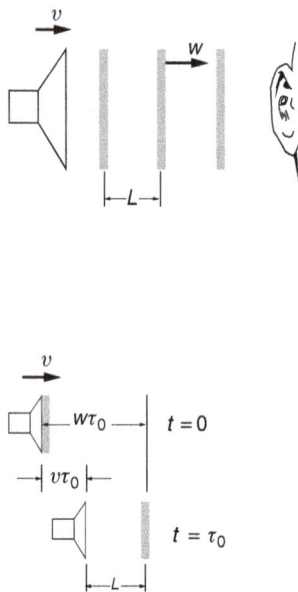

in principle be used be determine whether the Doppler shift is due to motion of the source or motion of the observer. The distinction is real because the motion is measured relative to a fixed medium such as air.

If these results were valid for light waves in space, we would be able to distinguish which of two systems is at absolute rest, which is not possible. To resolve this difficulty, we turn now to a relativistic derivation of the Doppler effect.

12.10.2 The Relativistic Doppler Effect

A light source flashes with period $\tau_0 = 1/\nu_0$ in its rest frame. The source is moving toward an observer with velocity v. Due to time dilation, the period in the observer's rest frame is

$$\tau = \gamma \tau_0.$$

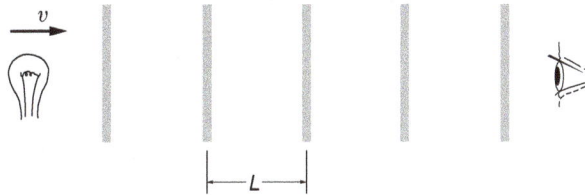

If the wavelength λ_D is the distance between pulses in the observer's rest frame, the frequency of the pulses is $\nu_D = c/\lambda_D$, where the wavelength λ_D is the distance between pulses in the observer's frame. Because the source is moving toward the observer this distance is

$$\lambda_D = c\tau - v\tau = (c - v)\tau$$

and

$$\nu_D = \frac{c}{(c - v)\tau}$$

$$= \left(\frac{1}{1 - v/c} \right) \frac{1}{\gamma \tau_0}$$

or

$$\nu_D = \nu_0 \frac{\sqrt{1 - v^2/c^2}}{1 - v/c}$$

which reduces to

$$\nu_D = \nu_0 \sqrt{\frac{1 + v/c}{1 - v/c}}. \tag{12.12}$$

ν_D is the frequency in the observer's rest frame and v is the relative speed of source and observer. As we expect, there is no mention of motion relative to a medium. The relativistic result plays no favorites with the classical results; it disagrees with both Eqs. (12.10) and (12.11) but treats

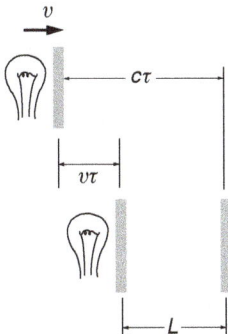

the case of moving source and moving observer symmetrically: it is the geometric mean of the two classical results.

12.10.3 The Doppler Effect Off the Line of Motion

We have analyzed the Doppler effect when the source and observer move along the line connecting them but this is not the most general situation. For instance, consider a satellite broadcasting a radio beacon signal to a ground tracking station that monitors the Doppler-shifted frequency.

We can readily generalize our method to find the Doppler effect for an observer in a direction at angle θ from the line of motion. We again visualize the source as a flashing light. The period of the flashes in the observer's rest frame is $\tau = \gamma\tau_0$, as before. The frequency seen by the observer is c/λ_D. The source moves a distance $v\tau$ between flashes and it is apparent from the sketch that

$$\lambda_D = c\tau - v\tau\cos\theta$$
$$= (c - v\cos\theta)\tau.$$

Hence

$$\nu_D = \frac{c}{\lambda_D}$$
$$= \frac{c}{(c - v\cos\theta)\tau_0\gamma}$$
$$\nu_D = \nu_0\frac{\sqrt{1 - v^2/c^2}}{1 - (v/c)\ \cos\theta} \tag{12.13}$$

where we have used $\tau_0 = 1/\nu_0$. In this result, θ is the angle measured in the rest frame of the observer. Along the line of motion, $\theta = 0$ and we recover our previous result, Eq. (12.12). At $\theta = \pi/2$ the relative velocity between source and observer is zero. The classical Doppler effect would vanish here, but relativistically there is a shift in frequency; ν_D differs from ν_0 by the factor $\sqrt{1 - v^2/c^2}$. This "transverse" Doppler effect is due to time dilation. The flashing lamp is effectively a moving clock and moving clocks run slow.

The relativistic Doppler effect agrees with the classical result to order v/c, so that any experiment to differentiate between them must be sensitive to effects of order $(v/c)^2$. Nevertheless, the relativistic expression was confirmed by H.E. Ives and G.R. Stilwell in 1938 by observing small shifts in the wavelengths emitted by fast-moving atoms.

A useful application of the Doppler effect is in navigational systems, as the following example explains.

Example 12.9 Doppler Navigation

The Doppler effect can be used to track a moving body, such as a satellite, from a reference point on the Earth. This provided the basis for a navigational system that was created when the first satellites were

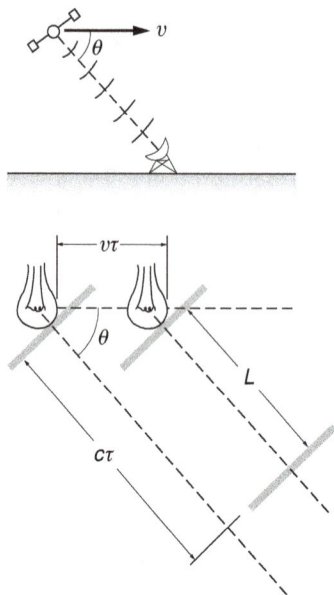

flown. Although it has been superseded by the Global Positioning System (GPS), the method is remarkably accurate; changes in the position of a satellite 10^8 m away can be determined to a fraction of a centimeter.

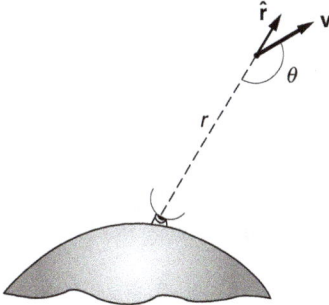

Consider a satellite moving with velocity $\mathbf{v}$ at some distance r from a ground station. An oscillator on the satellite broadcasts a signal with frequency ν_0. Since $v \ll c$ for satellites, we can approximate Eq. (12.13) by retaining only terms of order v/c. The frequency ν_D received by the ground station can then be written

$$\nu_D \approx \frac{\nu_0}{1 - (v/c)\cos\theta}$$

$$\approx \nu_0 \left(1 + \frac{v}{c}\cos\theta\right).$$

There is an oscillator in the ground station identical to the one in the satellite. At rest, both oscillators run at the same frequency ν_0 with corresponding wavelength $\lambda_0 = c/\nu_0$. In flight, the observed satellite frequency is different, and by simple electronic methods the difference frequency ("beat" frequency) $\nu_D - \nu_0$ can be measured:

$$\nu_D - \nu_0 = \nu_0 \frac{v}{c}\cos\theta.$$

The radial velocity of the satellite is

$$\frac{dr}{dt} = \hat{\mathbf{r}} \cdot \mathbf{v}$$

$$= -v\cos\theta.$$

Hence

$$\frac{dr}{dt} = -\frac{c}{\nu_0}(\nu_D - \nu_0)$$

$$= -\lambda_0(\nu_D - \nu_0).$$

ν_D varies in time as the satellite's velocity and direction change. To find the total radial distance traveled between times T_a and T_b, we integrate the above expression with respect to time:

$$\int_{T_a}^{T_b} \left(\frac{dr}{dt}\right) dt = -\lambda_0 \int_{T_a}^{T_b} (\nu_D - \nu_0)dt$$

$$r_b - r_a = -\lambda_0 \int_{T_a}^{T_b} (\nu_D - \nu_0)dt.$$

The integral is the number of cycles N_{ba} of the beat frequency that occur in the interval T_a to T_b. (One cycle occurs in a time $\tau = 1/(\nu_D - \nu_0)$, so that $\int dt/\tau$ is the total number of cycles.) Hence

$$r_b - r_a = -\lambda_0 N_{ba}.$$

This result has a simple interpretation: whenever the radial distance increases by one wavelength, the phase of the beat signal decreases one cycle. Similarly, when the radial distance decreases by one wavelength, the phase of the beat signal increases by one cycle.

Satellite communication systems operate at a typical wavelength of 10 cm, and since the beat signal can be measured to a fraction of a cycle, satellites can be tracked to about 1 cm. If the satellite and ground-based oscillators do not each stay tuned to the same frequency, ν_0, there will be an error in the beat frequency. To avoid this problem a two-way Doppler tracking system can be used in which a signal from the ground is broadcast to the satellite, which then amplifies it and relays it back to the ground. This has the added advantage of doubling the Doppler shift, increasing the resolution by a factor of 2.

We sketched the principles of Doppler navigation for the classical case $v \ll c$. For certain tracking applications the precision is so high that relativistic effects must be taken into account.

As we have shown, a Doppler tracking system also gives the instantaneous radial velocity of the satellite $v_r = -c(\nu_D - \nu_0)/\nu_0$. This is particularly handy, since both velocity and position are needed to check satellite trajectories. A more prosaic use of this result is in police radar speed monitors: a microwave signal is reflected from an oncoming car and the beat frequency of the reflected signal reveals the car's speed.

12.11 The Twin Paradox

Among the paradoxes that add to the fascination of special relativity, probably none has generated more discussion than the twin paradox. The paradox is simple to state: two twins, Alice and Bob, have identical clocks. Alice sets out on a long space voyage while Bob remains at home. Suppose that the spacecraft flies away in a straight line at constant speed v for time $T_0/2$ as measured by Alice using her onboard clock. She quickly reverses speed and heads back, returning home at T_0. Alice would observe that she has aged by time T_0, as measured with her onboard clock.

Because of time dilation, Bob would observe that the time for the journey is

$$T'_B = \gamma T_0 \approx T_0 \left(1 + \frac{1}{2}\frac{v^2}{c^2} \right).$$

Consequently, Bob concludes that because of the time dilation he is older than Alice by

$$\Delta T_{A,B} = \frac{1}{2} T_0 \frac{v^2}{c^2}, \tag{12.14}$$

or, equivalently, that Alice is younger than he.

If this same argument were applied by Alice, she would conclude that she is older than Bob by

$$\Delta T_{B,A} = \frac{1}{2}T_0 \frac{v^2}{c^2} \tag{12.15}$$

or that Bob is correspondingly younger than she.

Obviously they cannot both be right. Who is younger? Is there really any difference?

The paradox arises from ignoring the fact that the situations for the twins are not equivalent. Bob's system is inertial but for part of the time, Alice's is not. She must reverse her velocity in order to return to the starting point and while her velocity is changing, her system is not inertial. During this interval the situation becomes asymmetric: there is no question as to which twin is accelerating. If each were carrying an accelerometer such as a mass on a spring, Bob's would remain at zero while Alice's would show a large deflection as the spaceship reversed direction.

In principle, analyzing events in an accelerating system requires general relativity. Nevertheless, we can find the leading terms of the solution by invoking the equivalence principle and the analysis of the gravitational clock shift in Chapter 9. Recall that according to the principle of equivalence, there is no way to distinguish between an acceleration a and a uniform gravitational field $g = -a$. Due to the gravitational clock shift, Alice sees Bob's clock speed up during turnaround. We shall see that this time advance brings the two observers into agreement.

During turnaround, let us suppose that Alice experiences a uniform acceleration a applied for time τ_t. The time required to reverse the velocity is $a\tau_t = 2v$. During this time Alice experiences an effective gravitational field $g_{\text{eff}} = -a$ that points from Bob to her. As a result, Alice sees that Bob's clock has sped up due to the gravitational red shift (in this case, actually a blue shift). The fractional shift in the rate of the clock is $g_{\text{eff}}h/c^2$, where h is the "height" of the clock in the gravitational field. Turnaround occurs at time $T_0/2$, so $h = vT_0/2$. The total advance Alice measures in Bob's clock during turnaround is

$$\Delta T_{\text{grav}} = \frac{g_{\text{eff}}h}{c^2 \tau}.$$

Inserting the values $h = vT_0/2$, $g_{\text{eff}} = a = 2v/\tau$, and $\tau = v/a$, we have

$$\Delta T_{\text{grav}} = \frac{av^2 T_0}{c^2 a} = T_0 \frac{v^2}{c^2}. \tag{12.16}$$

Before taking the gravitational frequency shift into account, Alice believed that Bob was younger than she by $(1/2)T_0(v^2/c^2)$. However, when ΔT_{grav} is added to this time, she realizes that Bob is actually older by that amount. Both twins agree: at the end of the trip Alice is younger than Bob by $T_0 v^2/2c^2$. It appears that travel helps one stay relatively youthful.

Problems

*For problems marked *, refer to page 524 for a hint, clue, or answer.*

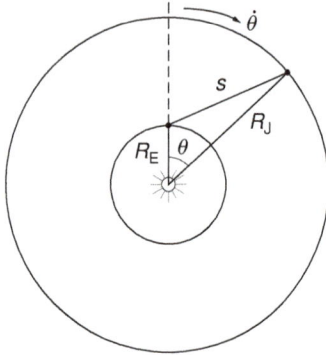

12.1 *Maxwell's proposal**

Section 12.3 mentioned Maxwell's proposal for measuring the effect of source motion on the speed of light, using Jupiter's moons as clocks. In the sketch (not to scale), the inner circle is the Earth's orbit and the outer circle is Jupiter's orbit. The angle θ is the position of Jupiter with respect to the Earth's position. Jupiter's period is 11.9 years and the Earth's period is 1 year, so that $\dot{\theta} = 2\pi(11.9 - 1)\,\text{rad/year} = 2.2 \times 10^{-6}\,\text{rad/s}$. The radius of Jupiter's orbit is $R_J = 7.8 \times 10^{11}$ m and the radius of Earth's orbit is $R_E = 1.5 \times 10^{11}$ m.

The problem is to find the time delay ΔT predicted by Maxwell's method. If s is the distance between Jupiter and Earth then

$$\Delta T = \frac{s}{(c - \dot{s})} - \frac{s}{(c + \dot{s})} \approx \frac{2s\dot{s}}{c^2}.$$

Calculate the maximum value of ΔT.

12.2 *Refined Michelson–Morley interferometer*

The improved apparatus used in 1887 by Michelson and Morley at the Case School of Applied Science (now Case-Western Reserve University) could detect a 0.01 fringe using sodium light, $\lambda = 590$ nm.

What is the upper limit to the Earth's velocity with respect to the ether set by this experiment? For comparison, the Earth's orbital velocity around the Sun is 30 km/s.

12.3 *Skewed Michelson–Morley apparatus*

In Section 12.3 arm A of the Michelson–Morley interferometer was assumed to be along the line of motion and arm B perpendicular, and the predicted time difference according to the ether theory was

$$\Delta\tau = \frac{l}{c}\left(\frac{v^2}{c^2}\right).$$

Calculate the expected time difference if arm A is at angle θ to the line of motion through the ether, as shown.

12.4 *Asymmetric Michelson–Morley interferometer*

If the two arms of the Michelson interferometer have different lengths l_1 and l_2, show that the fringe shift when the interferometer is rotated by 90° with respect to the velocity v through the ether is

$$N = \left(\frac{l_1 + l_2}{\lambda}\right)\left(v^2 c^2\right)$$

where λ is the wavelength of the source light.

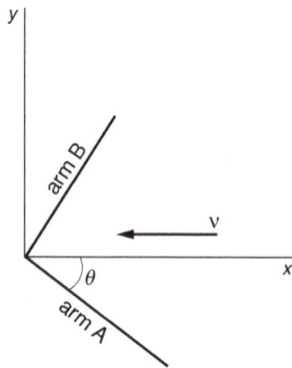

12.5 *Lorentz–FitzGerald contraction*

The Irish physicist G.F. FitzGerald and the Dutch physicist H.A. Lorentz tried to account for the null result of the Michelson–Morley experiment by the conjecture that movement through the ether sets up a strain that causes contraction along the line of motion by the factor $1 - \frac{1}{2}v^2/c^2$.

Show that this hypothesis can account for the absence of fringe shift in the Michelson–Morley experiment. (The hypothesis was disproved in 1932 by experimenters who used an interferometer with unequal arms.)

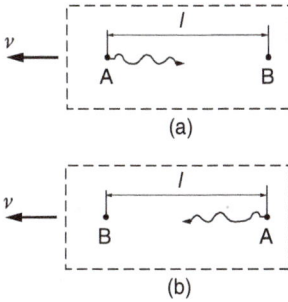

(a)

(b)

12.6 *One-way test of the constancy of c*

Light in a Michelson–Morley interferometer makes a roundtrip, and the predicted time delay is second order, proportional to v^2/c^2.

Here is an experiment that would give a first-order result proportional to v/c. Consider a laboratory moving through the ether with speed v in the direction shown. The observers have clocks and light pulsers. At time $t = 0$ A sends a signal to B a distance l away, sketch (a). B records the arrival time. The laboratory is then rotated $180°$, reversing the positions of A and B. At time $t = T$, A sends a second signal to B, sketch (b).

(*a*) Show that according to the ether theory, the interval that B observes between the signals is $T + \Delta T$, where

$$\Delta T \approx \frac{2l}{c}\frac{v}{c}$$

correct to order $(v/c)^3$.

(*b*) Assume that one clock in this experiment is on the ground and the other is in a satellite overhead. For a circular orbit with a period of 24 hours, $l = 5.6R_e$, where R_e is the Earth's radius = 6.4×10^6 m. Using an atomic clock stable to within 1 part in 10^{16}, what is the smallest value of v this experiment could detect?

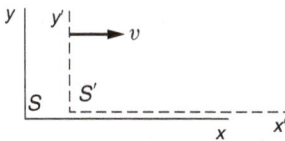

12.7 *Four events*

Note: S refers to an inertial system x, y, z, t and S' refers to an inertial system x', y', z', t', moving along the x axis with speed v relative to S. The origins coincide at $t = t' = 0$. For numerical work, take $c = 3 \times 10^8$ m/s.

Assuming that $v = 0.6c$, find the coordinates in S' of the following events

(*a*) $x = 4$ m, $t = 0$ s.

(*b*) $x = 4$ m, $t = 1$ s.

(*c*) $x = 1.8 \times 10^8$ m, $t = 1$ s.

(*d*) $x = 10^9$ m, $t = 2$ s.

12.8   Relative velocity of S and S'

Refer to the note and the sketch in Problem 12.7.

An event occurs in S at $x = 6 \times 10^8$ m, and in S' at $x' = 6 \times 10^8$ m, $t' = 4$ s. Find the relative velocity of the systems.

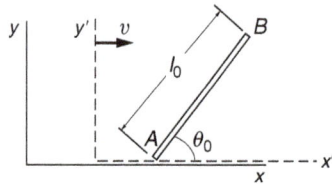

12.9   Rotated rod

A rod of length l_0 lies in the $x'y'$ plane of its rest system and makes an angle θ_0 with the x' axis. What is the length and orientation of the rod in the lab system x, y in which the rod moves to the right with velocity v?

12.10   Relative speed*

An observer sees two spaceships flying apart with speed $0.99c$. What is the speed of one spaceship as viewed by the other?

12.11   Time dilation

The clock in the sketch can provide an intuitive explanation of the time dilation formula. The clock consists of a flashtube, mirror, and phototube. The flashtube emits a pulse of light that travels distance L to the mirror and is reflected back to the phototube. Every time a pulse hits the phototube it triggers the flashtube. Neglecting time delay in the triggering circuits, the period of the clock is $\tau_0 = 2L/c$.

Now examine the clock in a coordinate system moving to the left with uniform velocity v. In this system the clock appears to move to the right with velocity v. Find the period of the clock in the moving system by direct calculation, using only the assumptions that c is a universal constant, and that distance perpendicular to the line of motion is unaffected by the motion. The result should be identical to that given by the Lorentz transformation: $\tau = \tau_0 / \sqrt{1 - v^2/c^2}$.

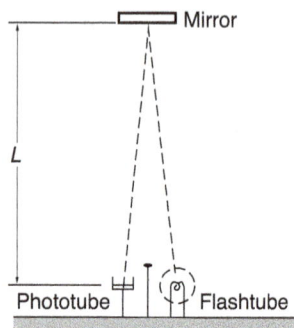

12.12   Headlight effect*

A light beam is emitted at angle θ_0 with respect to the x' axis in S'.

(a) Find the angle θ the beam makes with respect to the x axis in S.

(b) A source that radiates light uniformly in all directions in its rest frame radiates strongly in the forward direction in a frame in which it is moving with speed v close to c. This is called the headlight effect; it is very pronounced in synchrotron light sources in which electrons moving at relativistic speeds emit light in a narrow cone in the forward direction. Using the result of part (a), find the speed of a source for which half the radiation is emitted in a cone subtending 10^{-3} rad. (The sketch is considerably exaggerated, because 10^{-3} rad is only about 0.06 degree.)

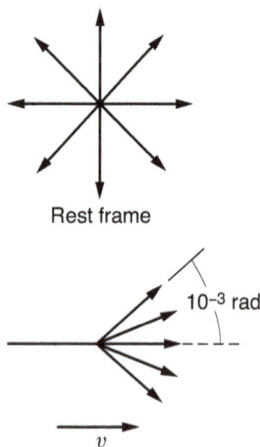

12.13 *Moving mirror*

The frequency of light reflected from a moving mirror undergoes a Doppler shift because of the motion of the image. Find the Doppler shift of light reflected directly back from a mirror that is approaching the observer with speed v, and show that it is the same as if the image were moving toward the observer at speed $2v/(1 + v^2/c^2)$.

12.14 *Moving glass slab**

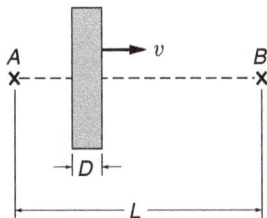

A slab of glass moves to the right with speed v. A flash of light is emitted from A and passes through the glass to arrive at B a distance L away. The glass has thickness D in its rest frame, and the speed of light in the glass is c/n. How long does it take the light to go from A to B?

12.15 *Doppler shift of a hydrogen spectral line**

One of the most prominent spectral lines of hydrogen is the H_α line, a bright red line with a wavelength of 656.1×10^{-9} m.

(a) What is the expected wavelength of the H_α line from a star receding with a speed of 3000 km/s?

(b) The H_α line measured on Earth from opposite ends of the Sun's equator differ in wavelength by 9×10^{-12} m. Assuming that the effect is caused by rotation of the Sun, find the period of rotation. The diameter of the Sun is 1.4×10^6 km.

12.16 *Pole-vaulter paradox**

The pole-vaulter has a pole of length l_0, and the farmer has a barn $\frac{3}{4}l_0$ long. The farmer bets that he can shut the front and rear doors of the barn with the pole completely inside. The bet being made, the farmer asks the pole-vaulter to run into the barn with a speed $v = c\sqrt{3}/2$. In this case the farmer observes the pole to be Lorentz contracted to $l = l_0/2$, and the pole fits into the barn with ease. The farmer slams the door the instant the pole is inside, and claims the bet. The pole-vaulter disagrees: he sees the barn contracted by a factor of 2, so the pole can't possibly fit inside. Let the farmer and barn be in system S and the pole-vaulter in system S'. Call the leading end of the pole A, and the trailing end B.

(*a*) The farmer in S sees A reach the rear door at $t_A = 0$, and closes the front door at the same time $t_A = t_B = 0$. What is the length of the pole as seen in S?

(*b*) The pole-vaulter in S' sees A reach the rear door at t'_A. Where does the pole-vaulter see B at this instant?

(*c*) Show that in S', A and B do not lie inside the barn at the same instant.

12.17 *Transformation of acceleration*
The relativistic transformation of acceleration from S' to S can be found by extending the procedure of Section 12.9. The most useful transformation is for the case in which the particle is instantaneously at rest in S' but is accelerating at rate a_0 parallel to the x' axis.

Show that for this case the x acceleration in S is given by $a_x = a_0/\gamma^3$.

12.18 *The consequences of endless acceleration**
The relativistic transformation for acceleration derived in Problem 12.17 shows the impossibility of accelerating a system to a velocity greater than c. Consider a spaceship that accelerates at constant rate a_0 as measured by an accelerometer carried aboard, for instance a mass stretching a spring.

(*a*) Find the speed after time t for an observer in the system in which the spaceship was originally at rest.

(*b*) The speed predicted classically is $v_0 = a_0 t$. What is the actual speed for the following cases: $v_0 = 10^{-3}c, c, 10^3 c$.

12.19 *Traveling twin*
A young man voyages to the nearest star, α Centauri, 4.3 light years away. He travels in a spaceship at a velocity of $c/5$. When he returns to Earth, how much younger is he than his twin brother who stayed home?

13

RELATIVISTIC DYNAMICS

13.1 Introduction

In Chapter 12 we saw how the postulates of special relativity lead to new kinematical relations for space and time. These relations can naturally be expected to have important implications for dynamics, particularly for the meaning of momentum and energy. In this chapter we examine the modifications to the Newtonian concepts of momentum and energy required by special relativity. The underlying strategy is to ensure that momentum and energy in an isolated system continue to be conserved. This approach is often used in extending the frontiers of physics: by reformulating conservation laws so that they are preserved in new situations, we are led to generalizations of familiar concepts. We can also be led to the discovery of unfamiliar concepts, for instance the concept of massless particles that can nevertheless carry energy and momentum.

13.2 Relativistic Momentum

To investigate the nature of momentum in special relativity, consider a glancing elastic collision between two identical particles A and B in an isolated system. We want the total momentum of the system to be conserved, as it is in non-relativistic physics. We shall view the collision in two frames: A's frame, the frame moving along the x axis with A so that A is at rest while B approaches along the x direction with speed V, and then in B's frame, which is moving with B in the opposite direction so that B is at rest and A is approaching. (The term "frame" is used synonymously with "reference system.") We take the collisions to be completely symmetrical. Each particle has the same y speed u_0 in its own frame before the collision, as shown in the sketches. The effect of the collision is to reverse the y velocities but leave the x velocities unchanged.

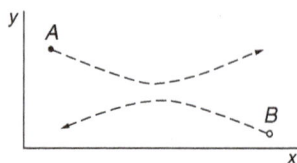

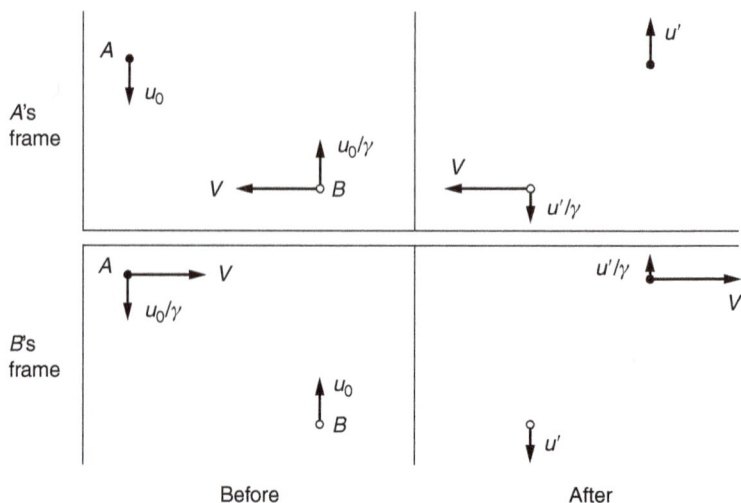

The relative x velocity of the frames is V. In A's frame, the y velocity of particle A is u_0, and by the transformation of velocities, Eqs. (12.8)

and (12.9), the y velocity of particle B is u_0/γ where $\gamma = 1/\sqrt{1 - V^2/c^2}$. The situation is symmetrical when viewed from B's frame.

After the collision the y velocities have reversed their directions as shown. The situation remains symmetric: if the y velocity of A or B in its own frame is u', the y velocity of the other particle is u'/γ.

Our task is to find a conserved quantity analogous to classical momentum. We suppose that the momentum of a particle moving with velocity $\mathbf{w}$ is

$$\mathbf{p} = m(w)\mathbf{w},$$

where $m(w)$ is a scalar quantity, yet to be determined, analogous to Newtonian mass but which could depend on the speed w.

The x momentum in A's frame is due entirely to particle B. Before the collision B's speed is $w = \sqrt{V^2 + u_0^2/\gamma^2}$ and after the collision it is $w' = \sqrt{V^2 + u'^2/\gamma^2}$. Imposing conservation of momentum in the x direction yields

$$m(w)V = m(w')V.$$

It follows that $w = w'$, so that

$$u' = u_0.$$

In other words, y motion is reversed in the A frame.

Next we write the statement of the conservation of momentum in the y direction as evaluated in A's frame. Equating the y momentum before and after the collision gives

$$-m_0 u_0 + m(w)\frac{u_0}{\gamma} = m_0 u_0 - m(w)\frac{u_0}{\gamma}$$

which gives

$$m(w) = \gamma m_0.$$

In the limit $u_0 \to 0$, $m(u_0) \to m(0)$, which we take to be the Newtonian mass, or "rest mass" m_0, of the particle. In this limit, $w = V$. Hence

$$m(V) = \gamma m_0 = \frac{m_0}{\sqrt{1 - V^2/c^2}}. \tag{13.1}$$

Consequently, momentum is preserved in the collision provided we define the momentum of a particle moving with velocity $\mathbf{v}$ to be

$$\mathbf{p} = m\mathbf{v} \tag{13.2}$$

where

$$m = \frac{m_0}{\sqrt{1 - v^2/c^2}} = \gamma m_0.$$

The quantity $m = \gamma m_0$ is referred to as the "relativistic mass" or more often simply as the mass of a particle. If the rest mass is intended, that needs to be made specific.

In relativity there is an upper limit to speed: the speed of light. However, there is no upper limit on momentum. Once a particle is moving with speed close to c, an increase in momentum comes about primarily through an increase in mass. High energy particle accelerators do not make particles go substantially faster and faster. A particle is quickly accelerated to speed close to c. After that, the accelerator principally makes the particle more and more massive with only a very small increment in speed.

The expression $\mathbf{p} = m\mathbf{v} = \gamma m_0 \mathbf{v}$ is sometimes taken as the starting point for developing relativistic dynamics, but in the early days of relativity attention was focused not so much on momentum but on the apparent dependence of mass on speed. Investigation of this problem provided the first direct experimental evidence for Einstein's theory.

Example 13.1 Speed Dependence of the Electron's Mass

At the beginning of the twentieth century there were several speculative theories based on various models of the structure of the electron that predicted that the mass of an electron would vary with its speed. One theory, from Max Abraham (1902), predicted $m = m(u_0)[1 + \frac{2}{5}(v^2/c^2)]$ for $v \ll c$ and another from Hendrik A. Lorentz (1904) gave $m = m_0/\sqrt{1 - v^2/c^2} \approx m(u_0)[1 + \frac{1}{2}(v^2/c^2)]$. The Abraham theory, which retained the idea of the ether drift and absolute motion, predicted no time dilation effect. Lorentz's result, while identical in form to that published by Einstein in 1905, was derived using the ad hoc Lorentz contraction and did not possess the generality of Einstein's theory. Experimental work on the effect of speed on the electron's mass was initiated by Kaufmann in Göttingen in 1902. His data favored the theory of Abraham, and in a 1906 paper he rejected the Lorentz–Einstein results. However, further work by Bestelmeyer (1907) in Göttingen and Bucherer (1909) in Bonn revealed errors in Kaufmann's work and confirmed the Lorentz–Einstein formula.

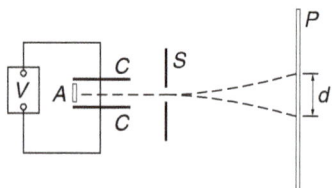

Physicists were in agreement that the force on a moving electron in an applied electric field $\mathbf{E}$ and magnetic field $\mathbf{B}$ is $q(\mathbf{E} + \mathbf{v} \times \mathbf{B})$ (the units are SI), where q is the electron's charge and $\mathbf{v}$ its velocity. Bucherer employed this force law in the apparatus shown in the sketch. The apparatus is evacuated and immersed in an external magnetic field $\mathbf{B}$ perpendicular to the plane of the sketch. The source of the electrons A is a button of radioactive material, generally radium salts. The emitted electrons ("beta-rays") have a broad energy spectrum extending to 1 MeV or so. To select a single speed, the electrons are passed through a "velocity filter" composed of a transverse electric field $\mathbf{E}$ (produced between two parallel metal plates C by the battery V) and the perpendicular magnetic field. $\mathbf{E}, \mathbf{B}$, and $\mathbf{v}$ are mutually perpendicular. The transverse force is zero when $qE = qvB$, so that electrons with $v = E/B$ are undeflected and are able to pass through the slit S.

Beyond S only the magnetic field acts. The electrons move with constant speed v and are bent into a circular path by the magnetic force $q\mathbf{vB}$. The radius of curvature R is given by $mv^2/R = qvB$, or $R = mv/qB = (m/q)(E/B^2)$.

The electrons eventually strike the photographic plate P, leaving a trace. By reversing $\mathbf{E}$ and $\mathbf{B}$, the sense of deflection is reversed. R is found from a measurement of the total deflection d and the known geometry of the apparatus. E and B are measured by standard techniques. Finding R for different velocities allowed the velocity dependence of m/q to be measured. Physicists believe that charge does not vary with velocity (otherwise an atom would not stay strictly neutral in spite of how the energy of its electrons varied), so that the variation of m/q can be attributed to variation in m alone.

The graph shows Bucherer's data together with a dashed line corresponding to the Einstein prediction $m = m_0/\sqrt{1 - v^2/c^2}$. The agreement is striking.

Today, the relativistic equations of motion are used routinely to design high energy particle accelerators. For protons, accelerators have been operated with m/m_0 up to 10^4, while for electrons the ratio $m/m_0 = 10^5$ has been reached. The successful operation of these machines leaves no doubt of the validity of relativistic dynamics.

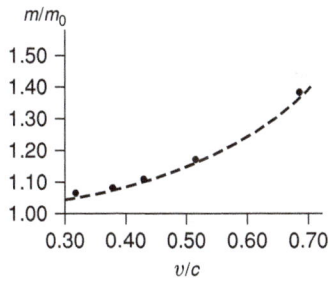

13.3 Relativistic Energy

By generalizing the Newtonian concept of energy, we can find a corresponding relativistic quantity that is also conserved. Recall the argument from Chapter 5: the change in kinetic energy K of a particle as it moves from $\mathbf{r_a}$ to $\mathbf{r_b}$ under the influence of force $\mathbf{F}$ is

$$K_b - K_a = \int_a^b \mathbf{F} \cdot d\mathbf{r}$$

$$= \int_a^b \frac{d\mathbf{p}}{dt} \cdot d\mathbf{r}.$$

For a Newtonian particle moving with velocity $\mathbf{u}$ the momentum is given by $\mathbf{p} = m\mathbf{u}$, where m is constant. Then

$$K_b - K_a = \int_a^b \frac{d}{dt}(m\mathbf{u}) \cdot d\mathbf{r}$$

$$= \int_a^b m\frac{d\mathbf{u}}{dt} \cdot \mathbf{u}\, dt$$

$$= \int_a^b m\mathbf{u} \cdot d\mathbf{u}.$$

Using the identity $\mathbf{u} \cdot d\mathbf{u} = \frac{1}{2}d(\mathbf{u} \cdot \mathbf{u}) = \frac{1}{2}d(u^2) = u\,du$, we obtain

$$K_b - K_a = \frac{1}{2}mu_b^2 - \frac{1}{2}mu_a^2.$$

It is natural to try the same procedure starting with the relativistic expression for momentum $\mathbf{p} = m_0\mathbf{u}/\sqrt{1 - u^2/c^2}$:

$$\begin{aligned}
K_b - K_a &= \int_a^b \frac{d\mathbf{p}}{dt} \cdot d\mathbf{r} \\
&= \int_a^b \frac{d\mathbf{p}}{dt} \cdot \frac{d\mathbf{r}}{dt}\,dt \\
&= \int_a^b \frac{d}{dt}\left[\frac{m_0\mathbf{u}}{\sqrt{1 - u^2/c^2}}\right] \cdot \mathbf{u}\,dt \\
&= \int_a^b \mathbf{u} \cdot d\left[\frac{m_0\mathbf{u}}{\sqrt{1 - u^2/c^2}}\right].
\end{aligned}$$

The integrand has the form $\mathbf{u} \cdot d\mathbf{p}$. Using the relation $\mathbf{u} \cdot d\mathbf{p} = d(\mathbf{u} \cdot \mathbf{p}) - \mathbf{p} \cdot d\mathbf{u}$ gives

$$\begin{aligned}
K_b - K_a &= (\mathbf{u} \cdot \mathbf{p})\big|_a^b - \int_a^b \mathbf{p} \cdot d\mathbf{u} \\
&= \frac{m_0 u^2}{\sqrt{1 - u^2/c^2}}\bigg|_a^b - \int_a^b \frac{m_0 u\,du}{\sqrt{1 - u^2/c^2}},
\end{aligned}$$

where we have again used the identity $\mathbf{u} \cdot d\mathbf{u} = u\,du$. The integral is elementary, and we find

$$K_b - K_a = \frac{m_0 u^2}{\sqrt{1 - u^2/c^2}}\bigg|_a^b + m_0 c^2\,\sqrt{1 - \frac{u^2}{c^2}}\bigg|_a^b.$$

Let point b be arbitrary and take the particle to be at rest at point a so $u_a = 0$:

$$\begin{aligned}
K &= \frac{m_0 u^2}{\sqrt{1 - u^2/c^2}} + m_0 c^2\,\sqrt{1 - \frac{u^2}{c^2}} - m_0 c^2 \\
&= \frac{m_0[u^2 + c^2(1 - u^2/c^2)]}{\sqrt{1 - u^2/c^2}} - m_0 c^2 \\
&= \frac{m_0 c^2}{\sqrt{1 - u^2/c^2}} - m_0 c^2
\end{aligned}$$

or

$$K = (\gamma - 1)m_0 c^2. \tag{13.3}$$

This expression for kinetic energy bears little resemblance to its classical counterpart. However, in the limit $u \ll c$, $\gamma = 1/\sqrt{1 - u^2/c^2} \approx \frac{1}{2}u^2/c^2$.

Using the expansion $1/\sqrt{1-x} = 1 + \frac{1}{2}x + \cdots$ we obtain

$$K \approx m_0 c^2 \left(1 + \frac{1}{2}\frac{u^2}{c^2} - 1 \right)$$

$$= \frac{1}{2} m_0 u^2.$$

The kinetic energy arises from the work done on the particle to bring it from rest to speed u. Using the relation $mc^2 = \gamma m_0 c^2$, we can rearrange Eq. (13.3) to give

$$mc^2 = K + m_0 c^2$$

$$= \text{work done on particle} + m_0 c^2. \qquad (13.4)$$

Einstein proposed the following bold interpretation of this result: mc^2 is the *total* energy E of the particle. The first term arises from external work; the second term, $m_0 c^2$, represents the "rest" energy the particle possesses by virtue of its mass. In summary,

$$E = mc^2. \qquad (13.5)$$

It is important to realize that Einstein's generalization goes far beyond the classical conservation law for mechanical energy. Thus, if energy ΔE is added to a body, its mass will change by $\Delta m = \Delta E/c^2$, irrespective of the form of energy. ΔE could be mechanical work, heat energy, the absorption of light, or any other form of energy. In relativity the classical distinction between mechanical energy and other forms of energy disappears. Relativity treats all forms of energy on an equal footing, in contrast to Newtonian physics where each form of energy must be treated as a special case.

The conservation of total energy $E = mc^2$ is a consequence of the structure of relativity. In Chapter 14 we shall show that the conservation laws for energy and momentum are actually different aspects of a single, more general, conservation law.

The following example illustrates the relativistic concept of energy and the application of the conservation laws in different inertial frames.

Example 13.2 Relativistic Energy and Momentum in an Inelastic Collision

Suppose that two identical particles each of mass M collide with equal and opposite velocities and stick together. In Newtonian physics, the initial kinetic energy is $2(\frac{1}{2}MV^2) = MV^2$. By conservation of momentum the mass $2M$ is at rest and has zero kinetic energy. In the language of Chapter 4 we say that mechanical energy MV^2 was lost as heat. As we shall see, this distinction between these different classical forms of energy does not occur in relativity.

Now consider the same collision relativistically, as seen in the original frame x, y, and in the frame x', y' moving with one of the particles. By the relativistic transformation of velocities, Eqs. (12.8) and (12.9), the relative velocity in the x', y' frame is

$$U = \frac{2V}{1 + V^2/c^2} \tag{1}$$

in the direction shown.

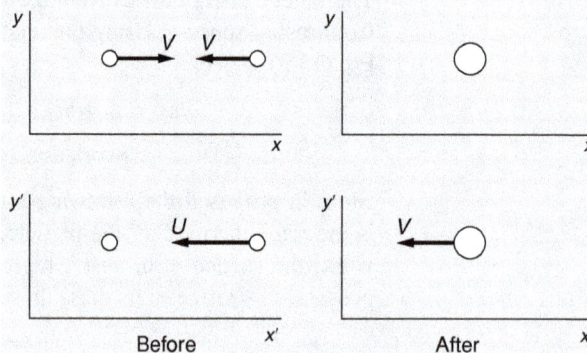

Let the rest mass of each particle be M_{0i} before the collision and M_{0f} after the collision. In the x, y frame, momentum is obviously conserved. The total energy before the collision is $2M_{0i}c^2/\sqrt{1 - V^2/c^2}$, and after the collision the energy is $2M_{0f}c^2$. No external work was done on the particles, and the total energy is unchanged. Therefore

$$\frac{2M_{0i}c^2}{\sqrt{1 - V^2/c^2}} = 2M_{0f}c^2$$

or

$$M_{0f} = \frac{M_{0i}}{\sqrt{1 - V^2/c^2}}. \tag{2}$$

Physically, the final rest mass is greater than the initial rest mass because the particles are warmer. To see this, we take the low-velocity approximation

$$M_{0f} \approx M_{0i}\left(1 + \frac{1}{2}\frac{V^2}{c^2}\right).$$

The increase in rest energy for the two particles is $2(M_{0f} - M_{0i})c^2 \approx 2(\frac{1}{2}M_{0i}V^2)$, which corresponds to the loss of Newtonian kinetic energy. Now, however, the kinetic energy is not "lost"—it is present as a mass increase.

By the postulate that all inertial frames are equivalent, the conservation laws must hold in the x', y' frame as well. Checking to see if our assumed conservation laws possess this necessary property, we have in the x', y' frame

$$\frac{M_{0i}U}{\sqrt{1 - U^2/c^2}} = \frac{2M_{0f}V}{\sqrt{1 - V^2/c^2}} \tag{3}$$

by conservation of momentum and

$$M_{0i}c^2 + \frac{M_{0i}c^2}{\sqrt{1 - U^2/c^2}} = \frac{2M_{0f}c^2}{\sqrt{1 - V^2/c^2}} \tag{4}$$

by conservation of energy.

The question now is whether Eqs. (3) and (4) are consistent with our earlier results, Eqs. (1) and (2). To check Eq. (3), we use Eq. (1) to write

$$1 - \frac{U^2}{c^2} = 1 - \frac{4V^2/c^2}{(1 + V^2/c^2)^2}$$
$$= \frac{(1 - V^2/c^2)^2}{(1 + V^2/c^2)^2}. \tag{5}$$

From Eqs. (1) and (5),

$$\frac{U}{\sqrt{1 - U^2/c^2}} = \frac{2V}{(1 + V^2/c^2)} \frac{(1 + V^2/c^2)}{(1 - V^2/c^2)}$$
$$= \frac{2V}{1 - V^2/c^2}$$

and the left-hand side of Eq. (3) becomes

$$\frac{M_{0i}U}{\sqrt{1 - U^2/c^2}} = \frac{2M_{0i}V}{1 - V^2/c^2}. \tag{6}$$

From Eq. (2), $M_{0i} = M_{0f}\sqrt{1 - V^2/c^2}$, and Eq. (6) reduces to

$$\frac{M_{0i}U}{\sqrt{1 - U^2/c^2}} = \frac{2M_{0f}V}{\sqrt{1 - V^2/c^2}},$$

which is identical to Eq. (3). Similarly, it is not hard to show that Eq. (4) is also consistent.

We see from Eq. (6) that if we had assumed that rest mass was unchanged in the collision, $M_{0i} = M_{0f}$, the conservation law for momentum (or for energy) would not be correct in the second inertial frame. The relativistic description of energy is essential for maintaining the validity of the conservation laws in all inertial frames.

Example 13.3 The Equivalence of Mass and Energy

In 1932 J.D. Cockcroft and E.T.S. Walton, two young British physicists, successfully operated the first high energy proton accelerator and succeeded in causing a nuclear disintegration. Their experiment provided one of the earliest confirmations of the relativistic mass–energy relation.

Briefly, their accelerator consisted of a power supply that could reach 600 kV and a source of protons (hydrogen nuclei). The power supply used an ingenious arrangement of capacitors and rectifiers to

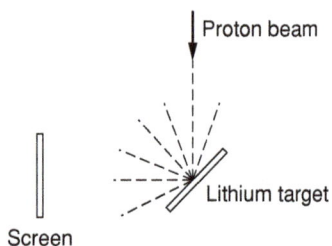

quadruple the voltage of a 150 kV supply. The protons were supplied by an electrical discharge in hydrogen and were accelerated in vacuum by the applied high voltage.

Cockcroft and Walton studied the effect of the protons on a target of ^{7}Li (lithium atomic mass 7). A zinc sulfide fluorescent screen, located nearby, emitted occasional flashes, or scintillations. By various tests they determined that the scintillations were due to alpha particles, the nuclei of helium, ^{4}He. Their interpretation was that ^{7}Li captures a proton and that the resulting nucleus of mass 8 immediately disintegrates into two alpha particles. We can write the reaction as

$$^7\text{Li} + {}^1\text{H} \rightarrow {}^4\text{He} + {}^4\text{He}.$$

The mass–energy equation for the reaction can be written

$$K_{\text{initial}} + M_{\text{initial}}c^2 = K_{\text{final}} + M_{\text{final}}c^2$$

where the masses are the particle rest masses. Applied to the lithium bombardment experiment, this gives

$$K(^1\text{H}) + [M\,(^1\text{H}) + M\,(^7\text{Li})]c^2 = 2K(^4\text{He}) + 2M(^4\text{He})c^2$$

where $K(^1\text{H})$ is the kinetic energy of the incident proton, $K(^4\text{He})$ is the kinetic energy of each emitted alpha particle, $M(^1\text{H})$ is the proton rest mass, etc. (The initial momentum of the proton is negligible, and the two alpha particles are emitted back-to-back with equal energy by conservation of momentum.)

We can rewrite the mass–energy equation as

$$K = \Delta M c^2,$$

where $K = 2K(^4\text{He}) - K(^1\text{H})$, and where ΔM is the initial rest mass minus the final rest mass.

The energy of the alpha particles was determined by measuring their range in matter. Cockcroft and Walton obtained the value $K = 17.2$ MeV (1 MeV = 10^6 eV = 1.6×10^{-13} J.

The relative masses of the nuclei were known from mass spectrometer measurements. In atomic mass units, amu, the values available to Cockcroft and Walton were

$$M(^1\text{H}) = 1.0072$$

$$M(^7\text{Li}) = 7.0104 \pm 0.0030$$

$$M(^4\text{He}) = 4.0011.$$

Using these values,

$$\Delta M = (1.0072 + 7.0104) - 2(4.0011)$$

$$= (0.0154 \pm 0.0030) \text{ amu}.$$

Proton beam

Lithium target

Screen

The rest energy of 1 amu is ≈ 931 MeV and therefore

$$\Delta M c^2 = (14.3 \pm 2.7) \text{MeV}.$$

The difference between K and $\Delta M c^2$ is $(17.2 - 14.3)$ MeV $= 2.9$ MeV, slightly larger than the experimental uncertainty of 2.7 MeV. However, the experimental uncertainty always represents an estimate, not a precise limit, and the result from these early experiments can be taken as consistent with the relation $K = \Delta M c^2$. It is clear that the masses must be known to high accuracy for studying the energy balance in nuclear reactions. Modern techniques of mass spectrometry have achieved an accuracy of better than 10^{-10} amu, and the mass–energy equivalence has been amply confirmed to within experimental accuracy. According to a modern table of masses, the decrease in rest mass in the reaction studied by Cockcroft and Walton is $\Delta M c^2 = (17.3468 \pm 0.0012)$ MeV.

13.4 How Relativistic Energy and Momentum are Related

Often it is useful to express the total energy of a free particle in terms of its momentum. In Newtonian physics the relation is

$$E = \frac{1}{2}mv^2 = \frac{p^2}{2m}.$$

To find the equivalent relativistic expression we can combine the relativistic momentum

$$\mathbf{p} = m\mathbf{u} = \frac{m_0 \mathbf{u}}{\sqrt{1 - u^2/c^2}} = \gamma m_0 \mathbf{u} \tag{13.6}$$

with the energy

$$E = mc^2 = \gamma m_0 c^2. \tag{13.7}$$

Squaring Eq. (13.6) gives

$$p^2 = \frac{m_0^2 u^2}{1 - u^2/c^2}.$$

We can solve for γ as follows:

$$\frac{u^2}{c^2} = \frac{p^2}{p^2 + m_0^2 c^2}$$

$$\gamma = \frac{1}{\sqrt{1 - u^2/c^2}}$$

$$= \sqrt{1 + \frac{p^2}{m_0^2 c^2}}.$$

Inserting this in Eq. (13.7), we have

$$E = m_0 c^2 \sqrt{1 + \frac{p^2}{m_0{}^2 c^2}}.$$

The square of this equation is algebraically simpler and is the form usually employed:

$$E^2 = (pc)^2 + (m_0 c^2)^2. \tag{13.8}$$

For convenience, here is a summary of the important dynamical formulas we have developed so far.

$$\mathbf{p} = m\mathbf{u} = m_0 \mathbf{u} \gamma \tag{13.9}$$

$$K = mc^2 - m_0 c^2 = m_0 c^2 (\gamma - 1) \tag{13.10}$$

$$E = mc^2 = m_0 c^2 \gamma \tag{13.11}$$

$$E^2 = (pc)^2 + (m_0 c^2)^2. \tag{13.12}$$

13.5 The Photon: A Massless Particle

In 1905, in the *annus mirabilis* when Albert Einstein published four papers each worthy of a Nobel Prize, the first paper, and the only one to actually receive the Prize, had the unlikely title *On a Heuristic Viewpoint Concerning the Production and Transformation of Light*. A heuristic theory is a theory based partly on guesswork, intended to stimulate thinking. The paper ostensibly provided an explanation for the photoelectric effect, the process by which electrons are ejected from a surface when it is irradiated with light. It is now recognized that the paper provided the foundation for the quantum theory of light, contributed significantly to the development of quantum mechanics, and made applications such as the laser possible.

Because light is inherently relativistic, Einstein's paper actually opened a chapter on relativity even before relativity had been announced. At the heart of his argument is a concept that makes little sense in Newtonian physics but perfect sense in relativistic physics: *massless particles that carry momentum*.

A little background is needed: In December, 1900, quantum physics was born when Max Planck proposed that the energy of a harmonic oscillator cannot be varied at will but only by discrete steps. If the frequency of the oscillator is ν, then the energy steps had size $h\nu$ where h is a constant, now called *Planck's constant*, $h \approx 6.6 \times 10^{-34} \, \mathrm{m^2 \, kg^2/s} = 6.6 \times 10^{-34}$ joule · second. Planck proposed this idea to solve the mystery of thermal radiation, often called *blackbody radiation*. The shape of the spectrum of radiation emitted by a warm body could not be accounted for by the known laws of physics, based on Newtonian mechanics and Maxwell's electromagnetic theory. Planck put forward his hypothesis more in the spirit of a mathematical conjecture than a physical theory, but in 1905 Einstein came to a similar conclusion, though by totally different reasoning, and his theory had some startling implications.

Einstein was thoroughly aware of the wave nature of light. He knew all about Maxwell's equations and how they predicted the existence of electromagnetic waves—light waves—that can travel through empty space with speed c. The wavelength of an electromagnetic wave λ and its frequency ν are related by $\lambda \nu = c$. There was a considerable body of experimental evidence that confirmed the wave nature of light, for instance the colors in soap bubble films that are a signature of light waves interfering, not to mention the fringes in Michelson's interferometer. Einstein, however, pointed out that these phenomena involve observations at the macroscopic (large scale) level. Macroscopic behavior results from the effect of many microscopic events. He pointed out that little was known about how light interacted with matter at the atomic or individual particle level. He went on to argue that light could also be understood from a particle point of view. He suggested that a light wave could behave as if it were a gas of particles, each possessing energy $\epsilon = h\nu = hc/\lambda$, where ν is the frequency of the wave. This particle hypothesis seemed to be in direct contradiction to the wave theory of light.

We now understand that light displays either wave-like or particle-like behavior depending on the situation. To understand light from the wave point of view, one starts by writing Maxwell's wave equations. Their solution reveals that time-varying electric and magnetic fields in space support each other to create an electromagnetic wave that travels at the speed of light. Furthermore, no matter which inertial coordinate system one chooses for describing the radiation process, the wave always propagates at speed c. In other words, Maxwell's equations are intrinsically relativistic. Einstein showed that we can also understand the relativistic behavior of light starting from a particle point of view, and that is the approach we now follow.

A startling consequence of the relativistic energy–momentum relation is the possibility of "massless" particles, particles that possess momentum and energy but have zero rest mass. The essential point is that a particle can possess momentum without possessing mass. This follows from the definition of relativistic momentum

$$\mathbf{p} = m_0 \mathbf{u} \left(\frac{1}{\sqrt{1 - u^2/c^2}} \right).$$

If we consider the limit $m_0 \to 0$ while $u \to c$, then $\mathbf{p}$ can remain finite. Evidently a particle without mass can carry momentum, provided that it travels at the speed of light. From Eq. (13.12),

$$E^2 = (pc)^2 + (m_0 c^2)^2,$$

and if we take $m_0 = 0$, then we have, denoting photon energy by the symbol ϵ,

$$\epsilon^2 = (pc)^2,$$
$$\epsilon = pc. \tag{13.13}$$

We have taken the positive square root because the negative solution would predict that in an isolated system the momentum of a photon could increase without limit as its energy dropped. Combining Eq. (13.13) with Einstein's relation $\epsilon = h\nu$, we find that a photon possesses momentum p of magnitude

$$p = \frac{h\nu}{c}. \tag{13.14}$$

The direction of the momentum vector is along the direction of travel of the light wave.

Einstein's quantum hypothesis was designed to solve a theoretical dilemma—the spectrum of blackbody radiation—but its first application was to a totally different problem—the photoelectric effect.

Example 13.4 The Photoelectric Effect

In 1887 Heinrich Hertz discovered that metals can give off electrons when illuminated by ultraviolet light. This process, the *photoelectric effect*, represents the direct conversion of light into mechanical energy (here, the kinetic energy of the electron). Einstein predicted that the energy a single electron absorbs from a beam of light at frequency ν is exactly the energy of a single photon, $h\nu$. For the electron to escape from the surface it must overcome the energy barrier that confines it to the surface. The electron must expend energy $W = e\Phi$ to escape from the surface, where e is the charge of the electron and Φ is an electric potential known as the *work function* of the material, typically a few volts. The maximum kinetic energy of the emitted electron is therefore

$$K = h\nu - e\Phi.$$

The work function depends on the poorly known chemical state of the surface, making the photoelectric effect difficult to investigate. Nevertheless, Robert A. Millikan overcame this problem in 1914 by working with metal surfaces prepared in a high vacuum system. He plotted the reverse voltage V needed to prevent the photoelectrons from reaching a detector as a function of the frequency of light. The voltage is given by

$$eV = K = h\nu - e\Phi.$$

The slope of the plot of V versus ν is

$$\frac{dV}{d\nu} = \frac{h}{e}. \tag{13.15}$$

The photoelectric effect: experimental results on the energy of photoelectrons and the frequency of light. The graph is from R.A. Millikan. From R.A. Millikan, *Physical Review* **7**, 355 (1916).

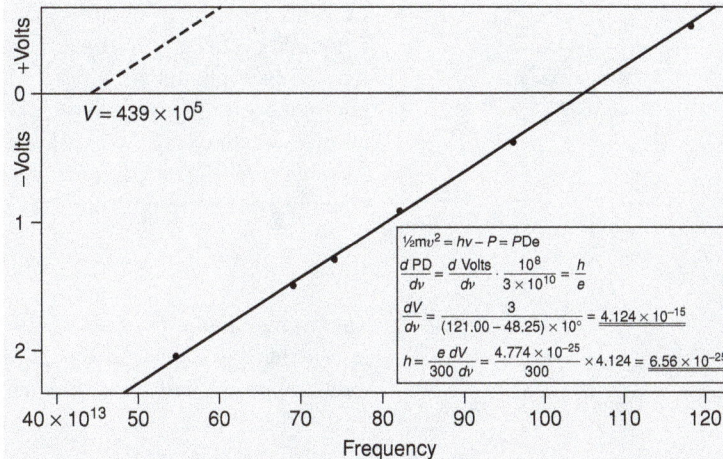

$V = 439 \times 10^5$

$\frac{1}{2}mv^2 = h\nu - P = PDe$

$\frac{d\,PD}{d\nu} = \frac{d\,\text{Volts}}{d\nu} \cdot \frac{10^8}{3 \times 10^{10}} = \frac{h}{e}$

$\frac{dV}{d\nu} = \frac{3}{(121.00 - 48.25) \times 10^{\circ}} = \underline{4.124 \times 10^{-15}}$

$h = \frac{e\,dV}{300\,d\nu} = \frac{4.774 \times 10^{-25}}{300} \times 4.124 = \underline{6.56 \times 10^{-25}}$

The graph of Millikan's results shows the linear relation between energy and frequency predicted by Einstein, and the slope of the line provides an accurate value for the ratio of two fundamental constants, Planck's constant and the charge of the electron.

The fact that light can interfere with itself, as in the Michelson interferometer, is compelling evidence that light has wave properties. Nevertheless, the photoelectric effect illustrates that light also has particle properties. Einstein's energy relation, $E = h\nu$, provides the link between these apparently conflicting descriptions of light by relating the energy of the photon to the frequency of the wave.

Example 13.5 The Pressure of Light

The photon picture of light provided an immediate explanation for a phenomenon that was also predicted by Maxwell's electromagnetic theory: the pressure of light. If a beam of light is absorbed or reflected by a body, it exerts a force on the body. The force per unit area, the *radiation pressure*, is too small to feel when we are in sunlight but it can have visible effects. Radiation pressure causes comets' tails to always point away from the Sun. On the astronomical scale, it helps prevent stars from collapsing under their gravitational attraction. In ultra-high intensity laser beams radiation pressure can be large enough to compress matter to the high density needed to initiate fusion reactions.

Energy flow in a light beam is often characterized by the beam's *intensity I*, which is the power per unit area of the light beam. If the number of photons crossing a unit area per second is $\dot{N}$ and each photon carries energy ϵ, then $I = \dot{N}\epsilon$.

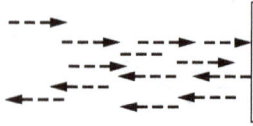

Consider a stream of photons in a monochromatic light beam striking a perfectly reflecting mirror at normal incidence. The initial momentum of each photon is $p = \epsilon/c$ directed toward the mirror, and the total change in momentum after the reflection is $2p = 2\epsilon/c$. The total momentum change per unit area per second due to the reflection is $2\dot{N}p = 2\dot{N}\epsilon/c$. This is the force on the light beam due to the mirror. The reaction force is the pressure P on the mirror due to the light. Hence

$$P = \frac{2\dot{N}\epsilon}{c} = \frac{2I}{c}.$$

The average intensity of sunlight falling on the Earth's surface at normal incidence, known as the *solar constant*, is ≈ 1000 W/m^2. The radiation pressure of sunlight on a mirror is therefore

$$P = 2I/c$$
$$= 7 \times 10^{-6}\,\text{N/m}^2$$

which is very small compared, for example, to atmospheric pressure 10^5 N/m^2.

Newtonian particles can be neither created nor destroyed. If they are combined, their total mass is constant. In contrast, massless particles can be created and annihilated. The emission of light occurs by the creation of photons, while the absorption of light occurs by the destruction of photons. The familiar laws of conservation of momentum and energy, as expressed in the theory of relativity, let us draw conclusions about processes involving photons without a detailed knowledge of the interactions, as the following examples illustrate.

Example 13.6 The Compton Effect

The photon description of light seemed so strange that it was not widely accepted until an experiment by Arthur Compton in 1922 made the photon picture inescapable: by scattering x-rays from electrons in matter, and showing that the x-rays scattered like particles undergoing elastic collisions, and that the dynamics were correctly described by special relativity.

A photon of visible light has energy in the range of 1 to 2 eV, but photons of much higher energy can be obtained from x-ray tubes, particle accelerators, or cosmic rays. X-ray photons have energies typically in the range 10 to 100 keV. Their wavelengths can be measured with high accuracy by the technique of crystal diffraction.

When a photon scatters from a free electron, the conservation laws require that the photon loses a portion of its energy due to the recoil of the electron. The outgoing photon therefore has a longer wavelength than the incoming photon. The shift in wavelength, first observed by Compton, is known as the *Compton effect*.

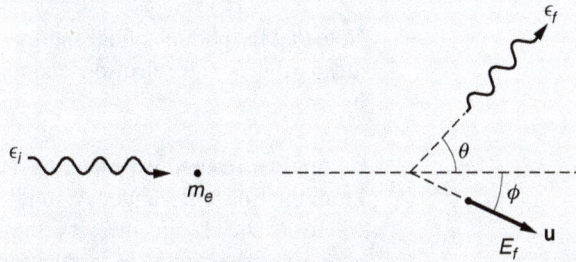

Suppose that a photon having initial energy ϵ_i and momentum ϵ_i/c is scattered at angle θ and has final energy ϵ_f. The electron has rest mass m_e and relativistic mass $m = \gamma m_e$. The electron is assumed to be initially at rest with energy $E_i = m_e c^2$. The scattered electron leaves at angle ϕ with momentum $\mathbf{p}$ and energy $E_f = mc^2$. Here $m = m_e \gamma = m_0/\sqrt{1 - u^2/c^2}$, where u is the speed of the recoiling electron.

The initial photon energy ϵ_i is known and the final photon energy ϵ_f and the scattering angle θ are measured. The problem is to calculate how ϵ_f varies with θ.

Conservation of total energy requires

$$\epsilon_i + m_e c^2 = \epsilon_f + E_f \tag{1}$$

and conservation of momentum requires

$$\frac{\epsilon_i}{c} = \frac{\epsilon_f}{c} \cos \theta + p \cos \phi \tag{2}$$

$$0 = \frac{\epsilon_f}{c} \sin \theta - p \sin \phi. \tag{3}$$

Because Compton detected only the outgoing photon our object is to eliminate reference to the electron and find ϵ_f as a function of θ. Equations (2) and (3) can be written

$$(\epsilon_i - \epsilon_f \cos \theta)^2 = (pc)^2 \cos^2 \phi$$

$$(\epsilon_f \sin \theta)^2 = (pc)^2 \sin^2 \phi.$$

Adding,

$$\epsilon_i^2 - 2\epsilon_i \epsilon_f \cos \theta + \epsilon_f^2 = (pc)^2. \tag{4}$$

To solve for ϵ_f, we introduce the energy–momentum relation in Eq. (13.12), which can be written $(pc)^2 = (mc^2)^2 - (m_e c^2)$. Combining this with Eq. (4) gives

$$\epsilon_f^2 - 2\epsilon_i \epsilon_f \cos \theta + \epsilon_f^2 = (\epsilon_i + m_e c^2 - \epsilon_f)^2 - (m_e c^2)^2,$$

which reduces to

$$\epsilon_f = \frac{\epsilon_i}{1 + (\epsilon_i/m_e c^2)(1 - \cos \theta)}. \tag{5}$$

Note that the photon's final energy ϵ_f is always greater than zero, which means that a free electron cannot absorb a photon, but can scatter it.

Compton measured wavelengths rather than energies in his experiment. From the Einstein frequency condition, $\epsilon_i = h\nu_i = hc/\lambda_i$ and $\epsilon_f = hc/\lambda_f$, where λ_i and λ_f are the wavelengths of the incoming and outgoing photons, respectively. In terms of wavelength, Eq. (5) takes the simple form

$$\lambda_f = \lambda_i + \frac{h}{m_e c}(1 - \cos \theta).$$

The quantity $h/m_e c$ is known as the *Compton wavelength* λ_C of the electron and has the value

$$\lambda_C = \frac{h}{m_e c}$$
$$= 2.426 \times 10^{-12} \text{ m}$$
$$= 0.02426 \text{ Å},$$

where 1 Å $= 10^{-10}$ m. (Å, called the angstrom, is a non-SI unit formerly used for wavelength measurements.)

The shift in wavelength at a given angle is independent of the initial photon energy:

$$\lambda - \lambda_0 = \lambda_C(1 - \cos \theta).$$

The figure shows one of Compton's results for $\lambda_0 = 0.711$ Å and $\theta = 90°$. The peak P is due to primary photons while the peak T is for photons scattered from a block of graphite. The measured wavelength shift is approximately 0.0246 Å and the calculated value is 0.02426 Å. The difference is less than the estimated uncertainty due to the experimental limitations.

We have assumed that the electron was free and at rest. For sufficiently high photon energies, this is a good approximation for electrons in the outer shells of light atoms. If the motion of the electrons is taken into account, the Compton peak is broadened or can have structure.

If the binding energy of the electron is comparable to the photon energy, momentum and energy can be transferred to the atom as a whole, and the photon can be completely absorbed.

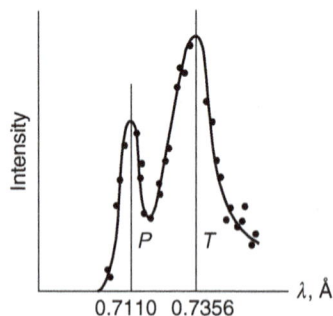

Intensity vs λ, Å. P at 0.7110, T at 0.7356.

Example 13.7 Pair Production

We have seen two ways by which a photon can lose energy in matter: photoelectric absorption and Compton scattering. If a photon's energy is sufficiently high, it can also lose energy in matter by the mechanism of *pair production*. The rest mass of an electron is $m_0c^2 = 0.511$ MeV. Can a photon of this energy create an electron? The answer is no, since this would require the creation of a single electric charge. As far as we know, electric charge is conserved in all physical processes. However, if equal amounts of positive and negative charge are created, the total charge remains zero and charge is conserved. It is therefore possible to create an electron–positron pair (e^-, e^+), two particles having the same mass but opposite charge.

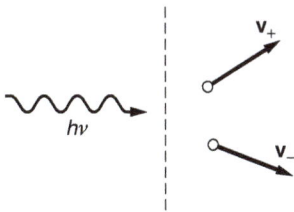

A single photon of energy $2m_0c^2$ or greater has enough energy to form an e^-, e^+ pair, but the process cannot occur in free space because it would not conserve momentum. To show why, imagine that the process occurs. Conservation of energy gives

$$hv = m_+c^2 + m_-c^2 = (\gamma_+ + \gamma_-)m_0c^2,$$

or

$$\frac{hv}{c} = (\gamma_+ + \gamma_-)m_0c,$$

while conservation of momentum gives

$$\frac{hv}{c} = |\gamma_+ \mathbf{v}_+ + \gamma_- \mathbf{v}_-| m_0.$$

These equations cannot be satisfied simultaneously because

$$(\gamma_+ + \gamma_-)c > |\gamma_+ \mathbf{v}_+ + \gamma_- \mathbf{v}_-|.$$

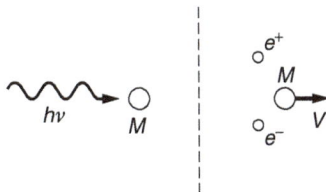

Pair production is possible if a third particle is available for carrying off the excess momentum. For instance, suppose that the photon collides with a nucleus of rest mass M_0 and creates an e^-, e^+ pair at rest. We have

$$hv + M_0c^2 = 2m_0c^2 + M_0c^2\gamma.$$

Since nuclei are much more massive than electrons, let us assume that $hv \ll M_0c^2$. (For hydrogen, the lightest atom, this means that $hv \ll 940$ MeV.) In this case the atom will not attain relativistic speeds and we can make the classical approximation

$$hv = 2m_0c^2 + M_0c^2(\gamma - 1)$$

$$\approx 2m_0c^2 + \frac{1}{2}MV^2.$$

To the same approximation, conservation of momentum yields

$$\frac{hv}{c} = MV.$$

Substituting this in the energy expression gives

$$hv = 2m_0c^2 + \frac{1}{2}\frac{(hv)^2}{Mc^2} \approx 2m_0c^2,$$

since we have already assumed $hv \ll Mc^2$. The threshold for pair production in matter is therefore $2m_0c^2 = 1.02$ MeV. The nucleus plays an essentially passive role, but by providing for momentum conservation it allows a process to occur that would otherwise be forbidden by the conservation laws.

Example 13.8 The Photon Picture of the Doppler Effect

In Chapter 12 we analyzed the relativistic Doppler effect from the standpoint of waves but we can also understand it from the photon picture. Consider first an atom with rest mass M_0, held stationary. If the atom emits a photon of energy hv_0, the atom's new mass is given by $M_0'c^2 = M_0c^2 - hv_0$.

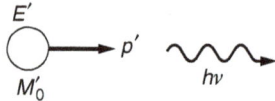

Next, we suppose that before emitting the photon the atom moves freely with velocity **u**. The atom's energy is $E = Mc^2 = \gamma M_0c^2$, where $\gamma = \sqrt{1 - u^2/c^2}$ and the atom's momentum is $p = Mu = M_0\gamma u$. After emitting a photon of energy hv the atom has velocity **u'**, rest mass M_0', energy E', and momentum p'. For simplicity, we consider the photon to be emitted along the line of motion.

By conservation of energy and momentum we have

$$E = E' + hv \tag{1}$$

$$p = p' + \frac{hv}{c}. \tag{2}$$

Rearranging Eqs. (1) and (2) gives

$$(E - hv)^2 = E'^2$$

$$(pc - hv)^2 = (p'c)^2.$$

Subtracting, and using Eq. (13.12), $E^2 - (pc)^2 = (m_0c^2)^2$, we have

$$(E - hv)^2 - (pc - hv)^2 = E'^2 - (p'c)^2 = (M_0'c^2)^2 \tag{3}$$

by the energy–momentum relation. Expanding the left-hand side and using $E^2 - (pc)^2 = (M_0c^2)^2$, with $M_0'c^2 = M_0c^2 - hv_0$, we obtain

$$(M_0c^2)^2 - 2Ehv + 2(pc)(hv) = (M_0'c^2)^2$$

$$= (M_0c^2 - hv_0)^2.$$

Simplifying, we find

$$v = v_0 \frac{(2M_0c^2 - hv_0)}{2(E - pc)}.$$

However,

$$E - pc = M_0 c^2 \gamma \left(1 - \frac{u}{c}\right)$$

$$= M_0 c^2 \sqrt{\frac{1 - u/c}{1 + u/c}}.$$

Hence

$$\nu = \nu_0 \left(1 - \frac{h\nu_0}{2M_0 c^2}\right) \sqrt{\frac{1 + u/c}{1 - u/c}}.$$

The term $h\nu_0/2M_0 c^2$ represents a decrease in the photon energy due to the recoil energy of the atom. Usually the recoil energy is so small that it can be neglected, leaving

$$\nu = \nu_0 \sqrt{\frac{1 + u/c}{1 - u/c}},$$

in agreement with the wave analysis that led to Eq. (12.12). However, the wave picture does not readily take into account the recoil of the atom. In modern experiments using high precision lasers and ultra-cold atoms, the recoil cannot be overlooked. On the contrary, it plays a crucial role in many studies.

Example 13.9 The Photon Picture of the Gravitational Red Shift

In Chapter 9 we derived an expression for the effect of gravity on time—the gravitational red shift—by invoking the equivalence principle. However, the effect of gravity on time can also be understood using the photon description of light and the conservation of energy.

Atoms can absorb or emit photons at certain characteristic frequencies. For a frequency ν_0, the atom loses energy $h\nu_0$ when it emits a photon, going from an upper energy state E_1 to a lower energy state, E_0, and it can gain energy $h\nu_0$ when it absorbs a photon, reversing the process.

Consider an atom with rest mass M_0 in its ground state with energy $E_0 = M_0 c^2$, in a gravitational field g. It absorbs a photon that increases its energy to $E_1 = E_0 + h\nu_0$. The mass of the atom is $M_1 = E_1/c^2 = (E_0 + h\nu_0)/c^2$. If we lift the atom to height H in a gravitational field g the work that we do is $M_1 gH$, so the final energy W_a of the atom is

$$W_a = E_1 + M_1 gH$$

$$= (E_0 + h\nu_0)(1 + gH/c^2)$$

$$= E_0 + h\nu_0 + h\nu_0 gH/c^2 + E_0 gH/c^2.$$

Consider an alternative scenario: the atom is first lifted to height H while it is in state E_0, and then a photon of energy $h\nu$ is radiated upward to put the atom in state E_1. The energy W_b of the atom with this procedure is

$$W_b = E_1 + M_0 g H = E_0 + h\nu + E_0 g H / c^2.$$

The final state of the system is the same in both scenarios. Consequently $W_a = W_b$ and it follows that

$$h\nu = h\nu_0 (1 + g H / c^2).$$

In fractional form, the gravitational red shift is

$$\frac{\nu - \nu_0}{\nu_0} = \frac{g H}{c^2}.$$

A word of explanation about the adjective "red." Our result reveals that if radiation travels outward from the Earth to a region of higher (less negative) gravitational potential, its energy decreases. Consequently, radiation emitted by a massive body such as the Sun is observed to shift to lower energy, equivalently to longer wavelengths, toward the red end of the spectrum. In contrast, radiation that comes down to the Earth from a satellite, for instance the signal from an atomic clock, is shifted to higher energy, which might be called a blue shift.

13.6 How Einstein Derived $E = mc^2$

Einstein's famous equation $E = mc^2$ is not to be found in his historic paper on relativity but only appeared a few months later in a short note titled *Does the Inertia of a Body Depend Upon Its Energy Content?* (translated from German). His argument was elegant in its simplicity, based entirely on elementary considerations of energy, momentum, and the Doppler shift.

Consider a body in system S at rest at the origin. The body has energy E_0 initially, and then sends out a pulse of light with energy $\epsilon/2$ in the $+x$ direction, and simultaneously a pulse with energy $\epsilon/2$ in the $-x$ direction. The body remains at rest after the emission by conservation of momentum, and its energy is then E_1

$$E_0 = E_1 + \frac{1}{2}\epsilon + \frac{1}{2}\epsilon.$$

In system S' moving with velocity v with respect to S, the initial energy of the body is H_0 and its energy after the emission is H_1. Taking the Doppler shift into account,

$$H_0 = H_1 + \frac{1}{2}\epsilon \left(\frac{1 - v/c}{\sqrt{1 - v^2/c^2}} \right) + \frac{1}{2}\epsilon \left(\frac{1 + v/c}{\sqrt{1 - v^2/c^2}} \right)$$

$$= H_1 + \frac{\epsilon}{\sqrt{1 - v^2/c^2}}.$$

The energy differences in the two systems are

$$(H_0 - E_0) - (H_1 - E_1) = \epsilon \left(\frac{1}{\sqrt{1 - v^2/c^2}} - 1 \right).$$

Einstein argued that the difference $H - E$ must equal the kinetic energy K of the body, to within an additive constant C that is independent of the relative velocity

$$H_0 - E_0 = K_0 + C$$
$$H_1 - E_1 = K_1 + C.$$

Thus

$$K_0 - K_1 = \epsilon \left(\frac{1}{\sqrt{1 - v^2/c^2}} - 1 \right)$$
$$\approx \frac{1}{2} \epsilon \frac{v^2}{c^2}.$$

Classically,

$$K_0 - K_1 = \frac{1}{2} \Delta m v^2.$$

Einstein then obtained his famous equation by comparing the two results for $K_0 - K_1$:

$$\Delta m = \frac{\epsilon}{c^2}.$$

Einstein concluded his brief paper by asserting that the equivalence of mass and energy must be a general law, holding for any form of energy, not just radiation.

Problems
*For problems marked *, refer to page 525 for a hint, clue, or answer.*

13.1 *Energetic proton*
Cosmic ray primary protons with energy up to 10^{20} eV (almost 10 J) have been detected. Our galaxy has a diameter of about 10^5 light years.

(a) How long does it take the proton to traverse the galaxy, in its own rest frame (proper time)? (1 eV = 1.6×10^{-19} J, $M_p = 1.67 \times 10^{-27}$ kg.) What is the proper time for a photon to traverse our galaxy?

(b) Compare the proton's energy to the kinetic energy of a baseball, mass = 145 g, traveling at 100 miles/hour.

13.2 *Onset of relativistic effects*
When working with particles it is important to know when relativistic effects have to be considered.

A particle of rest mass m_0 is moving with speed v. Its classical kinetic energy is $K_{cl} = m_0 v^2/2$. Let K_{rel} be the relativistic expression for its kinetic energy.

(a) By expanding K_{rel}/K_{cl} in powers of v^2/c^2, estimate the value of v^2/c^2 for which K_{rel} differs from K_{cl} by 10 percent.

(b) For this value of v^2/c^2, what is the kinetic energy in MeV of

(1) an electron ($m_0 c^2 = 0.51$ MeV)?
(2) a proton ($m_0 c^2 = 930$ MeV)?

13.3 *Momentum and energy*

In Newtonian mechanics, the kinetic energy of a mass m moving with velocity $\mathbf{v}$ is $K = mv^2/2 = p^2/(2m)$ where $\mathbf{p} = m\mathbf{v}$. The change in kinetic energy due to a small change in momentum is $dK = \mathbf{p} \cdot d\mathbf{p}/m = \mathbf{v} \cdot d\mathbf{p}$.

Show that the relation $dK = \mathbf{v} \cdot d\mathbf{p}$ also holds in relativistic mechanics.

13.4 *Particles approaching head-on**

Two particles of rest mass m_0 approach each other with equal and opposite velocity v in the laboratory frame. What is the total energy of one particle as measured in the rest frame of the other?

13.5 *Speed of a composite particle after an inelastic collision**

A particle of rest mass m_0 and speed v collides with a stationary particle of mass M and sticks to it. What is the final speed of the composite particle?

13.6 *Rest mass of a composite particle**

A particle of rest mass m_0 and kinetic energy $xm_0 c^2$, where x is some number, strikes an identical particle at rest and sticks to it. What is the rest mass of the resultant particle?

13.7 *Zero momentum frame**

In the laboratory frame a particle of rest mass m_0 and speed v is moving toward a particle of mass m_0 at rest.

What is the speed of the inertial frame in which the total momentum of the system is zero?

13.8 *Photon–particle scattering**

A photon of energy ϵ_i collides with a free particle of mass m_0 at rest. If the scattered photon flies off at angle θ, what is the scattering angle ϕ of the particle?

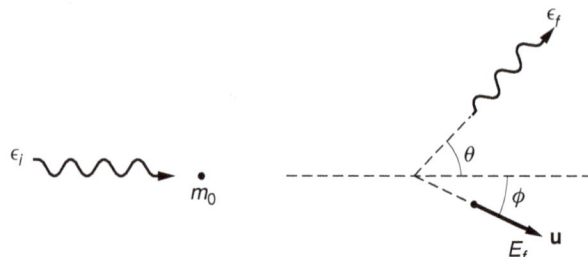

E_0

V m_0

13.9 *Photon–electron collision**

A photon of energy E_0 and wavelength λ_0 collides head-on with a free electron of rest mass m_0 and speed V, as shown. The photon is scattered at 90°.

(*a*) Find the energy E of the scattered photon.

(*b*) The outer electrons in a carbon atom move with speed $v/c \approx 6 \times 10^{-3}$. Using the result of part (*a*), estimate the broadening in wavelength of the Compton scattered peak from graphite for $\lambda_0 = 0.711 \times 10^{-10}$ m and 90° scattering. The rest mass of an electron is 0.51 MeV and $h/(m_0c) = 2.426 \times 10^{-12}$ m. Neglect the binding of the electrons. Compare your result with Compton's data shown in Example 13.6.

13.10 *The force of sunlight*

The solar constant, the average energy per unit area from the Sun falling on the Earth, is 1.4×10^3 W/m^2.

(*a*) How does the total force of sunlight compare with the Sun's gravitational force on the Earth?

(*b*) Sufficiently small particles can be ejected from the solar system by the radiation pressure of sunlight. Assuming a specific gravity of 5, what is the radius of the largest particle that can be ejected?

13.11 *Levitation by laser light*

A 1-kW light beam from a laser is used to levitate a solid aluminum sphere by focusing it on the sphere from below. What is the diameter of the sphere, assuming that it floats freely in the light beam? The density of aluminum is 2.7 g/cm^3.

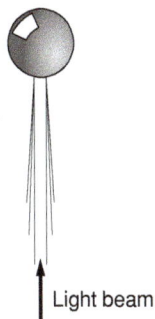

13.12 *Final velocity of a scattered particle*

A photon of energy $\epsilon_i = h\nu$ scatters from a free particle at rest of mass m_0. The photon is scattered at angle θ with energy $\epsilon_f = h\nu'$, and the particle flies off at angle ϕ.

Find an expression for the final velocity **u** of the particle.

Light beam

14 SPACETIME PHYSICS

14.1 Introduction

In 1908, three years after Einstein published the special theory of relativity, the mathematician Hermann Minkowski presented a geometrical formulation of Einstein's ideas based on the concept of a four-dimensional manifold that he called "spacetime." Minkowski famously asserted "Henceforth space by itself, and time by itself, are doomed to fade away into mere shadows, and only a kind of union of the two will preserve an independent reality." His claim may be a little exaggerated—we continue to move freely in a three-dimensional world while being swept forward relentlessly in time—but his point of view has been invaluable in extending the concepts of relativity to other areas of physics.

Special relativity provides an orderly procedure for relating the coordinates of events recorded by observers in different inertial systems. The essence of the theory is embodied by the Lorentz transformation. To set the stage for Minkowski's spacetime description of this transformation, let's briefly review how vectors transform in Newtonian physics.

14.2 Vector Transformations

We are interested here in the transformation properties of vectors, for instance some vector $\mathbf{A}$ that could represent a physical quantity such as force or velocity, or simply be an abstract mathematical quantity. To describe $\mathbf{A}$ in component form we introduce an orthogonal coordinate system S with coordinates (x, y, z) and unit base vectors $(\hat{\mathbf{i}}, \hat{\mathbf{j}}, \hat{\mathbf{k}})$. $\mathbf{A}$ can then be written

$$\mathbf{A} = A_x\hat{\mathbf{i}} + A_y\hat{\mathbf{j}} + A_z\hat{\mathbf{k}}. \tag{14.1}$$

The coordinate system is not fundamental but merely a construct that we introduce for convenience. We could use some other orthogonal coordinate system S' with coordinates (x', y', z') and base vectors $(\hat{\mathbf{i}}', \hat{\mathbf{j}}', \hat{\mathbf{k}}')$. If the two systems have the same origin, they must be related by a rotation. In the primed system,

$$\mathbf{A} = A'_x\hat{\mathbf{i}}' + A'_y\hat{\mathbf{j}}' + A'_z\hat{\mathbf{k}}'. \tag{14.2}$$

Because Eqs. (14.1) and (14.2) describe the same vector, we have

$$A'_x\hat{\mathbf{i}}' + A'_y\hat{\mathbf{j}}' + A'_z\hat{\mathbf{k}}' = A_x\hat{\mathbf{i}} + A_y\hat{\mathbf{j}} + A_z\hat{\mathbf{k}}. \tag{14.3}$$

To find the coordinates in S' given the coordinates in S, take the dot product of both sides of Eq. (14.3) with the corresponding unit vector:

$$A'_x = \mathbf{A} \cdot \hat{\mathbf{i}}' = A_x(\hat{\mathbf{i}} \cdot \hat{\mathbf{i}}') + A_y(\hat{\mathbf{j}} \cdot \hat{\mathbf{i}}') + A_z(\hat{\mathbf{k}} \cdot \hat{\mathbf{i}}') \tag{14.4a}$$

$$A'_y = \mathbf{A} \cdot \hat{\mathbf{j}}' = A_x(\hat{\mathbf{i}} \cdot \hat{\mathbf{j}}') + A_y(\hat{\mathbf{j}} \cdot \hat{\mathbf{j}}') + A_z(\hat{\mathbf{k}} \cdot \hat{\mathbf{j}}') \tag{14.4b}$$

$$A'_z = \mathbf{A} \cdot \hat{\mathbf{k}}' = A_x(\hat{\mathbf{i}} \cdot \hat{\mathbf{k}}') + A_y(\hat{\mathbf{j}} \cdot \hat{\mathbf{k}}') + A_z(\hat{\mathbf{k}} \cdot \hat{\mathbf{k}}'). \tag{14.4c}$$

The coefficients $(\hat{\mathbf{i}} \cdot \hat{\mathbf{i}}'), (\hat{\mathbf{j}} \cdot \hat{\mathbf{j}}')$, etc. are numbers that can be calculated for any given rotation. For instance, for rotation by θ around the z axis,

$$A'_x = A_x \cos\theta + A_x \sin\theta \tag{14.5a}$$

$$A'_y = -A_x \sin\theta + A_y \cos\theta \tag{14.5b}$$

$$A'_z = A_z. \tag{14.5c}$$

As an example, if we let $\mathbf{A}$ be the position vector $\mathbf{r}$ for a point that has coordinates (x, y, z) in system S and (x', y', z') in system S' rotated by angle θ around the z axis, then we have

$$x' = x\cos\theta + y\sin\theta \tag{14.6a}$$

$$y' = -x\sin\theta + y\cos\theta \tag{14.6b}$$

$$z' = z. \tag{14.6c}$$

Note that the x' and y' axes are both rotated in the same direction from the respective x and y axes. This is a trivial observation for rotations in three-dimensional space, but we will soon see that rotations in spacetime behave quite differently.

The transformation from S' back to S, by rotating the axes through angle $-\theta$, known as the *inverse transformation*, is

$$x = x'\cos\theta - y'\sin\theta \tag{14.7a}$$

$$y = x'\sin\theta + y'\cos\theta \tag{14.7b}$$

$$z = z'. \tag{14.7c}$$

Rotating axes through angle θ has the same effect on a vector's components as keeping the axes fixed and rotating the vector through angle $-\theta$. However, in this chapter we will always keep the vector fixed and rotate the axes.

14.2.1 Invariants and Scalars

Quantities that remain constant when a coordinate system changes are called *invariants*. Clearly, the components of a vector are not invariants but the vector itself is. Another invariant is the length of the vector $A = |\mathbf{A}|$ defined by

$$A = \sqrt{A_x^2 + A_y^2 + A_z^2} = \sqrt{A_x'^2 + A_y'^2 + A_z'^2}.$$

Quantities that do not change with a change in coordinate systems are called *scalars*. Simple numbers such as mass, temperature, and Avogadro's number are scalars. The lengths of vectors are also scalars. Many physical quantities are scalars, for instance mass, time intervals, and speed (as contrasted to velocity). Such scalars play important roles in Newtonian physics and they continue to play important roles in relativity, though their interpretation is different.

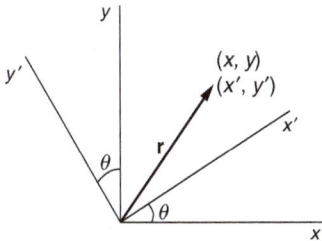

14.3 World Lines in Spacetime

To start our exploration of the spacetime world let's see how events evolve on a spacetime plot.

Motion in three-space is unrestricted. As time evolves, the point $\mathbf{r} = (x, y, z)$ can be anywhere. All of space is accessible to an observer, and all time in the future is accessible to observation, given enough time. Time, of course, moves in one direction only and is out of our control.

In spacetime physics, an event is a physical happening that can be specified by values for three spatial coordinates plus the time, (x, y, z, t). The emission of a pulse of light from the origin at $t = 0$ is for example an event with coordinates (0,0,0,0). Because coordinates with different physical dimensions are awkward to use, we will write time in units of length by multiplying it by a velocity. For this we naturally multiply time by the velocity of light so that $t \to ct$. Hence 1 μs corresponds to ≈ "300 m of time" and "1 m of time" corresponds to ≈ 0.0033 μs. With this convention we can speak of a four-dimensional spacetime with coordinates (x, y, z, ct).

Focusing on linear motion with coordinates x and ct, we can plot the evolution of an event on a spacetime diagram. By an awkward but universally accepted convention, time in spacetime diagrams is plotted vertically. As time evolves, a point in spacetime traces an upward path called a *world line*. The sketch shows two world lines: the vertical dashed line is the world line for a particle at rest (x constant), and the solid line is for a moving particle (x and t both increasing).

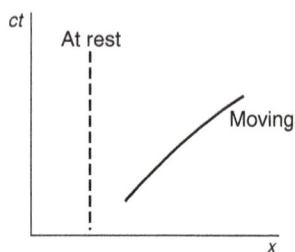

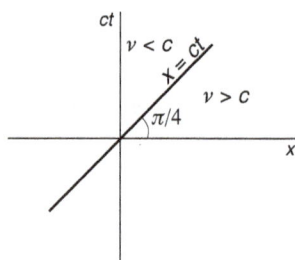

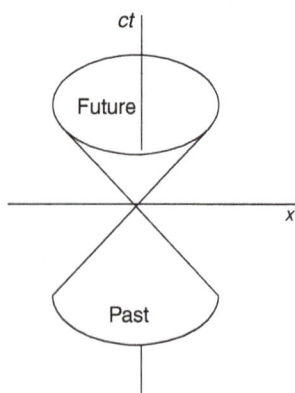

Note that with this convention, speed—in units of c—is given by the cotangent of the slope. A slope of ± 1 corresponds to a speed of $\pm c$. The fastest event is a pulse of light whose world line is given by $x = ct$ or $x = -ct$. This line is at an angle $\pi/4$ with respect to the horizontal, as shown. A world line at an angle less than $\pi/4$ would describe motion faster than light, which is prohibited. A three-dimensional plot that shows (x, y, ct) is in the form of two cones with their apexes at the origin, called *light cones*.

All future events lie in the upper light cone, past events in the lower light cone. The region of spacetime outside the light cones is physically inaccessible to an observer at the origin. Other events would be described by other light cones but for two events to be causally related their light cones must overlap.

We now turn to the question of how spacetime events appear to observers in different inertial systems. In Chapter 12 we derived the Lorentz transformation for changing coordinates between our standard inertial systems S and S'. The origin of S' moves at speed v along the x axis. Alternatively, the origin of S moves with speed $-v$ along the x' axis. Because time is now expressed in units of length ct, it is natural to express the relative velocity of the coordinate systems by the variable $\beta = v/c$. With this notation, the Lorentz transformation from S to S', Eqs. (12.3)

and (12.4), is

$$x' = \gamma(x - \beta ct) \tag{14.8a}$$
$$y' = y \tag{14.8b}$$
$$z' = z \tag{14.8c}$$
$$ct' = \gamma(-\beta x + ct). \tag{14.8d}$$

The reverse transformation from S' to S is given by

$$x = \gamma(x' + \beta ct') \tag{14.9a}$$
$$y = y' \tag{14.9b}$$
$$z = z' \tag{14.9c}$$
$$ct = \gamma(\beta x' + ct') \tag{14.9d}$$

where $\beta = v/c$ and $\gamma = 1/\sqrt{1 - \beta^2}$. The quantity β is always taken to be $\geq$ 0, with algebraic signs shown explicitly. Because the y and z coordinates are unchanged by the Lorentz transformation, we will concern ourselves only with x and ct.

There is a parallel between the rotation of coordinates in three-space and transformation in spacetime specified by Eqs. (14.8) and (14.9).

Rotations in three-space	Lorentz transformation
$x' = x\cos\theta + y\sin\theta$	$x' = \gamma(x - \beta ct)$
$y' = -x\sin\theta + y\cos\theta$	$ct' = \gamma(-\beta x + ct)$
$z' = z$	$y' = y$
$t' = t$	$z' = z$

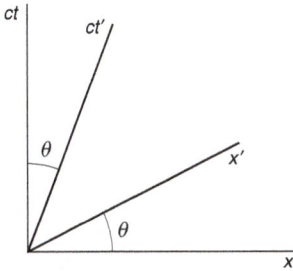

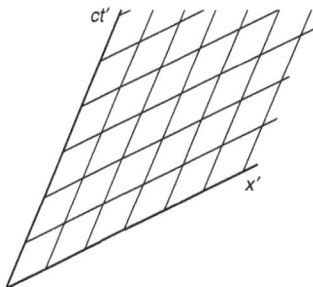

The Lorentz transformation is similar in form to the transformation of x and y due to a rotation around the z axis. Equation (14.8d) shows that the locus of the x' axis (for which $ct' = 0$), when plotted in the x-ct plane, is given by $ct = \beta x$. This describes a *counterclockwise* rotation of the x' axis from the x axis through angle $\theta = \arctan\beta$. In contrast, Eq. (14.8a) reveals that the ct' axis (for which $x' = 0$) is given by $x = \beta ct$, which describes a *clockwise* rotation of the ct' axis from the ct axis, through the same angle. In the limiting case $v = c$, $\theta = \pi/4$; the x' and ct' axes become coincident because $x' = ct'$.

According to these transformations, the axes are no longer orthogonal. The lines of constant x' and constant ct' in the figure form a grid of diamonds rather than squares. Furthermore, the scale of the coordinate axes is changed by the factor of γ. These are fundamental differences between the geometries of three-space and spacetime.

The loss of orthogonality is a characteristic feature of transformations in spacetime. Although we can describe time as a fourth dimension, time is fundamentally different from the spatial dimensions, and that difference is crucial to the geometry of spacetime. The length-scales in a spacetime diagram differ by a factor of γ and the appearance of a world

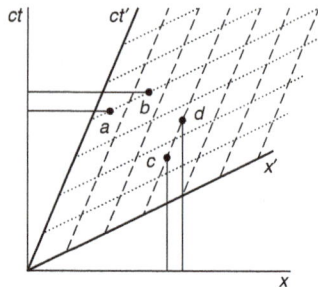

line depends on the relative motion of the observer. Consequently, every observer would describe events with a different diagram, so that care is needed in extracting geometrical relationships from a diagram.

The drawings show some events in spacetime in the systems $S(x, ct)$ and $S'(x', ct')$. In S', lines of constant time are dashed and lines of constant position are dotted. Notice that events a and b are coincident in S' but occur at different times in S. Similarly, events c and d occur at the same location in S' but at different locations in S.

14.4 An Invariant in Spacetime

In three-space, the length r of the position vector is $r = \sqrt{x^2 + y^2 + z^2}$, and is invariant under rotation. In spacetime, the quantity $x^2 + y^2 + z^2 - (ct)^2$ is invariant under the Lorentz transformation. To prove this, we have from Eqs. (14.8)

$$x'^2 + y'^2 + z'^2 - (ct')^2 = \gamma^2(x - \beta ct)^2 + \gamma^2(-\beta x + ct)^2 + y^2 + z^2$$

$$= \frac{1}{1 - \beta^2}\left[x^2(1 - \beta^2) - (ct)^2(1 - \beta^2)\right] + y^2 + z^2$$

$$= x^2 + y^2 + z^2 - (ct)^2.$$

Hence

$$x'^2 + y'^2 + z'^2 - (ct')^2 = x^2 + y^2 + z^2 - (ct)^2. \tag{14.10}$$

Consider a world line between an event starting at $\mathbf{R_1} = (\mathbf{r}, ct)$ and ending at $\mathbf{R_2} = (\mathbf{r} + \Delta\mathbf{r}, ct + \Delta ct)$. The displacement between the events is $\Delta\mathbf{R} = \mathbf{R_2} - \mathbf{R_1} = (\Delta\mathbf{r}, \Delta ct)$, where $\Delta\mathbf{r} = (\Delta x, \Delta y, \Delta z)$. The coordinates for these displacements when viewed by an observer in S' are $\Delta\mathbf{r}'$ and $\Delta t'$, which can be found using the Lorentz transformation. From Eq. (14.10) we have

$$\Delta r'^2 - \Delta(ct')^2 = \Delta r^2 - \Delta(ct)^2.$$

Note that by convention in relativity, Δx^2 is interpreted as $(\Delta x)^2$, not $\Delta(x^2)$. Because the Lorentz transformation depends on the relative velocity v/c, and because the relative velocity can be chosen arbitrarily, the only way that this equation can be satisfied is for the two sides to separately equal a constant. Denoting this constant by Δs^2, we have

$$\Delta s^2 \equiv \Delta r^2 - \Delta(ct)^2. \tag{14.11}$$

Consequently, Δs^2 is an invariant of the transformation. Δs^2 is often called the *separation* of the spacetime interval. Unlike the square of an ordinary number, $(\Delta s)^2$ can be negative.

14.4.1 Spacelike and Timelike Intervals

If $\Delta s^2 > 0$, we can always find a coordinate system that satisfies $\Delta r^2 = \Delta s^2$, with $\Delta(ct)^2 = 0$. Consequently, there is a coordinate system in which the events are simultaneous, but there is no frame in which the events are coincident in space. Such an interval is called *spacelike*. Similarly, if $\Delta s^2 < 0$, there is a coordinate system in which the events can coincide in space, but there is no frame in which they occur simultaneously. Such an interval is called *timelike*. If $\Delta s^2 = 0$, the events correspond to the emission and reception of a pulse of light. The world line obeys $\Delta r = \Delta ct$ and the separation is called *null*. A separation has the same value in all inertial systems.

For a given spacetime interval, the geometry of possible displacements looks entirely different from the geometry of a three-space interval, where the displacements of a given magnitude lie on a sphere. In two dimensions, the locus of possible displacement is a circle, $\Delta r^2 = \Delta x^2 + \Delta y^2$. In spacetime, the locus of an interval for the (x, y, ct) coordinates is a hyperbolic surface of revolution.

$$x^2 + y^2 - (ct)^2 = \pm |\Delta s^2|. \tag{14.12}$$

14.5 Four-Vectors

We developed the Lorentz transformation to satisfy the need for a spacetime transformation that preserves the speed of light as a universal constant. The reasoning behind the Lorentz transformation is more powerful than this application might suggest. The ideas can be extended to develop relativistic dynamics in a coherent fashion by general arguments that can apply to new systems, for example to the dynamics of electric and magnetic fields, though we will not pursue that development here. The starting point for this generalization is the concept of a spacetime vector, usually referred to as a *four-vector*.

A four-vector is a set of four numbers that transform according to the Lorentz transformation. Such vectors have the potential to be physically significant. Put another way, any set of four numbers that do not obey the Lorentz transformation cannot enter into a physical law because the law could not satisfy the principle of relativity. A good way to extend our understanding of physics in the relativistic world is by searching for four-vectors. A four-vector $\mathbf{A}$ has the general form

$$\mathbf{A} = (a_1, a_2, a_3, a_4)$$

where the components (a_1, a_2, a_3) are along the axes of the standard system $S = (x, y, z)$. Under a Lorentz transformation to a coordinate system S' moving in the x direction with velocity v, $\mathbf{A}_4 = (a'_1, a'_2, a'_3, a'_4)$, where

$$a'_1 = \gamma(a_1 - \beta a_4) \tag{14.13a}$$
$$a'_2 = a_2 \tag{14.13b}$$
$$a'_3 = a_3 \tag{14.13c}$$
$$a'_4 = \gamma(-\beta a_1 + a_4). \tag{14.13d}$$

As previously defined, $\beta = v/c, \gamma = 1/\sqrt{1-\beta^2}$. We shall use the convention that four-vectors are printed in upper case bold, while three-vectors are in lower case bold. Thus, the four-vector for four-position $\mathbf{R}$ can be written $\mathbf{R} = (\mathbf{r}, ct) = (x, y, z, ct)$.

14.5.1 Lorentz Invariants

The sum of the squares of the components of a four-vector, with a negative sign for the square of the time component, is called the *norm* of the four-vector. For example, the norm of $\mathbf{A} = (a_1, a_2, a_3, a_4)$ is $a_1{}^2 + a_2{}^2 + a_3{}^2 - a_4{}^2$.

A true four-vector transforms according to the Lorentz transformation, as in Eqs. (14.13). It is easy to show that the norm is the same in any frame, and the norm is therefore called a *Lorentz invariant*.

$$a_1'^2 + a_2'^2 + a_3'^2 - a_4'^2 = a_1{}^2 + a_2{}^2 + a_3{}^2 - a_4{}^2.$$

We showed this invariance for the four-vector $\mathbf{R}$ in Section 14.4.

14.5.2 Four-Velocity

The simplest kinematical quantity after position is velocity, and so it is natural to consider the rate of change of $\mathbf{R}$. However, the concept of *rate* requires the introduction of a clock, and the question then is whose clock? The only time on which all observers could agree is the proper time τ of a clock attached to the moving point. Of course, this would differ from the observer's time, but by a known amount.

Consequently, we can tentatively define the *four-velocity* $\mathbf{U}$ by

$$\mathbf{U} \equiv \frac{d\mathbf{R}}{d\tau} = \left(\frac{d\mathbf{r}}{d\tau}, \frac{d(ct)}{d\tau} \right). \tag{14.14}$$

To connect the spatial part with the familiar three-velocity of a Newtonian observer, we can rewrite this as

$$\mathbf{U} = \left(\frac{d\mathbf{r}}{dt} \frac{dt}{d\tau}, \frac{d(ct)}{d\tau} \right).$$

From the discussion of proper time in Section 12.8.3, the relation between a local time interval and the proper time interval is

$$d\tau = dt \sqrt{1 - v^2/c^2} = dt/\gamma. \tag{14.15}$$

Hence $dt/d\tau = \gamma$ and $\mathbf{U} = (\gamma\mathbf{u}, \gamma c)$, or

$$\mathbf{U} = \gamma(\mathbf{u}, c). \tag{14.16}$$

Consider the norm of the four-velocity $\mathbf{U} = \gamma(\mathbf{u}, c)$:

$$\mathbf{U} \cdot \mathbf{U} = \gamma^2(u^2 - c^2)$$
$$= -c^2.$$

The norm of $\mathbf{U} = -c^2$ is obviously a Lorentz invariant, the same in every reference frame.

Example 14.1 Relativistic Addition of Velocities

The relativistic law for the addition of velocities was derived in Section 12.9. Here we look at the same problem using the four-velocity approach.

Consider for simplicity velocity along the x axis. If the initial speed is u_1 along x, then we can take the initial four-velocity to be $\mathbf{U} = \gamma(u_1, c)$. If we observe an event with this velocity in system S_1 moving with speed v_1, then the Lorentz transformation Eqs. (14.8) shows that the components of the four-velocity are

$$u_1' = \gamma_1(u_1 - \beta_1 c) \tag{1a}$$

$$u_4' = \gamma_1(-\beta_1 u_1 + c), \tag{1b}$$

where $\beta_1 = v_1/c$ and $\gamma_1 = 1/\sqrt{1 - \beta_1^2}$.

If we now move into system S_2 moving with speed v_2 relative to S_1, we have

$$u_1'' = \gamma_1(u_1' - \beta_2 u_4') \tag{2a}$$

$$u_4'' = \gamma_1(-\beta_2 u_1' + u_4'), \tag{2b}$$

where $\beta_2 = v_2/c$ and $\gamma_2 = 1/\sqrt{1 - \beta_2^2}$. These two successive transformations are equivalent to a single transformation found by substituting Eq. (1) in Eq. (2). The result, after some rearranging, is

$$u_1'' = \frac{\gamma_2 \gamma_2}{1 + \beta_1 \beta_2}\left(u_1 - \frac{\beta_1 + \beta_2}{1 + \beta_1 \beta_2}c\right) \tag{3a}$$

$$u_4'' = \frac{\gamma_2 \gamma_2}{1 + \beta_1 \beta_2}\left(-\frac{\beta_1 + \beta_2}{1 + \beta_1 \beta_2}u_1 + u_4\right). \tag{3b}$$

Comparing Eq. (3) with Eq. (1), we see that the form is equivalent to moving into a coordinate system with speed β_3 given by

$$\beta_3 = \frac{\beta_1 + \beta_2}{1 + \beta_1 \beta_2}$$

or, in laboratory units,

$$v_3 = \frac{v_1 + v_2}{1 + v_1 v_2/c^2}.$$

This is the rule for the relativistic addition of velocities that we found in Section 12.9.

One might reasonably ask what purpose is served by introducing the four-velocity $\mathbf{U} = \gamma(\mathbf{v}, c)$, because the timelike component is simply a constant, c. But as we shall see in the next section, the constant plays an important role.

14.6 The Energy–Momentum Four-Vector

The natural next step is to define the four-momentum $\mathbf{P}$ of a particle of mass m moving with four-velocity $\mathbf{U}$. Using Eq. (14.16), we have

$$\mathbf{P} = m\mathbf{U} = \gamma m_0(\mathbf{u}, c). \tag{14.17}$$

Here m_0 is the mass observed when $u \ll c$ and $\gamma = 1/\sqrt{1 - u^2/c^2} \approx 1$. In Chapter 13 we employed the term *rest mass* for m_0. This quantity would more logically be called the *proper mass* since it is the mass measured in the particle's rest frame but because of historical circumstance the term *proper mass* is not used. In relativity, the simple term *mass* always refers to the *relativistic mass m* defined by

$$m \equiv \gamma m_0. \tag{14.18}$$

Because $\mathbf{U}$ is a four-vector and m is a scalar, $\mathbf{P} = m\mathbf{U}$ is also a four-vector. The four-momentum can be written

$$\mathbf{P} = (m\mathbf{u}, mc) = (\mathbf{p}, mc), \tag{14.19}$$

where $\mathbf{p} = m\mathbf{u}$ is the relativistic three-momentum. The norm of $\mathbf{P}$ is

$$\mathbf{P} \cdot \mathbf{P} = \mathbf{p} \cdot \mathbf{p} - (mc)^2 = C \tag{14.20}$$

where C is a constant. In the particle's rest frame $\mathbf{p} = 0$, $C = -(m_0 c)^2$. Thus the norm of the four-momentum is $-(m_0 c)^2$ and we can rewrite Eq. (14.20) as

$$p^2 = (mc)^2 - (m_0 c)^2. \tag{14.21}$$

To be useful, four-momentum must be conserved in an isolated system. Three-momentum vanishes in a system in which a particle is at rest. Consequently, in Eq. (14.20) $\mathbf{p}$ and mc must be separately conserved. To make connections with Newtonian concepts, recall that in Newtonian physics, in addition to the linear momentum of an isolated system, energy and angular momentum are also conserved. Four-momentum has no evident connection with angular momentum so let us guess that mc, the temporal component of the four-vector in Eq. (14.19), is related to the energy E. To be dimensionally correct, we tentatively set

$$mc = E/c \tag{14.22}$$

so that $\mathbf{P}$ takes the form

$$\mathbf{P} = (\mathbf{p}, E/c). \tag{14.23}$$

$\mathbf{P}$ is sometimes called the *energy-momentum four-vector*, but more often it is referred to simply as the *four-momentum*. Introducing Eq. (14.22) into Eq. (14.21), we have

$$(pc)^2 = E^2 - (m_0 c^2)^2 \tag{14.24}$$

or

$$E^2 = (pc)^2 + (m_0 c^2)^2, \tag{14.25}$$

a result we found in Chapter 13, Eq. (13.8). The term m_0c^2 is known as the *rest energy* of the particle.

For a massless particle, $m_0 = 0$, so $pc = E$ or $p = E/c$, a result we derived by different means in Chapter 13, Eq. (13.13). For massless particles (notably photons), the norm of the four-momentum is 0 and the particle must move at the speed of light. The four-momentum for a massless particle has the form

$$\mathbf{P} = \frac{E}{c}(n_x, n_y, n_z, 1), \tag{14.26}$$

where $\hat{n}$ is a unit vector in the direction of propagation.

For a particle of non-zero mass moving at low velocity

$$m = \gamma m_0 = \left(1 + \frac{1}{2}\frac{v^2}{c^2} + \cdots\right).$$

Consequently

$$E = m_0c^2 + \frac{1}{2}m_0v^2 + \cdots = m_0c^2 + K + \cdots \tag{14.27}$$

where K is the Newtonian kinetic energy. In the low velocity approximation, the total energy of a free particle is the sum of its kinetic energy and its rest energy. Whether this interpretation is reasonable depends on the experimental evidence.

From Eq. (14.25), we have

$$E^2 - (pc)^2 = (m_0c^2)^2. \tag{14.28}$$

The rest energy of a particle is evidently the relativistic invariant of the four-momentum.

Postulating $mc = E/c$ led us directly to $E = mc^2$, the most famous formula in science. Of course, the validity of this formula does not depend on its fame but rather on the role energy plays in relativistic dynamics, a role that can be made meaningful only by experience.

In Chapter 13 we derived expressions for relativistic energy and momentum by applying the Lorentz transformation to observations of collisions in an isolated system. Those arguments rested on our understanding of collision processes and our intuition about the symmetry of views by different observers. In this chapter we found the same results by totally different reasoning—by examining the transformation properties of four-vectors. This argument is mathematically elegant and also physically elegant, for it reveals a deep connection between energy and momentum. Understanding phenomena from totally different lines of argument adds to our confidence in the truth of the explanation and cannot fail to deepen one's pleasure in physics.

14.7 Epilogue: General Relativity

Einstein was dissatisfied with his 1905 paper on special relativity because it could not deal with gravity. For example, according to Newton's law of universal gravitation, changes in a gravitational field, due, for instance, to motion of the source mass, are felt instantaneously everywhere. According to special relativity, the effect could not propagate faster than the speed of light.

There is, however, a more profound dilemma. Newtonian gravity rests on the assumption that gravitational mass and inertial mass are identical. Einstein used this simple observation to motivate his new theory. He explained his line of reasoning with one of his famous "gedanken" (thought) experiments: the Einstein elevator. Because all things fall at the same rate, an observer in a stationary elevator in a gravitational field sees dropped objects accelerate downward, and would make the same observation if the elevator were accelerating upward in the absence of a gravitational field. Einstein pointed out that if the elevator were in space far from other bodies, then except for looking outside, there was no way to tell whether the elevator was accelerating up at rate g or at rest in a gravitational field g. He concluded that in a local region, there is no way to distinguish a downward gravitational field g from an upward acceleration of the coordinate system a. This observation is the principle of equivalence.

The principle of equivalence poses an immediate obstacle to the special theory of relativity. If a downward gravitational field is equivalent to an upward acceleration, then motion in a gravitational field is indistinguishable from motion in an accelerating coordinate system. Accelerating systems are inherently non-inertial. The crux of the dilemma is that the special theory of relativity is restricted to observations in inertial systems.

The special theory of relativity was published when Einstein was young and he presented it in a single paper that was comprehensible to all. In contrast, Einstein labored more than ten years to solve the problem of gravity, making a number of false starts. When he published his general theory of relativity in 1916 the paper was so complex that few readers could follow it. Furthermore, the only experimental confirmation was his theory's ability to account for the discrepancy of 43 arc-second/century in the precession of the perihelion of Mercury that we mentioned in Section 10.6.

Such a result might seem like a minor detail of planetary dynamics. However, general relativity also made a dramatic prediction: light is deflected by gravity. In particular, the path of light from a star would be slightly bent as it passed near our massive Sun. The deflection would be so small that it could be observed only during a solar eclipse when the Sun's brightness was blotted out for a short time, allowing stars in the part of the sky near the Sun to be seen. Observations were delayed by World War I but in 1919 two eclipse expeditions observed the effect. The theory of general relativity made the front page of newspapers, and the Einstein legend was born.

In 1919 cosmology was a topic for speculation but it was not part of science. Today we are in a golden age of cosmology and Einstein's general theory of relativity is its foundational paradigm. The expansion of the universe is described by Einstein's field equations and the signature of his theory is to be found everywhere. For example, he predicted gravitational lensing—the focusing of light from a distance source that passes around a star or through a gravitating medium such as a galaxy. Today, gravitational lensing is a well-established tool in astrophysics.

Perhaps the most dramatic experimental prediction of general relativity is the existence of black holes. If the bending of light by gravity is large enough, the light cannot propagate. This occurs when the energy required to overcome the gravitational attraction of a mass, GMm/r, exceeds the rest mass mc^2. Equating these gives the radius of the black hole as $R = GM/c^2$. The radius is typically only a few km, but the density is so great that the mass is huge. We saw in Example 10.9 that the mass of the black hole Sgr A* at the center of our galaxy is 4 million times the mass of the Sun. Not even light can escape from such an intense gravitational field—hence the name *black hole*. Fortunately, material falling inward radiates intensely and many black holes have been identified. Measured data for the orbit of a star around Sgr A* appear in Example 10.9.

Einstein launched two lines of progress in 1905: his photoelectric effect paper was seminal to the creation of quantum mechanics, and his paper on special relativity inspired general relativity. Quantum theory and gravitational theory are triumphs of twentieth century physics that changed our world view. These separate theories have yet to be reconciled: a quantum theory of gravity has yet to be formulated. Physics is never finished.

Problems

For problems marked *, refer to page 525 for a hint, clue, or answer.

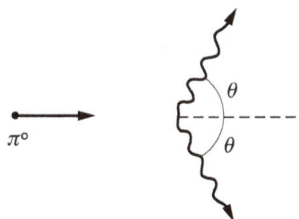

14.1 *Pi meson decay**

A neutral pi meson (π^0), rest mass 135 MeV, decays symmetrically into two photons while moving at high speed. The energy of each photon in the laboratory system is 100 MeV.

(*a*) Find the meson's speed V as a ratio V/c.

(*b*) Find the angle θ in the laboratory system between the momentum of each photon and the initial line of motion.

14.2 *Threshold for pi meson production**

A high energy photon (γ-ray) collides with a proton at rest. A neutral pi meson (π^0) is produced according to the reaction $\gamma + p \rightarrow p + \pi^0$.

What is the minimum energy the γ-ray must have for this reaction to occur? The rest mass of a proton is 938 MeV, and the rest mass of a π^0 is 135 MeV.

14.3 *Threshold for pair production by a photon*
A high energy photon (γ-ray) collides with an electron and produces an electron–positron pair according to the reaction $\gamma + e^- \to e^- + (e^- + e^+)$.

What is the minimum energy the γ-ray must have for the reaction to occur?

14.4 *Particle decay*
A particle of rest mass M spontaneously decays from rest into two particles with rest masses m_1 and m_2.

Show that the energies of the particles are $E_1 = (M^2 + m_1^2 - m_2^2)c^2/2M$ and $E_2 = (M^2 - m_1^2 + m_2^2)c^2/2M$.

14.5 *Threshold for nuclear reaction**
A nucleus of rest mass M_1 moving at high speed with kinetic energy K_1 collides with a nucleus of rest mass M_2 at rest. A nuclear reaction occurs according to the scheme $M_1 + M_2 \to M_3 + M_4$ where M_3 and M_4 are the rest masses of the product nuclei.

The rest masses are related by $(M_3 + M_4)c^2 = (M_1 + M_2)c^2 + Q$, where $Q > 0$. Find the minimum value of K_1 required to make the reaction occur, in terms of M_1, M_2, and Q.

14.6 *Photon-propelled rocket*
A rocket of initial mass M_0 starts from rest and propels itself forward along the x axis by emitting photons backward.

(*a*) Show that the four-momentum of the rocket's exhaust in the initial rest system can be written $\mathbf{P} = \gamma M_f v(-1, 0, 0, i)$, where M_f is the final mass of the rocket. (Note that this result is valid for the exhaust as a whole even though the photons are Doppler-shifted.)

(*b*) Show that the final velocity of the rocket relative to the initial frame is

$$v = \frac{\mu^2 - 1}{\mu^2 + 1}c,$$

where $\mu = M_0/M_f$ is the ratio of the rocket's initial mass to its final mass.

14.7 *Four-acceleration**
Construct a four-vector $\mathbf{A}$ representing acceleration. For simplicity, consider only straight line motion along the x axis. Let the instantaneous four-velocity be $\mathbf{U} = \gamma(u, 0, 0, c)$.

14.8 *A wave in spacetime*
The function $f(x, t) = A \sin 2\pi[(x/\lambda) - vt]$ represents a sine wave of frequency v and wavelength λ. The wave propagates along the x axis with velocity = wavelength × frequency = λv.

$f(x, t)$ can represent a light wave; A then corresponds to some component of the electromagnetic field that constitutes the light signal, and the wavelength and frequency satisfy $\lambda v = c$.

Consider the same wave in the coordinate system (x', y', z', t') moving along the x axis at velocity v. In this reference frame the wave has the form

$$f'(x', t') = A' \sin 2\pi \left(\frac{x'}{\lambda'} - v't' \right).$$

(a) Show that the velocity of light is correctly given provided that $1/\lambda'$ and v' are components of a four-vector $\mathbf{K}$ given in the (x, y, z, t) system by

$$\mathbf{K} = 2\pi \left(\frac{1}{\lambda}, 0, 0, \frac{v}{c} \right).$$

(b) Using the result of part (a), derive the result for the longitudinal Doppler shift by evaluating the frequency in a moving system.

(c) Extend the analysis of part (b) to find the expression for the transverse Doppler shift, by considering a wave propagating along the y axis.

HINTS, CLUES, AND ANSWERS TO SELECTED PROBLEMS

Chapter 1

1.1 *Vector algebra 1*
Ans. (a) $7\hat{\mathbf{i}} - 2\hat{\mathbf{j}} + 9\hat{\mathbf{k}}$; *(c)* 21

1.2 *Vector algebra 2*
Ans. (b) 101; *(c)* 2704

1.3 *Cosine and sine by vector algebra*
Hint: $\cos\theta = \mathbf{A} \cdot \mathbf{B}/AB$, $\sin\theta = |\mathbf{A} \times \mathbf{B}|/AB$.

1.7 *Law of sines*
Hint: Consider the area of a triangle formed by $\mathbf{A}, \mathbf{B}, \mathbf{C}$, where $\mathbf{A} + \mathbf{B} + \mathbf{C} = 0$.

1.9 *Perpendicular unit vector*
Ans. $\hat{\mathbf{n}} = \pm(2\hat{\mathbf{i}} - \hat{\mathbf{j}} + \hat{\mathbf{k}})/\sqrt{6}$

1.10 *Perpendicular unit vectors*
Hint: (a) The direction of a surface defined by two vectors is parallel to their cross product.

1.15 *Great circle*
Clue: If $\lambda_1 = 0$, $\phi_1 = 0$, $\lambda_2 = 45°$, $\phi_2 = 45°$, then $\theta = \arccos(0.5)$ and $S = (\pi/3)R$.

1.18 *Elevator and falling marble*
Clue: If $T_1 = T_2 = 4$ s, then $h = 39.2$ m.

1.19 *Relative velocity*
Ans. (a) $\mathbf{v_B} = \mathbf{v_A} - d\mathbf{R}/dt$

1.21 *Particle with constant radial velocity*
  Ans. (a) $v = \sqrt{52}$ m/s

1.23 *Smooth elevator ride*
  Clue: (d) If $a_m = 1.0$ m/s^2 and $T = 10.0$ s, then $D = 50.0$ m.

1.26 *Range on a hill*
  Hint: The rock hits the ground at the intersection of two curves.
  Clue: If $\phi = 60°$, then $\theta = 15°$.

1.27 *Peaked roof*
  Ans. $v = \sqrt{5/2}\,\sqrt{gh}$

Chapter 2

2.1 *Time-dependent force*
  Clue: (c) If $t = 1$ s, then $\mathbf{r} \times \mathbf{v} = 6.7 \times 10^{-3}\,\hat{\mathbf{k}}\,\mathrm{m}^2/\mathrm{s}$.

2.2 *Two blocks and string*
  Clue: If $M_1 = M_2$, then $x = gt^2/4$.

2.5 *Concrete mixer*
  Clue: If $R = 2$ ft, then $\omega_{\max} = 4\,\mathrm{rad/s} \approx 38\,\mathrm{revolutions/minute}$.

2.8 *Two masses and two pulleys*
  Clue: If $M_1 = M_2$, then $\ddot{x} = g/5$.

2.11 *Mass on wedge*
  Clue: If $A = 3g$, then $\ddot{y} = g$.

2.12 *Painter on scaffold*
  Clue: If $M = m$ and $F = Mg$, then $a = g$.

2.13 *Pedagogical machine*
  Clue: For equal masses, $F = 3Mg$.

2.14 *Pedagogical machine 2*
  Ans. $a_1 = -M_2 M_3 g/(M_1 M_2 + M_1 M_3 + 2M_2 M_3 + M_3{}^2)$

2.16 *Planck units*
  Hint: Write dimension equations of the form $[L_p] = [c]^a[h]^b[G]^c$.
  Replace each factor with its independent dimensions M, L, and T,
  and solve the three algebraic equations for a, b, c.
  Ans. (a) $L_p = 4.1 \times 10^{-35}$ m
  (b) $M_p = 5.4 \times 10^{-8}$ kg
  (c) $T_p = 1.3 \times 10^{-43}$ s

Chapter 3

3.2 *Sliding blocks with friction*
  Clue: If $F = 30$ N, $M_A = 5$ kg, $M_B = 6$ kg, then $F' = 25$ N.

3.4 *Synchronous orbit*
  Ans. $6.6R_e$

3.5 *Mass and axle*
  Clue: If $l\omega^2 = \sqrt{2}g$, then $T_{up} = \sqrt{2}mg$.

3.8 *Block and wedge*
  Ans. (a) $\tan\theta = \mu$
  Clue: (b) If $\theta = \pi/4$, then $a_{min} = g(1-\mu)/(1+\mu)$.
  Clue: (c) If $\theta = \pi/4$, then $a_{max} = g(1+\mu)/(1-\mu)$.

3.10 *Rope and trees*
  Clue: If $\theta = \pi/4$, $T_{end} = W/\sqrt{2}$, $T_{middle} = W/2$.

3.11 *Spinning loop*
  Ans. $T = M\omega^2 l/(2\pi)^2$

3.17 *Turning car*
  Clue: If $\mu = 1$ and $\theta = \pi/4$, all speeds are possible.

3.19 *Mass and springs*
  Clue: If $k_1 = k_2 = k$, then $\omega_a = \sqrt{k/2m}$, $\omega_b = \sqrt{2k/m}$.

3.22 *Mass, string, and ring*
  Clue: (a) If $Vt = r_0/2$, then $\omega = 4\omega_0$.

3.23 *Mass and ring*
  Ans. (a) $v_0/[1 + (\mu v_0 t/l)]$

3.24 *Retarding force*
  Ans. $v(t) = (1/\alpha)\ln[1/(\alpha bt/m + e^{-\alpha v_0})]$

Chapter 4

4.1 *Center of mass of a non-uniform rod*
  Ans. (a) $M = 2Al/\pi$
  (b) $X = l(1 - 2/\pi)$

4.11 *Freight car and hopper*
  Hint: There is a way to do this problem in one or two lines.
  Clue: If $M = 500$ kg, $b = 20$ kg/s, $F = 100$ N, then $v(10s) = 1.4$ m/s.

4.16 *Rope on table*
  Ans. (a) $x = Ae^{\gamma t} + Be^{-\gamma t}$, where $\gamma = \sqrt{g/l}$

4.20 *Reflected particle stream*
  Hint: The answer is *not* $\lambda(mv^2 + mv'^2)$.

4.23 *Suspended garbage can*
  Clue: If $\left(\frac{Mg}{2K}\right)^2 = v_0^2/2g$, $h = v_0^2/4g$.

Chapter 5

5.1 *Loop-the-loop*
  Ans. $z = 3R$

5.3 *Ballistic pendulum*
   Ans. (b) $v = [(m + M)/m] \sqrt{2gl(1 - \cos \phi)}$

5.4 *Sliding on a circular path*
   Clue: If $m = M$, then $v = \sqrt{gR}$.

5.6 *Block sliding on a sphere*
   Ans. $R/3$

5.7 *Beads on hanging ring*
   Clue: If $M = 0$, then $\theta = \arccos{(2/3)}$.

5.8 *Damped oscillation*
   Ans. (b) $n = \frac{k}{4f}(x_i - x_0)$

5.10 *Falling chain*
   Clue: The maximum reading is $3Mg$.

5.12 *Lennard-Jones potential*
   Hint: A power series expansion is useful. See Note 1.2.

5.16 *Snowmobile and hill*
   Ans. 45 mi/hr

5.17 *Leaper*
   Clue: > 1 hp, < 10 hp.

Chapter 6

6.4 *Bouncing ball*
   Clue: If $v_0 = 5$ m/s, $e = 0.5$, then $T \approx 1$ s.

6.13 *Nuclear reaction of α-rays with lithium*
   Ans. (a) neutron energy $= 0.15$ MeV

6.14 *Superball bouncing between walls*
   Ans. (a) $F = mv_0^2/l$
   Hint: (b) Find the average rate at which the ball's speed increases as the surface moves.
   Ans. (c) $F = (mv_0^2 l^2)/x^3$

6.16 *Converting between C and L systems*
   Ans. (a) $v_f = [v_0/(m + M)] \sqrt{m^2 + M^2 + 2mM \cos \Theta}$
   Clue: (b) If $m = M$, then $(K_0 - K_f)/K_0 = (1 - \cos \Theta)/2$.

Chapter 7

7.2 *Drum and sand*
   Clue: If $\lambda t = M_B$, $b = 2a$, then $\omega_B = \omega_A(0)/8$.

7.3 *Ring and bug*
   Clue: If $m = M$, then $\omega = v/3R$.

7.8 *Moment of inertia of a sphere*
    Ans. $I_0 = (2/5)MR^2$

7.10 *Cylinder in groove*
    Clue: If $\mu = 0.5$, $R = 0.1$ m, $W = 100$ N, then $\tau = 5.7$N $\cdot$ m.

7.11 *Wheel and shaft*
    Clue: If $F = 10$ N, $L = 5$ m, $\omega = 0.5$ rad/s, then $I_0 = 400$ kg $\cdot$ m^2.

7.14 *Stick on table*
    Ans. (c) $3g/4$
    Ans. (d) $Mg/4$

7.21 *Rolling cylinder*
    Ans. $\theta = \arctan(3\mu)$

7.23 *Disk, mass, and tape*
    Clue: (a) If $A = 2a$, then $\alpha = 3a/R$.

7.25 *Rolling marble*
    Clue: If $v_0 = 3$ m/s, $\theta = 30°$, then $l \approx 1.3$ m.

7.31 *Sliding and rolling cylinder*
    Clue: If $\omega = 3$ rad/s, then $\omega_f = 1$ rad/s.

7.34 *Marble in dish*
    Ans. $\omega = \sqrt{5g/7R}$

7.36 *Two twirling masses*
    Clue: If $m_A = m_B = 2$ kg, $v_0 = 3$ m/s, $l = 0.5$ m, then $T = 18$ N.

7.37 *Plank and ball*
    Clue: If $m = M$, then (a) $v_f = 3v_0/5$ and (b) $v_f = v_0/2$.

7.38 *Collision on a table*
    Ans. $\omega = (4\sqrt{2}/7)(v_0/l)$

7.39 *Child on ice with plank*
    Ans. (b) $2l/3$ from the child

7.41 *Leaning plank*
    Hint: Focus on the center of mass and make use of the energy equation.

Chapter 8

8.4 *Grain mill*
    Clue: If $\Omega^2 b = 2g$, then the force is twice the weight.

8.6 *Rolling coin*
    Ans. $\tan \alpha = 3v^2/2Rg$

8.12 *Euler's disk*
    Ans. (a) $\Omega_p = 2\sqrt{\frac{g}{R \sin \alpha}}$

Chapter 9

9.9 *Train on tracks*
   Ans. (a) approximately 300 lbs

9.10 *Apparent gravity versus latitude*
   Ans. $g = g_0 \sqrt{1 - (2x - x^2)\cos^2 \lambda}$, where $x = R_e \Omega_e^2 / g_0$

Chapter 10

10.2 r^3 *central force*
   Ans. (c) $r_0 \approx 2.8$ cm

10.8 *Projectile rise*
   Clue: If $\alpha = 60°$, then $r_{max} = 3R_e/2$.

10.10 *Satellite with air friction*
   Ans. (c) $\Delta K = +2\pi r f$ Note that friction causes the satellite to *increase* its speed.

10.14 *Speed of S2 around Sgr A**
   Ans. 7600 km/s

Chapter 11

11.8 *Scale spring constant*
   Ans. (a) 980 N/m
   Ans. (b) $\gamma = 2\omega_0 = 2\sqrt{k/(M + m)} = 18$ s^{-1}

11.9 *Velocity and driving force in phase*
   Hint: The necessary condition is $\sin(\omega t + \phi) = -\cos \omega t$ where ω is the driving frequency.

11.12 *Cuckoo clock*
   Ans. $Q \approx 68$, and the clock runs for 6 hours on a 1 J battery

11.13 *Two masses and three springs*
   Ans. (c) $\omega = \sqrt{k/M}$

Chapter 12

12.1 *Maxwell's proposal*
   Hint: Use law of cosines.

12.10 *Relative speed*
   Ans. $0.99995c$

12.12 *Headlight effect*
   Ans. (a) $\cos \theta = (\cos \theta_0 + v/c)/(1 + v/c \cos \theta_0)$
   Ans. (b) $v = (1 - 5 \times 10^{-7})c$

12.14 *Moving glass slab*
    Clue: If $v = 0$, then $T = [L + (n-1)D]/c$. If $v = c$, then $T = L/c$.

12.15 *Doppler shift of a hydrogen spectral line*
    Ans. (a) 662.7×10^{-9} m
    Ans. (b) 25 days

12.16 *Pole-vaulter paradox*
    Hint: Consider events at the ends of the pole from the point of view of each observer.

12.18 *The consequences of endless acceleration*
    Hint: $\displaystyle\int \frac{dx}{(1-x^2)^{3/2}} = \frac{x}{\sqrt{1-x^2}}$

Chapter 13

13.4 *Particles approaching head-on*
    Clue: If $v^2/c^2 = 1/2$, then $E = 3m_0c^2$.

13.5 *Speed of a composite particle after an inelastic collision*
    Ans. $v_f = \gamma vm/(\gamma m + M)$, where $\gamma = 1/\sqrt{1-v^2/c^2}$

13.6 *Rest mass of a composite particle*
    Clue: If $x = 7$, then $m = 4m_0$.

13.7 *Zero momentum frame*
    Clue: If $v^2/c^2 = 3/4$, then the speed is $2v/3$.

13.8 *Photon–particle scattering*
    Ans. $\cot\phi = (1 + E_0/m_0c^2)\tan\theta/2$

13.9 *Photon–electron collision*
    Ans. (a) $E = E_0(1 + v/c)/(1 + E_0/E_i)$, where $E_i = m_0c^2/\sqrt{1-v^2/c^2}$
    Ans. (b) $|\Delta\lambda| \approx 0.019\text{Å}$

Chapter 14

14.1 *Pi meson decay*
    Ans. (b) $\theta \approx 42°$

14.2 *Threshold for pi meson production*
    Ans. ≈ 145 MeV

14.5 *Threshold for nuclear reaction*
    Clue: If $M_1 = M_2 = Q/c^2$, then $K_1 = 5Q/2$.

14.7 *Four-acceleration*
    Ans. $\mathbf{A} = \gamma^4 a(1, 0, 0, u_x/c)$, where $a = du_x/dt$

APPENDIX A
MISCELLANEOUS PHYSICAL AND ASTRONOMICAL DATA

Speed of light, c	3.00×10^8 m/s
Gravitational constant, G	6.67×10^{-11} N m^2/kg^2
	6.67×10^{-11} m^3 kg^{-1} s^{-2}
Mass of proton, M_p	1.67×10^{-27} kg
Planck's constant, h	6.63×10^{-34} m^2 kg/s
Mean solar constant, S_{solar}	1.37×10^3 W/m^2
Mass of Sun, M_{sun}	1.99×10^{30} kg
Mass of Earth, M_{earth}	5.98×10^{24} kg
Mass of Moon, M_{moon}	7.34×10^{22} kg
Mean radius of Sun, R_S	6.96×10^8 m
Mean radius of Earth, R_e	6.37×10^6 m
Mean radius of Moon, R_M	1.74×10^6 m
Mean radius of Earth's orbit, $R_{E,orb}$	1.49×10^{11} m
Mean radius of Moon's orbit, $R_{M,orb}$	3.84×10^8 m
Period of Earth's rotation, T_{day}	8.64×10^4 s
Period of Earth's revolution, T_{year}	3.16×10^7 s

APPENDIX B
GREEK ALPHABET

A	α	alpha	N	ν	nu
B	β	beta	Ξ	ξ	xi
Γ	γ	gamma	O	o	omicron
Δ	δ	delta	Π	π	pi
E	ϵ	epsilon	P	ρ	rho
Z	ζ	zeta	Σ	σ	sigma
H	η	eta	T	τ	tau
Θ	θ	theta	Υ	υ	upsilon
I	ι	iota	Φ	ϕ	phi
K	κ	kappa	X	χ	chi
Λ	λ	lambda	Ψ	ψ	psi
M	μ	mu	Ω	ω	omega

APPENDIX C
SI PREFIXES

Factor	Name	Symbol	Factor	Name	Symbol
10^{24}	yotta	Y	10^{-1}	deci	d
10^{21}	zetta	Z	10^{-2}	centi	c
10^{18}	exa	E	10^{-3}	milli	m
10^{15}	peta	P	10^{-6}	micro	μ
10^{12}	tera	T	10^{-9}	nano	n
10^{9}	giga	G	10^{-12}	pico	p
10^{6}	mega	M	10^{-15}	femto	f
10^{3}	kilo	k	10^{-18}	atto	a
10^{2}	hecto	h	10^{-21}	zepto	z
10^{1}	deka	da	10^{-24}	yocto	y

INDEX